Postharvest Handling of Horticultural Crops

The present book covers the importance of post-harvest technology in horticultural crops, fruit growth, development and post harvest physiology, fruit maturity indices, harvesting of fruits and vegetables, initial handling of fruits and vegetable after harvesting, precooling of horticulture produce, transportation, marketing of fruits, vegetables and flowers, fruit ripening hormone and fruit ripening, packaging and storage of horticulture produce, postharvest disease and insect management, postharvest care and handling of cut flowers and spices. The book will guide food scientists, frontline extension officials, teachers, academic students, small entrepreneurs, investors, researchers, food engineers, quality control and assurance personnel, NGO's, Government agencies, bankers, journalist, etc.

R. L. Bhardwaj is working as Associate Professor (Horticulture) at College of Agriculture, Sumerpur-Pali (Rajasthan). Author has published 9 International, 34 National level journals research papers; 7 booklets, 16 folders, 10 popular article in English and 115 in Hindi, 14 Compendiums, 7 Books and guides, 3 book chapters, 1 practical manual and 1 chapter published in EOLSS from US. He have received more than 16 awards from different district, state and national level organisation for excellent work in field of horticultural development and farmer welfare.

Yogesh Kumar Sharma is presently working as Assistant professor (Horticulture) at Rajasthan Agriculture Research Institute, Durgapura (Jaipur), SKNAU, Jobner. He has an experience of 23 years of research and teaching. He has more than 20 research publication in reputed national journals to his credit and has published 125 popular article in different magazines, national and regional newspapers, 5 booklets and 10 folders.

Latika Vyas, is Professor of Home Science (Extension & Communication Management in the Directorate of Extension Education at Maharana Pratap University of Agriculture & Technology, Udaipur, Rajasthan, India. Dr. Latika successfully runs more than 10 researches and extension projects and organise 45 backstopping training for KVK scientists.

Postharvest Handling of Horticultural Crops

Dr. Raju L. Bhardwaj
Associate Professor (Horticulture-PHT)
Department of Horticulture
College of Agriculture, Sumerpur-Pali (Rajasthan)

Dr. Yogesh Kumar Sharma
Assistant Professor (Horticulture)
Division of Horticulture
Rajasthan Agriculture Research Institute, Durgapura (Rajasthan)

Dr. Latika Vyas
Professor (Home Science)
Maharana Partap University of Agriculture and Technology, Udaipur (Rajasthan)

NPH

NARENDRA PUBLISHING HOUSE

First published 2022
by CRC Press
4 Park Square, Milton Park, Abingdon, Oxon, OX14 4RN

and by CRC Press
6000 Broken Sound Parkway NW, Suite 300, Boca Raton, FL 33487-2742

CRC Press is an imprint of Informa UK Limited

British Library Cataloguing-in-Publication Data
A catalogue record for this book is available from the British Library

Library of Congress Cataloging-in-Publication Data
A catalog record has been requested

ISBN: 978-1-032-19955-9 (hbk)
ISBN: 978-1-003-26158-2 (ebk)

DOI: 10.1201/9781003261582

Contents

» Ethylene related plant growth regulators
» Ethephon
» Caution, etacelasil, haloethylsulfinic acids, 2-hydroxyethylhydrazine, others ethylene releasing compounds
» Ethylene management
» Prevention from sources of ethylene, ventilation, ethylene adsorbers, oxidation and catalysis, inhibitors of ethylene action
» Fruit ripening process
» Maturation, ripe, senescence, ripening of fruits, climacteric fruits, non-climacteric fruits, fruit respiration
» Fruits ripening facilities
» Ripening room, temperature, relative humidity, air flow
» Methods of fruit ripening
» Conventional methods of fruit ripening
» Chemical methods of fruit ripening
» Use of ethylene for fruit ripening
» Methods of applying ethrel, mode of action
» Chemical properties of ethylene
» Active ingredient, potential health hazard, ethylene biosynthesis
» Fruit ripening using calcium carbide
» Chemical properties of calcium carbide
» Chemical name
» Physical properties
» Hazard of calcium carbide
» Variations in the pattern of ripening within climacteric and non-climacteric groups
» Physiology of fruit ripening
» Stimulation of ripening
» Changes during ripening
» Changes in structure and composition, changes in cell structure during ripening, textural changes during ripening, change in taste of fruits, colour development in fruit, biochemical changes during ripening
» Hormones, respiration, climacteric fruits, non-climacteric fruits, carbohydrates, pectic substances, sugars, acids, pigments, lipids, protein and nucleic acids, amino acids, enzymes, vitamins, minerals, phenols, volatile compounds, changes in flavour of fruit
» After harvest changes in vegetables
» Interrelationships of the different aspects of the ripening process
» Biochemical control of fruit ripening
» External control of fruit ripening
» Effect of temperature, gaseous composition of storage atmosphere, effect of O2

- Transport packaging
- Types of transport packaging
- Wooden boxes

  Wooden crates, corrugated fibreboard boxes, plastic corrugated boxes, fibreboard boxes, plastic crates, sacks
- Unit load packaging/palletization
- Two principles are used in the assembly of pallet loads
- Pallet boxes
- Classification based on condition

  Field packing, shed packing, conventional packing, environment friendly packaging, modified atmosphere packaging (MAP), individual seal /shrink packaging technique, grape guard
- Classification based on packing materials
- Wooden packing material

  Pallet, slip sheets, wooden crates and lugs, wooden baskets and hampers, wooden field box, woven baskets, wire-bound crates, bulk bins
- Paper based packing material

  Corrugated fiberboard, waxed boxes, pulp containers, paper and mesh bags
- Plastics packing material

  Plastic bags, plastic field boxes, polystyrene boxes, shrink wrap, rigid plastic packages consumer packs

  Flexible plastic films, trays with overwrap, plastic punnets

  Plastic net bags, foam sleeve, light weight plastic crates, shrink wrap, corrugated boxes/cartons
- Transport packs

  Bamboo baskets, wooden boxes, corrugated fibre board/plastic boxes, plastic crates, sacks
- Precautions for packaging
- Quality of modern packaging
- Damages suffered by packed produce
- Prevention of injuries to the produce
- Standardization of packaging
- General operations in packing house
- Layout of packing house
- Need for a packinghouse
- Packing house operations
- Receipt of produce, sorting, cleaning and washing, fungicide treatment, quality selection and size grading, waxing, packaging, storage, dispatch
- Under or overfilled boxes

» Principles of storage of horticultural fresh produce

Control of respiration, control of transpiration, prolonging the dormancy period/ control of sprouting and rooting

» Methods of fruits and vegetable storage

» Traditional methods

In situ/ On site/ natural or field storage, sand and coir storage, bulk storage of dried bulb crops, clamp storage of root and tuber crops, pits or trenches, storage using evaporative coolers/ evaporative cooling, zero energy cool chambers (ZECC), windbreaks, cellars, night ventilation, natural ventilation, in-garden storage

» Improved storage methods

» Cold storage (Refrigeration, Chilling and Freezing)

» Status of cold storage and its potential in India

» Factors for effective cold storage of the produce

» Product quality

Temperature, relative humidity, air circulation and package spacing, respiration rates, heat evolution and refrigeration

» Weight loss in storage

» Sanitation and air purification

» Refrigeration cycle

Principle of refrigeration, components of refrigeration system, evaporator, compressor, condenser, expansion valve

» Low temperature (chilling) injuries

» Symptoms of chilling injury

» Skin pitting, browning or blackening of flesh tissues, rotting

» Management of chilling injury

Maintaining critical temperature, modified atmosphere storage, maintaining high rh, mechanism of chilling injury

» Primary events in chilling injury, secondary events in chilling injury

» Effect of cold storage on subsequent behaviour of horticultural produce

Temperature, relative humidity (RH), ethylene, odour volatiles, sulfur dioxide, smart fresh (1- methylcyclopropene), organic produce

» Cold chain

» Cooling concept

» Different methods of cooling

» Precooling

Room cooling, forced-air cooling, hydrocooling, ice cooling, top icing, package icing, slush ice, evaporative cooling, vacuum cooling, refrigeration

» Cold storage management

» Minimizing water loss during storage

» International standards bodies
» National standards bodies
» Aflatoxins (mycotoxins), trace metal contamination, pesticide residues
» Pre-harvest operations
» Maturity of spices crops
» Blanching and primary processing of black paper
» Continuous harvesting of spice crop
» Maturity indices in seed spices
» Parameters for deciding maturity indices
» Closing of flower initiation, type of the umbel, colour of the leaves, colour of the pods, colour of seeds
» Colour of fruits, stage of harvest
» Harvesting of spices
» Harvesting of leaves and stems
» Harvesting flowers and buds
» Harvesting roots and rhizomes
» Harvesting bark, wood and resins
» Post production operation
» Threshing
» Drying of produce
» Drying seeds and fruits
» Immediate drying
» Programmed drying the enzymatic development of flavour compounds
» Drying to retain colour
» Drying leaves and stems
» Drying flowers and buds
» Drying roots and rhizomes
» Cleaning/grading
» Packaging
» Packaging for seeds and fruits, Packaging for leaves and stems, Packaging for flowers and buds, Packaging for roots and rhizomes, Packaging for bark, wood and resins
» Transport of spices
» Storage
» Processing of spices
» Grinding and separating
» Extraction of essential oils
» Value addition in seed spices: An opportunity for high profit
» Value addition at farm level

Preface

National foods such as horticultural produces are among the most important foods of mankind at them are not only nutritive but are also indispensable of the maintenance of the health. The horticultural produce includes fruits, vegetables, flowers, ornamental plants, plantation crops, aromatic and medicinal plants and spices. During 2017-18, the production of horticulture crops was 311.71 Million Tonnes from an area of 25.43 million hectares. Fruits and vegetables account for nearly 90 per cent of total horticulture production in the country. Globally the cultivation and processing of fruits and vegetables are major industries. Besides meeting the increasing demand of the domestic population, which continues to grow, India exports some portion (4-5%) of its horticulture produce. At the same time, sizeable quantity of these is lost every year after harvest which is a significant economic loss. Postharvest losses occur in horticultural crops like fruits, vegetables and flowers, owing to the poor production planning, lack of suitable harvesting equipment, collection centres in major producing areas, ill equipped markets, unsuitable packing containers, insufficient storage capacity, be short of cold chain and substandard transport infrastructure, be deficient in organizational setup and poor state of processing industries. Losses in fruits, vegetables and flowers are estimated to vary between 20 to 30 per cent valued at nearly 8,000 to 12,000 crores annually depending upon the quality and market prices of produce. Post harvest management, is the process of the treating and handling of produce to stop or greatly slow down spoilage caused or accelerated by micro-organisms. The present book covers the importance of post-harvest technology in horticultural crops, fruit growth, development and post harvest physiology, fruit maturity indices, harvesting of fruits and vegetables, initial handling of fruits and vegetable after harvesting, precooling of horticulture produce, transportation, marketing of fruits, vegetables and flowers, fruit ripening hormone and fruit ripening, packaging and storage of horticulture produce, postharvest disease and insect management, postharvest care and handling of cut flowers and spices. The book will guide food scientists, frontline extension officials, teachers, academic students, small entrepreneurs, investors, researchers, food engineers, quality control and assurance personnel, NGO's, Government agencies, bankers, journalist, etc. It is a rich source of modern engineering technologies for income generating concept for agro based industries. The book is specially dedicated to the sub sector of the fruits and vegetables plants dealing with the fresh primary product from the product reception following the harvesting up-to the storage and before launches it to the market. We hope that this book will help as simple but comprehensive guide for all the persons that focuses on post harvest management skills and completing.

Authors

1 Importance of Horticultural Crop Production

The horticultural produce includes fruits, vegetables, flowers, ornamental plants, plantation crops, aromatic and medicinal plants and spices. Fruits and vegetables account for nearly 90 per cent of total horticulture production in the country. Globally the cultivation and processing of fruits and vegetables are major industries. Worldwide, over 866 million tons of fruit are produced annually with China, India, Brazil, and the United States counted among the top producers. India is the second largest producer of fruits (97.35 million tonnes) after China and rank first in production of mango, banana, papaya, sapota, pomegranate, acid lime and aonla. The diverse agro-climatic and edaphic conditions enable to produce variety of fruits (tropical, subtropical and temperate). However, mango and banana share more than half of the total fruit production in the country. Banana ranks first in production 30.80 million tonnes (31.64 %) followed by mango 21.82 million tonnes (22.41 %). Mango covers about 2.25 million hectares (34.70 %) of area under fruit crop in India. At present banana has the highest productivity (35.9 mt ha^{-1}) followed by papaya (33.4 mt ha^{-1}), grape (26.0 mt ha^{-1}) and pineapple (15.1 mt ha^{1}). India has been bestowed with wide range of climate and physic-geographical conditions and as such is most suitable for growing various kinds of horticultural crops such as fruits, vegetables, flowers, nuts, spices and plantation crops. India has increase in horticulture production over the last few years by significant progress in area expansion and increase in productivity of different fruit crops. Over the last decade, the area under horticulture grew by 2.6 per cent per annum and annual production increased by 4.8 per cent. During 2017-18, the production of horticulture crops was 311.71 Million Tonnes from an area of 25.43 million hectares. The production of vegetables has increased from 101.2 Million Tonnes to 184.40 Million Tonnes since 2004-05 to 2017-18 and production of fruits has increased from 50.9 Million Tonnes to 97.35 Million Tonnes since 2004-05 to 2017-18. In 2017-18, the total horticulture production was highest in case of Uttar Pradesh (392.48 Lakh Tonnes) followed by West Bengal (324.2 Lakh Tonnes). The total production of fruits is highest in case of Andhra Pradesh (152.15 Lakh Tonnes) with 15.63 per cent share, followed by Maharashtra (117.28 Lakh Tonnes) with 12.05 per cent share and Uttar Pradesh (105.39 Lakh Tonnes) with 10.82 per cent share of total fruit production. During 2017-18 the area under vegetables was 10.26 Million Hectares with a production of 184.40 Million Tonnes. For this period the total vegetable production was highest in case of Uttar Pradesh (283.16 Lakh Tonnes) with 15.4 per cent share followed by West Bengal (276.95 Lakh Tonnes) with 15.03 per cent share and Madhya Pradesh (175.45 Lakh Tonnes) with 9.52 per cent share of total production. The plantation crops production was highest in case of Kerala (60.54 Lakh Tonnes) with 33.48 per cent share followed by Karnataka (49.36 Lakh Tonnes) with 27.30 per cent share and Tamilnadu (42.34 Lakh Tonnes) with 23.42 per cent share of total production. The flower production was highest in Tamilnadu (4.83 Lakh Tonnes) with 17.33 per cent share followed by Andhra Pradesh (4.29 Lakh Tonnes) with 15.40 per cent share and Karnataka (3.24 Lakh Tonnes) with 11.63 per cent share of total production. The seed spices production was highest in Madhya Pradesh (11.92 Lakh Tonnes) with 14.67 per cent share followed by Rajasthan (11.44 Lakh Tonnes) with 14.09 per cent share and Andhra Pradesh (11.0 Lakh Tonnes) with 13.54 per cent share of total production.

Table 1: Crop wise area, production and productivity of major horticultural crops in India.
(Aera in 000 ha., Production in 000 MT and productivity in MT/ha.)

Fruits name	2015-16			2016-17			2017-18		
	Area	Production	PDY.	Area	Production	PDY.	Area	Production	PDY.
Fruit Crops									
Apple	277	2521	9.10	305	2265	7.43	301	2327	7.73
Banana	841	29135	34.64	860	30477	35.44	884	30808	34.85
Citrus	1024	11581	11.30	985	11419	11.59	1003	12546	12.51
Grapes	122	2590	21.22	137	2922	21.33	139	2920	21.01
Guava	255	4048	15.87	260	3826	14.72	265	4054	15.30
Litchi	90	559	6.21	93	568	6.11	92	686	7.46
Mango	2209	18643	8.43	2212	19506	8.82	2285	21822	9.55
Papaya	132	5667	42.93	134	5940	44.33	138	5989	43.40
Pineapple	110	1924	17.49	111	1861	16.77	103	1706	16.56
Pomegranate	197	2306	11.70	216	2613	12.10	234	2845	12.16
Sapota	107	1294	12.09	99	1236	12.48	97	1176	12.12

Fruits name	2015-16			2016-17			2017-18		
	Area	Production	PDY.	Area	Production	PDY.	Area	Production	PDY.
Others	937	9915	10.58	961	10285	10.70	965	10479	10.86
Fruit Total	6301	90183	14.31	6373	92918	14.58	6506	97358	14.96
Vegetable Crops									
Brinjal	663	12515	18.88	733	12510	17.07	730	12801	17.54
Cabbage	394	8806	22.35	395	8807	22.30	399	9037	22.65
Cauliflower	426	8090	18.99	454	8557	18.85	453	8668	19.13
Okra	511	5849	11.45	507	6003	11.84	509	6095	11.97
Onion	1320	20931	15.86	1306	22427	17.17	1285	23262	18.10
Peas	498	4811	9.66	530	5345	10.08	540	5422	10.04
Potato	2117	43417	20.51	2179	48605	22.31	2142	51310	23.95
Sweet potato	126	1454	11.54	128	1460	11.41	131	1500	11.45
Tapioca	204	4344	21.29	199	4171	20.96	173	4950	28.61
Tomato	774	18732	24.20	797	20708	25.98	789	19759	25.04
Other	3073	40115	13.05	3010	39579	13.15	3108	41590	13.38
Vegetable Total	10106	169064	16.73	10238	178172	17.40	10259	184394	17.97
Other Crops									
Aromatic	634	1022	1.61	664	972	1.46	720	866	1.20
Flower loose	278	1656	5.95	306	1699	5.55	324	1962	6.05
Flower Cut	-	528	-	-	693	-	-	823	-
Honey	-	88	-	-	95	-	-	105	-
Plantation Crop	3680	16658	4.52	3598	17972	4.99	3744	18082	4.82
Spices	3474	6988	2.01	3671	8122	2.21	3878	8124	2.09
Grand Total	24472	286188	11.69	24851	300643	12.09	25431	311714	12.25

TRENDS IN HORTICULTURE PRODUCTION AND PRESENT STATUS

Increase in horticulture produce has far outpaced the increase in area and productivity of horticultural crops. Comparison of 2013-14 with 1991-92 reveals that while the area has about doubled, the production has become three times of the level since 1991-92. The third advance estimate released by Ministry of Agriculture also corroborates the same as it indicates that the area under horticulture decreased from 24,198 thousand hectares in 2013-14 to 23,216 thousand hectares in 2014-15 whereas the production is estimated to have increased from 277,352 thousand MT in 2013-14 to 277,743 thousand MT in 2014-15. Despite of improvement in productivity, the same is yet to catch up with the world leaders, especially in case of vegetables where the productivity is lower than the world average.

Among fruits, banana and mango accounted for more than half (55%) of total fruit production both during 2013-14 and 2014-15, with production of banana alone accounting for about 33 per cent. Share of Citrus fruits was also significant and they accounted for about 13-14 per cent of overall fruit production. Among vegetables, potato production comprised the major share with 26 per cent and 28 per cent of overall vegetable production in 2013-14 and 2014-15 respectively. Together with tomato and onion (each accounting for 11-12 per cent of total vegetable production during 2013-14 and 2014-15) and brinjal (7-8% shares in vegetable production during 2013-14 and 2014-15). During 2013-14, the 0.4 per cent annual growth rate in production of vegetables was much less than 9.5 per cent growth rate in case of fruits. However, as per the third advance an estimate, the fruit production has declined by 3 per cent during 2014-15 even though the vegetable production has increased by 3 per cent during the period. Overall annual growth rate in case of horticulture produce is estimated to be merely 0.1 per cent during 2014-15. The same may be explained by about 4 per cent decline in area under horticulture crops with area under fruit crops declining by more than 13 per cent. Coconut continues to account for more than 90 per cent of plantation crop (areca nut, cashewnut, cocoa, coconut) production during 2013-14 and 2014-15. Chilies accounted for about a quarter of total spice production during 2013-14 and 2014-15 followed by garlic with 21 and 23 per cent share respectively. Among major spices, production of turmeric is estimated to decline steeply (by 28%) during 2014-15 compared to 2013-14. Under the horticulture crops, production of fruits is estimated to be around 97.38 million tonnes in 2018-19 compared to 97.36 million tonnes in the previous year. Vegetables production is estimated to rise 1.6 per cent at around 187.36 million tonnes. Among vegetables, onion production is estimated to be around 23.28 million tonnes, slightly higher than production in 2017-18. Potato production is estimated to be around 52.96 million tonnes, which is 3.2

per cent higher than 2017-18. Tomato production is estimated to be around 19.66 million tonnes, which is 0.5 per cent lower than 2017-18. As per the data, spices production is estimated to be around 8.61 million tonnes, which is 6.01 per cent higher than 2017-18. India is the largest producer of ginger and okra amongst vegetables and ranks second in production of potatoes, onions, cauliflowers, brinjal, cabbages, etc.

Table 2. Annual growth trends of area and production of horticulture crops (Growth in %)

Crops	2014-15 over 2013-14		2015-16 over 2013-14		2016-17 over 2015-16	
	Area	Production	Area	Production	Area	Production
Fruits	- 15.3	-2.7	3.1	4.1	2.8	3.0
Vegetables	1.6	4.0	5.9	-0.2	1.8	3.5
Flowers	-2.6	-6.7	11.6	1.9	11.4	2.9
Aromatics	34.0	12.0	-3.8	2.2	0.0	0.8
Plantation crops	-3.8	-4.5	4.1	7.0	-0.1	1.3
Spices	4.9	3.4	4.7	14.4	1.8	1.3
Total horticultural crops	-3.3	1.3	4.5	1.9	1.9	3.1

Source: Horticulture Statistics Division, Department of Agriculture, Coopn & Farmers Welfare

Exports

Besides meeting the increasing demand of the domestic population, which continues to grow, India exports some portion (4-5%) of its horticulture produce. During 2013-14 total exports of horticulture produce by India was 3.69 million metric tonnes which amounted to about Rs 143.6 billion. Even though the quantum of export decreased in comparison to the year before i.e. 2012-13 when it was 3.7 million metric tonnes, the value of export of horticulture produce increased by about 35.6 per cent from Rs 105.9 billion in 2012-13. Except for 2010-11 when the exports of horticulture declined by about 7 per cent, the export of horticulture produce has seen an increased during the last six years. During 2014-15, India exported fruits and vegetables worth Rs. 7474.14 crores which comprised of fruits worth Rs. 2771.32 crores and vegetables worth Rs. 4702.78 crores. During 2018-19, India exported fruits and vegetables worth Rs. 10236.93 crores/ 1,469.33 USD Millions which comprised of fruits worth Rs. 4817.35 crores/ 692.01 USD Millions and vegetables worth Rs. 5419.48 crores/ 777.25 USD Millions. Grapes, pomegranates, mangoes, bananas, oranges account for larger portion of fruits exported from the country while onions, mixed vegetables, potatoes, tomatoes, and green chilly contribute largely to the vegetable export basket. The major destinations for Indian fruits and vegetables are Bangladesh, UAE, Netherland, Nepal,

Malaysia, UK, Sri Lanka, Oman and Qatar. India's total export of floriculture was Rs. 571.38 Crores/81.94 USD Millions in 2018-19. The major importing countries were United States, Netherlands, United Kingdom, Germany, and United Arab Emirates. There are more than 300 export-oriented units in India. More than 50 per cent of the floriculture units are based in Karnataka, Andhra Pradesh and Tamil Nadu. With the technical collaborations from foreign companies, the Indian floriculture industry is poised to increase its share in world trade. India has also been exporting fruits and vegetable seeds and exports during 2018-19 stood at Rs. 849.27 crores/122.74 USD Millions. Netherland, United States, Pakistan, Bangladesh and Thailand were the major markets of India's fruits and vegetable seeds. Though India's share in the global market is still nearly 1 per cent only, there is increasing acceptance of horticulture produce from the country. This has occurred due to concurrent developments in the areas of state-of-the-art cold chain infrastructure and quality assurance measures. Apart from large investment pumped in by the private sector, public sector has also taken initiatives and with APEDA's assistance several Centres for Perishable Cargoes and integrated post harvest handling facilities have been set up in the country. Capacity building initiatives at the farmers, processors and exporters' levels has also contributed towards this effort.

The Indian food processing industry is primarily export orient. India's geographical situation gives it the unique advantage of connectivity to Europe, the Middle East, Japan, Singapore, Thailand, Malaysia and Korea. One such example indicating India's location advantage is the value of trade in agriculture and processed food between India and Gulf region. India's exports of processed food was Rs. 31111.90 crores in 2018-19, which including the share of products like mango pulp (Rs. 657.67 crores/ 93.97 USD millions), processed vegetables (Rs. 2473.99 crores/ 354.75 USD millions), cucumber and gherkins (Rs. 1436.08 crores/ 205.84 USD millions), processed fruits, juices and nuts (Rs.2804.97 crores/ 402.52 USD millions) pulses (Rs. 1680.18 crores/ 242.66 USD millions), groundnuts (Rs. 3298.33 crores/ 473.81 USD millions), guargum (Rs. 4707.05 crores/ 676.47 USD millions), jaggery and confectionary (Rs. 1606.32 crores/ 230.14 USD millions), cocoa products (Rs. 1350.86 crores/ 193.26 USD millions), cereal preparations (Rs. 3859.37 crores/ 553.17 USD millions), alcoholic beverages (Rs. 2103.97 crores/ 301.71 USD millions), miscellaneous preparations (Rs. 4072.98 crores/ 583.33 USD millions), and milled products (Rs. 1060.13 crores/ 151.85 USD millions).

Table 3: Export of horticulture produce from India (Value in lakh and quantity in mt)

Product	2015-16		2016-17		2017-18	
	Quantity	Value	Quantity	Value	Quantity	Value
Apple (Fresh)	20808.10	5980.18	22550.03	6309.53	12529.22	3916.59
Banana (Fresh)	95222.66	35767.20	110750.56	38759.67	101314.37	34877.39
Oranges (Fresh/Dried)	74339.13	15965.66	48111.64	11732.52	15840.48	3485.36
Grapes (Fresh)	132647.60	136225.58	198471.31	178171.38	188221.16	189994.86
Guavas (Fresh/Dried)	1913.73	713.29	1408.18	610.81	1229.75	553.26
Litchi	9.87	5.49	124.02	105.07	108.24	57.98
Mango (Fresh/ Dried)	36779.25	32063.91	52760.99	44366.03	49180.46	38234.02
Papaya (Fresh/ Dried)	13457.50	5717.04	12442.75	5118.78	9996.37	4150.44
Pineapple (Fresh/ Dried)	4187.53	1982.62	5204.56	2615.40	8339.82	3501.44
Sapota	978.61	967.16	1400.41	851.50	1602.60	890.70
Cabbage Lettuce (Fresh / Chilled)	349.33	65.38	286.02	42.65	527.28	84.45
Cauliflowers & Brocolli (Fresh/Chilled)	85.43	30.01	1338.76	227.35	177.86	95.23
Onion (Fresh/Chilled)	1382959.54	309720.86	2415739.05	310606.44	1588985.71	308882.23
Pea (Fresh/Chilled)	693.40	539.59	1073.68	567.10	410.11	179.41
Tomatoes (Fresh/Chilled)	160877.14	40412.58	267515.40	54815.65	47446.09	11421.91
Potatoes (Fresh/Chilled)	279650.47	36154.76	396340.67	66147.51	395748.12	41182.50
Sweet Potato	508.53	227.18	434.48	107.58	403.73	124.84
Floriculture	22691.66	48341.30	22020.35	54670.73	20703.47	50731.22
Walnut	3063.59	11359.06	1794.79	4825.33	3124.65	11927.03

Source: APEDA website accessed on 11.9.2018

Table 4: Import of horticulture produce in India (Value in lakh and quantity in mt)

Product	2015-16		2016-17		2017-18	
	Quantity	Value	Quantity	Value	Quantity	Value
Fresh fruits and vegetables	980271.00	1146601.60	1048743.00	1125207.10	1010357.00	1255016.61
Processed fruits and vegetables	54341.53	64682.10	56112.67	66418.21	68636.93	93718.56
Flowers	4768.81	11440.04	5560.35	13381.01	6243.17	13646.10

Source: APEDA website accessed on 12.9.2018

Indian Scenario of Fruits and Vegetables Availability

In India per capita availability of fruits and vegetables is quite low because of postharvest losses which account for about 20 to 30 per cent of production. Besides, quality of a sizable quantity of produce also deteriorates by the time it reaches the consumer. Most of the problems relating to the marketing of fruits and vegetables can be traced to their perishability. It is responsible for high marketing costs, market gluts, price fluctuations and other similar problems. There is a rise of about 4 per cent in the production of fruits and vegetables but there is significant increase in production area also which results in low per hectare production. This phenomenon is caused due to various factors in which economic and technological factors lead the race. State needs to educate the producers of the latest technology available and also help them by granting support to acquire them and proper supply chain management should be there in the country which would help prevention of exploitation of farmers and help increase their income.

Table 5: Projection regarding per capita availability in gms/day

Projections	2015-16	2016-17	2017-18
Production of fruits million MT	78.50	92.84	97.35
Production of vegetables million MT	176.79	175.0	184.39
Projected population (crores)	125.40	126.80	128.30
Per capita gross availability of fruits in gms/day	171.51	200.60	207.90
Per capita gross availability of vegetables in gms/day	386.25	378.13	393.76
Per capita net availability of fruits in gms/day (25% loss+ 5% exports and processing)	120.0	126.0	145.53
Per capita net availability of vegetables in gms/day (25% loss+ 5% exports and processing)	270.0	264.70	275.63

NUTRITIONAL VALUE OF FRUITS AND VEGETABLES

Fruits and vegetables remain an important source of nutrients in many parts of the world and offer advantages over dietary supplements because of low cost and wide availability. Most fruits and vegetables are available almost year-round in a wide variety and them not only tastety, but they also have favourable attributes of texture, colour, flavour and ease of use. The components obtained by the human body from fruits and vegetables include water, carbohydrates, fats, proteins, dietary fiber, minerals, organic acids, pigments, vitamins A, thiamine (B_1), niacin (B_3), pyridoxine (B_6), folacin (folic acid or folate), vitamins B_9, vitamins C (ascorbic acid), vitamins E and antioxidants. Legume vegetables, potatoes, and tree nuts, such as almond, filbert, pecan, pistachio, and walnut good source of proteins in the diet, and their proteins are of high quality as to their content of essential amino acids. Nuts are a good source of essential fatty acids, fiber, vitamin E, and minerals. Other important nutrients supplied by fruits and vegetables include riboflavin (B_2), zinc, calcium, potassium, and phosphorus. Maximum fruits and vegetables are relatively low in calories and fat (avocado and olives being the exceptions), free from cholesterol. Fruits and vegetables are relatively low in sodium and high in potassium. Ascorbic acid in fruits and vegetables enhances the bioavailability of iron in the diet. Because of all these characteristics, fruits and vegetables have a unique role in a healthy diet. Fruit and vegetable consumption is associated with reduced risk of major diseases, and possibly delayed onset of age-related disorders and promoting good health. However, in many cases fruit and vegetable consumption is still below the dietary guideline goal of consuming 5–10 servings each day. The nutritional value of fruits and vegetables depends on their composition, which shows a wide range of variation depending on the species, cultivar and maturity stage. The composition of fruits and vegetables includes a great number of metabolites however, it could be predicted that no single commodity might be rich in all these constituents.

General Composition of Fruits and Vegetables

Water: Most of the horticultural produce contains more than 80 per cent water, with some tissues, such as of cucumber, lettuce, marrow and melons, containing about 95 per cent water. The starchy tubers and seeds, for example yam, cassava and corn, contain less water, but even then they also contain more than 50 per cent water. The actual water content is dependent on the time of availability of produce will vary during the day if there are diurnal fluctuations in temperature. For most produce, it is desirable to harvest when the maximum possible water content is present as this result in a crisp texture. Hence the time of harvest can be an important consideration, particularly with leafy vegetables, which exhibit large and rapid variations in water content in response to changes in their environment.

Organic acids: Most of the fruits contain organic acids in excess of that required for the operation of the TCA cycle and other metabolic pathways. The excess acids are generally stored in the vacuole away from other cellular components. Lemon, lime, passion fruit and black currant often contain over 3 per cent of organic acids. There are two types of acids, namely aliphatic (straight chain) and aromatic acids. The most abundant acids in fruits and vegetables are citric and malic (both aliphatic) acids. However, large amounts of tartaric acid occur in grapes and tamarind, oxalic acid in spinach, ascorbic acid in aonla and isocitric acid in blackberries. Malic acid is the major component in apple, banana, cherry, melon and plum, whereas the citric acid is available in berries, citrus, guava, pear, pineapple and tomato. The acid content of fruits and vegetables generally decreases during maturation. For example, the citric acid content of clingstone peaches decreases faster than the malic acid content, while the malic acid content of apples and pears decreases faster than the citric acid content. Aromatic organic acids occur in several fruits and vegetables, but in very low concentrations. Benzoic acid occurs in cranberries, quinic acid in bananas and chlorogenic acid in potatoes. Organic acids play an important role in the sugar to acid ratio, which affects the flavour of fruits and vegetables.

Proteins: Fresh fruits and vegetables are not important contributors of protein to the human diet. Proteins represent less than 1 per cent of the fresh mass of fruit and vegetable tissues. Legume vegetables, potatoes, and tree nuts, such as almond, filbert, pecan, pistachio, and walnut good source of high-quality proteins. The proteins of fruits and vegetables are built from amino acids, but other related simple nitrogenous compounds also occur. The protein content of fresh fruits or vegetables is calculated by multiplying the total nitrogen content by a factor of 6.25. This calculation uses the fact that protein is comprised of about 16 per cent nitrogen, and the assumption that all nitrogen present is protein. The conversion ignores the fact that appreciable amounts of simple nitrogenous substances can be present in an uncombined form. In potatoes, 50 to 60 per cent of the nitrogen occurs in the form of simple soluble constituents, while in apples the estimates range from 10 to 70 per cent. Senescent tissues, such as those of overripe fruits, usually contain especially high proportions of non-protein nitrogen. Asparagine is abundant in potatoes and apples as non-protein nitrogen fractions. Pears and oranges are rich in proline, and black and red currants in alanine. The protein is mostly present as enzymes, rather than a storage pool as in grains and nuts.

Lipids and fatty acids: Lipids comprise less than 1 per cent in most of the fruits (grape 0.2%, banana 0.1% and apple 0.06%) and vegetables are associated with protective cuticle layers on the surface of the produce and with cell membranes. The avocado and olive are exceptions, having respectively about 35-70 per cent and 30-70 per cent fat content on a dry mass basis in the cells. Lipids are an energy source for plants during germination, forming components of cellular membranes and cuticular waxes, and they are mainly present as triglycerides (esters of glycerol and three fatty acids) or phospholipids (in which one fatty acid has been replaced by a phosphate group). Fatty acids are necessary for human physical functions, where they are used primarily to produce hormone-like substances that regulate a wide range of functions including blood pressure, blood clotting, blood lipid levels, the immune response and the inflammatory response. The human body can produce most fatty acids, except for linoleic acid and α -linolenic acid, which are widely distributed in plant oils.

Metabolizable carbohydrates: After water, carbohydrates are the most abundant constituents in fruits and vegetables, representing 50 to 80 per cent of the total dry weight. Carbohydrate functions include the storage of energy reserves and make-up of the structural framework of cells. Carbohydrates, like proteins, yield 4 kcal g^{-1}, while fats yield 9 kcal g^{-1}. Sugars are present mainly in ripe fruit, and starch occurs in both vegetables and unripe fruits. Glucose and fructose are the predominant forms of simple sugars found, especially, in fruits. Sucrose, the primary transport form of carbohydrate in most plants, is a disaccharide yielding glucose and fructose upon hydrolysis. Glucose, fructose and sucrose are water-soluble and together they comprise most of the sugars associated with the sweet taste of fruits and vegetables. In many fruits (e.g. apple, pear, strawberry, grape) glucose and fructose are present in greater amounts than sucrose, but in certain vegetables, such as parsnip, beetroot, carrot, onion, sweet corn, pea and sweet potato and in some ripe fruits such as banana, pineapple, peach and melon, the sucrose content is higher. Traces of other mono and disaccharide sugars such as xylose, arabinose, mannose, galactose and maltose may also be present in small amounts. Some fruits of the Rosaceae family could also have significant levels of the sugar alcohol sorbitol. Some non-starchy root vegetables, such as parsnip, beetroot and carrot, are relatively rich in simple sugars, containing between 8 to 18 per cent of total carbohydrates. Starch from plantain, cassava, yam, sweet potato and potato provides the bulk of energy in simple diets of subsistence groups in some developing countries.

Table 6. Sugar content of some ripe fruits (g/100g-1 fresh weight)

Fruit	Glucose	Fructose	Sucrose	Fruit	Glucose	Fructose	Sucrose
Apple	2	6	4	Orange (juice)	2	2	5
Banana	6	4	7	Peach	1	1	7
Cherry	5	7	0	Pear	2	7	1
Date	32	24	8	Pineapple	2	1	8
Grape	8	8	0	Tomato	2	1	0

Calorie value of fruits: Taking into consideration of the fruits calories, they seem to be the most preferred food choice, when it comes to weight loss program, eating fruits can give a boost to weight loss, by reducing appetite and the reason that can be attributed to this fact is that fruits make tummy feel full. The calories in Indian fruits are usually very low and owing to this reason, fruits serve as the perfect mid meal choice, when feel hungry. Fresh fruits also contain loads of water and fiber and thus are extremely beneficial for the human body.

Table 7. The calorie chart of Indian fruits (per 100 grams)

Fruits	Calories 100 g^{-1}	Fruits	Calories 100 g^{-1}
Dates	281	Plums	56
Avocado pear	190	Orange	53
Banana	95	Pears	51
Sapota	94	Peach	50
Pomegranate	77	Guava	49
Strawberries	77	Orange juice (100 ml)	47
Cherries	70	Pineapple	46
Mangoes	70	Grapes black	45
Guava	66	Kiwi fruit	45
Lychies	61	Papaya	32
Apple	56	Watermelon	16-26

Dietary fiber: A substantial proportion of carbohydrates are in attendance as dietary fibre, which helps passes through the intestinal system. Cellulose, pectic substances and hemicelluloses are the carbohydrate polymers that constitute fibre. Lignin, a complex polymer of aromatic compounds linked by proply units, is also a major component of fibre. Dietary fibre is not digested by people as they are not capable of secreting the enzymes necessary to break down the polymers to the basic monomeric units that can be absorbed by the intestinal tract. Starch and cellulose have the same composition, as they are synthesized from D-glucose units, but the bonding between the monomers differs. Starch comprises a-1, 4 linkages, which are hydrolyzed by a range of amylase enzymes secreted by human. Cellulose is a cell wall polymer of β -1, 4-linked glucose. In fruits and vegetables, the cell wall constitutes 1 to 2 per cent of the fresh weight, and cellulose could be as much as 33 per cent of that amount and little change in cellulose content occurs during ripening. Fruit tissues are particularly rich in pectins, which can account for up to 40 per cent of the total cell wall polysaccharides. Resistant starch consists of starch and its degradation products that are not digested in the small intestine. Green bananas and potato are also relatively rich in resistant starch. Non-digestible oligosaccharides are low molecular weight carbohydrates, intermediate in nature between simple sugars and polysaccharides. Several oligosaccharides might be hydrolyzed in the digestive tract, others might resist the digestive process. Fibre was once considered to be an unnecessary component in the diet, although it was thought to relieve constipation. Fibre deficiency in diet caused appendicitis, haemorrhoids, cancer of colon, hiatus hernia, constipation, ischaemic heart disease, deep vein thrombosis, obesity, diabetes, tumours of rectum, diverticulosis, varicose veins and gallstones. Fiber content of fruits and vegetables is usually in the range of 1 to 13 per cent (Almond 12.2, apple 2.4, asparagus 2.1, avocado 6.8, banana 2.6, broccoli 2.6, carrot 2.8, kiwifruit 3.4, lettuce 2.1, onion 1.7, orange 2.4, pea 2.6, peach 1.5, peanut 8.5, pear 3.1, pepper 2.1, pineapple 1.4, plum 1.4, potato 2.2, prunes 7.1, raisin 3.7, spinach 2.2, strawberry 2.0, tomato 1.2 and walnut 6.7 per cent). Nuts, legumes and dried fruits have higher levels of fiber than fruits and vegetables.

Vitamins: Vitamins are organic molecules required in trace amounts for normal development, which cannot be synthesized in sufficient quantity by the organism and must be obtained from the diet. The term "vitamin" derives from the words vital amine because the first vitamin discovered (thiamine) contained an amino group. They are usually grouped into fat-soluble (A, D, E and K) and water-soluble (B group and C) molecules. Today 14 vitamins are known vitamin A (retinol), B complex [vitamin B_1 (thiamine), vitamin B_2 (riboflavin), vitamin B_3 (niacin), vitamin B_5 (pantothenic acid), vitamin B_6 (pyridoxine), vitamin B_9 (folate/folic acid), biotin, choline and vitamin B_{12} (cyanocobalamine)] and vitamins C, D, E and K. The vitamins present in fruits and vegetables make an important contribution to human nutrition, as they have specific functions in normal body performance.

Vitamin A: Carotenoids are liposoluble pigments responsible for the yellow, orange and red colour of several fruits and vegetables. Carotenoids are terpenoids formed by eight isoprene units (2-methyl-1, 3-butadiene) and derived from isopentenyl diphosphate. The structural requirement for vitamin A is satisfied by around 60 carotenoids. Vitamin A plays an important role in vision, cell division and differentiation, bone development and reproduction. The average daily requirement for vitamin A for an adult is estimated at 5000 international units (1 IU = 0.3 μg retinol or 0.6 μg β carotene). Important source of carotene is mango 1800, cantaloupe 1000, papaya 810, guava 435, apricot 405, plum 295 and watermelon 230 μg in 100 g^{-1} fruit pulp, commonly supplying about 40 per cent of daily requirements. Dark-green vegetables (such as collards, spinach, and turnip greens), orange vegetables (such as carrots, pumpkin, squashes and sweet potato), orange-flesh fruits (such as apricot, cantaloupe, mango, nectarine, orange, papaya, peach, persimmon, and pineapple), watermelon, tomatoes and peppers can supply useful amounts of carotenoids. Only about 10 per cent of the carotenoids known to be present in fruit are precursors of vitamin A, all other carotenoids such as lycopene, the main pigment in tomato, have no vitamin A activity.

Vitamin B Complex: Thiamine is required in the human body for the metabolism of carbohydrates. A daily intake of 1–2 mg is generally considered as necessary for a normal adult. Legumes vegetables are especially rich in thiamine. Compared with ascorbic acid, thiamine is relatively stable at cooking temperatures, especially in a slightly acidic solution. However, losses of 25 to 40 per cent may occur during cooking. The average human requirement for riboflavin is estimated to be 1–2 mg day^{-1}. Green vegetables such as beans, beets, peppers, spinach are particularly rich in riboflavin. Starchy vegetables and fruits (except bael) are relatively poor sources of riboflavin. Niacin, also known as nicotinic acid, is a precursor to NADH, NAD and NADP, which play essential roles in living organisms. A daily intake of 10 mg to 15 mg niacin is recommended. There is evidence that niacin can be synthesized in the body from tryptophan. Almonds are a rich source, but no fruits or vegetables can be singled out as being rich in niacin except perhaps, cape gooseberry and avocado. Vitamin B_6 (pyridoxal phosphate) is a cofactor in many transamination, decarboxylation and deamination reactions. Vitamin B_6 is present in appreciable amounts in beans, cabbage, cauliflower, spinach, sweet potatoes, grapes, prunes, avocados and bananas. Pantothenic acid occurs widely in fruits, peas, beans, nuts, broccoli, mushrooms, potatoes and sweet potatoes. Symptoms of pantothenic acid deficiency in the diet include fatigue, headaches, sleep disturbance, tingling of hands and feet and lack of antibody production. Biotin is stable during cooking, processing and storage of fresh, canned and frozen fruits and vegetables. Deficiency leads to depression, sleeplessness and muscle pains. Folic acid is essential for reproduction, RNA synthesis and normal growth its deficiency will result in anaemia. The folic acid is present in banana fruits is 10g $100g^{-1}$ and in spinach, cabbage, mustard greens, butterhead, lettuce, broccoli, brussels sprouts, okra, green peas, asparagus and other green vegetables is $<5g\ 100g^{-1}$. The intensity of the green colour acts as a good guide to the folic acid content. Lack of folic acid in the diet can cause a red tongue, diarrhea and anemia. Choline is heat-stable and occurs in dried legumes and vegetables. Vitamin B_{12} does not occur in fruits and vegetables. Because vitamins of the B group are water-soluble, leaching losses occur during cooking and it not store in body so require in daily diet.

Vitamin C (Ascorbic acid): Ascorbic acid and its first oxidation product dehydroascorbic acid (which can be reduced in the human body) might be considered as vitamin C. Ascorbic acid is a water soluble carbohydrate derived compound showing antioxidant and acidic properties due to the presence of a 2,3-enediol moiety. The recommended dietary allowance of vitamin C for men is 75 mg and for young women is 90 mg daily. It helps prevent scurvy and cardiovascular disease aids wound healing and immune system. Aonla, raspberry, blackberry, red currant, gooseberry, cornelian cherry, rosehip, jujube, guava, broccoli, cabbage, cantaloupe, citrus fruits, kiwifruit, persimmon, pepper, pineapple, potato, strawberry, tomato, watermelon and leafy vegetables are rich in ascorbic acid. Ascorbic acid is highly susceptible to oxidation, either directly or through the enzyme ascorbate oxidase catalyzing the oxidation of ascorbic acid to dehydroascorbic acid, with the concomitant reduction of molecular oxygen to water. High storage temperatures, bruising and mechanical damage greatly increases the rate of ascorbic acid loss during storage. When vegetables are cooked before eating, high losses of vitamin C can occur. For instance, starchy vegetables may lose between 40 per cent and 80 per cent of their vitamin C during cooking, because of leaching and oxidation.

Table 8. Approximate level of vitamin C in different fruits and vegetables

Commodity	Vitamin C ($mg 100 g^{-1}$)	Commodity	Vitamin C ($mg 100 g^{-1}$)
Black currant, guava	200-210	Raw guava	184-190
Chili	150-160	Raw kiwi	118-120
Papaya	80-90	Raw litchi	72-75
Kiwi fruit	70-80	Raw papaya	62-65
Strawberry	40-45	Raw strawberry	57-60
Mango, carrot	30-35	Citrus fruits	31–53
Pineapple, potato,	25-30	Cantaloupe	42-45
Tomato, bean, cassava	20-25	Banana	25-30
Apple, peach	10-15	Beetroot, onion	5-10

Vitamin E: Vitamin E includes tocopherols and tocotrienols. They can be in eight different forms (four tocopherols and four tocotrienols). All the isomers have aromatic rings with a hydroxyl group that can donate hydrogen atoms to reduce reactive oxygen species (ROS). The different isomers are named alpha (α), beta (β), gamma (γ) and delta (δ), and this is related to the number and position of methyl groups in the ring. Each of the forms has its own vitamin E activity, α -tocopherol being the most active. Vitamin E deficiency results in stunted growth and decrease fertility. In general, vitamin E levels are more abundant in oily seeds, olives, nuts (such as almonds, cashew nuts, filberts, macadamias, pecans, pistachios, peanuts and walnuts) and avocados. Even though the levels of tocopherol in broccoli, corn, dry beans, lentils, chickpeas and dark green leafy vegetables are lower than in fat-rich products, but they are good sources compared to other fruits and vegetables. Vitamin E is highly susceptible to oxidation during storage and processing. It helps to prevent heart disease, LDL oxidation, cancer and diabetes.

Vitamin D and K: Vitamin D is a group of fat-soluble compounds. The main forms of vitamin D are ergocalciferol and cholecalciferol. It occurs only in trace amounts in fruits and vegetables. Vitamin K is essential for blood coagulation, but dietary deficiency is uncommon. The recommended daily intake is 120 μg. It occurs abundantly in lettuce, spinach, nuts, lentils, green onions, cauliflower, cabbage, broccoli, brussel sprouts, and leafy greens. As well as direct intake, it can also be produced by bacteria in the intestines.

Antioxidants in fruits and vegetables: Imbalance in the production of reactive oxygen species (ROS) leading to negative cellular alterations is known as oxidative damage, which is caused by several molecules. Reactive oxygen species are partially reduced forms of oxygen such as singlet oxygen, hydrogen peroxide (H_2O_2), superoxide or hydroxyl radical. Some, but not all of the components able to cause oxidative damage is free radicals (i.e. molecules with unpaired electrons, which determine their high reactivity). From a biological perspective, an antioxidant is considered as any compound able to oppose cellular oxidation. Diets rich in fruits and vegetables have been shown to reduce the incidence of cardiovascular disease and some chronic and degenerative diseases associated with oxidative damage. The incorporation of fruits and vegetables in the diet may also help to eliminate certain toxins. The protective effects have been associated with the presence of antioxidant compounds like ascorbic acid, carotenoids, vitamin E and phenolic compounds.

Phenolic Compounds

This group encompasses a great diversity of compounds derived from the aromatic amino acids phenylalanine and tyrosine. Their main functions are acting as deterrents of potential predators or antimicrobials, protecting against UV-radiation and contributing to the pigmentation of fruits and flowers. Phenolic compounds can contribute to the astringency and bitter taste of some products. They are generally present in low concentrations, but in certain cases, such as in blueberry, they can reach levels of more than 0.1 per cent. In general, they also accumulate in the peel more than in the pulp of fruits.

Flavonoids: Flavonoids represent a large group of phenolic compounds with two aromatic rings in their structure that are associated together by a 3C-oxygenated heterocycle. Phenolic compounds are usually present as glycosides, which reduce their activity against free radicals and increase their solubility.

Flavones and flavonols: Flavonols have a central ring of 3-hydroxypyran-4-one like rutin, luteolin and apigenin are common among flavones, while the most abundant flavonols are quercetin and kampferol. Onions are rich in these compounds. Blueberries also have high levels, especially in the peel, because synthesis is stimulated by exposure to light. Celery is a good source of flavones. Flavones are also present in citrus, but they are associated mainly with the fruit peel.

Flavanones and flavanols: Flavanones do not have the double bond in position 2, 3 of the central ring, while flavanols lack the carbonyl group at position 4. The genus citrus is characterized by the accumulation of flavanone glycosides. Orange juice is a source of the flavanone glycoside hesperidin. The flavanols catechin and epicatechin are common in grapes.

Isoflavones: Isoflavones are phytoestrogens present in legume vegetables. Soybean products are a good source of these compounds. The three most commonly found isoflavones are genistein, glycitein and daidzein.

Proanthocyanidins: Proanthocyanidins are oligomeric flavonoids (usually dimmers or oligomers of the flavanols catechin and epicatechin). They are common in the peel and seeds of grapes. Other sources of these compounds include apple, almond and blueberry.

Anthocyanidins: Anthocyanidins are pigments giving several fruits their characteristic red or purple colours, although in some conditions they can be uncoloured. Besides being pigments, anthocyanidins have great relevance due to their contribution to the antioxidant capacity of fruits and vegetables. There are six anthocyanidins more common in fruits and vegetables: pelargonidin, cyanidin, delphynidin, peonidyn, petunydin and malvidin. These compounds are usually present as glycosides associated with different sugars, since anthocyanidin glycosylation reduces antioxidant capacity relative to the free aglycons.

Table 9. Non-nutritive plant constituents (antioxidants) that may be beneficial to human health

Constituent	Compound	Sources	Effects on human wellness
Phenolic compounds			
Proanthoc-yanins	Tannins	Apple, grape, cranberry, pomegranate, aonla, okra, carrot, cabbage, cauliflower, lettuce, onions, spinach	Help prevent cancer
Anthocyanidins	Cyanidin, malvidin, delphinidin, pelargonidin, peonidin, petunidin	Red, blue, and purple fruits (such as apple, black berry, blueberry, cranberry, grape, nectarine, peach, plum, prune, pomegranate, raspberry and strawberry), carrot, tomato, turnip, beetroot, colour radish and pumpkins etc.	Help prevent heart disease, cancer initiation, diabetes, cataracts, and allergies, help lower blood pressure
Flavan-3-ols	Epicatechin, epigallocatechin catechin, gallocatechin	Apples, apricots, blackberries, plums, raspberries, strawberries, okra, carrot, cabbage, cauliflower, lettuce, onions, spinach, peas, tomatoes, amaranthus, fenugreek etc.	Help prevent platelet aggregation and cancer
Flavanones	Hesperetin, naringenin, eriodictyol	Citrus (oranges, grapefruit, lemons, limes, tangerines, etc.),	Help prevent cancer
Flavones	Luteolin, apigenin	Celeriac, celery, peppers, rutabaga, spinach, parsley, artichoke, guava, pepper, celeriae, spinach, leeks, beetroot, onion, snap beans etc.	Help prevent cancer, allergies, and heart disease
Flavonols	Quercetin, kaempferol, myricetin, rutin	Onions, snap beans, broccoli, cranberry, kale, peppers, lettuce	Help prevent heart disease and cancer initiation; are capillary protectants
Phenolic acids	Caffeic acid, chlorogenic acid, coumaric acid, ellagic acid	Blackberry, raspberry, strawberry, apple, peach, plum, cherry, ginger, turmeric, colocasia, okra, cabbage, cauliflower, lettuce, onions, spinach,	Help prevent cancer and lower cholesterol
Carotenoids			
Lycopene	-	Tomato, watermelon, papaya, Brazilian guava, autumn-olive, red grapefruit, pointed gourd, beetroots, leeks etc.	Help prevent cancer, heart disease, and male infertility

Constituent	Compound	Sources	Effects on human wellness
α-Carotene	-	Sweet potatoes, apricots, pumpkins, cantaloupes, green beans, lima beans, broccoli, brussel sprouts, cabbage, kale, kiwifruit, lettuce, peas, spinach, prunes, peaches, mango, papaya, squash, carrots	Help prevent/slow tumor growth
β-Carotene	-	Cantaloupes, carrots, apricots, broccoli, leafy greens (lettuce, swiss chard, etc.), mango, persimmon, red pepper, spinach, sweet potato	Help prevent cancer
Xanthophylls	Lutein, zeaxanthin, β-cryptoxanthin	Sweet corn, spinach, corn, okra, cantaloupe, summer squash, turnip greens	Help prevent macular degeneration
Monoterpenes	Limonene	Citrus (grapefruit, tangerine, etc.)	Help prevent cancer
Sulfur compounds	Glucosinolates, isothiocyanates, indoles, allicin, diallyl isulphide	Broccoli, brussel sprouts, mustard greens, horseradish, garlic, onions, chives, leeks	Help prevent cancer and diabetes; help lower cholesterol and blood pressure

Important Minerals

In general, vegetables are a richer source of minerals than fruits, but both vegetables and fruits are considered nutrient-dense foods in that they provide substantial amounts of micronutrients, such as minerals and vitamins, but relatively few calories. The major mineral in fruit is potassium and is present at more than 200 milligrams per 100 grams fresh weight in most produce. Many other essential minerals are present in fruit but their contribution to total dietary requirements is generally of minor importance. Although iron and calcium are present at nutritionally significant levels, but it is often in a form that is unavailable for absorption by human being. The nutritional value of various fruits depends not only on the concentration of nutrients in their produce but also on the amount of such produce consumed in the diet. Tomatoes and oranges are relatively low in concentration of nutrients but make the major contribution of all produce to Indian diets because of the large per capita consumption.

Potassium (K): Potassium-rich diet contributes to lower blood pressure, blunting the effects of salt. Inadequate levels of potassium intake have long been associated with higher blood pressure. Potassium also regulates heartbeat, assists in muscle contraction and is needed to send nerve impulses and to release energy from fat, carbohydrates and protein. Potassium is the most abundant individual mineral element in fruits and vegetables. It normally varies between 60 to 600 mg per 100 g^{-1} of fresh tissue. In fruits and vegetables, potassium occurs mainly in combination with various organic acids. Banana and plantains, leafy green vegetables, many dried fruits, oranges and orange juice, cantaloupes and honeydew melons, tomatoes and root vegetables are potassium-rich. Baked potato or sweet potato, banana and plantain, cooked dry beans, cooked greens, dried fruits (such as apricots and prunes), winter (orange) squash, cantaloupe are helps prevent hypertension (high blood pressure), stroke, and arteriosclerosis.

Calcium (Ca): Calcium is essential for bone and tooth formation. Because of this, calcium requirements are higher during adolescence. Calcium is also very important during later adulthood, and of great consequence from a public health perspective. Inadequate intake of calcium may increase the risk of osteoporosis, a condition in which decreased bone mass weakens bone. Dark green leafy cabbage, turnip greens and most green leafy vegetables are potential calcium sources because of their absorbable calcium content. Cooked vegetables (such as beans, greens, okra, and tomatoes), peas, papaya, raisins, orange, almonds, snap beans, pumpkin, cauliflower and rutabaga are good source of calcium. Regular consumption of these fruit and vegetables are helps in prevent osteoporosis, lower blood pressure, aids muscles, skeleton, and teeth.

Magnesium (Mg): Magnesium is important in protein synthesis, release of energy from muscle storage and body temperature regulation. It is critical for proper heart function and plays a role in bone formation. Magnesium activates over 100 enzymes. Generally, magnesium levels are significantly higher in vegetables than in fruits, but nuts are good sources of this nutrient. Dry fruits, spinach, lentils, okra, potato, banana, corn, cashews and legumes are rank higher in magnesium content. It helps to prevent osteoporosis, aids in nervous system, teeth and immune system.

Phosphorus (P): Inorganic phosphate is essential for skeletal mineralization and for multiple cellular functions, including glycolysis, gluconeogenesis, DNA and RNA synthesis, cellular protein phosphorylation, phospholipid synthesis and intracellular regulatory roles. Among tree fruits, nuts are natural sources of phosphorus.

Nitrogen (N): The largest requirement for nitrogen in eukaryotic organisms is the biosynthesis of amino acids, building blocks of proteins and precursors of many other compounds. Proteins represent a large percentage of the human body and carry out many different cell functions.

Sulfur (S): Sulfur is an essential nutrient required for growth, primarily used to synthesize cysteine and methionine. The sulfur-containing amino acids play pivotal roles in the structural and catalytic functions of proteins. Sulfur nutrition is important in the species within the order Brassicales (e.g. white cabbage, broccoli, cauliflower, capers) for the synthesis of anticarcinogenic glucosinolate compounds.

Manganese (Mn): Manganese is a key component of enzyme systems, including oxygen-handling enzymes. It supports brain function and reproduction and is also required for blood sugar regulation. Among horticultural crops, spinach is a good source of manganese.

Copper (Cu): Copper, a redox active metal, plays an important role in the oxidative defense system. Copper is necessary for the formation of hemoglobin and is required for the function of over 30 proteins, including superoxide dismutase, ceruloplasmin, lysyl oxidase, cytochrome c oxidase, tyrosinase and dopamine- β –hydroxylase. Amaranth and drumstick leaves, methi, turnip, mustard green, coriander, green peas, okra, tomato and caster apple etc. are very good source of copper.

Iron (Fe): Iron is required in numerous essential proteins, such as the heme-containing proteins, electron transport chain and microsomal electron transport proteins, and iron-sulfur proteins and enzymes such as ribonucleotide reductase, prolyl hydroxylase phenylalanine hydroxylase, tyrosine hydroxylase and aconitase. Almonds, pistachio nuts, walnuts, pecans, etc., are very good sources of iron. Different vegetables like parsley, broccoli, kale, turnip greens, collards, green peas, beans, karonda and banana are also considered good sources of iron.

Zinc (Zn): Zinc is a pervasive microelement that plays a catalytic or a structural role in more than 200 enzymes (e.g. carboxypeptidase, liver alcohol dehydrogenase and carbonic anhydrase) involved in digestion, metabolism, reproduction, and wound healing. Fruits are poor in zinc, but pecans and walnuts are good sources of this essential mineral. Parsley is also a good source of zinc.

Sodium (Na): It is important in electrolyte balance and essential in coregulating ATP with potassium. In addition, it has an important role in the regulation of blood pressure. Olives and spinach are horticultural sources of sodium. In general, fruits are poor in sodium, and are recommended for low-sodium dietary patterns.

Table 10. Nutritive constituents of vegetables and their impact on human wellness

Constituent	Sources	Effect on human wellness
Calcium	Curry leaves, amaranths leaves, drumstick leaves, methi, turnip, mustard green, coriander, green peas, okra, tomato, snap beans, pumpkin, cauliflower and spinach.	Osteoporosis, formation and maintenance of skeleton and teeth. It is required for normal contraction of muscle to make limbs move, heart for its normal function, nervous activity and blood clotting.
Iron	Drum stick leaves and fruits, amaranth, methi, mint, coriander, spinach, mustard green, lettuce, cabbage, peas, beans, tomatoes, carrot, bitter gourd, onion etc	Formation of haemoglobin of red cells of blood and plays an important role in the transport of oxygen. Required for tissues oxidation-reduction.
Magnesium	Spinach, okra, potato, green banana, corn etc.	It is required for osteoporosis, nervous system, teeth, immune system and cellular metabolism.
Potassium	Potato, sweet potato, green beans, peas, squash and cantaloupe etc.	Hypertension, stroke arteriosclerosis. Essential to maintain osmotic balance and keep cells in proper shape.
Phosphorus	Potatoes, carrot, tomatoes, cucumber, spinach, cauliflower, lettuce etc.	Deposited as calcium phosphate in the bone and the teeth. Phosphorus is also a component of nucleic acids.
Zinc	Okra, beans, carrot, cabbage, cauliflower, lettuce, onions, spinach, peas, tomatoes, amaranthus, fenugreek etc.	It is a co-factor for a number of enzymes. Zinc deficiency leads to growth failure and poor development of gonadal function.
Copper	Amaranth leaves, drumstick leaves, methi, turnip, mustard green, coriander, green peas, okra, tomato, etc.	It is involved in iron absorption, cross linking of connective tissues, neurotransmission and lipid metabolism.

Table 11. Predominating nutritional constituents of major vegetables

Vegetables	Energy (Kcals)	Moisture (g)	Protein (g)	Fat (g)	Mineral (g)	Carbohy-drate (g)	Fibre (g)	Calcium (mg)	Phosphorus (mg)	Iron (mg)	Carotene (µg)	Thiamine (mg)	Riboflavin (mg)	Niacin (mg)	Vit. C (mg)
Amaranths	45	85.7	4.0	0.5	2.7	6.1	1.0	397	83	3.49	5520	0.03	0.30	1.2	99
Ash gourd	10	96	0	0	0	2	1	30	20	1	0	0.06	0.01	0.4	1.0
Beans, scarlet	158	58	7	1	2	30	2	50	160	3	34	0.34	0.19	0.0	1.0
Bathua leaves	30	89.6	3.7	0.4	2.6	2.9	0.8	150	80	4.2	1740	0.01	0.14	0.6	12
Beet root	43	87.7	1.7	0.1	0.8	8.8	0.9	18.3	55	1.19	0	0.04	0.09	0.4	1.0
Bitter gourd	25	92	2	0	1	4	1	20	70	1	126	0.07	0.09	0.5	88
Bitter gourd, small	60	83	2	1	1	11	2	23	38	2	126	0.07	0.06	0.4	96
Bottle gourd	12	96	0	0	0	2	1	20	10	0	0	0.03	0.01	0.2	0
Brinjal	24	93	1	9	9	4	1	18	47	0	74	0.04	0.11	0.9	12
Brussels sprouts	52	85.5	4.7	0.5	1.0	7.1	1.2	43	82	1.8	126	0.05	0.16	0.4	72
Broad beans	48	85	4	0	1	7	2	50	64	1	9.0	0.08	-	0.8	12
Cabbage	27	91.9	1.8	0.1	0.5	4.6	1.0	39	44	0.8	120	0.06	0.09	0.4	120
Cauliflower	30	91	3	0	1	4	1	33	57	1	30	0.04	0.10	1.0	56
Carrot	48	86	0.9	0.2	1.1	10.6	1.2	80	530	1.03	1890	0.04	0.02	0.6	3
Cho-cho marrow	27	92	1	0	0	6	1	140	30	1	0	0	0.04	0.4	4.0
Cluster beans	16	81	3	0	1	11	3	130	57	1	198	0.09	0.03	0.6	49
Colocasia	97	73.1	3.0	0.1	1.7	21.1	1	40	140	0.42	24	0.09	0.03	0.4	0
Colocasia stem	18	94	0	0	1	4	1	60	20	0	104	0.07	0.07	0.1	3.0

Vegetables	Energy (Kcals)	Moisture (g)	Protein (g)	Fat (g)	Mineral (g)	Carbohy-drate (g)	Fibre (g)	Calcium (mg)	Phosphorus (mg)	Iron (mg)	Carotene (μg)	Thiamine (mg)	Riboflavin (mg)	Niacin (mg)	Vit. C (mg)
Cowpea pods	48	85	3	0	1	8	2	72	59	2	564	0.07	0.09	0.9	14.0
Cucumber	13	96	0	0	0	2	0	10	25	1	0	0.03	0	0.2	7.0
Double beans	85	74	8	0	1	12	4	40	140	3	-	-	-	-	22
Drumstick	26	87	3	0	2	4	5	30	110	0	110	0.05	0.07	0.2	120
Drumstick flowers	50	86	4	1	1	7	1	51	90	0	-	-	-	-	-
Field beans, tender	48	86	4	1	1	7	2	210	68	1	187	0.10	0.06	0.7	9
Figs, red	53	80	1	1	2	11	6	187	39	0	-	-	-	-	-
French beans	26	92	2	0	1	5	2	50	28	1	132	0.08	0.06	0.3	24
Ghosala	18	93	1	0	1	3	2	36	19	1	120	0.02	0.06	0.4	0
Giant chillies	24	93	1	0	1	4	1	10	30	0	427	0.55	0.05	0.1	137
Jack fruit, tender	51	84	3	0	1	9	3	30	40	2	0	0.05	0.04	0.2	14
Jack fruit seeds	133	65	7	0	1	26	2	50	97	2	-	-	-	-	-
Kankoda	52	84	3	1	1	8	3	33	42	5	1620	0.05	0.18	0.6	-
Karonda fresh	42	91	1	3	0	3	2	21	28	0	-	-	-	-	-
Karonda dry	364	18	2	10	3	67	0	160	60	39	-	-	-	-	-
Kheksa	29	90	1	0	1	6	2	27	38	0	-	-	-	-	-
Kovai	18	93	1	0	1	3	2	40	30	0	156	0.07	0.08	0.7	15
Knol-Khol	21	93	1	0	1	4	2	20	35	2	21	0.05	0.09	0.5	85
Ladies finger	31	90	2	0	1	6	1	66	56	0	52	0.07	0.10	0.6	13
Lettuce	21	93.4	2.1	0.3	1.2	2.5	0.5	50	28	2.4	990	0.09	0.13	0.50	10

Vegetables	Energy (Kcals)	Moisture (g)	Protein (g)	Fat (g)	Mineral (g)	Carbohy-drate (g)	Fibre (g)	Calcium (mg)	Phosphorus (mg)	Iron (mg)	Carotene (µg)	Thiamine (mg)	Riboflavin (mg)	Niacin (mg)	Vit. C (mg)
Lakuch, raw	73	90	2	1	1	14	3	67	25	0	-	-	-	-	-
Leeks	77	79	2	0	1	17	1	50	70	2	18	0.23	-	-	11
Lotus stem, dry	234	10	4	1	9	51	25	405	128	61	0	0.82	1.21	1.9	3
Mango, green	44	88	1	0	0	10	1	10	19	0	90	0.04	0.01	0.2	3
Onion stalks	41	88	1	0	1	9	2	50	50	7	595	0	0.03	0.3	17
Onion	50	86.6	1.2	0.1	0.4	11.1	0.6	46.9	50	0.60	0	0.08	0.01	0.4	11
Papaya, green	27	92	1	0	1	6	1	28	40	1	0	0.01	0.01	0.1	12
Parwar	20	92	2	0	1	2	3	30	40	2	153	0.05	0.06	0.5	29
Pink beans	44	87	3	0	1	7	2	54	70	2	454	0.06	0.02	0.6	12
Plantain flower	34	90	2	1	1	5	1	32	42	2	27	0.05	0.02	0.4	16
Plantain green	64	83	1	0	1	14	1	10	29	6	30	0.05	0.02	0.3	24
Plantain stem	42	88	1	0	1	10	1	10	10	1	0	0.02	0.01	0.2	7
Pumpkin fruit	25	93	1	0	1	5	1	10	30	0	50	0.06	0.04	0.5	2
Pumpkin flowers	39	89	2	1	1	6	1	120	60	0	-	-	-	-	-
Potato	97	74.7	1.6	0.1	0.6	22.6	0.4	10	40	0.48	24	0.10	0.01	1.2	17
Radish	17	94.4	0.7	0.1	0.6	3.4	0.8	35	22	0.4	3	0.06	0.02	0.5	15
Ridge gourd	17	95	1	0	0	3	0	18	26	2	33	-	0.01	0.2	5
Snake gourd	18	94	1	0	1	3	1	26	20	22	96	0.04	0.06	0.3	0
Spinach	26	92.1	2	0.7	1.7	2.9	0.6	73	21	1.14	5580	0.03	0.26	0.5	28
Sundakai, dry	269	12	8	2	5	55	18	390	180	2	450	-	-	-	-

Vegetables	Energy (Kcals)	Moisture (g)	Protein (g)	Fat (g)	Mineral (g)	Carbohy-drate (g)	Fibre (g)	Calcium (mg)	Phosphorus (mg)	Iron (mg)	Carotene (µg)	Thiamine (mg)	Riboflavin (mg)	Niacin (mg)	Vit. C (mg)
Sweet potato	120	68.5	1.2	0.3	1.0	28.2	0.8	46	50	0.21	6	0.08	0.04	0.7	24
Sword beans	44	87	3	0	1	8	2	60	40	2	24	0.08	0.08	0.5	12
Tapioca	157	59.4	0.7	0.2	1.0	38.1	0.6	50	40	0.9	-	0.05	0.10	0.3	25
Tinda, tender	21	93	1	0	1	3	1	25	24	1	13	0.04	0.08	0.3	18
Tomato, green	23	93	2	0	1	4	1	20	36	2	192	0.07	0.01	0.4	31
Turnip	29	91.6	0.5	0.2	0.6	6.2	0.9	30	40	0.4	0	0.04	0.04	0.5	43
Vegetable marrow	17	94	1	0	0	3	1	10	30	1	-	0.02	0	0.4	18
Water chestnut, fresh	115	70	5	0	1	23	1	20	150	2	12	0.05	0.07	0.6	9
Yam	111	69.9	1.4	0.1	1.6	26	1	35	20	1.19	78	0.07	-	0.7	-

Source: Gopalan. C, et al., 2004, Nutritive Value of Indian Foods, National Institute of Nutrition, ICMR, Hyderabad.

Table 12. Relative concentration of vitamins and minerals in fruit and vegetables, and the relative contribution of nutrients to diet

Nutrient concentration		Contribution of nutrients to diet	
Crop	**Rank**	**Crop**	**Rank**
Broccoli	1	Tomato	1
Spinach	2	Orange	2
Brussels sprout	3	Potato	3
Lima bean	4	Lettuce	4
Pea	5	Sweet corn	5
Asparagus	6	Banana	6
Artichoke	7	Carrot	7
Cauliflower	8	Cabbage	8
Sweet potato	9	Onion	9
Carrot	10	Sweet potato	10
Sweet corn	12	Pea	15
Potato	14	Spinach	18
Cabbage	15	Broccoli	21
Tomato	16	Lima bean	23
Banana	18	Asparagus	25
Lettuce	26	Cauliflower	30
Onion	31	Brussels sprout	34
Orange	33	Artichoke	36

Volatile compounds

All fruit and vegetables produce a range of small molecular weight compounds (less than 250) that possess some volatility at ambient temperatures. These compounds are not important quantitatively (normally less than 100 micrograms per gram fresh weight) but they are important to produce the characteristic flavour and aroma of fruit and to lesser extent, in vegetables also. Most fruits contain more than 100 different volatile compounds, mostly in minute amounts. The number of compounds present in produce is continually increasing, as the analytical techniques for their identification improves. The compounds are mainly esters, alcohols, acids and carbonyl compounds (aldehydes and ketones). Many of these compounds, such as ethanol, are common in all fruits. The threshold or minimum concentration, at which the odour of ethyl 2-methylbutyrate, the main characteristic odour of apple, can be detected organoleptically was found to be 0.001 microlitres per litre, that is, an apple of 100 grams is recognized if 0.01 micrograms of ethyl 2-methylbutyrate is present. For the characteristic desirable, detectable level of odour it is desirable that it must be present in the correct concentration. At different stages of maturation, different compounds become the dominant component of flavour so that a blindfolded subject would be able to detect the stage of development of a particular commodity by sniffing the aroma.

Table 13. Distinctive components of the aroma of fruits and vegetables

Fruit	Product	Compound
Apple	ripe	Ethyl 2-mehtylbutyrate
	green	Hexanal, 2-hexenal
Banana	green	2-Hexenal
	ripe	Eugenol
	overripe	Isopentanol
Grapefruit		Nootakatone
Lemon		Citral
Orange		Valencene
Rasberry		1-(p-Hydroxyphenyl)-3-butanone

Table 14. Predominating bio-chemical constituents of major fruits

S. No.	Name of fruit	Sugars	Organic acid	Pigments	Fat	Flavouring compound	Vitamins	Phenols	Mineral
1	Apple	Glucose & fructose	Malic acid	Anthocyanin-Idaein (cynidin-3-galactaside)	Stearic and arachidic acid	Ethyl 2- methyl-butyrate	Vitamin A & C	Quinic acid, epicatechin quercetin-3-o-β-D-galactopyranoside, phloretin-2-o-glucoside and cynidin-3-galacto-pyranoside	Potassium
2	Apricot	Sucrose	Citric acid	Carotenoids	-	Integrated response of myrcene, limonene, P-cymene, terpinolenetans-2-hexenol, α-terpined, geraniol, 2 methylbutyric and acetic acid, linalool and its 2 oxides α- octalactone and α-decalactone.	Vitamin C	Chlorogenic acid and P- coumaric acid derivatives (probably P-coumaryl quinic acid)	Potassium
3	Avocado	Glucose & Fructose	Malic acid	Carotenoids (α-carotene and cryptoxanthin)	Oleic, palmitic and linolenic acid	C 6 alcohol and aldehyde (trans-hex-2-en-1-ol, hexon-1- ol-and cis-hex 2- enol, hexenal)	Vitamin C	Catechin and flavone	Potassium
4	Banana	Sucrose	Malic acid	Carotenoid (β-carotene and lutein)	Olic, linoleic and linolenic	Amyl and isoamyl esters of acetic, propionic and butyric acids.	Vitamin C	Dopamine	Iron, Phosphorus

S. No.	Name of fruit	Sugars	Organic acid	Pigments	Fat	Flavouring compound	Vitamins	Phenols	Mineral
5	Ber	Glucose & Fructose	Malic acid	Carotenoid (β-carotene)	-	-	Vitamin A & C	Tannins	Phosphorus & iron
6	Cherry	Glucose & Fructose	Malic acid	Diglucoside of cynidin-keracyanin chloride	Phospholipid (45%) neutral (31%) glycolipid (24%)	Methylanthranilate and methyl salicylate	Vitamin C	Chlorogenic acid	Potassium
7	Citrus	Glucose & Fructose	Citric acid	Carotene and xanthophylls	Palmatic, palmitoleic, oleic, linoleic, and linolenic acids	Nootakatone (grapefruit), citral (lemon) and valencene (orange)	Vitamin C	Flavoniod (narirutin, narangin, hesperidin, neo-hesperidin, eriocitrin, diasmin) and triterpene (limonine & nomilin)	Potassium
8	Fig	Glucose & Fructose	Citric acid	Anthocyanin (Cynidin 3-glucoside & cynidin 3-rutinoside)	-	Ethyl acetate	Vitamin C	C-glycosylflavones	Potassium
9	Grape	Glucose & Fructose	Tartaric acid	Monoglucosides of malvidin (Black), Cyanidin (Red) and diglucoside (Muscadin cultivars)	-	Linalool and geraniol	Vitamin A	Tannins (complex esters of phenolic acids and sugars)	Potassium

S. No.	Name of fruit	Sugars	Organic acid	Pigments	Fat	Flavouring compound	Vitamins	Phenols	Mineral
10	Guava	Glucose & Fructose	Citric acid	Chlorophyll, carotene, xanthophylls and lycopene etc.	-	Methybenzoate, haxenol, P-phenyl ethyl acetate, methyl cinnamate & cinnamyl acetate	Vitamin C	(+) Catechin & (+) gallocatechin	Phosphorus
11	Kiwifruit	Glucose & Fructose	Citric acid	Chlorophyll (flesh)	-	Ethyl acetate, hexanal and hax-2-enal	Vitamin C	Tannins	Potassium & Nitrogen
12	Litchi	Glucose & Fructose	Malic acid	Cynidin-3-rutinoside and malvidin-3-acetylglucoside	-	-	Vitamin C	Genetisic, ferulic and P-hydroxy-benzoic acid	Phosphorus
13	Mango	Sucrose	Citric acid	Carotenoids (β-carotene and luteoxanthin)	Palmitoleate and linolenate acid	(z) Ocimene (E) Ocimene β- myrcene	Vitamin A & C	Tannic acid	Iron
14	Melons	Sucrose	Citric acid	β- carotene (muskmelon flesh) and lycopene (watermelon flesh)	-	Acetoin, 2,3-butyleneglycol	Vitamin C	-	Potassium
15	Papaya	Glucose & Fructose	Citric acid	Caricaxanthin (yellow skin) β-carotene (yellow flesh) and lycopene (red flesh)	C 16:1 fatty acid content of glycolioid and C 16:1 & C 18: 3 fatty acid content of phospholipids	Methyl butanoate (sweaty odour) and linalool (fresh papaya odour)	Vitamin C & A	-	Calcium & phosphorus

S. No.	Name of fruit	Sugars	Organic acid	Pigments	Fat	Flavouring compound	Vitamins	Phenols	Mineral
16	Peach	Sucrose	Malic acid	Anthocyanin (Cynidin-3-monoglucoside)	-	Lactones	Vitamin C	Chlorogenic acid, cafleic acid derivative epicatechin, catechin and cynidine	Potassium
17	Pear	Fructose	Malic acid	Carotenoids	-	Ethyl, propyl, butyl and hexyl acetates	Vitamin A	chlorogenic acid (fruit) and arbutin (seed)	Phosphorus & Calcium
18	Pineapple	Sucrose	Citric acid	Carotenoids	Phosphatidyl insositol, phosphatidyl chloline, phosphatidyl glycerol, phosphatidyl ethanolamine and phosphatic chic acid	Ethylacetate, butane-2-3 diol diacetate	Niacin	P-coumaric acid and ferulic acid	Potassium
19	Plum	Glucose & Fructose	Malic acid	Anthocynin (cynidin-3-rutinoside & peonidin-3-rutinoside)	-	Benzal-dehyde	Vitamin C	Neochlorogenic acid	Potassium
20	Strawberry	Glucose & Fructose	Citric acid	3-glucoside of cynidin and pelargonidin	Oleic acid	Ethyl hexanoate, ethyl butanoate, methyl hexanoate, hexyl acetate, ethyl propionate and 3-hexenyl acetate	Vitamin C	Chlorogenic acid, D-catechin and P-coumaricacid	Potassium

S. No.	Name of fruit	Sugars	Organic acid	Pigments	Fat	Flavouring compound	Vitamins	Phenols	Mineral
21	Tomato	Glucose & Fructose	Citric acid	Phytoent (colourless and lycopene)	-	Cis-3-hexenal, trans-2-2hexenal, hexenal β-ionone, 3-methyl butanol, cis-3-hexenol and methyl butanol	Vitamin C	Kaenperof-3-rutinoside, quercetin-3-glucosode and rutin	Potassium

Table 15. Proximate principles of different fruits

Fruit	Moisture (g)	Protein (g)	Fat (g)	Minerals (g)	Crude fibre (g)	Carbohy-drates (g)	Energy Kcal.	Calcium (mg)	Phosphorus (mg)	Iron (mg)	Carotene (μg)	Thiamine (mg)	Riboflavin (mg)	Niacin (mg)	Vit. C (mg)
Aonla	81.8	0.5	0.1	0.5	3.4	13.7	58	50	20	1.2	9	0.03	0.01	0.2	600
Apple	84.6	0.2	0.5	0.3	1.0	13.4	59	10	14	0.66	0	-	-	0	1
Apricot	19.4	1.6	0.7	2.8	2.1	73.4	306	110	70	4.6	58	0.22	-	2.3	2
Bael	61.5	1.8	0.3	1.7	2.9	31.8	137	85	50	0.6	55	0.13	0.03	1.10	8
Banana	70.1	1.2	0.3	0.8	0.4	27.2	116	17	36	0.36	78	0.05	0.08	0.5	7
Cherries	83.4	1.1	0.5	0.8	0.4	13.8	64	24	25	0.57	0	0.08	0.08	0.3	7
Dates dry	15.3	2.5	0.4	2.1	3.9	75.8	317	120	50	7.3	26	0.01	0.02	0.9	3
Dates fresh	59.2	1.2	0.4	1.7	3.7	33.8	144	22	38	0.96	-	-	-	-	-
Figs	88.1	1.3	0.2	0.6	2.2	7.6	37	80	30	1.0	162	0.06	0.05	0.6	5
Grapes (blue)	82.2	0.6	0.4	0.9	2.8	13.1	58	20	23	0.5	3	0.04	0.03	0.2	1
Grapes	79.2	0.5	0.3	0.6	2.9	16.5	71	20	30	0.52	0	-	-	0	1
Guava	81.7	0.9	0.3	0.7	5.2	11.2	51	10	28	0.27	0	0.03	0.03	0.4	212
Jack fruit	76.2	1.9	0.1	0.9	1.1	19.8	88	20	41	0.56	175	0.03	0.13	0.4	7
Lemon	85.0	1.0	0.9	0.3	1.7	11.1	57	70	10	0.26	0	0.02	0.01	0.1	39
Lichi	84.1	1.1	0.2	0.5	0.5	13.6	61	10	35	0.70	0	0.02	0.06	0.4	31

Fruit	Moisture (g)	Protein (g)	Fat (g)	Minerals (g)	Crude fibre (g)	Carbohy-drates (g)	Energy Kcal.	Calcium (mg)	Phosphorus (mg)	Iron (mg)	Carotene (μg)	Thiamine (mg)	Riboflavin (mg)	Niacin (mg)	Vit. C (mg)
Lime	84.6	1.5	1.0	0.7	1.3	10.9	59	90	20	0.30	15	0.02	0.03	0.1	63
Malta	90.3	0.7	0.2	0.4	0.6	7.8	36	30	20	1.00	0	-	-	0	54
Musambi	88.4	0.8	0.3	0.7	0.5	9.3	43	40	30	0.70	0	-	-	0	50
Loquat	88.2	0.6	0.3	0.5	0.8	9.6	43	30	20	1.30	559	-	-	0	0
Mango	81.0	0.6	0.4	0.4	0.7	16.9	74	14	16	1.30	2743	0.08	0.09	0.9	16
Musk-melon	95.2	0.3	0.2	0.4	0.4	3.5	17	32	14	1.40	169	0.11	0.08	0.3	26
Water-melon	95.8	0.2	0.2	0.3	0.2	3.3	16	11	12	7.90	0	0.02	0.04	0.1	1
Mulberry	86.5	1.1	0.4	0.6	1.1	10.3	49	70	30	2.30	57	0.04	0.13	0.5	12
Orange	87.6	0.7	0.2	0.3	0.3	10.9	48	26	20	0.32	1104	-	-	-	30
Papaya	90.8	0.6	0.1	0.5	0.8	7.2	32	17	13	0.50	666	0.04	0.25	0.2	57
Passion fruit	76.3	0.9	0.1	0.7	9.6	12.4	54	10	60	2.0	54	0.07	0.14	1.6	25
Peaches	86.0	1.2	0.3	0.8	1.2	10.5	50	15	41	2.4	0	0.02	0.03	0.5	6
Pears	86.0	0.6	0.2	0.3	1.0	11.9	52	8	15	0.5	28	0.06	0.03	0.2	0
Phalsa	80.8	1.3	0.9	1.1	1.2	14.7	72	129	39	3.1	419	-	-	0.3	22
Pineapple	87.8	0.4	0.1	0.4	0.5	10.8	46	20	9	2.42	18	0.20	0.12	0.1	39
Plum	86.9	0.7	0.5	0.4	0.4	11.1	52	10	12	0.60	166	0.04	0.1	0.3	5
Pomegranate	78.0	1.6	0.1	0.7	5.1	14.5	65	10	70	1.79	0	0.06	0.10	0.3	16
Pummelo	88.0	0.6	0.1	0.5	0.6	10.2	44	30	30	0.30	120	0.03	0.03	0.2	20
Respberry	84.8	1.0	0.6	0.9	1.0	11.7	56	40	110	2.30	1248	-	-	0.8	30
Sapota	73.7	0.7	1.1	0.5	2.6	21.4	98	28	27	1.25	97	0.02	0.03	0.2	6
Sheetha-phal	70.5	1.6	0.4	0.9	3.1	23.5	104	17	47	4.31	0	0.07	0.17	1.3	37
Straw-berry	87.8	0.7	0.2	0.4	1.1	9.8	44	30	30	1.80	18	0.03	0.02	0.2	52
Zizyphus	81.6	0.8	0.3	0.3	-	17.0	74	4	9	0.50	21	0.02	0.05	0.7	76

IMPORTANCE OF FRUITS AND VEGETABLES IN HUMAN NUTRITION

Fruits and vegetables are one of the oldest forms of food known to man. In fact, Adam the first man ate an apple the 'forbidden fruit' of heaven. There are many references to fruits in ancient literature. Vedas state that the fruits form the base of the "Food of Gods". According to the Quran fruits like grape, date palm, fig, olive and pomegranate are gifts and heavenly fruits of God. The people in prehistoric times regarded fruits as endowed with magic or divine properties. They gave them due reverence and dedicate them to their gods and goddesses. They also used their designs in decorating temples, vestment or ceremonial garments and sacred vessels.

In human nutrition fruits and vegetables play an important role for balanced diet. These provide not only energy rich food, but also vital protective nutrients like vitamins, minerals, dietary fiber and plant phytochemicals. Comparatively, fruits and vegetables are the cheapest sources of natural nutritive foods. Vitamins like A, B, C and minerals like calcium, magnesium and iron are abundantly available from these crops, which help in building resistance against diseases. Many fruits and vegetable like dates, banana, potato, sweet potato, colocasia, etc. are rich source of carbohydrates, while peas, beans, avocado etc. provide an ample supply of protein. Fruits are also a rich source of organic acids like citric acid in citrus fruits and tartaric acid in grapes, which stimulate appetite and help in digestion. Papaya contains protein digesting enzymes. There are many benefits of eating plenty of fruits, including weight control, increased energy levels, reduced risk of cardiovascular diseases, diabetes and cancers, control of hyper tension, high cholesterol levels and curing diseases like scurvy, night blindness, stomach disorders, ulcers etc. and slowing down of the ageing process. Realizing the worth of fruits and vegetables in human health ICMR recommend consumption of 120g of fruits and 280g of vegetables per capita per day. Many fruits and vegetables have neutraceuticals properties, they are highly beneficial natural effect on the human system. The main physiological actions of fruits and vegetables are as follows:

Weight loss and detoxification: The enzymes present in the fruits help digestion while the fiber in the fruits is beneficial for decontamination the colon, keeping it clean by getting rid of accumulated toxic material and additional fats, so fruit diets are also highly recommended for weight loss as well as detoxification programs especially. Because fruits and vegetables have lower calories than dairy products. The body continues to burn calories even if they are being consumed in lower quantities and this also helps the process of detoxification.

Hydrating effect: Taking of fruits and vegetables are the most pleasant way of hydrating the organism. The water absorbed by sick persons in this manner has an added advantage of supplying sugar and minerals at the same time.

Diuretic effect: Clinical observations have shown that potassium, magnesium and sodium contents of the fruit and vegetable act as a diuretic and dieresis, frequency of urination is considerably increased when fruits and vegetable juices are taken. They lower the urine density and thereby accelerate the elimination of nitrogenous waste and chlorides. As, however, fruits contain a very low level of sodium, they make a valuable contribution to a salt-free diet.

Alkalinizing effect: The organic acids of the salts in fruits produce alkaline carbonates, when transformed within the organism, which alkalinize the fluids. All fruits and vegetables promote intestinal waste elimination. This keeps the body free from toxic wastes, which creep into the blood from an overloaded, sluggish intestinal tract. The carbohydrates of fruits are chiefly in the form of sugar, dextrin and acids, which are easily digestible and are completely, absorbed. Hence, they are very useful for sick for quick energy and heat.

Mineralizing effect: Fruits and vegetables furnish minerals to the body. Dried fruits such as apricots, raisins and dates are rich in calcium and iron. These minerals are essential for strong bones and better blood formation respectively. Some fruits, like custard apple, furnish calcium to the extent of 800 mg fruit^{-1}, which is sufficient to meet our daily requirement of this mineral.

Laxative effect: The fibrous matter in fruits and vegetables, cellulose, aids in the smooth passage of the food in the digestive tract and simplifies bowel action. The sugars and organic acids contained in fruits also increase their laxative effect. Hence, regular use of fruits and vegetable prevents and cures constipation.

Tonic action: Fruits, as dependable sources of vitamins, exert a tonic effect in the body. Guavas, custard apples and citrus fruits, like lemons and oranges, are particularly valuable sources of vitamin C. These fruits are usually eaten fresh and raw, thus making the vitamins fully available to us. Several fruits and vegetables contain good amounts of carotene, which gets converted to vitamin A in the body. A medium sized mango (350-400g) can provide as much as 15,000 international units of vitamin A, that is sufficient for full one week and this vitamin can be stored in the body. The common papaya is an excellent source of vitamin C and carotene.

Medicinal effect: Fruits like apple, lemon, orange and pomegranate aid in the proper functioning of the heart. Nutrients in apples, dates and mangoes sharpen memory prevent exhaustion, hysteria, insomnia and mental tension. All forms of berries are rich in iron, phosphorus and sodium that are beneficial for blood building and nerve strengthening. Lemons are good for liver ailments, indigestion and rheumatism. Watermelons are good kidney cleansers. Pineapple and pomegranates help in curing inflammation of nose, throat, hay fever and other chronic or bronchial ailments. Grapefruit juice is an effective aid for common cold. Fresh and ripe fruits like grapes, apples, bananas and figs are good for all brain deficiencies. Even the kernel of walnut is a helpful remedy for weakness of the brain.

"Eating liberal amount of fruits and vegetables daily ensures a healthy energetic and active life throughout lifetime"

Nonfood Uses of Fruits and Vegetables

Many dry fruits are used as decorations or in dried flower arrangements, such as unicorn plant, lotus, annual honesty and milkweed. Ornamental trees and shrubs are often cultivated for their colourful fruits, including holly, pyracantha, viburnum, skimmia, beautyberry and cotoneaster. Fruits of opium poppy are the source of opium, which contains the drugs morphine and codeine, as well as the biologically inactive chemical thebaine from which the drug oxycodone is synthesized. Osage orange fruits are used to repel cockroaches. Bayberry fruits provide a wax often used to make candles. Many fruits provide natural dyes, e.g. walnut, sumac, cherry and mulberry. Dried gourds are used as decorations, water jugs, birdhouses, musical instruments, cups and dishes. Pumpkins are carved into jack-o-lanterns for halloween. The spiny fruit of burdock or cocklebur were the inspiration for the invention of velcro. Coir is a fibre from the fruit of coconut that is used for doormats, brushes, mattresses, floor tiles, sacking, insulation and as a growing medium for container plants. The shell of the coconut fruit is used to make souvenir heads, cups, bowls, musical instruments and birdhouses.

2 Importance of Postharvest Technology in Horticultural Crops

For the duration of the last 20 years, considerable emphasis has been given to the production of horticultural crops in India. During 2017-18, the production of horticulture crops was 311.71 million tonnes from an area of 25.43 million hectares. The production of fruits and vegetables has increased from 203.0 million tonnes to 281.75 million tonnes since 2004-05 to 2017-18. At the same time, sizeable quantity of these is lost every year after harvest which is a significant economic loss. These losses are known as post-harvest losses which occur at different stages in the chain of harvesting, packaging, storage, transport and marketing. Postharvest losses occur in horticultural crops like fruits, vegetables and flowers, owing to the poor production planning, lack of suitable harvesting equipment, collection centres in major producing areas, ill equipped markets, unsuitable packing containers, insufficient storage capacity, be short of cold chain and substandard transport infrastructure, be deficient in organizational setup and poor state of processing industries. Losses in fruits, vegetables and flowers are estimated to vary between 20 to 30 per cent valued at nearly 8,000 to 12,000 crores annually depending upon the quality and market prices of produce. Astonishingly more than three lakh tones of cashews and millions liters of coconut water currently go to waste. Pineapple and oranges produced in Assam are very poorly priced, owing to poor transportation and marketing facilities and to limited preservation capacity at the production site. A study jointly conducted by the management consultancy firm, McKinsey & Co. and the Confederation of Indian Industry, determined that at least 50 per cent of production of fruits and vegetables in the country is lost due to wastage and value destruction. The cost of wastage is estimated at Rs. 23,000 to 30,000 crores on an annual basis.

Table 16. Post-harvest losses in fruits and vegetables

Fruits	Post-harvest losses (%)	Vegetables	Post-harvest losses (%)
Guava	10-32	Cabbage	8-25
Papaya	20-35	Carrot	5-10
Jamun	30-60	Cauliflower	8-25
Mango	15-33	Chilies	5-35
Sapota	20-25	Cucumber	15-22
Pineapple	20-40	Leafy vegetables	30-40
Citrus	10-30	Onion	25-40
Ber	10-30	Pea	25-30
Banana	25-40	Potato	30-40
Phalsa	20-35	Radish	3-5
Grapes	10-25	Tomato	10-35
Average losses	20-25	Average losses	25-30

(Source: Prof. I.S.Singh, Post harvest handling and processing of fruits and vegetables, page 4)

Table 17. Fruit and vegetable wastes generated after processing, packing, distribution and consumption in organized sectors.

Country	Production (million tons)			Fruit & Vegetable processed (%)	Processed Fruit & Vegetable (million ton)	Losses and wastage (%)			Waste generated (million tonnes)
	Fruits	Vegetables	Total			Processing	Distrib-ution	Consu-mption	
India	97.35	184.40	281.75	2.2	6.20	25	10	7	2.60
China	122.19	473.06	595.25	23.0	136.91	2	8	15	31.98
Philippines	16.18	6.30	22.48	78.0	17.53	25	10	7	6.53
Malaysia	1.07	1.21	2.28	80.0	1.82	25	10	7	0.68
Thailand	10.27	3.81	14.08	30.0	4.22	25	10	7	1.57
USA	25.38	35.29	60.67	65.0	39.43	2	12	28	14.95

Table 18. Fruits and vegetables used for processing in different countries

S.No.	Countries	Per cent	S.No.	Countries	Per cent
1	Malaysia	83	5	Australia	60
2	Saudi Arabia	80	6	UK	50
3	Philippine	78	7	USA	40
4	Brazil	70	8	India	2.2

Fresh horticultural crops are varied in morphological structure like roots, stems, leaves, flowers and fruits. All horticultural crops fresh produce are high in water content and are subject to desiccation (wilting, shriveling) and mechanical injury. They are also susceptible to assault by bacteria and fungi, with pathological breakdown. High respiration rate, ethylene production and action, rates of compositional changes (associated with colour, texture, flavour and nutritive value), mechanical injuries, water stress, sprouting and rooting, physiological disorders and pathological breakdown are key causes of fruits and vegetable deterioration. The rate of biological deterioration depends on numerous environmental factors, including temperature, relative humidity, air velocity, and atmospheric composition (concentrations of oxygen, carbon dioxide, and ethylene), and sanitation procedures. Qualitative losses, such as loss in edibility, nutritional quality, caloric value, and consumer acceptability of fresh produce, are much more difficult to assess than are quantitative losses. Quality standards, consumer preferences and purchasing power vary greatly differ to across countries and cultures and these differences influence marketability and the magnitude of post-harvest losses. Lessening of post-harvest losses can increase fruit and vegetables availability to the growing world population, decrease the area needed for production, and conserve natural resources. Strategies for loss avoidance include, use of genotypes that have longer post-harvest life, employ of integrated crop management systems and Good Agricultural Practices (GAP) that result in good keeping quality of the commodity and exploit of proper post-harvest handling practices in order to maintain the quality and safety of fresh produce. Although minimizing post-harvest losses of already produced horticultural products are more sustainable than increasing production to compensate for these losses. Less than 5 per cent of the funding of agricultural research and extension programs worldwide is devoted to activities related to maintenance of produce quality and safety during post-harvest handling. This situation must be changed if success is to be achieved in reducing post-harvest losses of horticultural perishables.

ATTRIBUTE OF FRUIT AND VEGETABLE QUALITY

The word quality comes from the Latin qualitas, which represents attribute, property or basic nature of an object. However, it can be defined as the quality is a measure of the degree of excellence or acceptability by the consumer. It refers to the composite individuality of physical and chemical parameters of the fruits and vegetables. The characteristics that impart distinctive quality may be described by four different attributes like colour and appearance, flavour (taste and aroma), texture and nutritional value. As consumers, these four attributes typically affect in the order specified above, for example evaluate the visual appearance and colour first, followed by the taste, aroma and texture. The relative importance of each of these elements varies with the commodity, intended use and proposed consumer. Hence a green peel colour in orange fruit may be unacceptable in one market and yet be highly valued in another market. In case of potato tubers which contain a relatively high reducing sugar may be considered desirable for consumption after boiling but unacceptable for consumption after frying due to the occurrence of excessive maillard browning during the latter method of preparation. Each chemical component is influenced by both genetic and environmental factors before as well as after harvest. For example, most cultivars of Irish potato accumulate reducing sugars in response to exposure to below 5° C temperatures however, the degree to which cold sweetening occurs is quite variable among cultivars of potato.

Table 19. Quality components of fresh fruits and vegetables

Factor	Components
Appearance	Size: dimensions, weight, volume Shape and form: diameter, depth ratio, smoothness Colour: uniformity, intensity Gloss: wax Defects: external, internal
Texture	Firmness: hardness, softness Crispness, succulence, juiciness, mealiness, grittiness, toughness, fibrousness
Flavour	Sweetness, sourness, astringency, bitterness, aroma, off flavours
Nutrition	Energy, protein, vitamins, minerals, fiber
Safety	Naturally occurring toxicants, environmental contaminants Mycotoxins, microbial contamination

Appearance (visual) quality factors

Appearance is the first feeling that the consumer receives and the most important component of the acceptance and eventually of the purchase decision. These may include size, shape, colour, gloss, and free from defects and decay. Appearance is determined by physical factors including the size, shape, wholeness, presence of defects (blemishes, bruises, spots, etc.), finish or gloss and consistency.

Size and shape: Size and shape of fruits and vegetables make imperative involvement to the appearance of fresh produce. Size and shape may be influenced by cultivar, maturity, production inputs and the growing environment. It is important for fruits and vegetables to be of uniform size and attribute shape. Size and shape can be measured with manual operation (human judgment), simple scale, verniercalliper, micrometers, planimeters and machine.

Colour: Colour is caused by the spectral allocation of light. Surface colour is important for the fresh market and internal colour for the processing. Green colour is pinpointing of insufficient ripeness in fruit but it is desirable attribute in vegetables. Colour is derived from the natural pigments in fruits and vegetables, many of which change as the plant proceeds through maturation and ripening. The primary pigments imparting colour quality are the fat soluble chlorophylls (green) and carotenoids (yellow, orange and red) and the water soluble anthocyanins (red, blue), flavonoids (yellow), betalains and lycopine (red). In addition, enzymatic and non-enzymatic browning reactions may result in the formation of water soluble brown, gray and black coloured pigments. The enzymes involved in browning reactions include polyphenol oxidase, which catalyzes the oxidation of polyphenolic compounds, and phenylalanine ammonia lyase, which catalyzes the synthesis of precursors to phenolic substrates. The chlorophylls are responsive to heat and acid, but stable to alkali whereas their counterpart carotenoids are sensitive to light and oxidation but relatively stable to heat. Carotenoids may be bleached by an enzyme called lipoxygenase, which catalyzes the oxidation of lipid compounds. Anthocyanins are sensitive to both pH and heat, while the flavonoids are sensitive to oxidation but relatively stable to heat. When the unit size is large such as mango, papaya, guava, apple, melon, tomato, etc. individual units have to be measured. When the unit size is small such as karonda, strawberry, grape berries, phalsa, peas, etc. a representative sample can be measured. The instruments used for colour measurement are Hunter Colour Difference Meter Spectronic-20 (reflectance measurement) and spectrophotometer with a tristimulas integrator.

Compactness: Compactness is the most relevant feature in species where the inflorescence is the marketable part such as broccoli and cauliflower, and the heads such as lettuce, cabbage and endive. In general, it is not correlated to their organoleptic characteristics but rather is an indicator of the degree of development at harvest. Open inflorescences indicate that they were picked too late while non-compact heads are the consequence of a premature harvest. To a certain extent, compactness is also an indicator of freshness because it decreases with dehydration.

Uniformity: Uniformity is a concept applied to all the components of quality like size, form, colour, ripeness, compactness, etc. For the consumer it is a relevant feature that indicates that someone who knows the product has already selected and separated it into categories based on the official standards of quality. It is so important that selecting products for uniformity is the main activity of market preparation.

Defects: Defects can originate before harvest as well as after harvest by the result of damage by insects, diseases, birds, and hail develop various blemishes on surface of fruit such as scars, scabs, russeting, rind staining etc. Deformities caused by unfavourable environment or genetic abnormalities, mechanical injury caused during handling and transportation such as damage, bruising and crushing.

Textural (feel) quality factors

Texture includes firmness, crispness, juiciness, mealiness and toughness, depending on the commodity. Hand and mouth feel is judged by sense of contact. Hand finger feel as firmness (apple), softness (mango and plum) and juiciness (citrus, grapes and sweet corn). Mouth fell includes sensory characters such as chewiness, fibrousness, grittiness, mealiness and stickiness. Textural quality of horticultural produce is not only imperative for their eating and cooking quality but also for their shipping ability. Soft fruits cannot be shipped over long distances without substantial losses due to physical injuries. In many cases, the shipment of soft fruits necessitates that they be harvested at less than ideal maturity, from the flavour quality standpoint. Instruments developed for measuring the textural qualities of fruits and vegetables are succulometer and tenderometer. It applies compression and shear to measure the tenderness of peas, value of 90 to 95 indicates fancy quality peas and value over 100 indicates a lower grade standard. Shear press is used for many fruits and vegetables. Pressure tester is normally used for measuring the maturity of various fruits.

Pulp/fruit ratio: Physical quality also refers to the pulp/fruit ratio particularly in mango, litchi, custard apple and pomegranate. Higher pulps per cent indicate the better fruit quality.

Seed characteristics: Fewer numbers and small size of seeds is an advantageous character in guava. Softness of the seed is superiority attribute in pomegranate while seedlessness is in table grapes and water mellon.

Flavour (eating) quality factors

Flavour has been defined as a mingled but unitary experience which includes sensations of taste, aroma (odour), pressure, and often cutaneous sensations such as warmth, colour, ormild pain. Aroma compounds are volatile they are perceived primarily with the nose, while taste receptors exist in the mouth and are impacted when the produce is chewed. Flavour may have the major impact on acceptability and desire to consume it again. Taste has been divided into five primary tastes that is sweet, sour (acidity), salty, bitter, and umami (astringency). Umami can be described as a taste associated with salts of amino acids and nucleotides. In the evaluation of fruit and vegetable flavour, it is important to consider off-flavours as well as desirable ones. These off-flavours may be produced through the action of enzymes such as lipoxygenase or peroxidase, which form reactive free radicals and hydroperoxides that may catalyze the oxidation of lipid compounds. When these reactions occur, the result may be the development of undesirable flavourus described as rancid, cardboard, oxidized, or wet dog. Estimation of volatile acids, amines and succinic acid provides as indicators of off-flavour in stored fruits and vegetables. An objective analytical determination of critical components must be coupled with subjective evaluations by a taste panel to yield useful and meaningful information about the flavour quality of fresh fruits and vegetables. Gas chromatographic techniques have been developed to detect and estimate volatile flavouring components but it is very expensive. Quality acceptance is determined by organoleptic taste or sensory method. The two principle types of tests made by sensory panels are difference tests and preference tests. The difference test usually involves a small panel of trained people who determine whether a difference exists between two or more samples. Preference tests are designed to determine preference of representative populations, which are conducted by large panels of 100 or more untrained persons. Panel may consist of trained individuals but groups are better because differences of opinion average out.

Nutritional quality factors

Fruits and vegetables are of high food value as they are major source of both macro nutrients such as fiber and carbohydrates, and micro nutrients such as Vitamin C, B complex (thiamin, riboflavin, B_6, niacin, folate), A, E, minerals, and the lesser-studied polyphenolics, carotenoids, and glucosinolates. Most important vitamins of fruits and vegetables are, vitamin A (mango, papaya, carrot, spinach, etc.), vitamin B complex (cashew nut, papaya, peas, beans, etc.) and vitamin C (aonla, guava, citrus, tomato, leafy vegetables, etc.). Minerals, like calcium is considerably available in almond, litchi, karonda, ameranthus and fenugreek, while iron is available in karonda, dates, cashewnut, mint, litchi, garlic and potato. Proteins rich fruits are cashewnut, almond, walnut and vegetables are peas, beans, garlic and legumes. Reducing sugars i.e. glucose is available in dates, grape, fig etc. and fructose in dates, grapes, pear, etc., whereas non-reducing sugar i.e. sucrose is extensively found in mango, pine apple, banana, etc. Acids responsible for sourness of various fruits such as citric acid in citrus fruit, mango and guava, malic acid in apple and tartaric acid in grape and tamarind. Total soluble solids and acid ratio determines the quality of many fruits including melons. Other constituents of fresh fruits and vegetables are including carotenoids, flavonoids, isoflavones, phytosterols, and other phytochemicals (phytonutrients).

Safety factors for fruits and vegetables

A number of factors threaten the safety of fruits and vegetables. These include naturally occurring toxicants, such as glycoalkaloids in potatoes, natural contaminants, such as fungal toxins (mycotoxins), bacterial toxins, heavy metals (cadmium, lead, mercury), environmental pollutants and pesticide residues. Unless fertilized with animal and/or human waste or irrigated with water containing such waste, raw fruits and vegetables should normally be free of most human and animal enteric pathogens. Organic fertilizers, such as chicken manure and bone mill, should be sterilized prior to their application in fruit and vegetable production, so as to avoid the risk of contaminating fresh produce with Salmonella, Listeria, and other pathogens. Bacteria, like Shigella spp., Salmonella spp., Aeromonas spp., *Escherichia coli*, *Listeria monocytogenes* and the toxins produced by *Clostridium botulinum* and others, have been identified as responsible for illnesses associated with the consumption of fruits and vegetables. The hepatitis A virus has been detected on fresh produce as well as parasites like *Entomoeba histolyca*, and *Giardia lamblia*. Commodities that touch the soil are more likely to be contaminated than those that do not come in contact with the soil. The best approach to achieving and maintaining the safety of fresh fruits and vegetables is to focus on limiting potential contamination during their growth, harvesting, handling, treatment, packaging and storage. Strict devotion to Good Agricultural Practices, i.e. basic food safety principles associated with minimizing biological, chemical and physical hazards from the field throughout the distribution chain of fresh fruits and vegetables. Careful handling and washing of all produce to be consumed raw and the strict observance of proper sanitary measures are strongly recommended to reduce microbial contamination at the fruit and vegetables service, retail, and consumer levels.

Agrochemicals are the herbicides, insecticides, fungicides, fumigants, rodenticides, growth regulators, waxes, disinfectants, additives and all other products of a chemical nature used during production or postharvest handling. Their residues have always been a major concern, although advances in chemistry and biochemistry, better understanding of ecology as well as pests and diseases and the growing use of non-chemical control methods have made the present world a much safer place.

Table 20. Instrumental methods for determination of fruit and vegetable quality

Quality Attribute	Fruits	Vegetables	Objective method of measurement
Size	Diameter, drained weight	Sieve size, drained weight	Dimensions (scales, screens, sizing rings, micrometers, etc.), weight and volume
Shape	Height/weight ratio	Straightness	Dimension ratios, displacement, angles, diagrams and models
Wholeness	Cracked pieces	Cracked pieces	Counts, per cent whole, photographs, models
Gloss	Gloss, finish		Gloss meters, goniophotometers, wax platelets
Consistency	Consistency	Consistency	Consistometers, viscometers, flow meters, spread meters
Defects	Blemishes, bruises, spots, extraneous matter	Blemishes, bruises, spots, extraneous matter	Photographs, drawings, models, scoring systems, computer-aided visual techniques
Colour	Colour	Colour	Colour charts, dictionaries, reflectance and transmittance, colourimeters, pigment extraction and spectrophotometers
Texture	Texture, firmness, grit, character, mealiness	Texture, mealiness, succulence, fiber, maturity	Tenderometers, texture analyzers – compression, shearing, extrusion meters, analysis of solids, moisture, grit, fiber
Aroma	Aroma, ripeness, sweetness, fruitiness	Aroma	Gas chromatograph, enzymes
Flavour	Flavour, sweet, sour, bitter	Flavour, sweet, sour, salty, bitter	Hydrometer, refractometer, pH meter, determination of acidity and sugars, sodium chloride, enzymes, amines, bitter alkaloids or glucosides
Nutritional value	Vitamin A, B, C, E, polyphenolics, carotenoids	Vitamin A, B, C, E, polyphenolics, carotenoids, glucosinolates	HPLC and spectrophotometric methods

Pre-Harvest Factors affecting Post-Harvest Quality Of Horticultural Fresh Produce

The postharvest quality of fruits and vegetables are largely determined by preharvest factors such as production location, climate, nutrition and plant growth regulators and the secondary factors include soil quality and management, shading, rootstock, irrigation, pruning and crop load manipulation. The influences of important preharvest factors on postharvest quality of fruits and vegetables are described below.

Temperature: Atmospheric temperature has been found to influence fruit shape, size, colour and other quality parameters. Temperature variation during the early stage of fruit development caused variation in shape of orange fruit, undesirable thick peel and puffiness in citrus. Pineapple fruits grown in winter months or in cool growing areas had reduced eating qualities due to lower sugar/acid ratio. At high temperature, plant respires at faster rate and stored carbohydrates of fresh produce are quickly depleted and drastically reduce shelf life and quality. The produce which is having higher amount of stored carbohydrates show longer storage/vase life. For example high temperature during fruiting season of tomato leads to quick ripening of fruits on and off the plant, but quality is not fine. Orange grown in the tropics tend to have higher sugars and TSS than those grown sub tropics. Due to lower diurnal fluctuation in temperature, the tropical grown oranges tend to be green in colour and peel less easily. High temperature reduces the quality in citrus, radish, spinach, cauliflower, etc. and increased the quality in grapes, melons, tomato, etc.

Radiation: The duration, intensity and quality of light affect on fruits and vegetables quality at harvest. Most of the produce needs high light intensity (3000-8000 f.c.). Absorption of red light (625-700 nm) through pigments, phytochrome, is essential for carbohydrates synthesis which determines the shelf life of the produce. The vase life of the carnation and chrysanthemums is longer under high light intensity than low. Citrus and mango fruits produced in full sun generally had a thinner skin, lower weight, juice content and acidity but higher TSS. Citrus fruits grown in the shade may be less susceptible to chilling injury when subsequently stored in cold storage. In tomatoes, leaf shading of fruits produced a deeper red colour during the ripening than in the case of those exposed to light. The side of the fruit that have been exposed to sun will generally firmer than the non-exposed side. Exposure of potato to light causes greening, due to synthesis of solanine, which has toxic properties. High light intensity causes sunscald in citrus and tomatoes and reduces the pure white colour of cauliflower. In general, the lesser the light intensity the lower the ascorbic acid content of plant tissues. In low light intensity the leafy vegetables produce larger and thinner leaves. Under low light intensity boosts incidence of postharvest disease because of reduced physical and physiological integrity of the fruit, reduces fruit size and fruit elongation and decreases ripening characteristics. During storage high radiation interception cause chilling injury in grapefruit, this may be attributed to the increase in weight loss due to changes in wax and cuticular structures.

Seasons / Day and day length: Seasonal fluctuation and time of the day at harvest will significantly affect the postharvest quality of the produce. Synthesis of higher amount of carbohydrates during the day time and its utilization through translocation and respiration in the night is responsible for the dissimilarity in the longevity of the cut flowers. Roses and tuberose have been found to show longer keeping quality in the winter season under ambient condition than in the summer seasons. If long days onion (temperate) grown during short day (tropics) condition it leads to very poor storage quality.

Carbon dioxide: At higher level of CO_2 produce quality planting material, early and more flowering, rapid crop growth and development, increase yield of different plants. Production of chrysanthemum under green house at 1000 – 2000 ppm of CO_2 showed an increase in stem length, fresh weight, number of leaves and longevity of cut flowers.

Relative humidity: In general higher relative humidity around the fruit decreases water and Ca movement into the fruits. But in contrast, higher relative humidity increases Ca accumulation into the fruit by reducing leaf evapotranspiration, whereas decreases shelf life due to Ca toxicity. High relative humidity around the plant has also associated with increased maturity bronzing in banana fruit, thin rind and increased size in some other horticultural produce and this produce is more prone to high incidence of disease during post-harvest period. Low humidity may cause browning of leaf edge on plants with thin leaves or leaflets.

Nutrition of plants: Deficiencies, excesses or imbalances of various nutrients are known to result in disorders that can limit the storage life of many fruits and vegetables. Therefore, nutritional status is an important factor in quality at harvest and postharvest life of various fruits crops.

Macro nutrients: Elevated nitrogen negative in terms of postharvest quality attributes and increases fruit respiration, faster tissue decline thereby reducing their storage life. High nitrogen levels cause a reduction in ascorbic acid, TSS/acid ratio, sugar content, decrease flavour, TSS, firmness and colour of the fruits and in stone fruits, it increases physiological disorders, tissue softening and decrease fruit colour, whereas altered levels of essential amino-acids, thiamine, riboflavin and carotene. In green leafy vegetables such as spinach, lettuce, celery and cabbage, high nitrogen application under low light conditions can result in the accumulation of nitrates in plant tissues to unhealthy levels. Application of phosphorous minimizes weight loss, sprouting and rotting in bulb crops. Phosphorous nutrition can alter the post-harvest physiology of some produce by affecting membrane lipid chemistry, membrane integrity and respiratory metabolism. High phosphorus reduces size, weight and vitamin C in various fruits. The respiration rate of low-phosphorous fruits will be higher than that of high phosphorous fruits during storage. Its deficiency causes poor appearance of citrus, onion and carrot. High potassium has been found to improve post-harvest life and quality of anthurium, cut flowers and many horticultural produce. In bananas, elevated K increased total soluble solids and vitamin C and reduced acidity, whereas increased acidity in pineapples. Application of potassium in water melon tends to decline the post-harvest respiration rate. Potassium deficiency can bring about poor fruit development and abnormal ripening of fruits and vegetables. The storage potential of the fruits is largely dependent on the level of Ca and it is associated with produce texture. The application of Ca and high fruit Ca concentration resulted in increased firmness, reduced disease incidence, chilling injury, physiological disorders (skin splitting) and hindrance ripening. It also improved storability of many fruits e.g. apple, mango, guava, tomato, etc. The higher level of N, P and Mg and low levels of K and Bo lead to the Ca deficiency in fruits and reduce its storage life. The incorporation of 4 per cent Ca into proto pectin of middle lamella form bond with the cellulose of the cell wall and thus postponed softening in fruits. Infused Ca inhibits the internal browning, retarded respiration, and reduced the metabolism of endogenous substrates. Post climacteric respiration of apple decreased as peel Ca level increased from 400 to 1300 ppm. Physiological disorders of storage produce related to low Ca content of the tissue are bitter pit in apples, cork spot in pears, blossom end rot in tomato, tip burn in lettuce and hallow heart in potato and red blotch of lemons etc. Application of $CaCl_2$ delayed the accumulation of free sugars, decreased inorganic contents, mould development, softening and development of red colour in strawberry.

Micro-nutrients: Zinc is known to act as means of transportation for carrying ions across tissue and increase Ca content of the fruit. Zn deficiency causes straggled cluster in grape. The spraying of Mg decreased fruit Ca and increased bitter pits in apples. Magnesium increases size, weight and vitamin C in fruits such as citrus. Copper deficiency causes irregular blotch on citrus fruits and spoils the appearance. Adequate supply of Bo improves the mobility of Ca in the leaves and the fruits and subsequently increases fruit firmness, TSS, organic acids and reduce the incidence of the drought spot, bitter pit and cracking disorders. Boron deficiency cause flesh browning (e.g. aonla), gummy discoloration of albedo in citrus, lumpiness in papaya, hollow stem in cabbage, turnip and cauliflower, cracking of outer skin in beets and superficial rind pitting in oranges.

Table 21. Storage disorder and storage characteristic of Cox's Orange Pippin apple in relation to their mineral content

Disorder	Composition (mg 100 g^{-1})				
	N	P	Ca	Mg	K/Ca
Bitter pit	-	-	< 4.5	>5	>30
Break down	-	<11	< 5	-	>30
Lenticel blotch pit	-	-	<3.1	-	-
Loss of firmness	>80	<11	< 5	-	-
Loss of texture		<12		-	-

Irrigation: Stress due to unnecessary or insufficient water in the medium reduces the longevity of the produce. Extreme water stress reduces yield and quality, mild water stress decreases crop yield but may improve some quality attributes and no water stress increases yield but may reduce postharvest quality. Generally, the excessive availability of water can result in larger fruit, reduced firmness, flavour and increase disorders. Moisture stress boosts the rate of transpiration over the rate of absorption and irregular irrigation or moisture regime leads fruits and vegetable cracking (potato, lime, grape, pomegranate, dates, litchi, lemon, cherries, plums and tomatoes.). In case of mango, lowering irrigation during rapid fruit expansion stage reduced storage duration and fruit Ca concentration. Increased water stress condition in fruit crops increases TSS, firmness and yellow fruit colour, decreased fruit size, juice content and thick skin in citrus fruit. Keeping quality of bulb crops like onion and garlic will be meager if irrigation is not stopped before three weeks of harvest. Low irrigation reduced fruit growth rate and green life of bananas, whereas fully matured banana harvested soon after irrigation the fruit can easily split during handling procedure, allowing microorganism infection and post harvest rotting. If orange is too turgid at harvest (early morning) the flavdeo/oil gland in the skin can be ruptured during harvest, releasing phenolic compounds and causes Oleocellosis or oil spotting (green spot on the yellow or orange coloured citrus fruit after degreening). In green leafy vegetables, in excess of rain or irrigation can results in the leaves becoming harder and brittle, which can make them more susceptible to damage and decay during handling and transport. High humidity reduces the colour and TSS and enhances acidity in citrus, grapes, tomato, etc., but on other hand it is needed for superior quality of banana, litchi and pine apple.

Plant growth regulators (PGR): The application of plant growth regulators during development stage of fruit and vegetables has increase the size of the edible portion and to obtain early bearing of fruits. Pre-harvest application of plant growth regulators such as IAA, GA_3, growth retardants like B-9, BA, CCC, A-Rest and Phosphon-D are improve quality and longevity of flower crops. Application of gibberellins during fruit development increased fruit weight and berry quality in grapes. Cultar was found to increase the size and other quality attributes of avocado fruit. Cytokinin and auxin increase shelf life and reduced fruit splitting in persimmon and citrus. Application of GA_3 @ 50-100 ppm improves post-harvest quality of roses and enhance the longevity of tuberose flowers by anthocyanin development and it stimulate the accumulation of N, K, Mg and S. Pre-harvest spray with alar (1500ppm), maleic hydrazide (500ppm), and cycocel (500ppm) increased vase life of aster. Pre harvest application of maleic hydrazide (2000ppm) on onion field prevent them sprouting during storage.

Table 22. Effect of plant growth regulators on post harvest life and quality of fruits and vegetables

PGRs	Effects of plant growth regulators
Auxins	Increases size in loquat (2, 4, 5- TP), grape (IAA), mandarins (NAA) and TSS in mango (2, 4-D)
Gibberellic acid	Pre harvest application increases size and weight of grape, apricot, strawberry and fig. It causes parthenocarpic fruit in fig, guava, grape, tomato, etc. it reduces disorder of fruits e.g. water spot and corky spot in citrus.
Cytokinin	BA and PBA maintain green colour of leafy vegetables and cause parthenocarpic fruit in fig.
Ethylene	Ethphon increases anthocyonin (grape, plum, apple, chilies, and brinjal), carotenoid (mango, guava, papaya, citrus, tomato, etc.), ascorbic acid and TSS and reduces tannin (grapes, dates, etc.) and acidity (grape, mango, tomato, etc.). Pre-harvest treatment with ethephon reduces internal browning of pineapple.
Growth retardant	Pre-harvest application of Alar (B_9) increases colour in apple, cherry, apricot, etc. and maleic hydrazide inhibits sprouting in onion bulb. Foliar application of CCC before harvest resulted higher content of total sugars and TSS in banana fruits.

Pest and disease infestation: Infection by fungi, bacteria, mites and insects reduces the longevity as well as consumer acceptability. Tissue damage caused by them show wilting and produce ethylene leads to early senescence. Vascular diseases, stem rot, root rot of floral corps hinder the transport, affects the post-harvest life and quality. The potato tuber moth may infest tubers during growth if they are exposed above the soil and subsequently in the storage. Banana which suffers a severe infection with diseases of leaf spot may ripen pre maturely or abnormally after harvest. Pre-harvest application of systematic fungicide prevents quiescent infection like anthracnose of mango, papaya, crown rot of banana, stem end rot of citrus, lenticel rot of apple and brown rot of peaches. Pre-harvest spray of chitosan reduced the post-harvest fungal rot and maintains the keeping quality of strawberry.

Harvest maturity: For occurrence, when fruits and vegetables are harvested too late or too early in the season, overall taste, texture, and colour may be compromised. Generally, the fruits harvested at the advanced stage of maturity have increased fruit size and eating qualities (taste and aroma) but decreased shelf life. Harvesting of fruits and vegetables at immature stage leads to both qualitative and quantitative losses. Immature fruits fail to ripen normally with low nutritive values and have inferior flavour quality when ripe. On the other hand over mature fruits are likely to become soft and mealy with insipid flavour soon after harvest. Mango fruits harvested at more advanced stage of maturity showed uneven ripening of the flesh such as spongy tissue and soft nose. Mango fruits harvested 7 days before optimum maturity showed better storage performance but failed to arrest skin discolouration and caused uneven ripening. Many vegetables, in particular leafy vegetables, and immature fruit-vegetables such as cucumbers, green beans, peas, and okras, attain optimum eating quality prior to reaching full maturity. Most of the cut flowers are harvested at the immature stage. Roses are harvested at tight bud stage or cracked bud stage than the half open or full open stage.

Time of harvesting: Time of harvesting is advisable to harvest produce when temperature is mild as high temperature causes rapid respiration rate and excessive water loss. The recommended time for harvest of fresh horticultural produce is early morning or late evening hours exhibits longer post-harvest life than produce harvested during hot time of the day. The duration between harvesting and delivery to a market also can damage the quality of the fruit, vegetable and flower. If fresh produce isn't processed quickly, it may also lose nutritional value.

Methods of harvest: The method of harvesting (hand vs mechanical) can also have significant impact on the composition and post-harvest quality of fruits and vegetables. Sharp tools, secateurs, harvester, hand gloves, digger, vibrater, trimmer and any such items should always be used to detach the fruits, vegetable and flowers from the mother plant. Mechanical injuries (such as bruising, surface abrasions and cuts) can accelerate loss of water and vitamin C resulting in increased susceptibility to decay causing by pathogens. Cut flowers with long stem have higher post-harvest life than short stem because shorter stem have less carbohydrate reserves. Slant cut helps in facilitating the maximum surface area to absorb water at rapid rate during vase life, where as crushed stem block water absorption.

Cultivar and rootstock genotype: Cultivar and rootstock genotype have an important role in determining the taste quality, nutrient composition and postharvest life of fresh commodities. In citrus, Troyer and Carrizo (Citrange) rootstock produce the fruit of excellent quality of oranges, mandarins and lemons. In guava *Psidium pumilum* rootstock increases sugar and *Psidium cujavillus* ascorbic acid content of fruit. Varieties with thick peel, high firmness, low respiration rate and low ethylene production rates would usually have longer storage life. The cultivars that have ability to withstand the rigors of marketing and distribution will have lesser losses after harvest. Varieties with resistance to low temperature disorders and/or decay-causing pathogens can be stored well for longer duration with minimum storage losses.

Foliar nutrient sprays: Calcium is often considered to be the most important mineral element in determining fruit quality, especially in apples and pears, where it has been demonstrated to reduce metabolic disorders, maintain firmness and reduce decay. In apples, to significantly affect fruit firmness or decrease fruit decay caused by postharvest wound pathogens, it is necessary to raise the level of flesh calcium (dry weight basis) to 800 to 1000 $\mu g\ g^{-1}$.

Cultivation practices: High-quality production process is important in achieving good yields and quality of fresh produce. Certain aspects are mainly important, such as, weed control, crop hygiene and proper plant protection. Fruit thinning increases fruit size and helps in obtaining better quality produce. Girdling 30 to 40 days before harvest can enhance fruit size, advance and synchronize maturity in peach and nectarine. Girdling can also cause the pits of peach and nectarine fruits to split, especially if it is done too early during pit hardening. In grapes, the girdling makes balance between vegetative and fruit growth and it may improve bunch shape, berry size, colour and sugar.

Canopy manipulations: Appropriate pruning and fruitlet thinning make a balance between yield and fruit size. Pruning affects the berry size, colour, acidity and sugar content of grape, phalsa, ber, peach, apple, etc. Fruits which are exposed to high light environment (outside canopy) acquire higher TSS, acidity, fruit size, aroma, and shelf life compared to which lies inside the canopy. The use of more efficient training systems that allow sunlight penetration into the center and lower canopy areas is recommended to reduce the number of shaded fruit. An open canopy lowers the relative humidity around the plant, reduces wetting periods, and improves spray penetration, which reduces disease and insect problems and improves foliar nutrient applications. An open canopy also enables the harvesters to pick fruits and vegetables more rapidly, decreasing the likelihood of overripe commodities. Summer pruning and leaf pulling around the fruit increases fruit light exposure it can increase fruit colour.

Miscellaneous: The locations of the branch and crop load manipulation have also been found to influence fruit qualities. In the case of apple, the fruits located in the centre of the cluster had higher sugar content than those of terminal ones. The mango fruits on the longer panicles are generally larger and remain on the tree until maturity. The fruit count: leaf count ratio (F: L) influences high bush blueberry fruit quality more than mineral nutrition. A high F: L results in later ripening, lesser TSS and smaller berries. Low leaf: fruit ratio in banana caused reduced green life resulted from taking longer time to attain minimum marketable size. On the other hand, high leaf: fruit ratio increased sugars and soluble solids concentrations because of reduced competition for carbohydrate but increased fruit disorders and disease severity. Fruit thinning reduced the competition for nutrients, light and space, which resulted in attractive shape and size of the product. Thinning in grapes, dates, peaches, plum, etc. increases size, colour, acidity and sugar content of fruits. Bunch covering with plastic bags produced banana fruits of better colour and quality. Fruit injury will reduce appearance and may be source of infection for fungus diseases. Expedited and careful handling, immediate cooling after harvest, maintenance of optimum temperatures during transit and storage, and effective decay control procedures are important factors in the successful post-harvest handling of fruits and vegetables.

Post-Harvest Factors Affecting Quality of Horticultural Fresh Produce

Post-harvest handling practices also affect on the quality of various stored fruits, vegetables and flowers by different factors such as temperature, light, relative humidity, water quality, chemical treatment, packing and packaging etc.

Temperature: High temperature causes off flavour, weight loss and wilting particularly in leafy vegetables and decreases vitamin C content. Low temperature reduces the appearance quality of fruit by checking the carotenoid development. Temperature lower than the optimum end result in chilling and freezing injury of various fruits and vegetables. Senescence accelerate at higher temperature, whereas at low temperature, respiration rate comes down and lesser amount of ethylene produce by fruit, vegetable and flowers and the multiplication of microorganism does not take place at faster rate. However, the effect of dropping temperature on maintaining produce quality is not uniform over the normal temperature range i.e. 0 - 30° C for non chilling sensitive produce, 7.5 - 30° C for moderately chilling sensitive produce, 13 - 30° C for chilling sensitive plants. Normal ripening occur at temperature range of 10-30° C, but best quality fruit develops ripening at 20-23° C.

Heat of respiration: High heat of respiration deteriorates the quality of produce and speed up the growth of microorganisms during storage and transport. Provision for fresh air circulation must be helpful to remove respiration heat and reduce ethylene concentration.

Relative humidity: Low humidity leads to weight loss and wilting whereas high tends to encourage growth of microorganisms. Humid atmosphere may cause the development of fungal and bacterial diseases, which damages produce during storage and transport. Injured produce remove water quickly and emit a larger quantity of ethylene than healthy ones. Many horticultural produce should be kept at 80-95 per cent relative humidity for maintenance of freshness and turgidity. Produce should not be stores in dry atmosphere because they become less turgid through quick transpiration and start showing wilting symptoms when they have lost 5-15 per cent of their fresh weight. The rate of transpiration from the produce is reduced with the increases relative humidity, although to retain high relative humidity coupled with high temperature results in faster infection by pathogens.

Cleaning and washing: Cleaning fruits and washing root vegetables for remove dirt and spray residue and provides good appearance or freshness to the produce and minimize microbial load from surface.

Trimming: Removal of damaged, diseased and discoloured parts improves the appearance of horticultural fresh produce, it helps in reduce ethylene production and growth of microorganisms during storage and transport.

Chemical treatment: Treating fruits and vegetables with oil (groundnut / mustard) and waxol emulsion reduces weight loss and maintain freshness. Delighting fruits with ethephon and alar increases colour and reduces astringency. Sulphur is used on banana as a paste (0.1% active ingredients) to control crown rot fungi. Sodium or potassium bisulphite are used in a saw dust mixture (with in a pad that can be placed inside a carton) to release SO_2 for control of spoilage by mold in grapes. Fruits and vegetables can be dipped in hypochlorite solution (50-70 ppm available chlorine) then, rinsed with tap water for control of bacteria, yeast and molds which affect the quality.

Pre-cooling: Pre-cooling maintains the freshness or appearance and reduces weight loss in various fruits and vegetables. Pre cooled products are allowed to be packed, but there should not be any direct contact between product and the containers.

Hot water treatment: Increases carotene and total sugar in fruits e.g., mango. It also helps in control post-harvest spoilage in papaya, mango, oranges, beans, pepper, etc.

Packing: Loose packing causes more damage in fresh produce than tight packing. Wrapping fruits (apple, mango, etc.) in tissue paper reduces weight loss and maintain good colour. Similarly packing fruits and vegetables in ventilated polythenes bag retain colour and trim down weight loss.

Light: Potted flowering plants and cut flowers, it is sensible to illuminate the plants with 2 - 3 k lux (200-300 f.c.) with fluorescent and incandescent to create illumination of red and blue light.

Water quality: Water quality relates to pH, EC values, hardness contents of phytotoxic elements and microorganism causing vascular obstruction affecting longevity of the produce particularly cut flowers. Longevity of flowers reduced when salts concentrations reaches 200 ppm (roses, chrysanthemum and carnation) and 700 ppm (gladiolus). Basic ions like Ca++ and Mg++ present in hard water are less harmful than soft water containing sodium ions. Use of de-ionised water is better than ordinary tap water in enhancing vase life and even use of boiled water containing less air then tap water is readily absorbed by stem. Use of millipore filter water enhances flow rate of water through cut stem and reduction of air blockage from vessel. Acidification of alkaline water with H_2So_4, HCL and citric acid has been found to increase the vase life of cut flowers. At low pH, microbial population in stem of the flowers decreases. The optimum pH for extending the vase life of flowers varies from 4.0 - 5.0. Wetting agents or surfactants like Tween -20®/APSA® at 0.1 -0.01 per cent (1.0 ml - 0.1 ml L^{-1}) decrease the surface tension of water, increase the lateral water flux which removes air bubbles and helps to maintain a continuous xylem water column in cut flowers.

Plant hormones: Use of cytokinin (Kinetin, BA and B-9), auxin (IAA), maleic hydrazide, cycocel (CCC) and gibberellins (GA_3) are delay senescence of the produce and are known to be ethylene inhibitors. Abscisic acid hastens the developmental process associated with aging and increase sensitivity of the tissue to the ethylene production. Abscisic acid is also involves in senescence to increase the permeability of the tonoplast leading to cell disorganization, resulting in decreased water uptake and development of water stress effects.

Use of preservatives: A mixture of chemicals such as sugars, germicides, salts, growth regulators, etc. is being used to extend the vase life of the flowers. Sugars, biocides, anti-ethylene compounds (1-MCP, potassium permanganate) and hydrated compound are used for conditioning. All sugars used in holding solution make excellent media for the growth of micro-organism causing stem plugging. Therefore, sugars must be used in the combination with germicides in the vase solution. Metallic salts like silver nitrate, cobalt chloride, $AlSo_4$, $ZnSo_4$, calcium nitrate and nickel chloride are used to extend the vase life of flowers.

Ventilation and spacing: Provision for air exchange must be maintained to remove respiration heat and ethylene. Sufficient commodity spacing should be provided so that at least one side remains exposed for air circulation to prevent heat generation.

Packing and packaging of fresh produce: Packing and packaging methods can greatly influence air flow rates around the commodity, thereby affecting temperature and relative humidity management of produce while in storage or in transit. Packaging protects the produce from mechanical injury, and contamination during marketing. Packaging accessories such as trays, cups, wraps, liners, and pads may be used to help immobilize the produce within the packaging container while serving the purpose of facilitating moisture retention, chemical treatment and ethylene absorption.

Length of storage: One of the most significant factors that affect the quality of fresh produce is storage. Making sure that fresh produce is stored at optimum conditions is a key to retain their quality. If it is stored in poor storage conditions such as high temperatures, it will lose its nutritional value or spoil quickly. Storing fresh produce beyond the recommended periods even at optimum temperature can still cause loss of nutritional value.

Nature of Fruit and Vegetable Losses

Postharvest losses in fruits, vegetables and flowers both qualitative and quantitative occur between harvesting to consumption. The important sites where post-harvest losses occur in India are 15-20 per cent at farmers field, 15-20 per cent in packaging, 30-40 per cent in transportation and 30-40 per cent in marketing. The need of the hour is to minimize these losses. Nature of losses observed in fruits and vegetables is as under:

Direct losses: Those losses caused by waste or consumption by non-human agents, such as insects, rodents, birds, fungi, bacteria and others.

Indirect losses: Those due to deterioration in quality or acceptability of the product up to the point of complete rejection by the consumer, e.g. changes in its appearance, texture, and colour caused by climate, improper handling, transportation or infrastructure.

Economic losses: Those losses brought about by changes in market conditions and expressed in economic terms, e.g. losses due to changes in demand and supply.

Loss in weight: Most of the fruits, vegetables and flowers cell are turgid at time harvest with respect to time from get reduced in weight due to transpiration of water.

Shrinking and wilting: Water loss through transpiration from outer surface of fresh produce by stomata. Most of the fruits shriveled when 5 to 6 per cent loss in weight. Tender fruits (Jamun), flowers and leafy vegetables get flaccid due to loss of water shortly.

Change in colour: Many fruits and vegetables lose their natural colour due to high temperature and dry conditions during storage and transporation. Litchi becomes brown, spinach and lettuce become pale yellow and white curd of cauliflower becomes yellow.

Change in texture: Many fruits including tomato, mango, banana, jamun become very soft due to over ripening and leafy vegetables become very hard and brittle.

Change in taste: Some fruits (mango, banana and sapota) give flat, sour and unpleasant odour during transportation in airtight vehicles. Potato tends to become sweet, when store at low temperature and tasteless carrot due to high temperature storage.

Bruising: Fresh fruits and vegetables get bruised due to defective harvesting, handling, packing, transport and storage. It causes high respiration and transpiration rate and produce catch quick deterioration.

Sprouting: It is a great problem in vegetables likes onion, garlic, ginger, yam and potatoes if not stored at proper temperature and humidity. The stored food material of produce translocated to sprouts and produce get loss in quality and shelf-life.

Toughening: Vegetables like green beans and sweet corn become hard due to development of spongy tissue in prolonged storage.

Greening: Exposure of potatoes and sweet potatoes to light during storage cause green tissue. These portions contain solanine which has toxic properties.

Rotting: Rotting of fruits and vegetables is caused by pathogens and microorganisms. Fungi are the main contributory factor to post-harvest losses. High water containing fruit and vegetable become tender and favourable for microbial infestation.

Causes of Post-Harvest Losses

Significant postharvest losses occur in horticultural fresh produce, owing to the lack of suitable harvesting equipment, collection centers in major producing areas, suitable packing containers, commercial storage facilities, a cold chain and proper transportation system.

Several factors are responsible for the losses of fruits and vegetables during various stages of handling. These can be broadly classified in to three groups, i.e., physical injuries, physiological disorders and pathogenic disorders.

Table 23. Factors determining postharvest losses of horticultural fresh produce

Stage / phase	**Factors**
Pre-harvest	Genetic, agronomic practice, environmental conditions
Harvest	Maturity, physical damage
Immediate Postharvest	Pre-cooling, sorting, curing, washing, chemical treatments
Postharvest storage	Temperature, modified atmosphere, ventilation, physical protection
Transportation and handling	Packaging, temperature, atmosphere
Non-Technical	Managerial skills, administrative skills, inadequate extension service, shortage of capital, shortage of foreign exchange

Physical or mechanical losses

It can be caused by inappropriate harvest methods, poor packaging and transportation resulting in cuts, abrasions, bruises, breakage or leakage. Physical damage can be normal or abnormal. Normal deterioration is due to the natural aging process or senescence of the products. Abnormal decline is that which occurs due to adverse conditions such as unfavourable temperature at either extremes, i.e. too low but above freezing (above 0° C but below $10\text{-}12^{o}$ C) resulting in chilling injury or too high (above $35\text{-}38^{o}$ C) resulting in heat injury. Other examples of physiological damage include sprouting in onion, potato and yams during storage, blossom end rot in tomatoes, internal rind spot on watermelons, tip burn in lettuce, shrinking in flowers and leafy vegetables etc. Injury caused in fruits or vegetables during harvesting, grading, packing, transportation and storage is the source for entrance of micro-organisms that causes spoilage or losses. Improper physical handling can result injuries due to followings:

Abrasion: It can occur during harvesting, packing a transport of the produce. Rough containers or loosely packed containers allow the individual fruits to rub against each other or against containers surfaces and causes abrasion.

Compression: Packing, more produce into little container or stacking of containers higher than their design strength cause compression damage. Compression spot encourage the pathogen those responsible for fruit rotting.

Impact: Dropping of crates and bags of fruits or vegetables on a concrete floor or careless dumping of produce from containers on to retail display causes impact damage, i.e., bruise or crack.

Punctures: Punctures are sometimes caused by stem or pedicel in eggplant, cucumber, tomato, chilies etc. It provides a way to penetrate the microorganisms in fruits and which are responsible for decaying of internal tissue of produce.

Tears: This is most common in leafy vegetables and tissue exposed, lead to rapid dehydration, discoloration and decay.

The important diseases which enters through wounds or injury are stem end rot of mango, crown rot of banana, green and blue mold of citrus, pedicel rot of pineapple, rhizopus rot of sweet potato, bacterial soft of potato etc. Moreover the injured fruits and vegetables faced serious problem in marketing.

Control measures physical or mechanical losses

Careful handling: Preventative measure should be taken during harvesting, grading, packing, storage, transportation and marketing to prevent injury. Harvesting of produce at appropriate stage, grade on basis of market requisite and use protective and safe packing material and store or transport in low temperature with refrigerated vans.

Removal of injured fruits: Injured fruits by any physical injury must be removed during grading, packing and storage to avoid the infection of various diseases. Injured fruits are providing better environment of fast multiplication of pathogens.

Chemical protection: Injuries incurred by fruits and vegetables during harvest and handling are the major sites of infection by wound pathogens. Chemical treatment to control these infections should be applied as soon as possible after harvest. Post-harvest treatment of fruits with pyrimethanil (0.1%) before storage for minimizes the problems.

Physiological disorders

Physiological disorders are abnormal non-parasitic metabolism that usually shortens useful post-harvest life of fresh fruits and vegetables. Some disorders inherited at harvest and other develop later. These disorders cause losses and affect marketability of fruits and vegetables.

Pre-Harvest Period Causing of Post-Harvest Physiological Disorders

High temperature: High temperature during growth period causes ohanez spot of grapes, kelsey spot of plums, scald in fig, pomegranate and potato, failure of ripening in avocado (26-30° C) and tomato (above 30° C), browning in litchi, side cracking of prunes, blackening of peaches, water core of apple and pit burning of apricot. Excessive exposure to sunrays caused sunburn in fig, citrus, peach and sun scald in apple.

Low temperature: Low temperature causes cold injury in terms of chilling injury in chill sensitive crops including tomato, pea and freezing injury in citrus fruits during winter. Black heart, collar rot and frost croach in apple also caused due to low temperature.

Mineral nutrition deficiency: Several physiological disorders of fruits and vegetables are attributed to the deficiency of mineral nutrients, calcium and boron being the most important. Bitter pit in apple and blossom end rot in tomato, calyx end rot in persimmon and grape due to calcium deficiency, internal cork of apples and internal necrosis in aonla and mango, hen and chicken, millerandage and coluere in grape, fruit cracking in pomegranate to boron deficiency and exanthema (rind browning) in citrus, wither tip, rough bark and cracking in apple due to copper deficiency are some of the example of physiological disorders caused by mineral deficiency.

Water relations or imbalance: Fluctuation of moisture during growth and ripening causes fruit splitting of lime, lemons, grapes, litchi, fig, pomegranate and tomato. Black end of pear, cork spot of apple, endoxerosis, jonathan spot of apple, black heart of celery and tip burn of lettuce also caused by moisture imbalance.

Post-Harvest Period Causes of Post-Harvest Physiological Disorders

Senescent period (Ageing): Senescent fruits and vegetables are highly susceptible to micro-organisms due to reduced level of microbial inhibitors such as tannin, 3, 4 dehydroxy benzol dehyde, 6-methyl mullein, etc. A loss in weight of only 5-6 per cent causes shrinkage in many fruits and vegetables. Such fruits and vegetables are also highly susceptible to micro-organisms.

High temperature injury: Heat or high temperature during storage accelerates the physiological process in stored fruits and vegetables. It also causes excessive transpiration with resultant of shriveling or wilting, cracking of stored fruits and vegetables. Water loss is a loss of saleable weight and thus is a direct loss in income. The direct effects of excessive heat in storage are blemishes such as spotting, russeting and scald in apple and potato. High temperature also causes failure of ripening in banana (above 30° C) which cease ethylene production. Tomatoes do not develop proper colour (red/pink) due to high temperature (above 32° C).

Low temperature injury: Low temperature break down is the major physiological disorders of tropical fruits. Extremely low temperature (chilling injury) reduces the quality of the product and shortens shelf life. Chilling injury differs from freezing injury and it occurs at temperature, which are low but above the freezing point of the tissue. It occurs at 0-12.8° C in many fruit such as mango, banana, citrus, apple and vegetables such as potato, onion and tomato. Various disorders caused by chilling are pitting in rind, discolouration of rind, failure of ripening in banana, and browning of flesh in apple and plum, rusty brown specks, spots in lima beans and surface scald, blackening of seeds in brinjal. Freezing injury results when temperature goes below the freezing point of the fruits and vegetables in storage. It occurs at -1.5° C to -1.0° C in many fruits such as mango, banana, citrus, grapes, apple and vegetables such as potato, onion, tomato, etc. It causes watery break down, soft break down and browning or discolorations of fruits and vegetables.

Humidity: Most of the fruits and vegetable require 85-90 per cent relative humidity during storage. High humidity causes growth of fungus that reason of spoilage, e.g., Botrytis rots in grapes. Low humidity causes shrinkage and loss in weight in many fruits and vegetable. Dehydration injury occurs when the humidity of storage falls below normal. When this occurs, potato, sweet potato, apples and stone fruits become soft, flabby and wrinkled with consequent reduction in quality.

Volatile compounds: They are ethylene, esters, ketone, hydrocarbons, etc. produced in fruits and vegetables during storage. Esters accumulation causes scald in apple (brown discolouration of skin followed by internal browning) which is a serious problem in storage. Ethylene produced by many fruits during storage and its accumulation increases the rate of respiration, faster ripening, fruit softening and spoilage. Such fruits are highly susceptible to micro-organisms, stem end rot and anthracnose in citrus.

High CO_2 and low O_2: Respiration of fruits and vegetables is affected by level of CO_2 and O_2. If CO_2 level is high during storage starts anaerobic respiration, which leads to many physiological disorders such as off flavour, internal discolouration in citrus and brown heart in apple, internal browning of citrus, black heart of potato, black speck of crucifers and red heart of lettuce are typical examples of physiological disorders induced by unfavorable oxygen relationships.

Refrigerant (chemical) injury: Sulphur dioxide (SO_2) causes bleaching of skin pigments in apple, banana, mango, grape and citrus. Surface pitting and water break down in potato, discolouration of pigment in onion, bleaching and skin break down in tomato caused by higher concentration of sulphur dioxide during storage. Ammonia causes discolouration, browning and break down in apple, banana, grape, citrus, mango, potato, onion, tomato, etc.

Control measures post-harvest physiological disorders

For minimizing post-harvest physiological disorders maintain proper moisture condition during growth and maturation stage by quantify irrigation. Control of high field temperature by application of fogger irrigation and low temperature by spraying of low concentration (0.1%) sulphuric acid (H_2So_4). Nutritional balance with proper application of mineral nutrients and organic manure. Harvesting of produce at right stage of maturity, because over mature stage causes fast spoilage of fruits and vegetables, whereas early mature harvesting reduces development of qualitative parameters in fruits but it is appropriate in vegetables. Harvested produce should be kept under shaded tree or using tarpaulins or shade nets for avoiding microbial contamination. Delaying of senescence with ethylene absorbents, growth regulator treatments and skin coating is necessary for extend shelf life of fruits and vegetables. For minimizing internal and field heat of fruits by pre-cooling, packing in boxes with ice and bags or cartons keept in cold storage. High humidity during storage can be reduced by opening of ventilation window and keeping of calcium chloride in store, which absorbs moisture. Low humidity in storage can be increased with lowering the temperature and water spray or fog or moist on storage floor. Control of volatiles with esters absorbents such as oil treated paper for wrapping of fruits and as liners or cushioning materials and ethylene absorbents such potassium permanganate. Refrigerant injury is best controlled by prevent leakage of SO_2 and ammonia and using proper concentration during cold storage. Control high CO_2 level by providing proper ventilation. Injury due to low temperature can control by maintaining optimum temperature of storage, emulsion treatment of fruits before storage and wrapping of fruits in tissue paper or polythene bags.

Spoilage caused by physiological disorders in fruits and vegetables

When fruit are put into storage, they are on a slow path to senescence and death, and a number of disorders can arise during that time. Several storage disorders have physiological origins, which may be chilling related, and are often highly specific to species, cultivar, season and even growing region. Fruit maturity at picking is one of the important factor, with less mature fruit generally being more susceptible to chilling injury. Sensitivity to storage disorders depends on many factors, including maturity at harvest, lack or imbalance of nutrients and adverse growing conditions. Even if fruit are susceptible at harvest, the expression of disorder symptoms is dependent on storage conditions and duration, and symptoms may not always develop. The development of chilling injury is often described as a time by temperature relationship, i.e. it develops sooner at lower temperatures. This is true for damage that is a direct result of exposure to low temperature and which is seen almost immediately after exposure. However, many chilling disorders develop only after long periods in storage and are associated with an inability of fruit to ripen correctly at low temperatures (e.g. kiwifruit, peach, avocado). It seems that at low temperatures the natural highly coordinated process of ripening is disrupted by an element that is temperature sensitive. If removed from storage early enough, no symptoms of chilling develop when the fruit ripens at higher temperature. Thus far, chilling damage has been described as a single disorder, yet there are numerous symptoms that may develop in the fruit flesh or skin that differ among species and cultivars. In addition, there are disorders that develop as fruit start to senesce, irrespective of storage duration or temperature, and that may have similar symptoms to chilling injury in the fruit flesh.

Apple

Brown heart: Develop brown spot near the core part of the fruit. Discolouration of flesh due to high CO_2 concentration (above 0.5%) during storage. Make sure that the CO_2 concentration is below 0.5 per cent during storage.

Bitter pit: Green fruits are more susceptible than mature or ripe fruits. Golden delicious groups are also more prone to bitter pit. Build up brown spot close to the blossom end and it occurs as sunken discoloured pits in the skin with spongy, dry brown flesh beneath. It is primarily a response to inadequate calcium content. Pre-harvest spray of calcium (0.5%) greatly reduced.

Internal cork: Boron deficiency and extensive water stress condition during maturation caused cork browning. Pre-harvest spray of boron (0.5%) and proper irrigation decrease the effects.

Internal browning: Develop brown strips in the flesh up to a greater depth. Fruits give alcoholic smell due to internal tissue fermentation. Store fruits in CA storage at 4.5° C temperature.

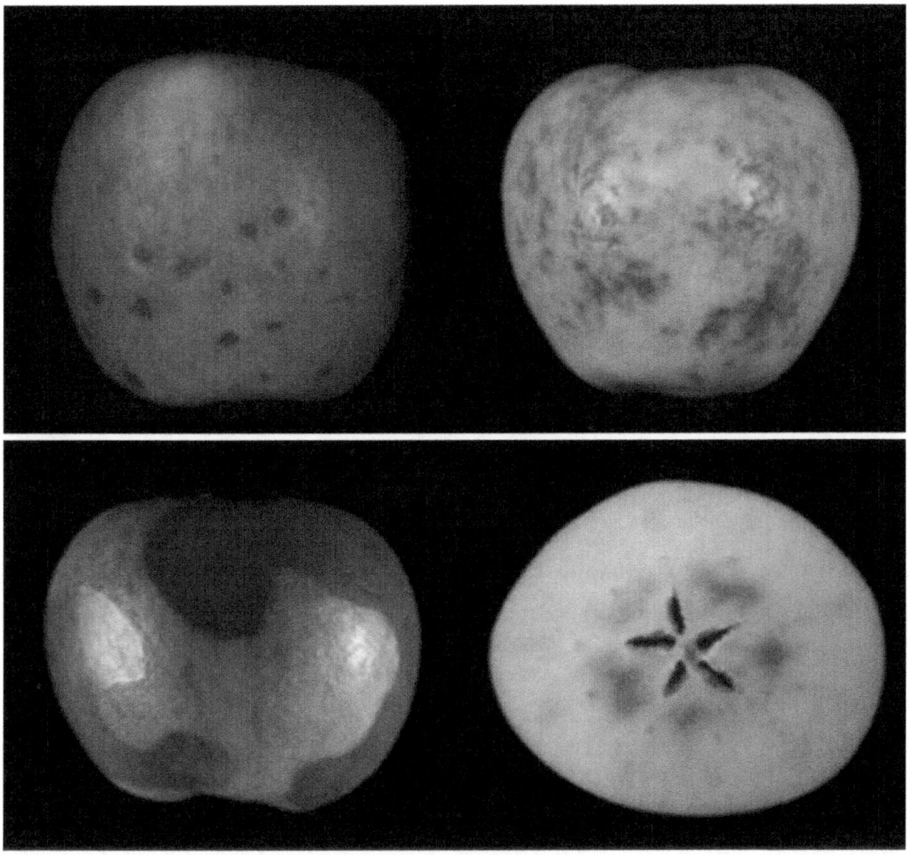

Figure 1. Physiological disorders of apple fruit. The top left panel shows bitter pit, a disorder associated with calcium deficiency. It can be partially controlled by preharvest sprays of calcium salts directly onto the fruit. The top right panel shows superficial scald, a low temperature disorder of the skin that can be controlled by 1-MCP treatment prior to cool storage. The bottom left panel shows soft scald, a low temperature disorder with symptoms of brown lesions that extend into the flesh. Incidence can be increased by over-maturity of the fruit at harvest and preharvest climatic conditions. The bottom right panel shows core flush, a browning within the core line that is a form of senescent breakdown.

Soggy break down: Flesh surrounding the core becomes soft and black. Storing the fruits at 2.2° C temperature produces this disorder. Wax treatment of the fruits and store first at 35° C for 24 hours and then store at 1° C temperature.

Water core (core flesh): It is a situation where there are glassy, waterlogged sections of tissue towards the centre of the fruit, typically centered in the region of the vascular bundles. Severe watercore leads to anaerobiosis, development of fermentation aromas and core browning similar to core flush. Fuji is an especially susceptible cultivar. Watercore is more severe in sweet fully mature fruit and involves a breakdown in transport of sorbitol across cell membranes. Application of balance dose of nutrients and harvest the fruit at fully mature stage.

Storage break down: Flesh becomes soft and breaks and peels split due to long storage. Pre-harvest spray of 2.5 per cent $CaCl_2$ before two weeks of harvest. Post-harvest treatment with 2 per cent $CaCl_2$ + 0.1 per cent azoxystrobin + 0.5 per cent difenconuzole and then store.

Chilling injury: Development of surface discoloration, pitting and brown core due to long time storage of the fruits below 0° C temperature. Keep the fruits at lowest safe temperature (2-3° C) of storage.

Freezing injury: Formation of soft brown blemish and break down due to temperature below 1.5° C. Maintain safe temperature (2-3° C) of storage.

SO_2 injury: Bleaching of pigments due to higher concentration of SO_2 (above 0.5%) during storage. Use safe concentration (0.1%) of SO_2.

Ammonia injury: Brown/discolored of fruits and internal break down of fruit tissues due to high concentration 1 per cent of ammonia. Check leakage of ammonia during storage and maintain ammonia level below 0.1 per cent.

Superficial scald: Superficial scald is a brown discolouration of the skin surface, particularly in cultivars like GrannySmith. It becomes visible to be connected with the accumulation of the hydrocarbon α–farnesene insusceptible cultivars, the oxidation products of which are brown and may cause cell collapse. Superficial scald can be reduced by a postharvest dip in an antioxidant free-radical scavenger like diphenylamine (DPA), harvesting at full maturity, wrapping of fruits with antioxidant dip tissue paper and store at 7.7 to 15° C temperature.

Soft scald: Soft scald or deep scald develops as sharply defined brown lesions on the skin that usually extend into the flesh. Soft scald is a low temperature disorder, partially avoided by slow (delayed) cooling or by storing at slightly warmer temperatures.

Core flush: It is most serious in McIntosh, is a browning of internal fleshy tissues surrounding the core of a fruit, and may have more than one cause. One reason seems to be the O_2 supply to the core, since conditions potentially causing anaerobiosis (large size, a closed and airtight calyx and a low O_2 atmosphere) increase incidence. It is most serious in fruit stored for long periods at around 0° C, and may be greatly reduced by storage at 4° C under CA.

Mango

Spongy tissue: A segment of fruit flesh becomes soft or spongy with large air pockets. Heat radiation is the cause earlier known as calcium deficiency. Harvesting at 3/4th maturity and treating fruits with ethephon (500 ppm). Organic mulching on plant canopy area also minimizes the effect of radiation.

Chilling injury: Pitting and discolouration of flesh and skin when store at temperature below 10° C. Wrapping of fruits in tissue paper and stored at 12° C and above.

Freezing injury: Discolouration and water break down of tissues when fruits are stored at 10° C and below temperature. Wrapping of fruits in tissue paper and stored at 12° C and above.

SO_2 injury: Bleaching and surface discolouration due to SO_2 concentration of 0.5 per cent or above. Use of proper concentration of SO_2 (0.1%) during storage and provide appropriate ventilation.

Ammonia injury: Surface pitting and rupture of tissue. Ammonia injury occurs if concentration is 1 per cent or above. Proper control on leakage of ammonia during storage.

Banana

Shriveling and poor ripening: Irregular ripening with shriveled fruits is caused by high temperature and low humidity. Pre-cooling the fruits and storing at 90-95 per cent humidity.

Chilling injury: Black peels, placentas become hard and stop ripening due to store of the fruits at temperature below 12° C. For control the fruits are storing in polythene covering/ bag at high temperature.

Freezing injury: Dark peel and brown pulp soft break down of peel at storage temperature below 1° C. Storing fruits in polythene covering/bag at higher temperature.

SO_2 injury: High SO_2 concentration (above 0.5%) is cause bleaching of skin pigments. Use proper concentration of SO_2 and store the produce at appropriate ventilated place.

Ammonia injury: In storage ammonia concentration above 1 per cent cause black peels and breaks down of tissue. Control on ammonia leakage during storage and proper ventilation.

Citrus

Oleocelosis: Rupturing of oil glands of peel in lime and lemon, which results brown raised and spots on peel. Low temperature and relative humidity or injury during harvesting and handling are the major causes. Harvest fruits at full colour development stage and avoid injury by using harvesting tools.

Black button: Colour of button becomes black in sweet orange. It caused by high humidity when fruits are stored in polythene. Post-harvest treatment with 100ppm 2, 4-D solution.

Red blotch: Black irregular blemish on lemon fruits. Harvesting at green mature stage and storing in perforated polythene bag.

Splitting: Cracking of lime and lemon fruits due to boron deficiency and moisture fluctuations. Maintain moisture condition appropriately throughout the growing period and apply boron by foliar spry.

Granulation (Sclerocitosis / crystallization): Juice vesicles become elongate, hard and rough in mandarin and sweet orange. Pre-harvest spray of 2, 4-D and early harvesting.

Stylar end break down: Dry spot at stylar end in lime and lemon due to mechanical injury. Careful harvesting and treating fruits with 100 ppm 2, 4-D.

Chilling injury: Rind pitting and grayish scald like discolouration due to temperature below -1.5° C. Storing fruits in polythene bag and at lowest safe temperature 10-13° C.

Freezing injury: Watery break down spots on fruits due sore at temperature below 3° C. Store of fruits at 10-12° C temperature.

SO_2 injury: Bleaching of pigments due to high SO_2 concentration in storage condition. Use proper ventilation and reduce SO_2 concentration.

Ammonia injury: Formation of dark brown spots and discoloration. Cause is high concentration (above 0.5%) of ammonia. Control on leakage of ammonia during storage.

Grape

Berry drop: Dropping of berry is a serious problem during storage and transportation due to weak pedicel. Pre-harvest spray with GA_3 100 ppm or 4 per cent wax emulsion 3-7 days prior to harvest.

Fruit cracking: Round crakes on apex of fruits. Storage after pre-cooling of fruits and maintain moisture balance during crop production.

Guava

Chilling injury: Pulp injury and decay due to low temperature below 4-5° C. Fruits are store at 9-10° C temperature with 90-95 per cent relative humidity.

Papaya

Chilling injury: Pitting, failure to ripen, off flavour and decay when temperature below 7° C. Store at protected temperature (7-10° C) with news paper wrapping.

Litchi

Browning of fruits: Pink colour changed to brown during storage. Dipping fruits in sodium-hypo-chloride solution, air dried and storing fruits by packing in polythene.

Pineapple

Endogenous brown spot (Black heart): Brown spot in the flesh due to abnormal storage temperature and high CO_2 concentration. Storage at 7-10° C temperature with proper ventilation.

Chilling injury: Browning, dryness of crown and development of water soaked area in pulp. Cause is low temperature below 7° C. Store the fruits at 7-10° C temperature with paper wrapping.

Potato

Sun scald: Due to high temperature and exposed potato comes in direct contact of solar radiation. Proper curing is necessary and produce store in shade area during harvesting and at the time of storage and control on temperature.

Chilling injury: Surface pitting and discolouration due to temperature below 0° C. Control temperature of storage up to 4° C.

Freezing injury: Produce store on temperature below 1.5° C tissue damage due to watery break down. Control storage temperature above 4° C and pack the produce with jute bags.

SO_2 injury: Bleaching, pitting, watery break down due to high concentration of SO_2. Use safe concentration of SO_2 (below 0.01%)

Ammonia injury: Dark brown pock marks internal discolouration and watery break down due high concentration of Ammonia (above 0.5%). Check leakage of ammonia during storage with proper ventilation.

Sprouting: Potato tuber sprouts during storage. Treat potato with 25 mg CIPC dust (1% active ingredient) per tonne of tubers. In case of liquid formulation CIPC can be applied in the form of a fog using a fogging machine 35-45 ml of commercial liquid is required for a tonne of potato. After treatment the store should be kept airtight for 24-48 hours.

Greening: Develop green colour compound solanin on the surface of the potato due to direct contact with sunlight. Protect produce from light during cultivation as well as storage.

Sweet Potato

Chilling injury: Decay, pitting internal browning hard core when stored below 30° C. Maintain lowest safe temperature at 4-5° C.

Onion

Chilling injury: Mottled patches on scales due to temperature below 0° C. Maintain optimum temperature at 12-15° C.

Freezing injury: Soft water breaks down take place when produce store below 1° C temperature. Keep up most favorable temperature at 12-15° C during storage.

SO_2 injury: Discolouration of pigments due to high concentration of SO_2 (above 0.5%). Control on SO_2 concentration with proper ventilation.

Ammonia injury: Black discolouration and break down due to high concentration of ammonia. Check leakage of ammonia during storage and facilitate proper ventilation.

Sprouting and rotting: Bulb sprouting due to presence of moisture and high temperature. Pre-harvest spray of 2000 ppm maleic hydrazide.

Garlic

Shriveling: Shrivelled due to loss of water. Protect from water loss and store at cool and dry place.

Browning and drying of clove: Due to higher water losses during storage the clove became brown, hard and dry. Stored at cool and dry place with proper ventilation.

Tomato

Chilling injury in ripe fruit: Surface pitting with brown spotting and poor colour development due to temperature below 8° C. Store at lowest safe temperature of storage 8-10° C with perforate polyethylene packing.

Chilling injury in mature green fruit: Water soaking, softening with whites spotting and decay due to temperature below 10° C. Store at protected temperature 13° C with high relative humidity in perforate polyethylene packing.

Freezing injury: Soft brown watery break down due to temperature below 0.5° C. Maintain safe temperature in storage 13° C.

SO_2 injury: Inhibition of colour development, bleaching of skin and watery breaks down due to high SO_2. Use safe SO_2 concentration (below 0.1%).

Ammonia injury: Due to high concentration of ammonia (<1%). Check leakage of ammonia during storage and provide adequate aeration.

Improper ripening: Ripening not uniform and develop superficial greenish white spots due to high temperature (<38° C). Control on temperature during fruit development and ripening at lower than 32° C temperature.

Okra

Chilling injury: Discolouration and build up water soaked areas, pitting and decay when stored below 7° C temperature. Maintain lowest safe temperature of 7° C during storage.

Sweet pepper

Chilling injury: Sheet pitting, alternaria rot and darkening of seeds due to temperature below 7° C. Fruits store at safe temperature at 8° C to 12° C.

Pathological damage

The maximum loss of fruits, vegetables and flower during storage is due to of attack of pathogenic organisms causing rot. It can be to fungal, bacterial and viral infections, e.g. Anthracnose in tomato, pepper, cucumber and watermelon, bacterial soft rot in brinjal, mango, papaya and cucumber, gemini virus in pepper, brinjal and tomato. Different species of Penicillium, Aspergillus and Cladosporium normally causes moulds on infected hosts in storage. Distinctive storage moulds are blue moulds and green moulds (*Penicillium* sp.) causing rotting in apple and citrus, gray moulds (*Botrytis cinerea*) of grapes, citrus and pears, gray moulds (*Rhizopus nigricans*) of straw berry, brown rot (*Sclerotinia fructicola*) of stone fruits and vegetables such as peas, beans, cabbage, carrot, turnip and lettuce, late blight of potato and tomato rot (*Phytophthora* sp.) and fusarium rots of stored fruits and vegetables such as onion, tomato, beans, melons, cucumber and potato. Numerous bacteria naturally cause spots, pits, discolouration and minor abnormalities on surface of fruits and vegetables. Examples are bacterial spot of tomato (*Xanthomonas vesicatoria*), bacterial blight or pod spot of peas (*Pseudomonas* sp.), fruit spot of stored fruits (*Xanthomonas* sp.) cucumber spot (*Pseudomonas* sp.) and black pit of citrus (*Xanthomonas citri*). Mosaic symptoms on peas, celery, tomato, beans and cucumber in storage, irregular destruction of chlorophyll or other pigments which shows mottled appearance, internal cork in sweet potato, net necrosis in potato and stone pit in pear are some of the examples of viruses causing spoilage.

Infection process of pathogenic diseases during storage

There are several way of infection process of pathogen but maximum possibility of entrance of microorganisms in to the host tissue by wounds injury examples diplodia stem end rot of mango, crown rot of banana, green and blue mold of citrus, pedicel rot of pineapple, rhizopus rot of sweet potato and bacterial soft rot of potato, and by the stomata like brown rot of plum and late blight of potato, and as a result of lenticels such as blue mold of apple. In some fruits the pathogen also enters through hair sockets like, brown rot of peaches and by means of cuticle like botrytis rot of grapes and brown rot of cherries. The external factors responsible for infection to the host are temperature and water relation to spore germination and mycelia growth and nutrient availability.

CONTROL OF POST-HARVEST DISEASES

A number of approaches for the control of post-harvest diseases are feasible for example prevention (use of disease free material, resistant varieties, etc.), eradication (culling, trimming, remove diseased fruit and vegetables, hot water treatments of some fruits and vegetables), retard growth of pathogen (temperature control, fumigation, modified atmospheres, radiation, etc.), chemical control (pre-harvest and post-harvest level), and combination of treatments (hot water dip + fungicides, wax + fungicides and low temperature + radiation). Pre-harvest application of chemicals reduces post-harvest decay to a substantial degree in many fruits and vegetables. Such application of systematic fungicide prevents quiescent infection like anthracnose of mango and papaya, crown rot of banana, stem end rot of citrus, lenticels rot of apple and brown rot of peaches. Botrytis rot of grapes can be controlled by spraying the fruits with captan. Rhizopus rot of peaches after harvest may be significantly reduced by spraying the fruits one week before harvest with carbendazim 0.2 per cent. Orchard spray with pyrimethanil 1000 ppm reduced the incidence of brown rot of peaches after harvest. Bordeaux mixture application before harvest reduces the incident of late blight in potato. The spray of growth regulator such as GA_3 before harvest on citrus check the infection of blue mould. GA_3 makes the peal firm and hard which avoids the fungus infection. Dormant infection can be eradicated in many cases by the application of systematic fungicide after harvest as the fungus pathogen is restricted to the epidermal or lenticels regions by the resistance of the host and so the chemical is able to penetrate in to these outer layers and inhibit the fungal development. Dichloram is highly effective treatment for control of Rhizopus rot of sweet cheery, nectarines, peaches and botrytis rot of several vegetables during long term cold storage. Sodium orthophenylphenate is very useful for control of post-harvest diseases of citrus fruits, apples, pears, sweet potato and other highly perishable fruits and vegetables. Azoxystrobin, fludioxomil and pyrimethanil are extensively used against stem end rot of citrus, lenticels rot of apples, stem end rot and crown rot of banana, latent infection of anthracnose on banana and citrus green and blue mold. Sulphur is used on banana as a paste (0.1% active ingredient) to control crown rot of fungi. Sodium or potassium bisulphite are used in a sawdust mixture (with in a pad that can be placed inside a carton) to release SO_2 for control of spoilage causes by mold in grapes.

Post Harvest Diseases of Important Fruits and Vegetables

Apple

Blue mold (*Penicillium italicum*): Affected portion become soft and watery and fruit produces off flavour. Treat the fruits with azoxystrobin or fludioxomil at1000 ppm for 5 minutes.

Black mold (*Aspergillus niger*): The pulp of affected parts becomes soft and watery. Treat fruits with sodium bicarbonate solution of 1.25 per cent before storage).

Bitter rot (*Glomerella cingulata*): Black small lesions are formed below the epidermis. This lesion produces the spores, which breaks the epidermis and cause spoilage. Golden delicious is highly affected. Treat fruits with trisodium phosphate solution (2%).

Pink rot (*Trichothecium roseum*): This disease is related with scab and cause heavy loss during storage. First brown spots are formed then conidia of pink colour are appeared and fruits give off flavour. Fruits which have scab it has also pink rot. Avoid storage of scab affected fruits and treat the fruits with 2 per cent solution of trisodium phosphate or fludioxomil at 100 ppm for 5 minutes.

Aonla

Blue mold (*Penicillium islandicum*): Affected portion cover with blue mycelium and become soft and watery. Treating fruits with azoxystrobin or fludioxomil or Thiabendozole at 1000 ppm solution for 5 minutes before storage.

Mango

Alternaria rot (*Alternaria alternate*): The symptoms are consist of either small black spots 0.5-1.0 mm in diameter with a dark centre and diffuse borders or dark lenticels resulting in a darkening of intercellular spaces and cell collapse. Treat the fruits with azoxystrobin or fludioxomil at 1000 ppm for 5 minutes.

Anthracnose (*Colletotrichum gelosporoides*): Large, spreading round brown or black lesions on surface of fruit. Pre-harvest spray of 0.2 per cent carbendazim and post-harvest treatment with azoxystrobin or fludioxomil or Thiabendozole at 1000 ppm solution for 5 minutes before storage. Hot water treatment at 46° C for 60-120 minute or at 48° C for 60 minute or irradiation of mangoes at 1000 Gy for the control of anthracnose.

Stem end rot (*Botryodiplodia theobromae*): Brown or black colour of fruit surface near the stalk. Treating fruits with hot solution (43° C) of 6 per cent borax for 3 minutes. Post-harvest dip with azoxystrobin or fludioxomil or thiabendozole 1000 ppm solution for 5 minutes before storage for effectively controlled.

Soft rot (*Rhizopus arrhizus*): Rotting of fruit surface with watery blemish. Store fruits at

7° C treatment with tissue paper wrapping. Treating with mixed solution of 5 per cent 2-aminothiazole + 5 per cent 2-aminopyridin.

Black mold rot (*Aspergillus niger*): It is a important postharvest disease of mango in India. The affected fruits show yellowing at the base and develop irregular, hazy, greyish spots which coalesce into dark brown or black lesions. Immerse the fruits with azoxystrobin of 1000 ppm solution before storage.

Bacterial rot (*Pseudomonas mangiferae indica*): Develop brown or black spot on fruit surface. Treating the infested fruits with Agromycin-100 before storage.

Banana

Anthracnose (*Gloeosporium musarum*): Black spot shriveled and fruits become black. Store fruits after treating with 0.1per cent azoxystrobin or with 2 ppm of sodium hypochlorite for 2 minutes.

Crown rot or finger stalk rot (*Gloeosporium musarum*): Stem infected with anthracnose causing separation of fruit from cluster (hand). Treat the hands with mixture of 0.5 per cent sodium orthophenyl phenate + 6 per cent waxol. After treatment apply antifungal paste on cut surface then pack and store at 10° C temperature.

Botryodiplodia finger rot (*Botryodiplodia theobromae*): Peels turn black, become soft and shriveled and pulp starts rotting with alcoholic smell. Store banana at 10° C temperature because it spreads very fast at 25-30° C. Post-harvest dip with azoxystrobin or fludioxomil 1000 ppm solution for 5 minutes before storage for effectively controlled.

Soft rot (*Rhizopus oryzae*): Rotting of fruits with watery blotch. Treatment of fruits with diphenylamine at 1000 ppm or 2-Aminopyridin before storage.

Citrus

Green mold (*Penicillium digitatum*): Rind turns out to be watery and breaks easily with pressing. Watery portion covered with green mycelium. Treating the fruits with mixture of pyrimethanil (0.1 to 0.2%) and waxol (6 to 9%).

Blue mold (*Penicillium italicum*): Peels become watery and on watery portion there is growth of blue colour mycelium. Treating the fruits with mixture of carbendazim (0.1 to 0.2%) and waxol (6 to 9%).

Aspergillus rot (*Aspergillus niger*): Brown colour water soaked blemish build up on the outer surface of the fruits. Prevent injury during harvesting and treating the fruits with 0.2 per cent fludioxomil solution.

Stem end rot (*Botryodiplodia theobromae*): Sweet orange is affected seriously. Brown or black colour of fruit surface near the stalk. Pre harvest spray of azoxystrobin or pyrimethanil (0.1 to 0.2%) and treats the fruit with 6 to 9 per cent borax.

Alternaria (black) rot (*Alternaria citri*): The fruits infected by alternaria rot produce blackish-green mycelium on the surface of satsuma mandarins and damage are serious in the long term storage. Treat the fruits by dipping process with 1000 ppm azoxystrobin or fludioxomil for 5 minutes.

Grey-mould rot (*Botrytis cinerea*): Grey-mould rot is common on lemons and satsuma mandarins in cold storage under high humidity. Grey-mould rot produces characteristic grey spores which enter through wounds or injured rind. Treating the fruits with mixture of carbendazim (0.1 to 0.2%) and waxol (6 to 9%).

Green-mould (*Penicillium digitatum*) **and blue mould** (*P. italicum*) **rot**: Green mould and blue mould are wound parasites, but will spread by contact of fruits on the tree or in the packing house, transit, storage and market. Particularly careful handling during harvesting to marketing is essential to avoid wound infection. Treat the fruits with mixture of carbendazim or thiophanate methyl (0.1 to 0.2%) and waxol (6 to 9%) before storage.

Stem-end rot (*Diaporthe citri, Diaporthe medusaea and Diplodia natalensis*): Stem-end rot appears as water-soaked spots near the stem end of the fruit, which then turn tan to blackish-brown. Postharvest treatment with imazalil (0.2%) and etaconazole (0.2%) and low temperature (4-5° C) with high relative humidity (80-88%) all help to reduce the occurrence of stem-end rot. Green mould, blue mould, sour rot and stem-end rot in citrus fruit can be partly controlled by biocontrol agents such as *Bacillus subtilis*, *Pichia guilliermondii* and *Trichoderma* spp.

Anthracnose (*Colletotrichum gloeosporioides*): Superficial leathery appearance, infected areas changes from a light brown water soaked appearance to a dark bronze almost black colour as it ages. Application of bordeaux mixture or copper oxychloride at early March and December. Postharvest treatment with carbendazim (0.2%) and thiophanate methyl (0.2%) fungicides may reduce fruit losses.

Custard apple

There are several pathogenic fungi to attack on fruit include *Botrytis cinerea* Pers., *Rhizopus stolonifer* (Fr.) Lind., *Penicillium expansum* (Link.) Thorn., *Phomopsis vexans* (Sacc. and Syd) Harter, Alternaria sp., Fusarium sp., botrytis, phomopsis and rhizopus are the most important pathogens while penicillium and alternaria are secondary pathogens. Penicillum, botrytis and rhizopus rot cause soft and wet rots, while phomopsis rot causes spots on the epidermis that give the skin a leather-like and finally scaly appearance. These diseases have been controlled by using immersion treatment with azoxystrobin (0.1%) and pyrimethanil (0.1%) solution. Treating with sodium-o phenyiphenate (SOPP) solution (0.25%), air dried and packed in perforated polythene bag for storage has been shown to be effective.

Guava

Macrophoma rot (*Macrophoma allahabadensis*): Browning of skin then it becomes water soaked. Development of orange or dark green mycelium which turns black and covers the whole fruits. Avoid injury during harvesting, packing and store fruits in cold storage. Treat the fruits with mixture of carbendazim or thiophanate methyl (0.1 to 0.2%) and waxol (6 to 9%).

Anthracnose (*Colletrotrichum gloeosporioides* Penz., *Botryodiplodia theobromea* Pat. and *Pestalotiopsis psidii*): The symptoms develop as the fruit matures, initially as tiny brown spots on the fruit surface, which later grow bigger and develop into sunken patches. Pre-harvest control the disease by spraying the plants several times with Topsin M or Indofil M-45. To manage *Colletrotrichum gloeosporioides* Penz. by using antagonistic fungi such as *Aspergillus pullulans*, *Cladosporium cladosporioides*, *Epicoccum purpurascens*, *Fusarium oxysporum* and *Trichoderma harzianum*.

Fruit rots (*Rhizopus stolonifer*, *R. arrhizus*, *Botryodiplodia theobromea*, *Botryophaeria ribes*, *Lasiodiplodia theobromea*, *Pestalotia psidii*, *Cylindrocladium parvum* and *Aspergillus niger*): The fruit decay problem may start as a latent infection in the field. Pre-harvest spraying the plants with Topsin M or Indofil M-45 was effective in controlling the disease. Apply dipping treatment on fruits with mixture of carbendazim or thiophanate methyl (0.1 to 0.2%) and waxol (6 to 9%) before storage.

Stylar end ring rot (*Phomopsis psidii*): This disease too may originate in the field and appears as very small, discoloured, necrotic lesions on the surface. These lesions later develop into a rough surface ring towards the stylar end of the fruit. Application of azoxystrobin (0.1-0.2%) or fludioxomil (0.1%) or thiabendozole (0.10-0.12%) solution gives better control if applied as hot water dip (48-50° C, 5 min). Gamma irradiation between 0.15 and 0.45 kGy appears promising in controlling disease. Biological agents such as bacteria (*Bacillus subtilis*), antagonistic fungi (*Trichoderma* spp.) or yeast (*Debaryomyces hansenii*) are also claimed to be effective in controlling diseases of harvested guava fruits.

Litchi

Blue (*Penicillium* sp.) **and Black mold** (*Aspergillus* sp.) **rots:** Browning of skin then it becomes water soaked spots and rotting of fruits. Treating with sodium-o- phenyiphenate (SOPP) solution (0.25%), air dried and packed in perforated polythene bag for storage. The postharvest diseases of litchi is controlled by the use of fungicides (azoxystrobin 1000 ppm, fludioxomil 1000 ppm, thiabendozole 1200 ppm), maintaining the cool-chain, refrigeration storage (0° C to 5° C), heat treatments (42° C), packing shed and orchard hygiene, correct packaging materials (perforated polythene bag) and pre-harvest insect control.

Papaya

Fruit-surface rots: It caused by anthracnose, chocolate spot, cercospora black spot and phytophthora on green/immature fruit and mycosphaerella, phomopsis, alternaria, stemphylium, fusarium and guignardia on ripe fruit infect through wounds occurring before or during harvest. Pre-harvest control by spraying the plants several times with Topsin M or Indofil M-45 and post harvest dipping with azoxystrobin (1000 ppm) or fludioxomil (1000 ppm) or Thiabendozole (1200 ppm) for effective manage the disease.

Anthracnose (*Colletotrichum capsici, C. circinans, C. dematium, C. gloeosporioides* and *C. papayae*): Out of all these species, *C. gloeosporioides* is the most commonly reported pathogen. As infected fruits ripen, latex is exuded at the fruit surface and small water-soaked spots appear. As the infection advances, circular sunken lesions with light orange spore masses form. These infections have been controlled by spraying carnendazim (0.2%) during fruit development stages and immersion treatment with azoxystrobin (0.1%) and pyrimethanil (0.1%) solution in storage.

Chocolate spot (*Colletotrichum gloeosporioides*): Minute, superficial reddish-brown lesions are the initial symptoms of this disease. As the fruit ripens, lesions may remain superficial or they may enlarge and become sunken with water-soaked margins. Application of carbendazim or copper oxychloride (0.2%) at early maturing stage of fruits. Postharvest treatment with carbendazim (0.2%) and thiophanate methyl (0.2%) fungicides may reduce fruit losses.

Dry rot (*Mycosphaerella* spp.): This organism is unable to penetrate the cuticle enzymatically and thus is associated with mechanical injuries. Small wrinkles in the fruit surface are the first symptoms at later stage lesions with brown, translucent margins develop. Treatment of fruits with diphenylamine (1000 ppm) or difenconuzole (0.2%) for 5 minutes before storage.

Wet rot (*Phomopsis* spp.): The entire infected area is soft and translucent and black pycnidia may form at the central portion of the lesion. A wet rot proceeds rapidly from the surface into the fruit cavity. Pre-harvest manage the disease by spraying the thiabendazole (0.2%) or imazalil (0.2%) and post harvest dipping with azoxystrobin (1000 ppm) or fludioxomil (1000 ppm) or thiabendozole (1200 ppm) is useful for effective control.

Alternaria fruit spot (*Alternaria alternata*): This disease is characterized by circular or oval black lesions that become covered with black spore masses. Lesions are usually restricted to the surface of the fruit and do not cause extensive rotting of parenchyma tissues. Treat the fruits with 1000 ppm azoxystrobin or fludioxomil for 5 minutes before storage.

Fusarium rot (*Fusarium solani*): It occurs sporadically on fruits after harvest and is characterized by small, dry lesions which later become covered by a white mycelial mat. Water-soaked lesions are also produced by this organism. Before storage the fruits treat with azoxystrobin (1000 ppm) or fludioxomil (1000 ppm) for 5 minutes and covered with potassium permanganate treated news paper cuttings.

Guignardia spot (*Guignardia* spp.): Sunken greenish-black lesions occasionally observed on fruit surfaces. This disease is often seen when papayas are pretreated in hot water (42° C) for 40 minutes during postharvest fruit fly disinfestation. The fruits immerse with azoxystrobin (1000 ppm) or fludioxomil (1200 ppm) for 5 minutes before storage.

Stem-end rots (*Mycosphaerella* spp.): Initially characterized by a translucent zone around the peduncle. As the infection advances the diseased tissue becomes black, wrinkled and dry and a white mycelium is formed at the stem-end. Postharvest treatment with imazalil (0.2%) and etaconazole (0.2%) and low temperature (4-5° C) with high relative humidity (80-88%) help to decrease the incidence of stem-end rot.

Internal fruit infections (*Erwinia herbicola*, *Enterobacter cloacae*): External symptoms are absent, however the diseases can be observed only after the fruit is cut open. Two such diseases are known: purple-stain and internal yellowing disease. Purple-stain, caused by pigment-producing strains of *Erwinia herbicola*, is characterized by violet-to-purple streaks in the vascular tissue and latex ducts surrounding the seed cavity. Internal yellowing disease is caused by *Enterobacter cloacae*. The infected fruit flesh is translucent with a bright yellow or lime green discolouration. These infections have been controlled by spraying thiophanate methyl (0.2%) during fruit development stages and dipping treatment with azoxystrobin (0.1%) and pyrimethanil (0.1%) solution in storage.

Pineapple

Black rot or stem end rot (*Ceratocystis paradoxa*): Black rot is widespread post harvest disease of pineapple, affecting 11 to 50 per cent fruits during transport and storage. Infection starts on stem end and affected fruits become black and rots. Pre-harvest spray of 0.2 per cent carbendazim and post-harvest dipping treatment with fludioxomil (1000 ppm) or thiabendazole (900 ppm) for 5 minutes.

Black rot (*Chalara paradoxa* (De Seynes) Sacc. (syn. *Thielaviopsis paradoxa* (De Seyn.) *Ceratocystis paradoxa* (Dade): Black rot, also called Thielaviopsis fruit rot, water blister, soft rot or water rot, is a universal fresh fruit problem characterized by a soft watery rot. Black rot is commercially controlled in fresh fruit by minimizing bruising of fruit during harvest and handling, store in refrigeration chambers and treat with fungicides. The fruits must be dipped with azoxystrobin (1000 ppm) or fludioxomil (1200 ppm) or sodium -o-phenyl-phenate (0.2%) for 5 minutes within 6-12 hours following harvest prior to packing and shipping.

Fruitlet core rot : This disease is caused by a complex involving the fungi *Penicillium funiculosum* Thorn, *Fusarium moniliforme* Sheldon var. subglutinas and the round yeast, (*Candida guilliermondi*), the pineapple fruit mite (*Steneotarsonemus ananas* Tryon) and pineapple red mite (*Dolichotetranychus floridanus* Banks). Fruitlet core rot, black spot, fruitlet brown rot and eye rot are all terms used to describe the brown to black colour of the central part of an individual fruitlet. Application of carbendazim or copper oxychloride (0.2%) at early maturing stage of fruits. Postharvest treatment with sodium -o-phenyl-phenate (0.2%) and thiophanate methyl (0.2%) fungicides may decrease fruit losses.

Yeasty fermentation: In damaged and overripe fruits and in fruits with interfruitlet cracking already present, yeasts start growing and dividing or new yeasts invade. This growth leads to fermentation, with bubbles of gas and juice escaping through cracks in the skin. The skin turns brown and leathery and the fruit becomes spongy with bright yellow flesh. For effective control of the disease pre-harvest spray with thiabendazole (0.2%) or imazalil (0.2%) and post harvest dipping with azoxystrobin (1000 ppm) or fludioxomil (1000 ppm) or thiabendozole (1200 ppm).

Pomegranate

Grey-mould rot (*Botrytis cinerea*): Fruit rotting usually starts at the calyx, the skin becomes light brown, tough and leathery. The pulp-bearing seeds disintegrate into a dark mass in advanced infections. Treating the fruits with mixture of carbendazim (0.1 to 0.2%) and waxol (6 to 9%).

Heart rot (*Aspergillus niger* and *Alternaria* spp.): Affected fruits show a slightly abnormal skin colour but internally a mass of blackened arils becomes prominent. Usually there is a black line of decay expanding from the calyx to the interior part of the fruit for the period of storage and transport. Application of carbendazim or copper oxychloride (0.2%) at the time of maturation and post harvest dipping with fludioxomil (1000 ppm) or thiabendozole (1200 ppm).

Penicillium rot (*Penicillium expansum* and *Penicillium* spp.): Produces watery areas at the infection site followed by masses of blue or green spores. Dipping treatments with aqueous topsin-M (0.1 %) and bavistin (0.05 to 0.1 %) inhibited the growth pathogen.

Brinjal

Alternaria rot (*Alternaria tenuis*): Numerous brown, sunken, circular spots with definite margins when small and after they coalesce to form irregularly shaped areas. The decay can be controlled by holding fruits at about 10° C and pre harvest spray with topsin-M (0.1 %) and bavistin (0.05 to 0.1 %).

Bacterial soft rot (*Erwinia carotovora*): Lesions are greyish brown, the skin is wrinkled and readily punctured and the underlying tissue is watery. Very careful handling and rapid cooling at 10° C retard this decay even after weather favouring this disease.

Phompsis rot (*Phomopsis vexans*): This disease originates before harvest particularly during warm wet weather. Cooling to 10° C and pre harvest spray with topsin-M (0.1 %) and bavistin (0.05 to 0.1 %) will considerably decrease fruit rot.

Bell pepper

Alternaria rot (*Alternaria tenuis*): Alternaria rot can infect peppers through intact skin, but only affects tissue that has been physiologically injured by too much sun or too low temperature. Treat the fruits by 1000 ppm, azoxystrobin or fludioxomil for 5 minutes and holding peppers at 7 to 10° C temperature.

Bacterial soft rot (*Erwinia carotovora*): The bacteria enter via wounds, either before or after harvest. The skin over the rot is sunken wrinkled and slightly discoloured. Immersing the peppers for 1-1½ minutes in 52.5 to 53.5° C water significantly decreases soft rot.

Garlic

Blue mold (*Penicillium corymbiferum*): Rotting of cloves. Avoid injury to cloves at harvest.

Onion

Bulb rots (*Aspergillus niger, Rhizopus nigerians, Botrytis cinerea*): Cause rotting of bulb with black and brown outer growth of fungi. Spray bavistin 0.1 or carbendazim 0.2 per cent after harvest.

Bacterial rot: Rotting of onion bulbs with yellow oozing. Treat the bulb with 2000 ppm thiabendazole before storage.

Potato

Fusarium rot (*Fusarium oxysporum*): Discoloration followed by water soaked lesions. Aerial mycelium and spores are formed profusely on the surface of the lesions and expansion of decay. Sanitary measures and optimum storage conditions with 4° C temperature and 85-90 per cent relative humidity. Before storage the fruits treat with azoxystrobin (1000 ppm) or thiabendazole (0.2%) for 5 minutes.

Bacterial soft rot (*Pseudomonas solanacearum*): Rotting of tubers with watery spots. Adopt proper sanitary measures, surface disinfection and maintenance of low temperature (4° C) in storage.

Late blight of potato (*Phytophora* sp.): Tubers rotting at time of storage. Control disease infield by application of fungicide spray bavistin 0.1 or carbendazim 0.2 per cent. Store only clean tuber at low temperature (4° C).

Tomato

Brown spots on fruits (*Macrosporium* sp., *Phoma* sp., *Cladosporium* sp.): Brown and black spots are formed at upper surface and the fruit rot. Apply streptocycline 0.02 per cent as post-harvest treatment.

Alternaria rot (*Alternaria alterata*): The lesion extends into the flesh where it produces a firm, dry, blackened mass of tissue thoroughly raised by mycelium of the fungus. The dramatic loss of resistance after fruits has been held for extended periods at less than 12.8° C. Treat the fruits with 1000 ppm azoxystrobin or fludioxomil for 5 minutes and holding fruits at 7 to 10° C temperature.

Grey mould rot (*Botrytis cineria*): The most common symptom of the disease is the dirty white colour of the mycelium over the lesion. Dipping the fruits with mixture of carbendazim (0.1 to 0.2%) and waxol (6 to 9%) solution for 5 minutes.

Primary Events in Postharvest Horticultural Produce

Bio-chemical changes: The fruits, vegetables and flowers are a living, biological entity when it is attached to the growing plant in its agricultural environment, but even after harvest the commodity is still living as it continues to perform most of its metabolic reactions and maintain the physiological systems that were present when it was attached to the mother plant. An important characteristic of fruits, vegetables and ornamentals, is that they transpire, and respire by taking up oxygen (O_2) and giving off carbon dioxide (CO_2) and heat. Although attached to the plant, losses due to respiration and transpiration are replaced from the flow of sap, which contains water, photosynthates and minerals. Respiration and transpiration continue after harvest, and since the produce is dependent completely on its own food reserves and moisture content. Therefore, losses of food and moisture are not replenished and decline begins very past due high water contents in all fresh horticultural produces. Formerly it was believed that the chemical transformations which occur in postharvest fruits and vegetables were the consequence of incompetence and de-compartmentation of the cellular setting. Further it was assumed by some that chemical transformations were strictly catabolic a falling apart of the tissue components. Know it recognize that both anabolic and catabolic reactions play a role in postharvest biochemical transformations, and that these changes are not the significance of disorganization but are the consequence of a biological clock, i.e., are under some form of genetic control. The links between cellular respiration and gene expression and between these primary events and secondary metabolism is by no means clear. Temperature is the ecological factor that most influences the deterioration rate of harvested commodities. For each increase of 10° C above optimum, the rate of deterioration increases by two to three folds. Exposure to undesirable temperatures results in many physiological disorders such as chilling injury and heat injury. Temperature also influences the effect of ethylene, reduced O_2 and elevated CO_2, those are responsible for either delay or accelerate the deterioration of fresh horticultural crops. The magnitude of these effects depends on the commodity, cultivar, physiological age, O_2 and CO_2 levels, temperature and duration of storage.

Table 24. Metabolic changes in postharvest fruits and vegetables

Degradative	Synthetic
Destruction of chloroplast	Synthesis of carotenoids and anthocyanins
Breakdown of chlorophyll	Synthesis of flavour volatiles
Starch hydrolysis	Synthesis of starch
Organic acid catabolism	Synthesis of lignin
Oxidation of substrate	Preservation of selective membranes
Inactivation of phenolic compounds	Interconversion of sugars
Hydrolysis of pectin	Protein synthesis
Breakdown of biological membranes	Gene transcription
Cell wall softening	Formation of ethylene biosynthesis pathway

Respiration of produce: The loss of food reserves during respiration results in, accelerate of senescence or aging, as this energy supplies are exhausted to maintain the commodity's living status, reduced nutritional value, loss of flavour quality, especially sweetness and loss of salable weight. The energy released as heat, known as vital heat, affects postharvest technology consideration, such as estimations of refrigeration and ventilation requirements. The rate of perishability of harvested commodities is generally proportional to the respiration rate. Based on their respiration and ethylene production patterns during maturation and ripening, commodities are classified either as climacteric or non-climacteric. Climacteric commodities, such as hot pepper, tomato, bitter gourd, mango, banana, passion fruit etc., show a large increase in CO_2 and ethylene production rates coincident with ripening, while non-climacteric commodities, such as brinjal, cucumber, yard long beans, okra, lettuce etc., show no change in their respiration rate and ethylene production during ripening. Figure 2 shows the relative changes during growth and development associated with the climacteric and non-climacteric patterns of respiration.

There is a contrary relation between respiration rate and storability of postharvest crops. Hence, those crop which exhibit a relatively high respiration rate tend to deteriorate rapidly whereas, crops which respire slowly can be stored for longer periods of time. In addition, reduction in the respiration rate by physical or chemical techniques is usually accompanied by a corresponding extension in storage life. The benefits of lowering storage temperature vary from commodity to commodity for at least two reasons. First, the temperature coefficient of respiration, normally about 2.0, can vary considerably and may range from 1 to 7 for different crops hence, reducing the temperature in a storage facility from 30° C to 20° C may have a more profound influence on both respiration and storability for certain crops than for others. Some crops are sensitive to chilling temperatures and the resulting anomalous metabolism can lead to a wide array of physiological disorders. Similarly, respiration rate and storability respond to modification of the storage atmosphere by reducing oxygen and increasing carbon dioxide partial pressure.

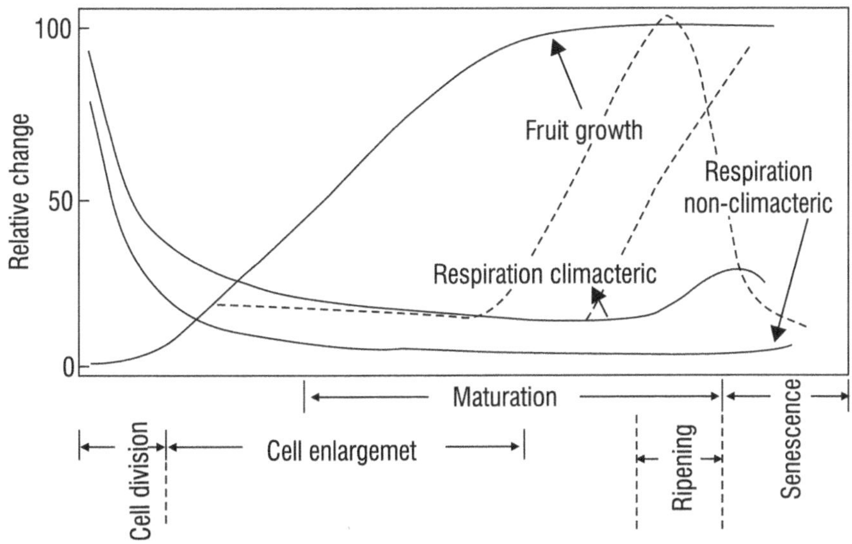

Figure 2. Growth and respiration patterns during development (Biale, 1964)

Table 25. Susceptibly of various crops to extremes in CO_2, and O_2 concentrations

Commodity	Minimum O_2 tolerance (%)	Maximum CO_2 tolerance (%)	Injury Symptoms
Lettuce	0.50	2-4	Brownish-red discolouration of midrib
Broccoli	0.25	20	Tolerant
Cauliflower	-	5	Excessive softening, discolours on cooking
Potatoes	5	10	Prevents wound healing

Genetic expression: Various lines of evidence now indicate that the metabolic transformations associated with detached plant parts are not simply the result of cellular disarray but are an organized, directed set of events under genetic control. Mutant selection studies have demonstrated that fruit ripening is under the control of nuclear genes. Inhibitors of mRNA synthesis and protein synthesis can block chemical transformations which are associated with characteristic quality changes in ripening and senescing plant tissues. New enzymes, such as the syntheses involved with ethylene biogenesis and polygalacturonase in tomato, are formed during the period of physiological deterioration. Protein turnover remains high during plant senescence, at least in part due to the formation of new mRNA. In developing tomato fruit, there is a burst in mRNA synthesis just prior to the onset of ripening.

Table 26. Possible control sites of gene expression in postharvest commodities

Event	Control	Evidence
DNA ⇩	Transcription	Mutant studies indicate nuclear genes control ripening. Incorporation of uridine into RNA increases. Ethylene stimulates RNA synthesis.
m RNA protein ⇩	Past-transcription protein translation	Incorporation 01 amino acids into protein increases. New enzymes (e.g., polygalacturonase and ACC synthase) appear. Cyclaheximide retards or prevents ripening
Active protein amino acid	Post-translation degradation	Certain proteins are turned over rapidly in ripening stags.

Synthesis of plant hormones: The plant hormone appears to play a role in the orchestration of the myriad of chemical transformations in the postharvest cell. Plant hormone, abscisins and ethylene are appearing to promote senescence in postharvest crops. The presence of certain hormones (auxins, cytokinins, and gibberellins) appears to maintain the tissue in a juvenile state that is, these hormones appear to prevent ontogenic changes associated with senescence. Ethylene (C_2H_4) a naturally occurring, gaseous plant hormone, is plays a particularly important role in postharvest behaviour determination and more significantly recognized as the ripening hormone. Storage of high ethylene commodities such as mango, banana, tomatoes or bitter melons can be detrimental to the quality of ethylene sensitive commodities such as yard long beans, seime and okra resulting in chlorophyll degradation (green to yellow colour and eventual darkening of skin), toughening of tissues, poor flavour and rapid deterioration. ACC synthesize as the enzyme responsible for the biogenesis of ethylene and its inhibition by compounds such as AVG has broadened the ability of the postharvest physiologist to manage ethylene formation and hence ethylene mediated changes in the crop. Thus, ethylene generating compounds, such as ethephon, ethrel, and ACC (1-amino-cyclopropane-l-carhoxylaic acid) can be used to trigger certain physiological processes. Inhibitors of ethylene biogenesis such as AVG (aminoethoxwinvlelvcine) can be used to retard other physiological events which are mediated by ethylene. Blueberries, in which postharvest losses are minimized and quality attributes are enhanced or maintained, provide an example of a benefit arising from the prevention of endogenous ethylene formation. Apples having an internal ethylene concentration near 1 ppm should be shipped immediately to market by the grower whereas fruit harvested at a time when internal ethylene is about 0.1 ppm are suitable for long-term storage in refrigerated, controlled atmosphere with little incidence of storage disorders and physiological deterioration. Ethylene may act at the level of direct enzyme control, influence energy metabolism and the redox potential of the cell, alter the selective permeability of cellular membranes, or act by controlling gene expression by some direct or indirect mechanism. Hormonal control of gene expression may be one of the most important mechanisms in senescing plants.

Table 27. Management of ethylene and its action

Application	Effects
Ethylene generating compounds	Desirable
Ethephon (2-chloroethyl phosphonic acid)	Accelerate ripening
ACC (1-amino-cyclopropane-1- carboxylic)	Uniform ripening for mechanical harvest De-green certain fruits or vegetables Enhance flavour development Undesirable Die-back of parent plant, excessive softening Accelerate lignification's e.g., asparagus
Inhibition of ethylene formation	Desirable
AVG (aminoethoxyvinylglycine) Inhibition of ethylene action	Delay senescence and extend storage life
Ag+	Effect quality indices in desirable way
CO_2	Undesirable Reduces aroma development in apples

Moisture loss of produce: Moisture loss is influenced by the difference between vapour pressures inside and outside of the commodities. Since water vapour moves from areas of higher to areas of lower concentration, water is lost from commodities when they are subjected to most marketing conditions. Water loss can be reduced by maintaining high moisture content in the air around the fresh produce, apply precooling, reducing air movement, protective packaging, applying a surface coating or waxing, trimming and curing. Outstanding water loss was generate wilting and shrinking of produce and loss of snapping quality of tomato, mango, leafy vegetables, flowers, yard long beans and okra. Excess water loss effect on softening, flaccidity, limpness, loss of crispness, juiciness and nutritional quality of fresh produce. For most of the produce, it is desirable to harvest when the maximum possible water content is present as these results in a crisp texture. In order to minimize produce water loss, both temperature and moisture in the surrounding air of the produce must be controlled. Water loss is generally highest in freshly harvested vegetables and will continue as long as the commodity temperature or vapour pressure is higher than that of the surrounding air.

Compositional changes: There are so many desirable or undesirable compositional changes take place throughout storage of fruit and vegetables. Loss of chlorophyll (green colour) is desirable in fruits but not in vegetables. Development of carotenoids (yellow and orange colours) is desirable in fruits such as citrus and papaya. Red colour development in tomatoes is due to a specific carotenoid (lycopene), beta-carotene (provitamin A) is very important in nutritional quality. The development of anthocyanins (red and blue colours) is desirable in fruits such as cherries. Changes in anthocyanins and other phenolic compounds may result in tissue browning which is undesirable from appearance quality standpoint. Starch to sugar conversion is undesirable in potatoes, peas and desirable in fruits and sweet corn. Breakdown in pectins and other polysaccharides results in softening of fruits and consequent increase in susceptibility to physical injuries. Loss of vitamin content, especially vitamin C, is detrimental to nutritional quality.

Changes associated with physical damage: Physical injuries during harvest are not only unsightly but also accelerate water loss, provide locations for fungal infection and stimulate CO_2 and ethylene production by the commodity. Chilling injury depending on the commodity could cause the undesirable changes like surface and internal discoloration, pitting, water-soaking, uneven ripening or failure to ripen, off-flavour development, and accelerated incidence of surface moulds and decay. The high temperatures can cause bleaching, surface burning or scalding, uneven ripening, excessive softening and desiccation.

Changes associated with pathological breakdown: In some cases pathogens can infect apparently healthy tissues and become the primary cause of deterioration. The onset of ripening in fruits, and senescence in all commodities, results in their becoming susceptible to infections by pathogens. Stresses such as physical, chilling and sunscald lower the resistance of the commodity to pathogens.

Criteria for the Selection of Appropriate Post-Harvest Technologies

The basic recommendations for maintaining post-harvest quality and safety of produce are the same regardless of the distribution system (direct marketing, local marketing, export marketing). In situations where the point of sale is only a matter of hours away from the site of harvest, careful harvesting and handling and the observance of proper sanitation practices are adequate measures for assuring the quality and safety of fruits and vegetables targeted for the fresh market. Pre-cooling, refrigeration and packaging however become essential when fresh produce must be moved over long distances. Furthermore, over-investment in handling facilities can result in economic losses, if consumers in the target market are unable to take up these added costs. Proper education of all stakeholders along the post-harvest chain (growers, harvesters, handlers and those involved in marketing) is more critical than the level of sophistication of the equipment used in post-harvest handling. Commodity requirements can be met through the use of simple and inexpensive methods in many cases. Mechanical injuries are major causes of losses in the quality and quantity of fresh horticultural commodities in all handling systems. The incidence and severity of mechanical injury can be greatly minimized by reducing the number of steps involved in harvesting and handling and by educating all personnel involved, about the need for careful handling. Post-harvest horticulturists therefore need to coordinate their efforts and to cooperate with production horticulturists, agricultural marketing economists, engineers, food technologists, microbiologists, and others who may be involved in various aspects of the marketing systems. In most cases, solutions to existing problems in the post-harvest handling system require the use of existing information rather than new research.

Importance of Post Harvest Management of Fruits and Vegetables

Importance of post harvest technology lies in the fact that it has capability to meet fruit and vegetable requirement of growing population by eliminating avoidable losses of nutritive perishable items of fruit and vegetables. Post-harvest technology of fruit and vegetables has potential to create rural industries. In India, where 65 per cent of people live in the villages and 60 per cent depend on agriculture have experienced that the process of industrialization has shifted the food, feed and fibre industries to urban areas. This process has resulted in capital drain from rural to urban areas, decreased employment opportunities in the rural areas, balance of trade in favour of urban sector and mismatched growth in economy and standard of living including the gap between rural and urban people. It is possible to evolve appropriate technologies, which can establish fruit and vegetable processing based rural industries. A process by which fruits and vegetables are prevented from getting spoilt for a long period of time. The colour, taste and nutritive value of the fruits and vegetables are also preserved. This means that the colour and taste of fruit and vegetables which is present at the time of preservation should not change, example of preserving grapes as raisins by sun drying process. During the drying process the grapes change colour from green to brown, and taste sweeter. However once they are dried and stored they do not change colour or taste e.g., mango juice, murraba, squash, pulp, chutney, pickle, raw mango powder etc. Mango is a summer fruit and grows in large quantities during the months of April to August. Usually all the quantity grown in a region cannot be consumed by the people staying there. The farmers make arrangements to transport the excess quantity to regions where either mango is not grown or where that particular variety of mango is not available. If does not do this, the excess produce will rot and go waste and the growers will then lose money. There is still some quantity which is left after the fresh fruit is consumed by people has to be preserved for consumption during the months when fruits are not available. Stored of fresh fruits and preservation of the fruit and vegetable is done during the month when these are available in large quantity and therefore at low cost.

Need for Post Harvest Technology

Fruits, vegetables and ornamentals are ideally harvested based on optimum eating or visual quality. However, since they are living biological entities, they will deteriorate after harvest. The rate of deterioration varies greatly among products depending on their overall rate of metabolism, but for many it can be rapid. For example marketing chains where produce is transported from farm to end user within a short time period, the rate of post harvest deterioration is of little consequences. However, with the increasing remoteness of production areas from population centers, the time lag from farm to market is considerable. The deliberate storage of certain produce to capture better return adds to this time delay between farm and end user, by extending the marketing periods into times of shorter supply. Thus a modern marketing chain puts increasing demands on produce and creates the need for the post harvest techniques that allows retention of quality over an increasingly longer period.

3 Fruit Growth, Development and Postharvest Physiology

Fruit evolved as means for production and spreading of seeds. Humans then imposed further selection pressures to develop products for our use. Our concept of a fruit as a sweet and fleshy object for eating is really quite recent in evolutionary terms.

ORIGIN OF FRUIT TISSUES AND FRUIT SET

Pollination, followed by pollen tube growth and fertilization, initiates fruit enlargement. If pollination does not occur, flowers are shed (only rare exceptions). However, a developmental series of gene expression for fruit growth has already been established well ahead of floral biology. Primordia may have been initiated up to 6 months previous to a particular flower opens, and ovary development carry on during flower growth with ovary tissues forming late in this process. As part of that outcome, homology between leaves and sepals is noteworthy and evident in many fruits. Sepals show leaf-like cell layers, with stomata and chloroplasts. The generic term 'fruit' covers a wide range of structures, all supporting and protecting seeds, but where the various parts have developed from the original fertilized flower in various distinguishing ways. In the simplest form, ovary walls grow along with seeds, and as they develop, the ovary walls dry out to become a pod (legume) or capsule (poppy). In others (particularly fleshy fruit), the main structure can arise by exaggerated development of a particular part of the original floral unit. These include ovary wall or central axis, the receptacle that supports anthers and ovary, or even petals and sepals. In morphological terms, fruit are structures that develop from fertilized or stimulated ovules, plus associated floral parts that originate from the parent plant. Mechanistically, a fruit is a single dispersal unit that includes seeds and associated tissues, developed as a single body. This broad description includes structures derived from a single ovary (as in simple fruit such as apple, avocado and mango) as well as compound fruit where separate ovaries are joined (an aggregate fruit such as blackberry and cherimoya) or where separate flowers are collected into a single structure (pineapple and breadfruit). During fruit development, an ovary wall becomes a pericarp, either dry as in a dehiscent pea pod and the indehiscent caryopsis of barley or fleshy as in berries (grape). Three morphologically distinct strata are present and developed to varied degrees such as exocarp (fruit skin), mesocarp (fruit flesh) and endocarp (inner cell layers). An exocarp will develop a cuticle and may exhibit a variety of morphological features such as coarse hairs (kiwifruit) or fine hairs (peach). The exocarp plus cuticle restricts gas exchange, and determines the general appearance of ripening fruit. Most cuticles are highly impermeable to gases, so that water vapour, O_2 and CO_2 diffuse mainly via either stomata or lenticels or by mass flow through cavities at the calyx and stem ends of fruit. Mesocarp tissues usually represent the fleshy part of a fruit, and commonly hold chloroplasts and starch grains. In fleshy fruit such as berries (e.g. tomato, kiwifruit and grape) tissue typically comprises large parenchyma cells and contains the main vascular network. Endocarps are less common, but typically develop as a dense hard case around a seed, as in peach, apricot or macadamia. An ovary must be stimulated in some way for fruit growth to occur, this is normally by pollination and fertilization. Gibberellins and auxins are involved in the pollination stimulus, and subsequent hormone production by the fertilized ovary is critical to stimulating fruit development. By implication, a suitable balance of growth regulators applied to unpollinated fruitlets can result in fruit set, and in practice gibberellins GA_4 and GA_7 are very effective in setting parthenocarpic (seedless) grape and cucumber fruits. By contrast, parthenocarpy is rare in kiwifruit, although repeated applications of naphthaleneacetic acid (NAA) with benzyladenine (BA) and gibberellin have been successful. The growth regulators alone or in combination can trigger cell division in ovaries or related tissues that ultimately become fruit. Seedless fruit have arisen via human selection of genotypes in which ovaries produce an adequate supply of growth regulators without any stimulation from the germinating pollen and developing seed (triploid banana), or where fertilisation is closely followed by seed abortion (stenospermocarpic, as in sultana grape). In the absence of pollination, levels of endogenous hormones such as auxins and gibberellins normally fall markedly and flowers abscise or fruitlets stop growing.

Dynamics of Fruit Growth

Fruit can augment in mass or volume by 100-fold or more from fertilization to maturity and such changes commonly follow a sigmoid curve. Interpretation of such growth curves is complex because a single variable (mass, length, volume) is commonly applied to an object that contains several organs and different tissue types, each developing at their own rate and in accordance with their own programme. Moreover, at a cellular level, comparative levels of division and expansion change with ontogeny, while shifts in airspace percentage also play a part in volume increases. Added to this, changes in storage products (oil, starch and sugar) and structural carbohydrate (endocarp thickening) influence dry matter content.

Cell division and enlargement: In spite of complexities of fruit growth and development, there are some overall consistencies in patterns of cell division and enlargement, as well as tissue differentiation and fruit enlargement. During the first 1–4 weeks, fleshy tissue volume increases rapidly because of cell division and embryo volume remains small. In many commercial fruit (e.g. apple, kiwifruit, tomato and peach), cell division may cease a few weeks after of anthesis, and fruit growth slows down, reflected as an inflection in the growth curve, and signalling an end to the first sigmoid phase. During early growth, the fertilized embryo and endosperm develop and seeds start to form. A second phase begins where the pericarp resumes growth and continues to enlarge until slowing for a second time as fruit mature. This second phase in fruit growth is mainly accomplished by cell expansion in longitudinal, radial and tangential planes. Longitudinal growth, where cells enlarge parallel to the long axis of the fruit, will often be a big factor for development of elongated fruit such as cucumber and marrow. Radial growth increases diameter as in some pumpkins. Mature watermelons end up with some of the largest parenchyma cells in the plant kingdom, about 0.7 mm in diameter. In contrast to this general pattern where cell division come to a close after a few weeks, pericarp cells of avocado fruit continue to divide over the whole growth period so that cells in mature fruit are still relatively small. Cell enlargement is a continues process but not homogeneous. Cells in various regions of a fruit often enlarge at different rates and in different planes, so that many mature fruit show strong gradients in cell size from their surface to the centre. In apple fruit, cells closest to the core are smallest, with cell size increasing towards the fruit surface. Conversely in many berries, such as cucumber, kiwifruit and grape, the smallest cells are found in outer regions of the pericarp, with size increasing progressively towards inner regions.

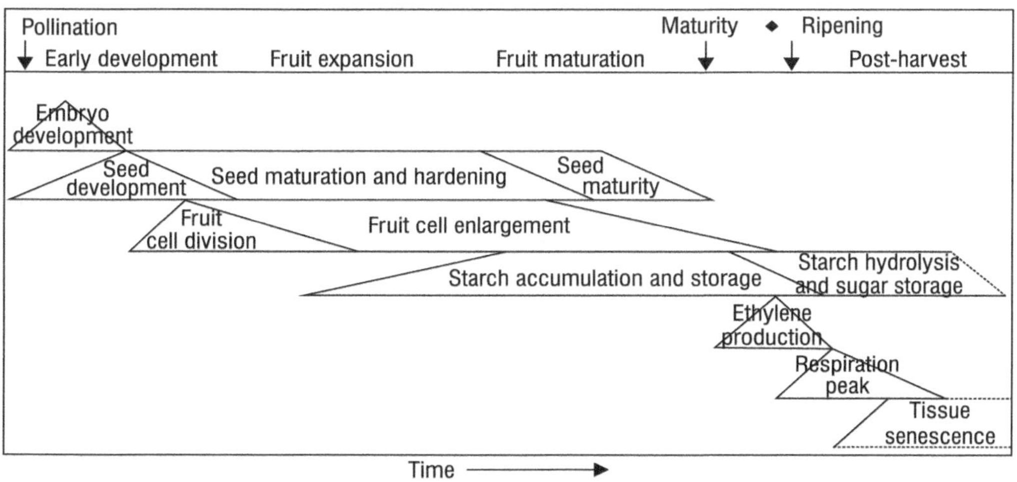

Figure 3. A large number of complex processes are integrated in space and time during seed development and fruit growth and are shown here schematically. In broad terms, embryo differentiation and seed development are already well advanced as pericarp enlargement gets underway, and seed maturation

usually precedes onset of ripening; consequently fruit ingested prematurely still represent vehicles for seed dispersal. A phase of carbohydrate accumulation during fruit maturation gives way to starch hydrolysis and sugar storage during maturation, accompanied by a peak in ethylene output and respiratory activity as fruit ripen. (Original diagram courtesy I.B. Ferguson)

Cell differentiation: Patterns of cell enlargement and differentiation in cell layers can influence the quality of mature fruit. For example, pepino fruit with a compact exocarp composed of tightly packed cells are less likely to bruise during postharvest handling than cultivars having large intercellular airspaces. As cell size increases during development, other accompanying characteristics also change, such as cell wall thickness, differentiation of specific cell types (e.g. sclereids) and the formation of cell inclusions such as oil droplets or calcium oxalate crystals (raphides). In Feijoa and pear, development of sclereids in the mesocarp provides the characteristic gritty texture. As another example, juiciness of orange depends on prior differentiation of juice sacs in the endocarp. The extent and distribution of airspaces are particularly important, affecting both fruit texture and physiological properties. For instance, in apple, airspace relative to fruit volume can double during development, while cell wall thickness and relative cell surface area both decline. Such changes affect gas exchange and diffusion of solutes through pericarp tissues due to increased tortuosity. Cell division ceases first in the exocarp and last in the innermost regions of the central core. The outer pericarp is first seen as a homogeneous population of cells but by after pollination two cell types become visible, namely small isodiametric parenchyma cells full of starch grains, and much larger heavily vacuolate ovoid cells in which the frequency of starch grains per unit volume is low.

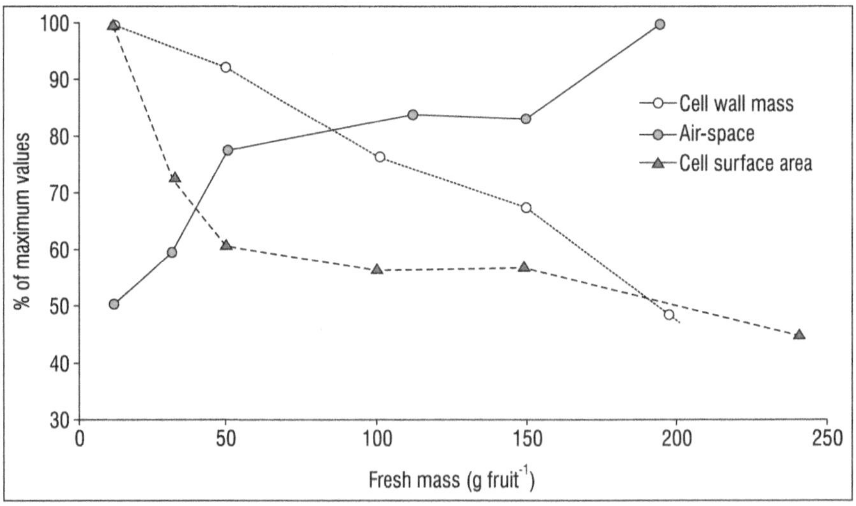

Figure 4. Physical characteristics of apple fruit change during growth and development with a notable increase in gas space to a maximum of around 0.1 mL gas mL^{-1} tissue at maturity, and a corresponding decrease in cell wall mass to around 15 mg g^{-1} fresh mass. Cell surface area shows an early rapid decrease from around 340 cm^2 mL^{-1} tissue. (Based on Harker and Ferguson (1988) Physiol. Plant. 74, 695-700)

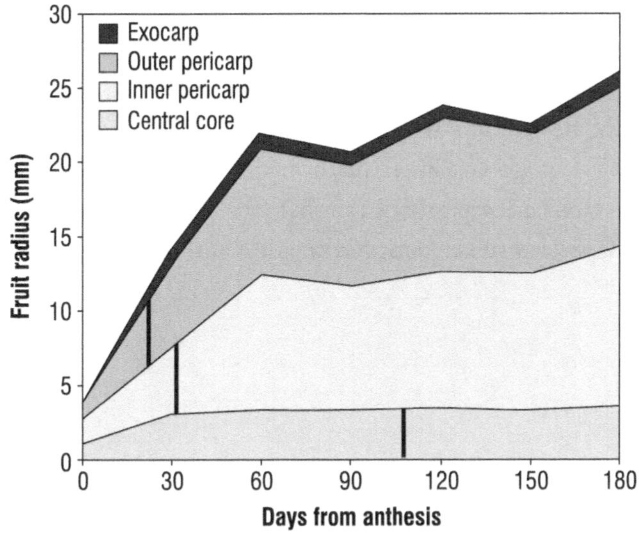

Figure 5 Radial growth in kiwifruit is due mainly to enlargement of outer and inner pericarp. Vertical lines indicate cessation of cell division in each tissue. (Unpublished data courtesy K. Gould and I.B. Ferguson)

Seed development and fruit growth: Fertilization is normally crucial for fruit set and pericarp development. As fertilized ovules develop into seeds, this influence on pericarp growth continues where production of hormones by the endosperm and developing embryo promotes pericarp growth. The importance of seeds as sources of hormones for initiation and stimulation of fruit growth is implied by fruit response to exogenous hormones in parthenocarpic systems (development of fruit without seeds). Applying auxin and gibberellins to unfertilized embryos is one way of achieving parthenocarpy, another is to use auxin transport inhibitors such as chlorflurenol to prevent loss of auxin from embryos so that a threshold level for pericarp response is exceeded. Parthenocarpy in tomato and cucumber indicate that high auxin levels enhance embryo cell division, and this cell division phase seems to be more critical than subsequent cell expansion in determining final fruit size. Such results imply a cooperative mode of action where gibberellins combine with auxins to begin cell division. Seed cytokinins and cell division are similarly related because tomato seeds accumulate cytokinins that subsequently influence cell division in surrounding pericarp tissue. Such interdependence between seed development and fruit growth shows up in final fruit size. Parthenocarpic fruit have reduced auxin content and are generally smaller than wild-type fruit. In apple fruit, seed numbers frequently correlate with fruit growth or with shape and size of fruit. Inadequate pollination of kiwifruit results in distortion and a curvilinear relationship emerges between seed number and fruit weight. A similar response is obtained when young seeds are surgically removed from immature strawberry fruit, causing a corresponding distortion in flesh development. Despite ample evidence that natural control of fruit shape is primarily exerted by plant hormones originating from seeds and stimulating growth to varying degrees, this is not true for all fruit. In banana, fertile seeds actually suppress development of the fleshy pulp. In this anomalous case, fertilization failure allows an ovary to grow. In marrow, tomato and kiwifruit, ovary shape dictates spatial distribution of seeds. They in turn influence pericarp growth, so that fruit size and shape then become a function of initial ovary shape plus subsequent fertilization and seed development.

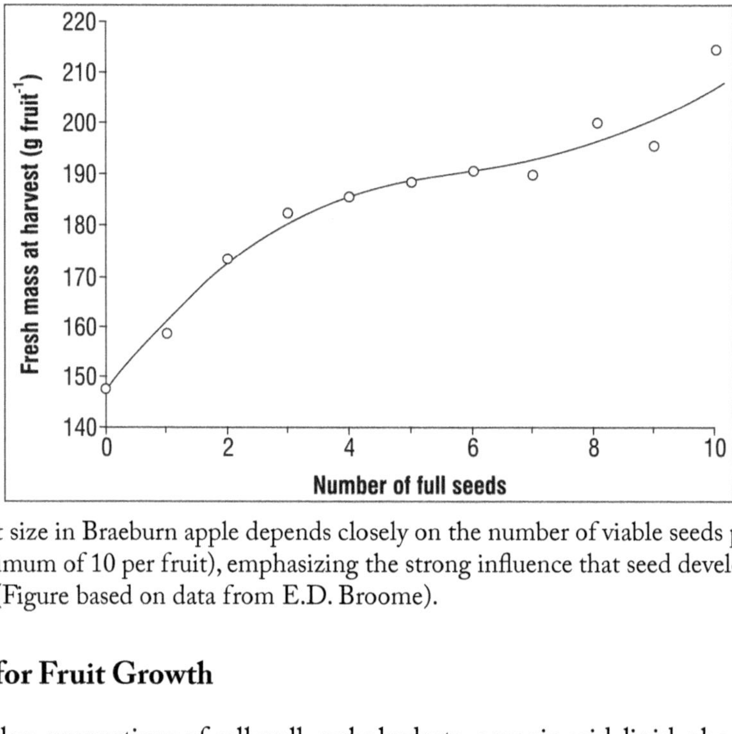

Figure 6. Fruit size in Braeburn apple depends closely on the number of viable seeds per fruit (up to a normal maximum of 10 per fruit), emphasizing the strong influence that seed development has on fruit growth. (Figure based on data from E.D. Broome).

Resources for Fruit Growth

As fruit develop, proportions of cell wall, carbohydrate, organic acid, lipid, phospholipid and volatile (aroma) compounds change dramatically and within each of those groups there are changes in the proportion of individual group members. In developmental process of fruit comprise with rate of growth, attainment of maturity, gain final fruit size, develop aroma, flavour and texture in ripe fruit. Enlarging fruit require carbohydrate to sustain cell division, enlargement and tissue specialization. Only in later stages are carbohydrates typically retained as either starch or soluble sugars. Soluble carbohydrate is mainly imported as photoassimilate, with only a minor contribution from local CO_2 fixation, and re-assimilation of respiratory CO_2. During peak fruit expansion, usually early summer, there is an intense flow of photoassimilate from mature leaves (sources) into rapidly enlarging fruit (sinks). Sugars generated by photosynthesis, along with amino acids and phosphate within the plant's vascular network, move via the phloem into enlarging fruit.

Photoassimilate distribution: Sources of photoassimilate can be identified by providing individual leaves with $^{14}CO_2$ and following the pattern of labelled material into neighbouring organs. Leaves typically begin to show a net export of photoassimilate at about 50–60 per cent of full size. Distribution patterns of $_{14}C$-labelled products relate to developmental morphology of fruiting shoots. Typically, source leaves are nearby on the same lateral branch, both above and below the fruit. In apple, fruiting spurs may develop primary leaves (emerging soon after budburst), then spur leaves (in a rosette at the base of the flower), then bourse leaves (growing on spur bourse shoots). Each in turn provides assimilate for the next phase of leaf growth (primary → spur → bourse) then as leaf expansion come to an end, all provide assimilate to the developing fruit. Leaves on adjacent extension shoots can provide some photoassimilate to fruit, but if indeterminate growth continues the furthermost leaves become progressively less important as suppliers, and more significant as competitors. If the normal suppliers are removed, carbohydrate can come from longer distances, sometimes from leaves more than a meter away. The relative strength of source and sink is a major factor for distribution patterns, but transport options are dictated by vascular connections. During plant growth, development occurs in an orderly and patterned manner, creating separate files of leaves. This pattern (phyllotaxis) is accompanied by a matching pattern of vascular connections. Photoassimilate tends to move along a pathway of least resistance, following these direct vascular connections where they exist, hence distribution patterns generally follow phyllotaxis. This importance of phyllotaxis in carbohydrate allocation to the fruit is well shown in kiwifruit, where specific leaf–fruit connections exist. Patterns of assimilate distribution from leaf to fruit have been studied by taking a number of matched lateral fruiting branches of kiwifruit vines, then supplying $^{14}CO_2$ to one leaf on each, at various nodal positions along the stem, from node 1 (base) to node 10 (tip). Each lateral also had one fruit each on nodes 1 and 2, while the remaining nodes had leaves only. Distribution of $_{14}C$-labelled photosynthate was allowed to proceed for 6 d, and the total radioactivity in each leaf and fruit on the lateral was then measured. Specifically, node 1 fruit received assimilate from their own subtending leaf (node 1 leaf) and from leaves on nodes 6 and 9. Node 2 fruit was supplied by its subtending leaf and leaves on nodes 7 and 10. Assimilate from remaining leaves was distributed generally within the main body of the plant. However, if the apex of the lateral was removed to stop extension growth, fruit then drew assimilate from all leaves. By implication, a drastic change in source–sink relationships can override restriction on carbon transport imposed by vascular patterns in intact plants.

Composition of photoassimilates: Radiolabelling of photoassimilates has also been used to identify which compounds are transported into storage organs. Analyses of phloem tissues and phloem sap show that in most plants carbohydrate enters fruits primarily as sucrose.

In the woody Rosaceae (apple, pear, stone fruit), the sugar alcohol sorbitol is the major photosynthetic product at 60–85 per cent of transported carbon, the remainder being mainly sucrose. Regardless of transport form, photoassimilate arriving in fruits is rapidly converted to the storage products characteristic of the fruit in question (principally starch, glucose, fructose and sucrose). Thus the identity of labelled sugars in fruit often differs markedly from the form transported. For example, sorbitol concentration is high during early development of apple fruit and more or less reflects the composition of photoassimilate in transit. By maturity, sorbitol content will typically decline to below 5 per cent of the total soluble carbohydrate. If sorbitol reaching fruit is not fully metabolized, apoplastic accumulation results and pericarp tissues become glassy in a disorder called watercore. This is a common problem with some apple cultivars such as Fuji. Sugar transport and accumulation can thus have economic importance both in terms of desired taste characteristics and postharvest fruit quality. Photoassimilate is commonly transported from leaves (sources) to fruit (sinks) as sucrose, and most agricultural plants fit this model. Even genera as diverse taxonomically as yucca and vitis transport sucrose almost exclusively, but stone fruit and pome fruit of the woody Rosaceae use sorbito. Ash trees translocate most of their photoassimilate as mannitol. As a further variant, some cucurbits (e.g. squash) use stachyose and related compounds. In all cases, translocated sugar is an energy rich source of carbon, but sucrose is not the universally translocated sugar, as indicated below:

Table 28. Composition of photoassimilate sugars in different fruits

Plant species	Photoassimilate sugars					
	Sucrose	Glucose	Fructose	Sorbitol	Mannitol	Other
Apple	22	4	3	71	-	-
Ash	11	1	1	-	65	22
Grapevine	93	4	3	-	-	-
Yucca	97	2	1	-	-	-

FRUIT COMPOSITION AND SENSORY ATTRIBUTES

Carbon transport and subsequent metabolism in developing fruit cannot be viewed in isolation, particularly when aspects of fruit quality, such as taste and flavour, are directly dependent on such processes. In particular, sugar acid balance and contents are primary determinants of the taste attributes of fruit, and so are of major significance for consumers. Too much acid and the fruit are tart and unpalatable too little and the fruit is insipid and bland. In horticultural terms, acid levels are often expressed as titratable acidity and this is used as one indicator of taste. Another indicator used is the refractive index of the expressed sap (recorded as °Brix). This is a measure of the soluble solids concentration (SSC %) of expressed juice and represents the sum of organic acid, salts and sugar contents. Several organic acids may be present, but certain ones are characteristic of particular species or cultivars. For example, malic acid predominates in pome fruit, citric acid is dominant in citrus, while tartaric acid is dominant in grape. In kiwifruit, malic, citric and quinic acids are the major ones, and in total may exceed 1.5 per cent of the fresh weight. Acids are not transported into fruit via phloem connections, but are synthesized in situ. Part of the acid component comes from metabolism of the sugar imported through the phloem, but part can be synthesised by local fixation. In citrus, dark fixation of CO_2 by mature fruit makes a meager contribution to acid balance, but inter-conversion of imported carbon is of more consequence. In that case, citrate synthase and subsequent enzymes in the citric acid cycle appear to determine whether imported carbon (as sucrose) is transformed into other sugars or is metabolized further to organic acids.

Starch sugar balance is a major factor in consumer perceptions of fruit quality. In many fruit and vegetables, including apple, banana and kiwifruit, starch accumulates throughout development, being laid down as granules in plastids. In kiwifruit, starch may reach 50 per cent of the total dry matter towards the end of fruit growth (at about 15 weeks after pollination). As fruit approach maturity (17–20 weeks after pollination), there is a rapid onset of starch hydrolysis. Starch content at the onset of this conversion is not enough to account for all the sugar present in ripe fruit, and this implies that maturing fruit continue importing sugar up to harvest. Continuing import of $_{14}C$-labelled photoassimilate into maturing peach and apricot fruit confirms that pattern. The dynamic between starch breakdown and soluble sugar increase can be a critical index of fruit maturity. Starch content is strongly linked with sugar content of ripe fruit and hence with consumer perceptions of fruit taste. Starch content of fruit at harvest (commonly estimated from fruit dry matter content) is used commercially as a proxy for potential fruit taste. As an additional factor in their dietary appeal, fruit are rich sources of vitamins, particularly vitamin C (L-ascorbic acid). Moreover, vitamin C can be a major metabolite (greater than 2g kg^{-1} fresh weight) in fruit such as acerola, rosehip, quandang, kiwifruit, citrus, blackcurrant and guava, and has strong anti-oxidant properties. This may account for a notable absence of browning in kiwifruit and citrus when sliced (in conjunction with relatively low levels of polyphenols and polyphenol oxidase in those tissues). Vitamin C levels increase in the fruit during early growth, and tend to be stable through to maturity. The B group vitamins such as B_1, B_2, pantothenic acid and biotin are present in both fruit and seeds, while B_3 and B_6 are particularly abundant in seeds. The vitamin A precursor β-carotene is found in useful quantities in some fruit, for example peach, apricot, melon and cherry. Phenolics such as anthocyanins and tannins are also important in fruit and are responsible for much of the visual appeal of intact fruit (e.g. tamarillo), exposed flesh (e.g. cherry) or extracted juice (e.g. guava). They also contribute to flavour characteristics, adding a slight and pleasing astringency (as with the dessert apple) or a more aggressive one (as with cider apple and green banana). Tannins in persimmon fruit are a special feature of that fruit and provide an interesting example of the potential dominance of a single quality characteristic in determining how a given fruit is used.

MINERAL NUTRIENTS OF FRUITS AND VEGETABLES

Mineral nutrients move into the fruit most rapidly during the early stages of development at a time when xylem water flow dominates. As fruit approach maturity, surface to volume ratio declines, the skin becomes less permeable to water loss, and large amounts of photoassimilate are imported via phloem connections. As a result, a significant part of the water reaching fruit now enters through the phloem and is accompanied by photoassimilate. Mobile ions such as K^+ and HPO_4^{2-} are loaded into the leaf veins along with the photoassimilate, travel in the phloem and so reach fruit over the whole growing season. In contrast, less mobile nutrients such as Ca^{2+} fail to reach fruit during later stages, so that Ca^{2+} concentration remains steady or even declines slightly. Nutrient deficiencies in fruit are relatively uncommon, except for those associated with calcium. Calcium deficiencies are expressed in the form of blossom end rot in tomato, and bitter pit plus lenticel blotch in apple fruit. These apple disorders tend to be expressed during postharvest storage, but symptom expression is somehow related to the previous ripening environment. These disorders show up as a pitting of flesh and skin, reducing fruit value or even rendering those commodities unmarketable. Such commercial penalties have resulted in development of preharvest sprays and postharvest dips of calcium salts that diminish bitter-pit incidence in harvested fruit. Where there is little or no calcium recycling via phloem, calcium needs to be applied directly to fruit to have a beneficial effect.

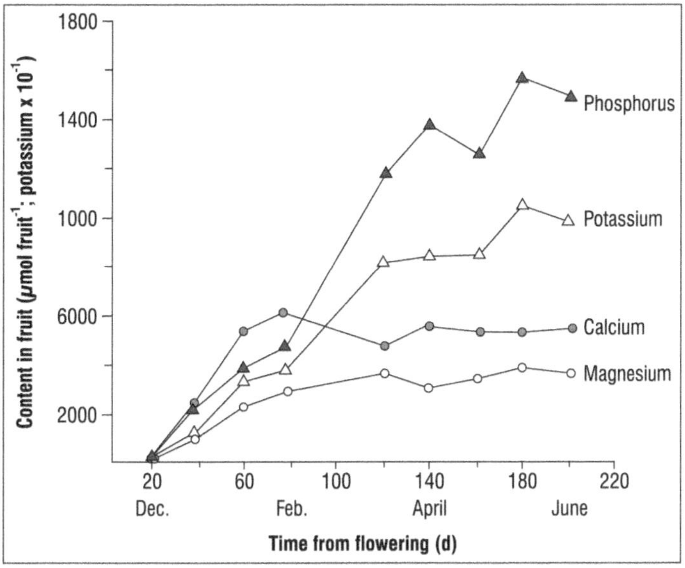

Figure 7. In kiwifruit, as in most fruit, accumulation of calcium is confined to early stages of development that coincide with cell division. By contrast, phosphorus and potassium move into fruit over the whole growing season and are able to enter via either the xylem or phloem. Magnesium import is meagre but progressive. Note the expanded scale for potassium. Fruit content of nutrient ions at maturity would be as follows: phosphorus 1600, calcium 600, magnesium 400 and potassium 1000 μmol per fruit. (Based on Ferguson (1980) N. Z. J. Agric. Res. 23, 349-353)

Carbon accumulation

During development, photoassimilate is stored in fruit (and other sink organs such as root vegetables, seeds and flowers) in either a soluble or insoluble form. Fruit that store carbon in a soluble form (e.g. berry fruit, peach, persimmon, melon, grape, citrus) need to remain on the plant until nearly ripe if they are to survive postharvest storage and meet customer expectations. In most cases the major and rapid increase in soluble sugar content does not occur until late in development, signalling the beginning of ripening. Because the sugar source is the parent plant, harvesting such fruit too early reduces their final sugar content to unacceptable levels. In contrast, there are other fruit that store their carbon in insoluble forms, particularly starch. This allows greater efficiency in accumulating carbon, as the storage product is more compact, osmotically inactive and better segregated from metabolic processes. Examples are avocado, which stores carbon as starch and lipid, and kiwifruit, apple, pear, mango, papaya and banana, all of which store carbon as starch. In fruit that store carbon as sugars and organic acids during development, colour changes followed by the initiation of softening signal that fruit are becoming mature and ready to harvest. The challenge facing postharvest physiologists is to assess when such fruit are still sufficiently firm to allow easy handling and storage, but have enough sugar for a true ripe flavour to develop. A developing crop is typically monitored with a simple refractometer test (soluble solids concentration based on refractive index of expressed juice). In grape, a rapid rise in sugar content (beginning at 'veraison') may need to reach SSC values of around 20 per cent (with, say, 17% sugar and 2% acid as the main soluble components) before a successful harvest can be assured. There is more to good flavour than high sugar content, with sugar to acid ratio being particularly important, so that a combination of SSC to acid ratio with a minimum SSC level may be required. In fruit that store carbon as starch, time of picking is less dependent on sugar content, since a doubling or more of sugar concentration by starch hydrolysis can still take place after the fruit are picked. Once starch utilization has started, fruit can be picked without much detriment to final eating quality. SSC measurements alone are then insufficient, and measurements of total solids, dry matter, oil content and starch concentration are used as well. For example, in kiwifruit sequential measurements of SSC are combined with dry matter quantity, in apple and pear, the pattern of starch distribution within the fruit is recorded along with SSC and ethylene measurements, in mango total solids, fruit shape, flesh colour and firmness are recorded, in papaya colour changes and sequential SSC measurements are recorded, in avocado dry matter combination with oil content is used. In banana, shape or fullness of fruit is an important criterion rather than starch level because bananas can be harvested over a remarkably wide range of maturities and still ripen satisfactorily. Thus, when fruit store carbon in insoluble forms, they can often be harvested while still hard, lending greater flexibility to postharvest handling and ensuring a longer storage life.

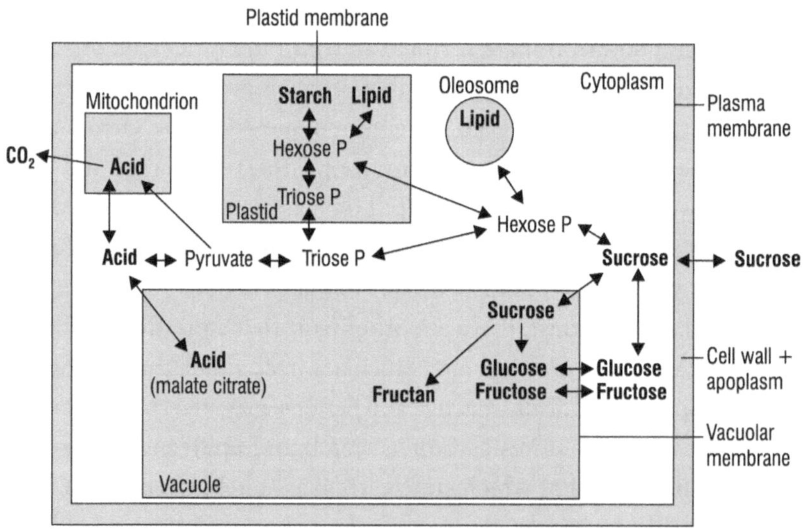

Figure 8. Carbohydrate economies in developing fruit are derived by import and recycling of photoassimilates between different metabolic compartments. Sucrose or less commonly other forms of translocated carbon arrive via phloem conduits and are loaded into cytoplasmic compartments. As

shown schematically, carbohydrate can then be partitioned to vacuolar storage or converted to other sugars and, in the form of hexose phosphate, transferred to plastids where it is used to synthesize starch. Each step is reversible, and as happens during ripening, starch in plastids is transformed back to sugars that subsequently accumulate in vacuoles as indicated. Other specialized organelles (oleosomes) store lipids while mitochondria draw upon imported and locally fixed carbon for ATP generation. (Original diagram courtesy R.L. Bieleski)

Sugar storage

In sugar storing fruit a major shift in metabolism generally takes place when fruit expansion is almost complete, heralding a rapid increase in sugar content. Unloading of sugars from the phloem usually occurs by a symplastic route, but in some species is interrupted by an apoplastic step. Control points for sugar entry and accumulation by fruit include:

- » Rate of sugar production by leaves and delivery to transport pathways.
- » Reallocation of sugar from supporting vegetative growth towards fruit growth.
- » Enhanced unloading of sugar from transport streams into fruit.
- » Enhanced transfer of sugar across plasma membranes into cells or through plasmodesmatal connections between cells.
- » Onward metabolism of sugar in the cytoplasm, or transfer to storage in vacuoles.
- » Increased respiratory utilization of sugar to provide energy for metabolic processes.

As with any biological system, multiple controls operate concurrently to drive a given pattern of maturation. Such events lead us to more robust indicators of ripeness, and improved ways of manipulating maturation to yield higher sugar content and better handling properties. In tomato, there are different genotypes that accumulate either hexoses or sucrose. Most cultivars are hexose accumulators, in which acid invertase is active during growth and ripening. In melon, where sucrose is the main sugar to increase, there is a corresponding decrease in acid invertase and an increase in sucrose phosphate synthase activity (this synthesizes sucrose from hexose phosphate and adenylated precursors). However, in grape berries where hexoses begin to accumulate at veraison and reach very high levels, sucrose phosphate synthase, sucrose synthase and hexokinase activities all increase, but acid invertasem RNA abundance and activity both peak just prior to or at veraison and then decline.

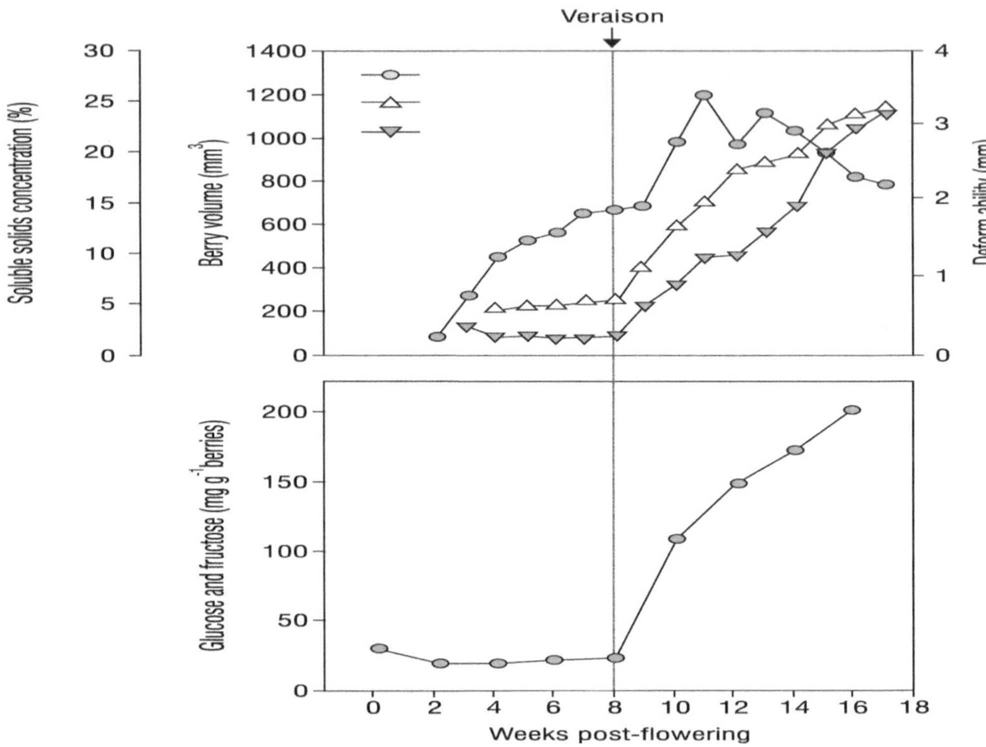

Figure 9. Grape undergoes an abrupt change in physiology midway through development. For about 8 weeks after flowering, berry volume increases steadily but fruit are hard (low deformability) and sugar content low. At veraison invertase activity rises abruptly and reducing sugar content increases rapidly, reaching about 20 per cent of fresh weight when ripe. Berries attain full size by 10–12 weeks, and approach an asymptote in sugar content 2–3 weeks later. (Based on Davies and Robinson (1996) Plant Physiol. 111, 275-283)

Starch storage

Fruit that store starch switch from starch synthesis during development to starch hydrolysis during ripening. Starch sugar interconversion involves a larger number of enzymes and a greater complexity of control than is required for sugar storage alone. No transgenic plants have yet been reported in which a starch storing fruit or lipid storing fruit has been altered to store only sugars, or vice versa, but results with potato suggest it could be possible. In potato, control of both insoluble carbon storage and sugar to starch conversion has been attained using sense and antisense constructs to alter the expression of specific carbohydrate enzymes. Kiwifruit provide an example of biochemical changes in a starch-storing fruit. They are normally harvested with starch contents ranging from 4 to 10 per cent (dry matter concentrations between 14 to 20%, SSC between 6.2 to 12%). At harvest, the main sugars are sucrose, glucose and fructose. As fruit ripen after harvest, sucrose content increases only slightly, while fructose and glucose increase in parallel to become the predominant sugars in ripe fruit. Labelling with radioactive precursors indicates that all three sugars are actively synthesized during ripening, and there are increases in the activities of a number of sucrose-metabolizing enzymes, particularly sucrose phosphate synthase and invertase. The starch-degrading enzyme α-amylase increases two-fold. At room temperature malate decreases and citrate increases while quinate remains unchanged. However, during cool storage (0–4° C), malate increases. Such changes are enough to alter the flavour balance in the ripe fruit. Where fruit store carbon in an insoluble form, there are several potential control points for sugar metabolism take place. In addition to sugar metabolism, hydrolysis of starch to glucose, transfer of sugar precursors from starch-containing plastids (amyloplasts), synthesis or degradation of sucrose and synthesis of hexoses, transfer of sugar to vacuoles or export from cells, carbon flow between sucrose and malate or citrate, production of CO_2 from sugar or acid precursors and transfer of malate or citrate across the vacuolar or mitochondrial membranes activities are accomplished during storage and ripening of fruits. Interference in any of these processes should affect ripening and/or flavour development after harvest. Such interference may be physical (as in storage temperature), chemical (as in atmosphere composition) or genetic (by modifying activities of specific proteins which control flow between particular metabolites).

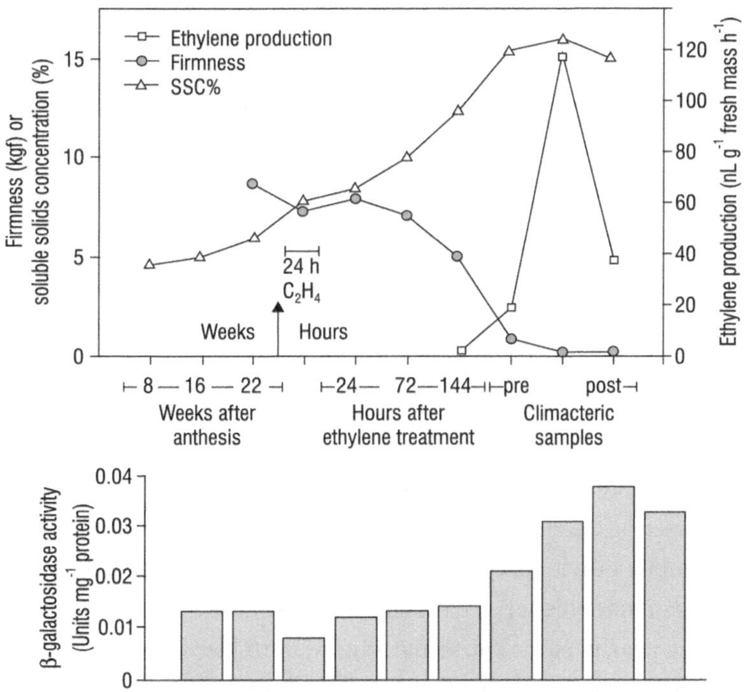

Figure 10. Kiwifruit show some dramatic changes in physiological status during development. For 24 weeks after flowering, fruit are hard and sugar content is low. Meanwhile starch content (not shown)

rises to about 12% of fresh mass. Ripening in fruit harvested around 24 weeks can be triggered by exposure to an external source of ethylene even though the fruit are not yet producing ethylene by themselves, and are incapable of an autocatalytic response. Within a week of such treatment starch becomes hydrolysed and sugar concentration rises from 7% to 15%. Fruit then soften rapidly and several cell wall-modifying enzymes increase in activity. Fruit do eventually show a peak in ethylene production, but not until ripening is well under way. (Figure based on data from J. Win)

Fruit Ripening

Several processes take place since fruit ripen as well as become edible, and then senesce. These changes may take place while fruit are still attached to the plant or after harvest. Tomato, banana and avocado are examples of fruit that at harvest can be at a mature green but unripe stage and are inedible until subsequent ripening processes have occurred. In contrast, strawberry, orange, pomegranate and grape are examples of fruit that need to stay on the tree or vine until ready to eat in order to have their desired eating characteristics. Several major changes take place as fruits ripen, and in ripening processes like, changes in carbohydrate composition resulting in sugar accumulation and increased sweetness, change in colour, texture and flesh softening, formation of aroma volatiles and accumulation of organic acids with associated development of flavour. These changes make the ripe fruit attractive to animals, which in eating the fruit will disperse the seeds and enlarge the range and improve the survival chances of the next generation of the plant. Lignified pits and seeds encased in a fibrous core might be discarded after eating the flesh, whereas smaller seeds might pass through the animal's digestive system and be deposited with the animal's excrement.

Ethylene and regulation of ripening process

Ethylene emission is closely associated with fruit ripening in many species and it is the plant hormone that regulates and synchronizes the different aspects of the ripening process like colour development, aroma production and texture of the fruits. Historically fruit have been categorized into two classes (climacteric and non-climatric) of behaviour with respect to ethylene physiology and respiratory pattern. In the first category (climacteric fruit), as fruit progress towards edibility the respiratory rate increases sharply to a maximum (climacteric rise) followed by a decline as fruit senesce. The major rise in ethylene production may take place before, just after or close to the respiratory peak and show a single phase respiratory response to exogenous ethylene. Such fruit are classed as climacteric fruits, like apple, avocado, apricot, banana, blueberry, cantaloupe, cherimoya, feijoa, fig, kiwifruit, mango, papaya, passionfruit, pear, persimmon, tomato and watermelon. Climacteric fruit ripen after harvest, and need not remain on the tree or vine. The second category of fruit, exemplified by blueberry, cherry, citrus, cucumber, grape, mandarin, pineapple, sweet orange and strawberry.

Respiration rate either remains almost unchanged or shows a steady decline until senescence intervenes, with little or no increase in ethylene production these are called nonclimacteric fruit. Fruits can show a multiple phase response to exogenous ethylene and ripen only if they remain attached to the parent plant.

In some species, particularly pepper and melon, different cultivars or genotypes exhibit characteristics typical of climacteric or non-climacteric behaviour. It has been shown that in climacteric fruit some ripening changes occur independently of ethylene, and that some non-climacteric fruit have ethylene-requiring changes during ripening. With the development of more sensitive ethylene measuring devices, many non-climacteric fruit appear to show an increase in ethylene evolution at previously undetectable levels upon ripening. Some other fruit, such as kiwifruit, a hybrid ripening pattern is seen, with most of the ripening changes occurring in the absence of any detectable rise in ethylene and CO_2 production, a climacteric response occurs only towards the end of ripening. Exposure to exogenous ethylene promotes ripening of kiwifruit, but if exposure to ethylene is insufficient, or fruit are too immature, then removal of ethylene results in non-climacteric behaviour. Ethylene as a ripening trigger is used commercially with banana, avocado and early-season kiwifruit to ensure that fruit are at optimum ripeness when eaten.

The pathway of ethylene biosynthesis is as follows:

i. Methionine (a sulphur containing amino acid also important in protein synthesis) is converted to SAM (S adenosyl methionine) through the action of SAM synthase.

ii. SAM is converted to ACC (1-amino cyclopropane-1-carboxylic acid) through the action of ACC synthase (ACS).

iii. ACC is converted to ethylene through the action of ACC oxidase (ACO).

In fruit with a climacteric behaviour, ethylene biosynthesis occurs at very low and basal levels during fruit development prior to ripening. This ethylene production is auto-inhibitory, and has been termed System 1 ethylene. At the initiation of ripening there is a change in the regulation of ethylene biosynthesis, which become auto-stimulatory and involves the induction of specific ripening-related ACO and ACS genes different from those that are responsible for System 1 ethylene. Production of ethylene greatly increases due to the autocatalytic regulation, and is known as System 2 ethylene. At the point at which the fruit becomes competent to ripen, there is a transition from System 1 to System 2 ethylene that may be regulated by developmental genes such as RIN.

Colour development, fruit softening, the accumulation of sugars and organic acids, and the production of aroma volatiles could in some cases be separated. For example, in antisense ACO melon fruit, degreening of the rind, softening and the accumulation of organic acids were sensitive to different levels of ethylene, and flesh pigmentation was ethylene independent. Ripening is a series of parallel processes involving both ethylene-independent and ethylene-dependent pathways, the latter requiring different sensitivities to ethylene to proceed. The other important factor in the regulation of fruit ripening is the way in which plants perceive ethylene and the signal transduction pathway that leads to the ethylene response. In summary, ethylene is perceived by receptors that are negative regulators of the signalling pathway. In the absence of ethylene the receptors actively suppress ethylene responses, but when these receptors bind ethylene they undergo a conformational change, leading to removal of the suppression and this allows de-repression of the signalling pathway. The signal is transduced through a MAP kinase pathway that ultimately leads to the stabilisation of a class of EIN3 (ethylene insensitive 3) transcription factors. The EIN3 name originated from the ethylene insensitive phenotype observed in Arabidopsis mutants. The stabilisation of EIN3 leads to an increase in the transcription of genes associated with each ripening trait. Ethylene receptors are multi-gene families (six genes in tomato) encoding two types of closely related proteins, one sub-family with a histidine kinase domain and the other sub-family with aserine/ threonine kinase function. The importance of receptors in tomato fruit ripening has also been shown by the semi-dominant mutant Never-ripe (Nr), the fruit of which are unable to ripen. This was found to be due to a mutation in the ETR3 receptor, making the fruit impaired in its ability to perceive ethylene. During ripening some ethylene receptors increase in transcription and it appears that receptor expression is used to restore the ability to respond to ethylene, implying that there is a corresponding loss of receptor protein. This suggests model in which during climacteric fruit ripening there is an increase in receptor turnover, allowing the fruit the ability to rapidly turn off ripening if ethylene is removed from the system. This is observed in tomato, apple and kiwifruit suppressed in ACO expression, which require continuous exposure to exogenous ethylene for ripening.

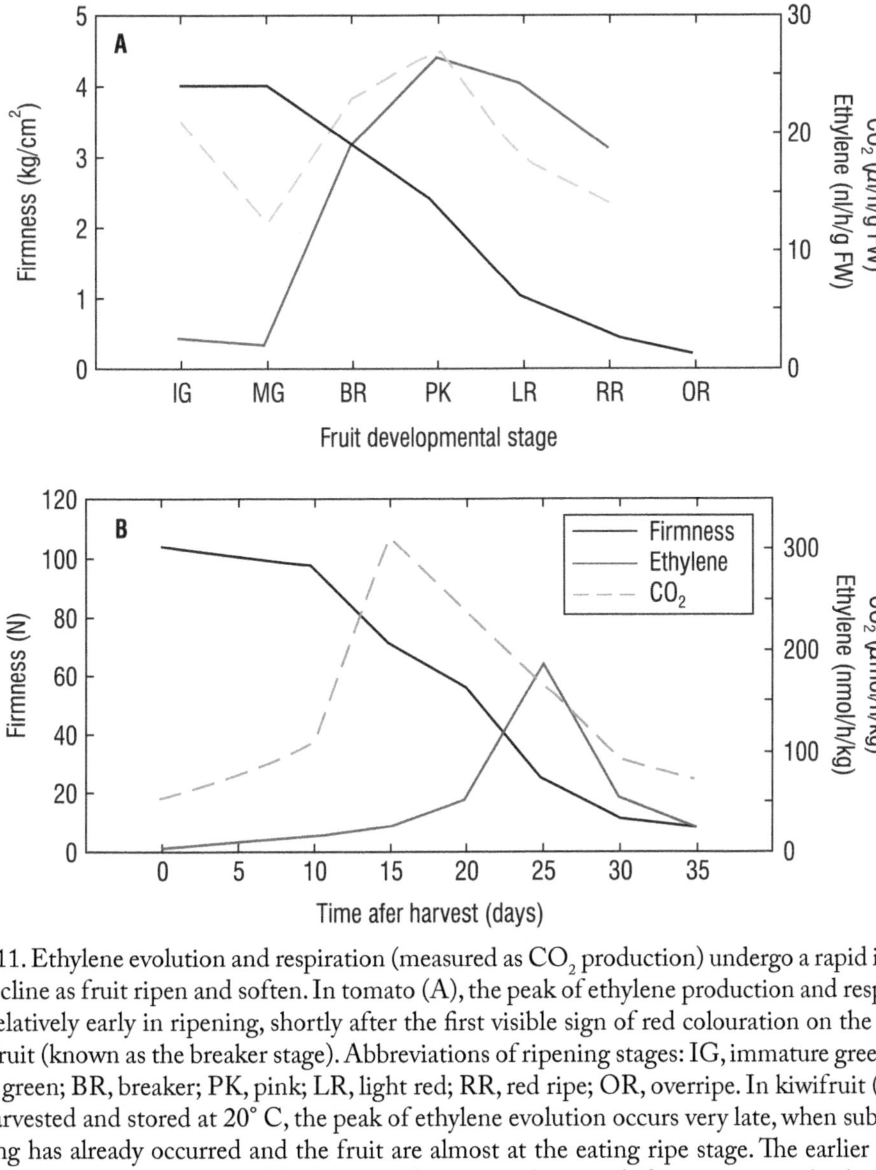

Figure 11. Ethylene evolution and respiration (measured as CO_2 production) undergo a rapid increase then decline as fruit ripen and soften. In tomato (A), the peak of ethylene production and respiration occur relatively early in ripening, shortly after the first visible sign of red colouration on the outside of the fruit (known as the breaker stage). Abbreviations of ripening stages: IG, immature green; MG, mature green; BR, breaker; PK, pink; LR, light red; RR, red ripe; OR, overripe. In kiwifruit (B) that were harvested and stored at 20° C, the peak of ethylene evolution occurs very late, when substantial softening has already occurred and the fruit are almost at the eating ripe stage. The earlier peak in respiration may have been caused by harvest. The two studies use different units so absolute values cannot be compared between the species. (Data taken from MacLauchlan and Brady (1994) Plant Physiol. 105, 965-974; Rothan and Nicolas (1989) HortScience 24, 340-342; Taglienti et al. (2009) Food Chem. 114, 1583-1589)

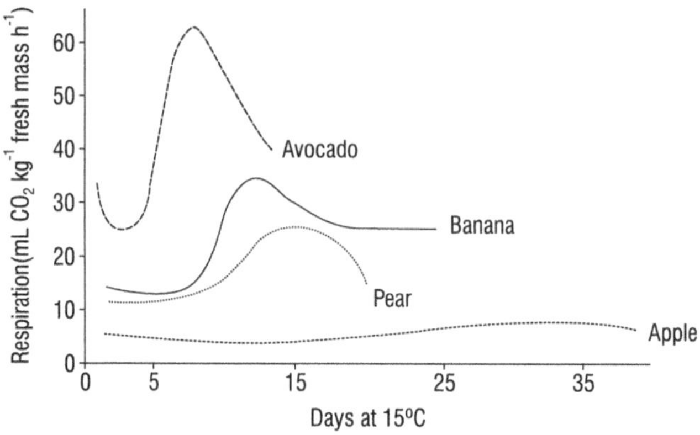

Figure 12. Respiratory output of CO_2 can undergo dramatic change as fruits ripen. Early research on apple and pearled to a classic model of a postharvest climacteric rise associated with ripening and linked in time with ethylene production. Studies with tropical fruits such as avocado and banana then revealed characteristic 'waveforms' of even wider amplitude. (Based on Biale (1950) Annu. Rev. Plant Physiol. 1, 183-206)

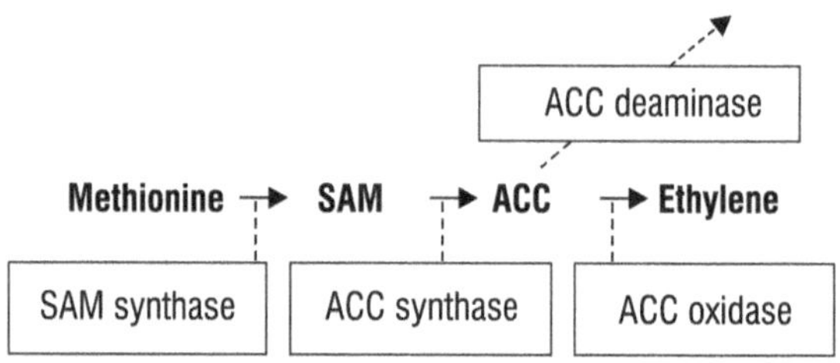

Figure 13. The pathway of ethylene biosynthesis in plants. The two genes controlling the committed steps to ethylene biosynthesis, ACC synthase (ACS) and ACC oxidase (ACO) are highly transcriptionally regulated. In one series of experiments, the biochemical precursor of ethylene, ACC, was depleted by expression of a bacterial ACC deaminase transgene, but this reaction does not normally occur in plants.

Developmental control of ripening

Ethylene play important role in fruit ripening, but is not the only regulator. A cold treatment can trigger ripening in detached apple and kiwifruit, acting either independently of ethylene or by increasing sensitivity to existing very low levels of ethylene. Other hormones also appear to play important roles particularly, declining levels of auxin and increasing levels of abscisic acid may control the onset of ripening in non-climacteric species such as grape, pomegranate and strawberry. Abscisic acid may also play a role in controlling the onset of ripening of climacteric species. The ripening-inhibitor (rin), colourless non-ripening (cnr) and non-ripening (nor) mutants all have fruit that fail to ripen, though with different characteristics. Although these fruit do not produce elevated levels of ethylene and will not ripen in response to exogenous ethylene, they are not completely insensitive and some ethylene-responsive genes (but not the whole ripening process) can be induced by ethylene treatment. The products of these three genes are transcription factors and are thought to be key developmental genes that control ripening progression, apparently acting upstream of the ethylene production pathway. RIN encodes a MADS-box gene that clusters in the SEPALLATA clade. CNR encodes a SQUAMOSA promoter binding protein. Both proteins are necessary for the induction of ripening related increases in respiration and ethylene biosynthesis, although since they are important in the ripening of both climacteric and non-climacteric fruit they appear to be more global regulators of ripening with some functions that are ethylene independent. Transgenic tomato fruit that had been suppressed in the ethylene signalling pathway and treated with 1-MCP showed an ethylene independent increase in the expression of ripening-related ACS genes and ethylene production. This is apparently controlled by developmental factors, and would be sufficient to induce the autocatalytic increase in ACS expression and ethylene production typical of tomato ripening.

Texture and softening

During fruit ripening, softening and textural changes are components of the suite of modifications that make ripe fruit attractive to animals that might disperse the seeds. The texture of ripe fruit differs drastically between species, with crisp, hard apple and deformably soft avocado representing the extremes. The characteristic textures of different fruit and their manner of softening can be linked both with anatomical features and with changes that occur to the cell wall during ripening. Some fruit that are picked while hard, such as kiwifruit and tomato, will subsequently soften markedly as a result of extensive modifications to the cell wall structure that include substantial swelling. Other fruits, such as apple or watermelon, remain crisp and soften only slightly. Their thin cell walls remain relatively unaltered. Both types of softening occur in the pear family, Asian pear (Nashi) shows a crisp apple-like texture, whereas many European pears soften to give ripe fruit with a melting texture. Many textural characteristics relate to the fate of fruit flesh when it is fractured and crushed in the mouth. For example, a ripe apple has large (0.1–0.3 mm diameter), turgid, thin-walled cells that are loosely packed (airspace 20% of fruit volume). When that flesh is chewed, cells fracture and release their sugary contents as free juice. In contrast, ripe kiwifruit has minimal airspace (2% of fruit volume) and cell walls are thick and hydrophilic. Such cells tend to pull apart when the flesh is chewed, resulting in a paste moistened by liquid held in cell walls or released by damaged cells. Avocado also has cells with walls that are thick and soft and which tend to pull apart, but also has a high proportion of oil that gives the pulp an oily quality in the mouth. Softening in kiwifruit occurs over a period of weeks, and can be divided into a number of phases. Modification of the cell wall plays an important part in determining fruit texture and ripening characteristics. The plant primary cell wall consists of a network of strong, rigid cellulose microfibrils held together by a complex matrix of polysaccharides consisting of two types: the hemicelluloses (composed mainly of neutral sugars) and the pectins (rich in galacturonic acid), together with smaller amounts of structural proteins. The outer part of each cell wall, which abuts and provides attachment to neighbouring cells is composed mainly of pectins and is called the middle lamella. Fruit softening involves alterations to various pectin and hemicelluloses polysaccharide wall components, and changes to the bonding between some polymers. Chemical analyses of cell wall components in a range of species, show some consistent changes during the early stages of ripening like increase solubilisation of pectin (but without further degradation), the cell wall swells and become more hydrophilic, loss of galactose from pectins and de-esterification of some pectins, these are responsible for commence rapid softening to ripeness. Softening is also associated with a further increase in pectin solubilisation, loss of galactan and arabinan side chains from pectic polymers, and more cell wall swelling. After the softening progresses depolymerisation (a reduction in size) of the hemicellulosic polysaccharide xyloglucan takes place, this is associated with a reduction in cell wall strength. Depolymerisation of pectin, which is associated with dissolution of the middle lamella and reduced intercellular adhesion. All these changes have been observed in a wide range of fruit types, although the extent and relative timing varies somewhat between species. Such observations indicate that pectin solubilisation and cell wall swelling are important events in the control of softening in kiwifruit and probably most other species with a melting texture. Cell wall modification is much less extensive in fruit with acrisp, fracturable texture such as apple and capsicum pepper. As fruit become fully ripe, dissolution of the middle lamella means that it eventually virtually disappears as a visible structure under the microscope. Dissolution of the middle lamella results in a great reduction in intercellular adhesion, and cells now have fewer regions of attachment to each other and become more rounded in appearance as they pull away from neighbouring cells. The primary walls are also weakened by the various changes that have occurred, and cells easily rupture when bitten or chewed, releasing the cell contents as juice. Many of the cell wall-modifying enzymes produced during ripening are regulated by ethylene. Because transgenic fruits retained more firmness and left on vines for longer time, resulting in more carbohydrate accumulation prior to harvest. Moreover, fruit could be harvested partially coloured rather than mature green, thereby allowing ripening processes to progress more naturally and yielding fruit with better flavour and appearance.

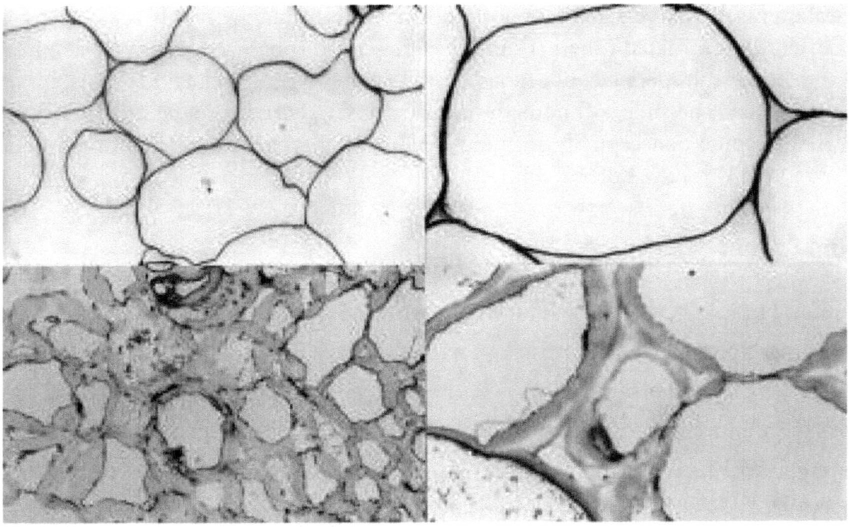

Figure 14. Anatomical features such as cell size, wall thickness and the distribution of intercellular gas spaces greatly influence our perception of fruit texture and eating pleasure. Cross-sections of ripe apple flesh (top pair) and ripe kiwifruit flesh (bottom pair) at low (x120) and high (x310) magnification. The apple tissue (top left) shows cells having densely staining thin walls. Tissue from kiwifruit (bottom left) shows cells with thick, swollen and weakly staining walls. The two right-hand figures give views of individual cells, corresponding to the tissue views. (Photomicrographs courtesy I. C. Hallett, E.A. MacRae and T.F. Wegrzyn).

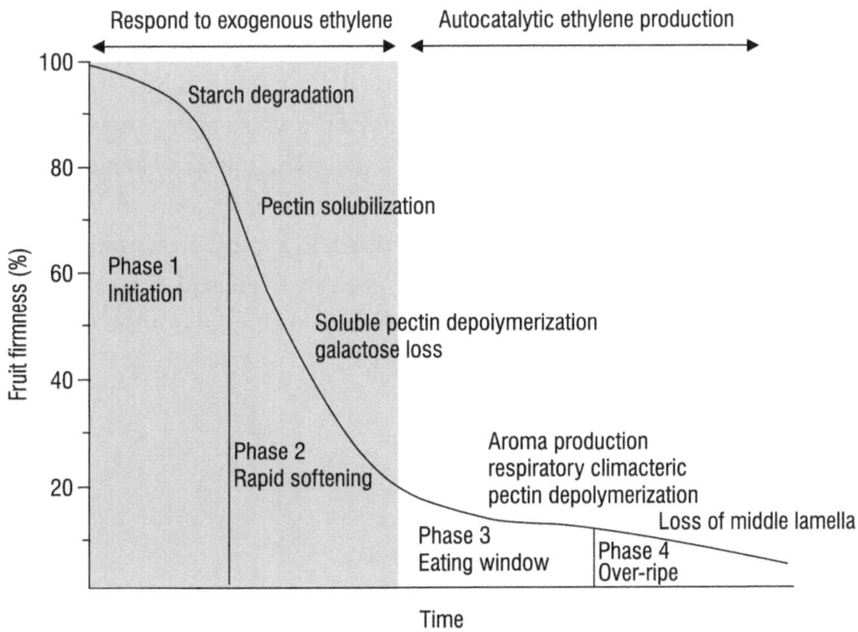

Figure 15. Schematic representation of postharvest ripening in kiwifruit, showing the timing of key

physiological events. At harvest, fruit do not produce ethylene but are highly sensitive to exogenous ethylene. Softening is initiated (phase 1) and becomes rapid (phase 2). Relatively late in softening, compared with other fruit species, endogenous autocatalytic ethylene production begins, aroma volatiles are produced and fruit become soft enough to eat (phase 3). If fruit progress to the over-ripe stage (phase 4), they become unacceptably soft and exhibit 'off-flavour' notes. (Figure reproduced from Atkinson *et al.* (2011) J. Exp. Bot. 62, 3821-3835)

Colour and flavour

Throughout the ripening process several fruit change colour and it is important visible indicator of maturity and ripeness of the fruits e.g. bananas, berry and stone fruit. Colour change in fruits typically involves chlorophyll loss and an increase in production of yellow, orange, red and purple pigments by enzymes in the carotenoid biosynthetic pathway. The hydrocarbon carotenes, e.g. lycopene (red) and β-carotene (orange) or the oxygen-containing xanthophylls, e.g. lutein (yellow). Besides providing highly attractive pigmentation, carotenoids protect the plant's photosynthetic apparatus from excessive light energy and are essential requirements for human and animal nutrition. Red and purple pigments of the type seen in grapes and boysenberries are anthocyanins, which are products of the phenylpropanoids pathway. Anthocyanin pigments are water-soluble, synthesized in the cytosol and localised in vacuoles. Their basic ring structure can be modified by hydroxylation, methylation or glycosylation and their specific colour is modified by pH, metal ions and co-pigments to produce the subtlety of colours seen in nature. Like carotenoids, anthocyanins have many human health benefits and are widely used as natural food colourants. The accumulation of anthocyanins is regulated by transcription factors of two classes (R_2R_3 MYB and basic helix loop helix), regulatory proteins that co-ordinate gene expression of the whole phenylpropanoids pathway. In fruit, this regulation system has been well characterised in grape and apple. In white berry grapes, $VvMYBA_2$ is inactivated by mutations in the coding region and $VvMYBA_1$ has a retrotransposon in the promoter and is not transcribed. In apple fruit, a mini-satellite repeat structure in the promoter region of the MYB_{10} gene up-regulated the expression of this regulatory gene, which increased the level of anthocyanin throughout the plant producing a fruit with striking red colour in all over the flesh.

Figure 16. Colour diversity in ripe kiwifruit and apple is determined by the presence or absence of chlorophyll, carotenoid and anthocyanin compounds in different fruit tissues. (Photographs courtesy Plant & Food Research)

Flavour is the imperative factor determining if purchasers will repurchase a particular fruit. Therefore, all varieties are produced and stored to maintain the greatest flavour and aroma properties. Two main factors determine a fruits characteristic flavour the correct sugar/ acid balance and the production of aroma by volatile compounds. These volatile compounds can include amixture of volatile acids, aldehydes, alcohols, esters, terpenoids and aromatics. Human taste sensations and experiences play an important part in characterizing volatile compounds in fruit and wine, so a vocabulary has been developed to describe their sensory nature like woody, grassy, floral, spicy and citrus. With advanced GC-MS techniques many ripe fruit can be shown to contain >100 volatile compounds that contribute to their flavour and aroma. However, the absolute concentration of a volatile compound itself does not determine how important it is to perceived flavour and aroma. Sometimes compounds found at very low levels, e.g. parts per billion, are required to give a fruit its characteristic aroma. Compounds are given odour activity values (compound concentration divided by the minimum concentration that can be detected by the human nose) to show their importance to aroma. Only, one or two key volatile compounds can be regarded as characteristic for fruit of a given species or cultivar and are used in synthetic mixtures to represent that commodity. Specific volatiles are especially important in wine grapes where an individual volatile can become the dominant characteristic used in marketing a specific wine type. Examples include the grassy character of methoxypyrazine in Sauvignon Blanc, the richness of b-damascenone in red wine, or the foxy character of methyl anthranilate produced by *Vitis labruscana*. Other examples of important volatile sin fruit be fond of, raspberry – raspberry ketone, Hort16A kiwifruit – ethyl butanoate and 1, 8-cineole, hayward kiwifruit – (E)-2-hexenal and hexanal, apple - 2-methyl butyl acetate and a-farnesene and strawberry furaneol.

Aldehydes, acids and esters are derived from fatty acids and branched chain amino acids. The first committed step in straight-chain ester production is performed by lipoxygenase that produce13-hydroperoxide linoleic acid from linoleic acid. These compounds are converted by cytochrome p-450 lyases to aldehydes such as hex-3-enal and hex-2-enal. Alcohol dehydrogenases (ADHs) can then transform aldehydes to the corresponding alcohols, which contribute green aromas. Alcohol acceptor substrates are then esterified with coenzyme acid donors by alcohol acyl transferases (AATs) to form esters, which generally contribute fruity and sweet characteristics. In apple, the enzyme $MpAAT_1$ can produce a range of esters in ripe fruit and is up-regulated by ethylene during ripening. Branched chain esters are produced from isoleucine by branch chain aminotransferases and then pyruvate decarboxylase to produce aldehydes. These aldehydes are then available for ADHs and AATs to form branched chain alcohols and esters, respectively. Sesquiterpenes and monoterpenes also contribute to fruit flavour and aroma profiles, often by adding floral or spicy top notes. In apple, the most important ripe-fruit terpene is a-farnesene produced by the sesquiterpene synthase AFS_1. Hort16A kiwifruit produce 1,8-cineole and *A. arguta* (baby kiwifruit) produce a-terpinolene, which add spicy/minty notes to these fruit. Terpenoid compounds are produced by terpene synthase enzymes, using farnesyl diphosphate (FPP) and geranyldiphosphate (GPP) as substrates. In the plastid, the MEP (2-C-methyl-D-erythritol 4-phosphate) pathway leads to the production of GPP, while in the cytoplasm the mevalonate pathway provides precursors for FPP formation. The primary terpenoid skeletons can subsequently be modified further, e.g. by oxidation, hydroxylation, glycosylation or methylation by a range of other enzymes to increase terpenoid diversity.

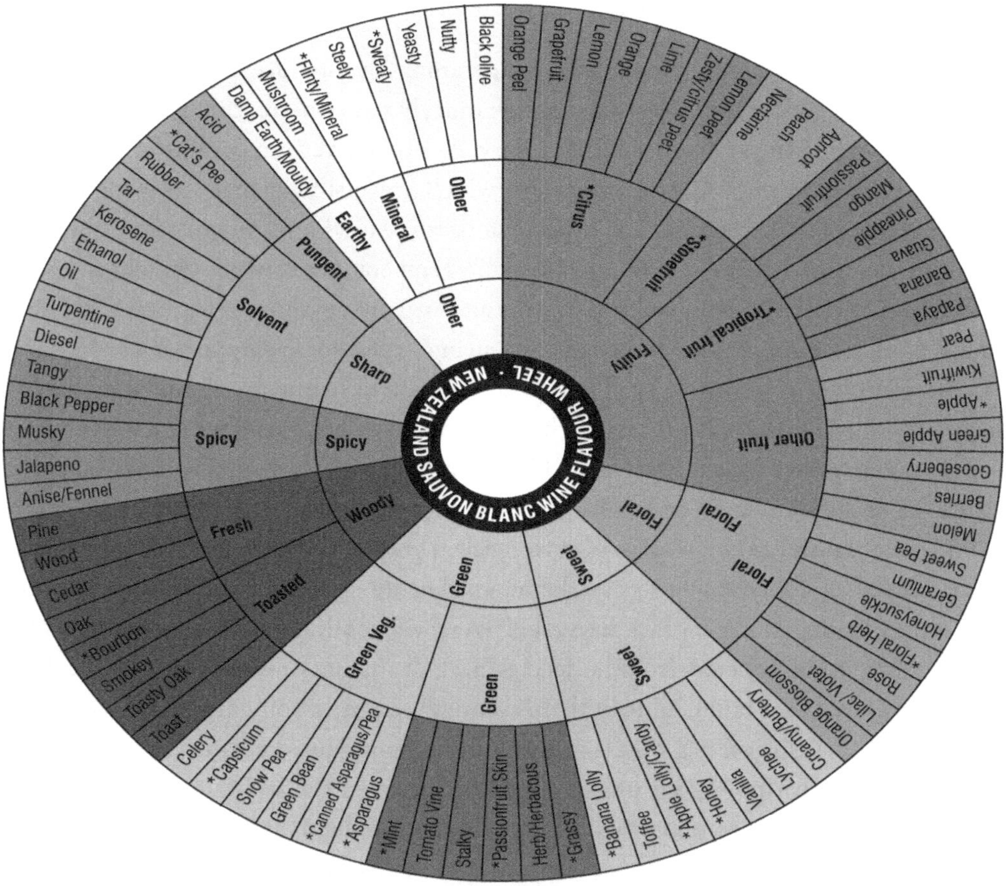

Figure 17. Flavour wheels that is used to systematically categories and define sensory characteristics of wine (Courtesy F.R. Harker).

Extending Storage Life of Fruits and Vegetables

The main method used to prolong the storage life of fruit is through reducing the fruit temperature to slow metabolism. Refrigerated storage deliberates the rate of ripening and senescence of the fruit, and also measureds the development of any rots. The way in which temperature management is implemented after harvest can significantly affect the quality of the fruit at the end of storage, both in the amount of ripening retardation and also the presence or absence of disorders. The basic effect of refrigerated storage on fruit can be supplemented by modification of the atmosphere in the cool store, by reducing oxygen and increasing carbon dioxide concentrations. More recently, the application of the inhibitor of ethylene action 1-methylcyclopropene (1-MCP) has become common to slow the ripening of fruits, particularly certain cultivars of apple. The way in which all these technologies impact on the fruit is dependent on the physiological state, or maturity, of the fruit at harvest.

Temperature and relative humidity

To obtain the maximum benefit from cold temperatures, the produce must be store at lowest safe temperature. Below the lowest safe temperature, but at non-freezing temperatures, the fruit may develop symptoms of chilling injury. At even lower temperatures, generally in the range –0.5° C to –1.5° C, freezing occurs which irreversibly damages a living product. Because of this, –0.5° C is usually the lowest temperature used for storage of fruit, including some apple cultivars, berries or Hayward kiwifruit. Temperatures at which chilling symptoms occur are around 8° C for subtropical species and may be anything up to 14° C for some tropical fruit, for example unripe banana and mango need to be shipped at 13–14° C. At 0° C, respiration is reduced to a level that is just enough to maintain cell function. Sugar is slowly consumed during this process so that fruit with low sugar content at harvest are less durable. Commodities such as kiwifruit, which are picked with large supplies of carbohydrate in the form of starch, have an additional source of sugar to utilize, giving longer storage lives than those entirely reliant on soluble reserves, such as grapes. Temperature management may also be viewed as a two-stage process, the removal of the field heat and then temperature maintenance during storage. Once harvested, fruit will continuously lose water to a point where quality will be affected. In some species, a small amount of water loss may accelerate ripening (e.g. avocado), but in all fruit there eventually comes a point at which loss of water, usually first seen as shrivelling, results in the fruit becoming unacceptable. Water loss from the fruit is driven by the vapour pressure gradient between the fruit and the surrounding environment. While the capacity for air to hold water is reduced at low temperatures, there is always gradient driving water from the fruit into the cool store atmosphere.

Controlled and modified atmospheres

The storage life achievable by refrigerated storage can be extended by modifying the store atmosphere. Elevated CO_2 and reduced O_2 used either separately or together, can delay ripening and slow the onset of senescence. These methods were originally developed on a commercial scale for apple, but have been progressively applied to many other fruits. Container shipping helped the introduction because a sealed container made it easier to maintain the required temperature and atmosphere regimes. Initial low O_2 concentrations may be achieved through the use of nitrogen generators or O_2 scrubbers, or the fruit may be allowed to reduce the O_2 concentration through respiratory activity. To prevent from becoming too low O_2 concentration air can be exchanged with the atmosphere. CO_2 accumulates from respiration, but can be avoided from increasing excessively by absorbing it with lime, or removal with an activated carbon scrubber or by purging from the store with nitrogen. In a closed controlled atmosphere system it is also possible to scrub ethylene. The removal of ethylene is particularly important for ethylene sensitive fruit such as kiwifruit, where even low levels (e.g. 30ppb) in the store atmosphere can reduce the storage life of the fruit.

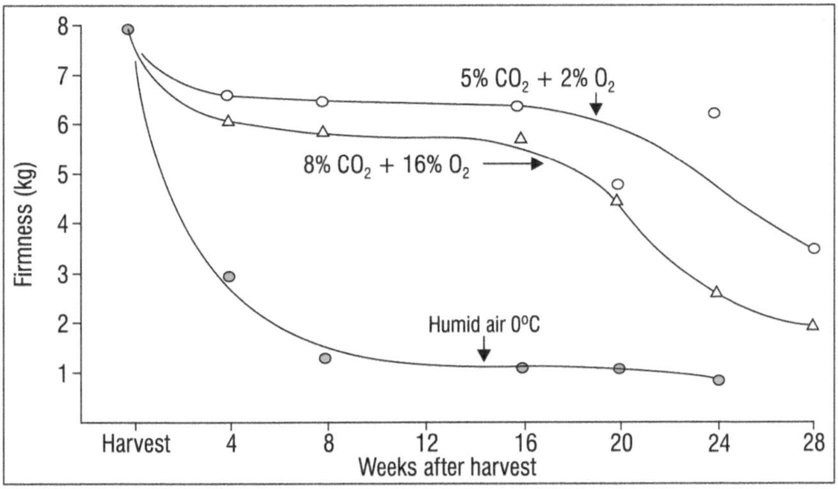

Figure 18. Storage life of kiwifruit can be greatly extended by controlled atmospheres. Under standard conditions (humidified air, 0° C) firmness declines exponentially over time, reaching limited acceptability by 8 weeks. Storage life ends once fruit firmness drops below about 0.9 kgf. Fruit are then soft enough to eat. Softening in cold store was slowed and storage life greatly extended by holding fruit in atmospheres containing either 5% CO_2 + 2% O_2 (top curve) or 8% CO_2 + 16% O_2 (middle curve). (Based on McDonald and Harman (1982) Sci. Hort. 17, 113-123)

For optimizes the CA process, since using a predetermined atmosphere tends to err on the side of safety by setting the O_2 concentration well above the lowest safe level to allow for the variability in low O_2 tolerance amongst fruit from different orchards or seasons. Although this eliminates the risk of fruit becoming anaerobic, it also diminishes the potential benefit. While early attempts at dynamic CA utilized ethanol sensors to detect if fruit metabolism was becoming anaerobic, it was the development of a fluorescence sensor that could give a rapid measurement of the fruit response to low O_2 stress that allowed the commercialization of dynamic CA. The sensor is placed over a sample of the fruit in the store, the O_2 concentration is decreased until a response is detected from the fruit and then the O_2 concentration is increased slightly above the low O_2 stress point. The procedure can be repeated throughout the storage period so that the O_2 concentration can be continually matched to the capacity of the fruit to withstand low O_2. An alternative way of utilizing the beneficial effects of low O_2 and high CO_2 is termed modified atmosphere (MA) storage. In this system, fruit respiration is used to reduce the concentration of O_2 and increase that of CO_2 inside an enclosed space, usually the export box or retail packs. The fruit is prevented from becoming anaerobic by making such enclosures out of plastic films that are partially permeable to O_2 and CO_2. Both gases come to an equilibrium based on respiration rate, the specific permeability of the film, the surface to volume ratio of the package and the amount of fruit in the package. Hence, this form of storage is highly dependent on being able to control the fruit temperature, since this determines the rate of respiration. The independence of having fruit in smaller packages that can be moved intact throughout handling and retailing suggest that MA may be more versatile than CA, although in practice any inability to maintain adequate cold-chain conditions can result in fruit spoilage as packages turn anaerobic at higher than desired temperatures. Coating fruit in waxes or other compounds may act in a similar way to MA, by modifying the gas permeability of the fruit skin, thereby reducing the flow of O_2 in and CO_2 out of the fruit. As with MA, if the restriction of oxygen flow into the fruit is too huge, the fruit may turn anaerobic and ferment. There are several possibilities, mostly involving fruit respiration and ethylene metabolism. In general the fruit respiration is suppressed in response to the changed atmosphere. This could occur via acidification of the cytosol, resulting from an elevated CO_2 concentration redirecting metabolism towards alcohol or lactate/succinate or malate production rather than CO_2 production. Another alternative is a direct effect of ultra-low O_2 concentrations (<2%) on cytochrome *c* oxidase in the mitochondrial electron transfer pathway, preventing that enzyme from functioning properly. Fruit differ with respect to critical values for tolerance to low O_2 or high CO_2 concentrations, and ideally might make a model for predicting the tolerance limits for a new cultivar or fruit from specific background information on its physiological behaviour. However, there is a key problem in manipulating atmospheres by static modelling approaches. The critical gas composition exists within the flesh of a fruit, not in the environment around it, while differences in genetic background cause each cultivar to behave differently with respect to metabolism and thus internal gas composition. Species vary in their response to the altered atmospheres of CA, and can even differ according to cultivar and harvest. This variation is seen in both the final concentration of CO_2 and O_2 within stored fruit, and in the time taken to equilibrate. Normally, an internal 0.5 per cent (0.5 kPa) partial pressure is the minimum O_2 level tolerable, and 10 per cent (10 kPa) is the maximum for CO_2. Conditions during storage are especially critical because optimum levels of CO_2 and O_2 are on the threshold between aerobic respiration (desirable) and anaerobic respiration (undesirable). Fruit differ in their sensitivities to anaerobic respiration, but are normally intolerant of prolonged periods (>3 days), after which disorders and off-flavours come out. Yet the anaerobic metabolites ethanol and acetaldehyde are common volatiles of many ripe fruit, and treatment with these metabolites, or short anaerobic periods before storage, can have beneficial effects on storage life in some fruit, although they are not used commercially. Fruit in which ethanol and acetaldehyde have been induced are able to metabolize these compounds without tissue damage. Both low O_2 (2 %) and high CO_2 (5 %) can independently improve firmness retention during storage, with a synergistic effect when used in combination. However, whilst effective in retarding ripening, there are risks to the fruit. The greatest firmness retention is achieved by a rapid establishment of the CA, although too rapid establishment of 5 per cent CO_2 can result in increased physiological disorders and rots. Higher, concentrations of CO_2 at about 10-15 per cent can result in a differential softening of the fruit flesh and core, resulting in a core that is firm relative to the pericarp tissues.

Blocking ethylene action

Ethylene having a pivotal role in the ripening of many fruit, the use of the ethylene action inhibitor 1-MCP has been considered for extending the storage life of a wide range of species through retarding fruit ripening and softening. 1-MCP is usually applied after harvest as a gas treatment in a sealed store, container or tent, with the active ingredient released from a powder by dissolving in water. Successful use of 1-MCP to delay ripening depends on the physiology of the fruit, most likely on the natural rate of replacement of the ethylene receptors that are blocked by 1-MCP. Since binding of 1-MCP to existing ethylene receptors is irreversible, a single period of exposure can delay ripening for several to many days, depending on the rate of synthesis of new receptors. There has been a rapid uptake of 1-MCP use for commercial storage of some apple cultivars, although for other cultivars the treatment has little effect on fruit softening. The rapid uptake for apple is associated with the way in which apple fruit ripen, which involves only limited softening and with firmness retention being a key quality component, i.e., people like crisp apples. This dissimilarity with the physiology of other species in which ripening involves a softening of the fruit coordinated with changes in flavour and colour. In stone fruit such as peach, the ripening inhibition is rapidly overcome, and repeated exposure to 1-MCP may be necessary, which can be commercially unfeasible.

4 Fruit Maturity Indices

Maturity at harvest is the most important factor that determines storage-life and final fruit quality. The phase of fruit and vegetable development at which a product is regarded as mature depends on its final use. Fruit and vegetables are eaten at all stages of development, like sprouted seeds, vegetative leaves, stems, roots and flowers, whole fruit as well as seeds and nuts. Maturity must be defined for each product in some cases for each variety of a particular product. The employ of maturity standards supply consumers with a minimum level of quality assurance. Another cause for establishing maturity standards is that the majority horticultural products are harvested by hand. A simple colour guide and size can assist pickers harvest produce at the correct stage of development. Fruits harvested premature may lack flavour and may not ripen properly, while produce harvested belatedly may be fibrous or have very limited market life. Immature fruits are more subject to shrivelling and mechanical damage, and are of inferior flavour quality when ripe. When mango picked at immature stage develop white patches or air pockets during ripening and lacking in normal brix acid ratio or sugar acid ratio, taste and flavour. Overripe fruits are likely to become soft and mealy in texture with insipid flavour, soon after harvest susceptible to microbial spoilage and physiological disorder. The storage life of over mature fruits is considerably reduced than fruits picked at the proper maturity. Similarly, vegetables are harvested over a wide range of physiological stages, depending upon which part of the plant is used as food. For example, small or immature vegetables possess better texture and quality than mature or over-mature vegetables. All fruits, with a few exceptions (such as pears, avocados, and bananas), reach their best eating quality when allowed to ripen on the plant. However, some fruits are usually picked mature but unripe so that they can withstand the postharvest handling system when shipped long-distance. Most currently used maturity indices are based on a compromise between those indices that would ensure the best eating quality to the consumer and those that provide the needed flexibility in marketing.

A group of fruits that are not capable of continuing their ripening process once removed from the plant like berries, cherry, citrus fruits, grape, litchi, pineapple, pomegranate, and tamarillo. These fruits produce very small quantities of ethylene and do not respond to ethylene treatment except in terms of degreening (removal of chlorophyll); these should be picked when fully-ripe to ensure good flavour quality. Another group of fruits that can be harvested mature and ripened off the plant like apple, apricot, avocado, banana, cherimoya, guava, kiwifruit, mango, nectarine, papaya, passion fruit, pear, peach, persimmon, plum, quince, sapodilla, sapota. These fruits produce much larger quantities of ethylene in association with their ripening, and exposure to ethylene will result in faster and more uniform ripening.

Mature: It is derived from Latin word 'Maturus' which means ripen. It is that stage of fruit development, which ensures attainment of maximum edible quality at the completion of ripening process.

Maturation: It is the developmental process by which the fruit attains maturity. It is the transient phase of development from near completion of physical growth to attainment of physiological maturity. There are different stages of maturation e.g. immature, mature, optimally mature, over mature. Maturity at harvest is the vital factor that determines postharvest-life and final quality such as appearance, texture, flavour, nutritive value of fruits and vegetables. On the basis of consumption the fruits and vegetables take in two groups:

Immature fruits and vegetables: The optimum eating quality is reached before full maturity and delayed harvesting results in lower quality at harvest and faster deterioration rate after harvest. Such as green bell pepper, green pepper, cucumber, summer squash (soft-rind), chayote, lima beans, snap beans, sweet pea, edible-pod pea, okra, brinjal, and sweet corn etc. All are usually produce extremely low quantities of ethylene, but very responsive to ethylene and can be damaged by exposure to 1 ppm or higher concentrations. Ethylene exposure accelerates chlorophyll degradation, induces yellowing of green tissues, encourages calyx abscission (brinjal), and accelerates fruit softening.

Mature fruits and vegetables: In this group the optimum eating quality is reached after full maturity such as red tomato, red peppers, muskmelons, watermelon, pumpkin, winter squash (hard-rind) and maximum fruits. Most of the fruits in this group produce larger quantities of ethylene in association with their ripening and exposure to ethylene treatment will result in faster and more uniform ripening as indicated by loss of chlorophyll (green colour), increase of carotenoids (red, yellow, and orange colour), flesh softening and increased intensity of characteristic aroma volatiles.

Therefore, harvesting of fruits and vegetables at proper stage of maturity is of paramount importance for attaining desirable quality. The level of maturity actually helps in selection of storage methods, estimation of shelf life, selection of processing operations for value addition etc. The maturity has been divided into five categories as below:

Horticultural maturity: It refers to the stage of development when plant and plant part possesses the pre-requisites for use by consumers for a particular purpose i.e. ready for harvest. Example: Fresh vegetables for canning/ dehydration/ IQF – Individual Quick Frozen/ harvesting for local or distant market.

Physiological maturity: It refers to the stage in the development of the fruits and vegetables when maximum growth and maturation has occurred. It is usually associated with full ripening in the fruits or ripening when it is harvested i.e. ready for eating or processing or seed production. The physiological mature stage is followed by senescence.

Commercial maturity: It is the state of plant organ required by a market. It commonly bears little relation to physiological maturity and may occur at any stage during development stage.

Harvest maturity: It may be defined in terms of physiological maturity and horticultural maturity, it is a stage, which will allow fruits / vegetables at its peak condition when it reaches to the consumers and develop acceptable flavour or appearance and having adequate shelf life.

Shipping maturity: The necessity of shipping mature fruits and vegetables long distances has often been encouraged harvesting them at less than ideal maturity, resulting in sub-optimal taste quality to the consumer.

Vegetables are harvested at harvest maturity stage, which will allow it to be at its peak condition when reaches to the consumer, develop an acceptable flavour or appearance, size required by the market and should have sufficient shelf life. Most of the fruits are harvested when they are mature and allowed to ripen subsequently during transit, storage or when displayed in retail shops. The optimal stage of maturity of most of the fruits is transient and delay in harvesting by a day or two may render them more mature. So, maturity indices help in deciding that when a given commodity should be harvested to provide some marketing flexibility and to ensure the attainment of acceptable eating quality to the consumer. It also enhances freshness/appearance /elegance and ensures an adequate postharvest shelf life. It facilitates scheduling of harvest, packing and long distance transportation of the produce. It needs to evaluate the time of harvest for a given fruit. The grower should know the margin time of safety for marketing and to supply to the consumer with fruit of good eating quality.

MATURITY INDICES

Maturity indices have been developed to determine harvest time for several fruits and vegetables. These indices are based on the changes that occur during the development of the fruit or vegetable. A holistic approach using multiple criteria for measure optimal fruit maturity. Measuring maturity and ripeness in fruit provide reasonable prediction tools that can be applied commercially. However, there are limitations that can lead to imperfect prediction with consequent variation in quality. Some of this variation can be managed by postharvest treatments such as heat, phytohormones, anaerobic metabolites and 1-methylcyclopropene. Nonetheless, application of knowledge of biological variation and uncertainty analysis can improve the outcome. The molecular biology tools like analysis of differential protein expression can potentially provide a powerful alternative assessment of maturity, particularly in stone fruit, which are very difficult to evaluate using existing methods. For this purpose identification of the components of the complex changes in gene and protein expression during ripening is essential because these changes are responsible for the quality or metabolic changes in fruit. There is a strong correlation between metabolic component shifts which provides the basis for relating to gene expression to actual and specific quality shifts during ripening and in response to postharvest treatments. The measures of molecular events may be more reliable than current measures of maturity and ripeness since they are less affected by short term fluctuations in the growing and handling environments. Therefore, the determination of harvest time is based on a compromise between flexibility in marketing and best eating quality for the consumer. Both subjective and objective indices are available for use in different fruits and vegetables are included:

Table 29. Maturity indices for fruits and vegetables

Maturity indices	Fruits and vegetables
Calendar date	All fruits and vegetables
DFFB	All fruits and radish
Mean heat units	Apple, pear, grape, mango, ber, litchi, datepalm, sweet corn and many vegetables
T-Stage	Apple
Size	All fruits, beans, carrot, cucumber, cherry, asparagus, zucchini and cauliflower.
Shape	Compactness in cabbage, cauliflower and broccoli, angularity in banana, full shoulder development in mango

Maturity indices	Fruits and vegetables
Surface morphology	Grape (cuticle formation), mango, sapota, tomato, banana and litchi (glossyness), netting on some melons, oil glands on citrus, development of wax on pome fruit and some cucurbits
Specific gravity	Cherries, mango, ber, watermelons, potatoes
Fruit retention strength	Apple and some cucurbits
Colour (Surface) (Seed) (Flesh)	All fruits, papaya, tomato, watermelon, jamun and vegetables Apple and pears Mango, papaya, watermelon, muskmelon, strawberry
Firmness (Penetrometer)	Pome and stone fruits, beans, lettuce, melons and vegetables
Tenderness (Tenderometer)	Pea
Abscission layer	Melons, apples, feijoas
Touch/Finger Squeezing	Beans, okra, peas
Solidity-Bulk density/X/ Gamma rays	Lettuce, cabbage, brussels sprouts
Tight bud/ bud crack/ flower opening	Rose and many cut flowers, loose flower- crossandra, marigold etc
Total solids: Dry weight	Potato, avocado, kiwifruit etc
TSS	All fruits, tomato, water melon
Starch content -Iodine test	Apple, banana, pear etc
Sugar content (Refractrometer)	All fruits, specially in grapes, pome and stone fruits
Juice content	Citrus fruits
Acidity and Sugar/ acid ratio	Pomegranate, citrus, papaya, grape and kiwi fruit
TSS/acid ratio	Grape and citrus

Maturity indices	Fruits and vegetables
Sugars	Pome, stone fruits and grape
Astringency (Tannin)	Persimmon, dates, aonla, bael, cucurbits and many vegetables
Oil content	Avocado
Respiration and ethylene evolution rates	Apple, pears and many fruits
Optical methods (380-730 nm)	Apricot, banana, orange, papaya
Aroma	Many fruits
Fruit opening	Nutmeg, chow chow (over mature), ackee
Acoustic / Vibration	Melons, apple, tomato (unripe 110- ripe 80 Hz)
Electrical Characteristics	Peach (unripe 550, ripe 150 Hertz)
Electormagnetic – Nuclear magnetic resonance (NMR)	Apple, banana, avocado, peach, pear, onion
Near-Infrared reflectance (400-2500 nm)	Mango, pineapple
Radiation (X-rays & gamma - rays)	Lettuce, potato and some exotic vegetables

It is advisable to use different parameters concurrently to improve the accuracy of the determinations. The final decision on harvesting will take account of the current market value of the expected yield, and also the time during which the crop will remain in marketable condition. With seasonal crops, growers are often tempted to harvest too early or too late in order to benefit from higher prices at the beginning and end of the season. Furthermore, one maturity index cannot be applied across all cultivars of the same crop. There are slight differences among the cultivars and it is important to only use the maturity indices that are relevant to the specific cultivar in question.

Methods for Judging Maturity

The principles dictating at which stage of maturity a fruit or vegetable should be harvested are crucial to its subsequent storage and marketable life and quality. Fruits picked at the wrong stage of maturity may develop physiological disorders in storage and may exhibit poor dessert quality. It has been stated that apples should be harvested at a stage of maturation and ripening which is less than ripe. For example apple picked too early may not ripen properly in storage and may develop superficial scald, bitter pit and extreme shrivelling. They are also likely to be of low dessert quality from the stand point of colour, flavour and texture. The ripe fruits are prone to bruising and decay. It has been suggested that if pears are permitted to ripen on the tree, the dessert and storage quality become drastically reduced through the occurrence of mealiness and core breakdown with excessive accumulation of acetaldehyde. Harvesting, storage, marketing and canning experiences have shown that this type of fruits should be removed from the tree while still in a firm hard green condition. For selecting the harvest maturity of fruits or vegetables it should be kept in mind that harvested commodity should have its peak acceptable quality (non-toxic, size, appearance and flavour with adequate shelf life).

Table 30. Methods and parameters to determine harvest indices

Methods	Parameters to determine harvest indices
Computational methods	Calendar date, days from full bloom to harvest (DFFB), mean heat units, T-stage, days to fruit set.
Physical methods	Fruit retention strength, fruit size and surface morphology, weight, specific gravity, colour (skin, flesh and seed), flesh firmness, firmness, ease of separation, brittleness of the floral part, juice content, bulk density (cole crops/lettuce), structural properties (soft/rough), development of abscission layer (melons)
Chemical methods	Total soluble solids (TSS), titrable acidity, TSS/acid ratio, sugars (total and reducing), sugar/acid ratio, bioelectrical conductance, starch content, tannin content (dates, persimmon and litchi), oil content, juice content
Physiological methods	Respiration rate, ethylene evolution rate, transpiration, production of volatiles.
Geo metrical methods	Particles size and shape of the produce, particle composition and orientation in a given tissue or food moisture content of produce
Other method	Brittleness, elasticity and hardness

Computational Methods

Calendar date: It is one of the commonly used indices of maturity and is practically accurate provided flowering if weather during growing season is normal. Every crop required specific period for maturation in definite climate and soil condition. On the basis of calendar dates calculate the required period for maturation and date of harvesting by simple mathematical adding of days.

Table 31. Harvesting periods of important fruits and vegetables

Fruits	Harvesting period	Vegetables	Harvesting period
Apple	August- November	Brinjal	January- December
Banana	January- December	Broccoli	December-January
Grapes	January-April or December- January	Cabbage	October- February
Guava	January-March or November- December	Cauliflower	October- February
Jamun	May- July	Carrot	November – February
Kinnow	November-January	Cucurbits	March- May or July – September
Lime/ Lemon	July- September or Year round	Chilli (Rabi)	November- February
Litchi	May-June	Cowpea pod	July – September
Mandarin	November-January	Okra	February- April or July – September
Mango	April-July	Onion	January- December
Mosambi	January- December	Peas	January- April
Orange	January- March or November- December	Potato	November- February
Papaya	January-December	Radish	November- March
Pineapple	January- December	Sweet potato	October- December
Sapota	January- December	Tomato	January- December

Days from full bloom to harvest (DFFB): Predicting date of first acceptable harvest using elapsed time from full bloom is the method which shows the correction in the standard figure on the basis of variation in the full bloom date. For every 5° C variation from an average night temperature, a correction of one day is made in the standard figure from full bloom. For every 2 days difference in average full bloom date, an adjustment of 1 day is required.

Suppose full bloom date of a cultivar is May 23 (the optimum date of picking is 120 days from full bloom) but in a particular season it was recorded on May 29 then first acceptable harvesting date will be = 120 + 3 = May 26 + 123 days = September 26. In case of Mango 110 -125 day (Var. Alphonso and Pairi) and Banana 99 - 107 days (var. dwarf cavendish) is required for maturation.

Table 32. Approximate number of days from planting to market maturity under optimum growing conditions

Name of crop	Early variety	Late variety	Name of crop	Early variety	Late variety
Beans, bush	46	65	Muskmelon	75	90
Beans, pole	56	72	Mustard	40	60
Lima beans	65	78	Okra	50	60
Beets	50	80	Onions	85	120
Broccoli	70	150	Parsley	70	85
Brussels sprouts	90	100	Parsnips	100	130
Cabbage	62	110	Peas	58	77
Carrots	60	85	Potatoes	90	120
Cauliflower	55	65	Pumpkin	110	120
Chinese cabbage	70	80	Radishes spring	22	40
Chives	85	90	Radishes winter	50	60
Corn	70	100	Rutabagas	85	95
Cucumber	60	70	Spinach	40	50
Eggplant	70	85	Winter squash	50	68
Kohlrabi	55	65	Squash summer	80	120
Lettuce, head	60	85	Tomatoes	65	100
Lettuce, leaf	40	50	Turnips	40	75
Watermelon winter	100	110	Watermelon spring	65	95

Mean heat units: The apparent effect of temperature on flowering and fruit development advises the use of degree days or of heat summation for the more accurate prediction of harvest date. The time required for fruits to reach maturity is determined mainly by the total amount of heat received which can be expressed interms of, temperature-time value called degree days or heat units. The number of degree-days to maturity is determined over a period of several years. Heat units are not useful for photoperiod sensitive species. A Heat unit is calculated by (daily mean temp – base temp) x Number of Days (flowering to harvest) base temperature for apple, tomato, spinach and pumpkin is 10° C, 15° C, 20° C, 13° C, respectively.

The number of heat units required for a growing area can be estimated as follows:

$$H = T - TL \times D$$

Where,

H	=	Heat units or degree days
T	=	Mean monthly temperature
TL	=	Base line temperature
D	=	Number of days in a month.

Example: The mean monthly temperature of February 20° C, March 25° C, April 30° C, May 35° C and June 40° C. Khadravi datepalm bloom about 15th February and ripe about 30th June, calculate the total number of degree days. Base line temperature is 10° C.

Months	Temperature		Degree days
Degree days for the month of February	20–10 x 13	=	130
Degree days for the month of March	25–10 x 31	=	465
Degree days for the month of April	30–10 x 30	=	600
Degree days for the month of May	35–10 x 31	=	775
Degree days for the month of June	40–10 x 30	=	900
Total number of degree days			2870

Table 33. Mean heat units requirement of various fruits

Fruits	Cultivar	Cut out or Base Temperature	Degree days
Apple	Red Delicious	10° C/50° F	1659 - 1705
Pear	Bartlett	7.25° C/45° F	2240 - 2296
Grapes	Thompson Seedless	10° C/50° F	1600 - 2000
	Bhokri	-	3284
	Gulabi	10° C/50° F	3508
	Banglore Blue	10° C/50° F	3562

Mango	Banganapally	18° C/64.5° F	1426 - 1445
Ber	Gola	7.25° C/45° F	1980 - 2236
	Kaithli	-	2236 - 2566
	Umran	-	2516 - 2920
Litchi	Shahi	15° C/59° F	813
Banana	-	9.8° C/49.5° F	1930
Asparagus	-	10° C/50° F	120-410
Peas	Early Wisconsin	4.4° C/40° F	1319
	Alaska	-	1200

Number of days from fruit set: Fruit set refers to the conversion of a flower to fruit after fertilization. In some fruits, the time taken between fruit set until the fruit starts showing signs of maturity has been recorded and this can be used to determine harvest time. For instance, in alpanso and pairi mango varieties, it takes about 110 to 125 days after fruit set for surface colour to change from dark green to olive-green and flesh colour from white to pale yellow.

Table 34. Time taken from pollination to horticultural maturity

Vegetables	Time to harvest maturity (days)	Vegetables	Time to harvest maturity (days)
Ridge gourd	5-6	Pepper (red stage)	60-70
Squash	7-8	Pumpkin (mature)	65-70
Brinjal	25-40	Tomato (mature green)	35-45
Okra	4-6	Tomato (red ripe stage)	45-60
Pepper (green stage)	45-55	Peas	30-35

T-Stage: The T-Stage is the points at which the angle formed by the fruit receptacle and the pedicel reaches 90° and its timing is obtained by inter polation. The angle between pedicel and receptacle is greater than 90° early in the growing season and less than 90° late in the season. The number of days between the T-Stage and the date of optimal harvesting time of apple showed a remarkable consistency for a given location. It was determined that the number of days from the T-Stage to harvest for apple varied from 120 -130 days.

Physical Methods

Fruits remain attached with the main plant due retention strength. It is the force required to pull fruit from the tree (at the natural abscission site) which indicates the maturity status of the fruit. Fruit with best storage quality could be picked from the spurs using a slight rolling action. Physical methods are found as temperature dependent indices of maturity. Fruit size and weight are also the measures of fruit maturity, which dependent upon a number of variables such as soil, cultivar, fertilizer application and climatic conditions which vary from place-to-place and season-to-season. In banana, litchi, nuts, mango, pineapple, etc. changes in the surface morphology is the reliable maturity index and determine harvest maturity. The other features of maturity are specific gravity (by water displacement method), colour of skin, flesh and seed (by horticulture colour chart), flesh thickness (by penetrometer) and TSS (by hand refractometer) are also used frequently in different fruit crops as indices of maturity.

Size and shape: Maturity of fruits and vegetable are evaluated by their final shape and size at the time of harvest. Fruit shape may be used in some instances to decide maturity. The visual appearance of fruits and vegetables is the most important quality factor, which decides its price in the market. The consumer (wholesaler or retailer) observes the quality of fresh fruits and vegetables with their visual or external appearance. Medium size produce is always preferred by the consumers, because they tend to view large fruits as more mature. The appearance of the product is the most critical factor in the initial purchase, while subsequent purchase may be more related to texture and flavour. For example, the fullness of cheeks adjacent to pedicel may be used as a guide to maturity of mango and some stone fruits.

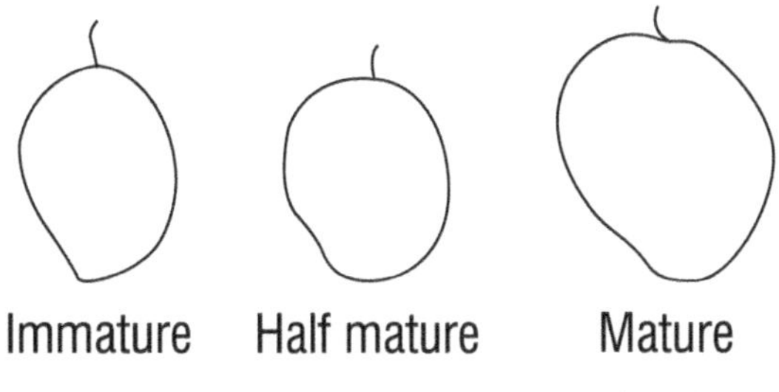

Figure 19. Evaluation of mango harvest maturity on the basis of its shape of shoulder (Source: Von Loesecke, 1949).

In the same way, a few cultivars of banana are less angular in cross section during its developmental and maturation phase. Banana size is usually has limited value with reference to its maturity index. Though size of the fruit is often specified as a quality standard. For example a large size is the indication of commercial over-maturity and under-sized produce indicates immature state. This assumption is not always reliable for all-purpose.

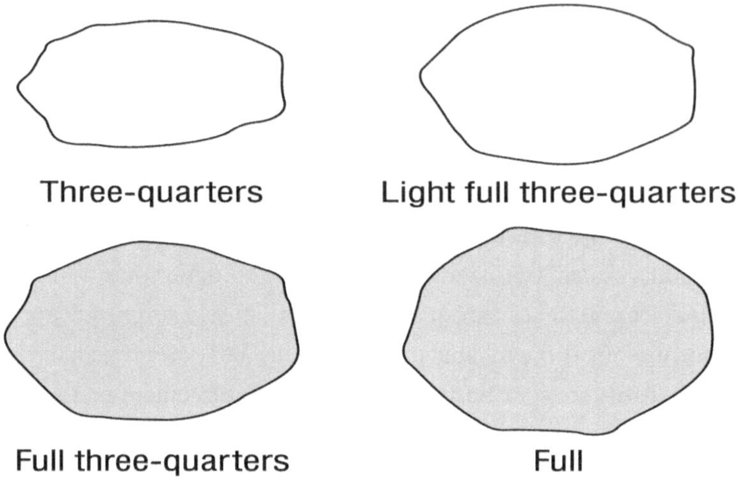

Figure 20. Cross section of the middle banana fingers showing the changes in angularity as they mature on the plant (Source: Von Loesecke, 1949).

The changes in size of crop as it is growing are frequently used to determine when it should be harvested. Litchi, avocado, green beans, okra, asparagus and potato related to size at maturity. In banana, width of individual fingers can be used for determining their harvest maturity. In baby corn, more immature and smaller cob is marked for maturity. Avocado industry is using minimum fruit weight and diameter for each commercial variety in conjunction with picking dates.

Skin colour: The loss of green colour of many fruits is a precious guide to maturity. There is initially a gradual loss in intensity of colour from deep green to lighter green and with many commodities, a complete loss of green colour with the development of yellow, red or purple pigments. Colour changes may vary from cultivars, seasons, site, light etc. In most of the fruits green colour changes to light green/ yellow/ red/ purple /violet during ripening after the optimum maturity. When it is still green it may be possible to develop the colour after harvest but not all the flavour characteristics. Ground colour as measured by colour charts, is useful index of maturity for apple, pear and stone fruits, but is not entirely reliable as it is influenced by factors other than maturity. For some fruits, as they mature on the tree, development of blush colour, that is additional colour superimposed on the ground colour, can be a good indicator of maturity. Examples are red or red-streaked apple cultivars and red blush on some cultivars of peach. In papaya, apples and pears, one can judge ripeness by the seed colour. If the seed coat colour has turned completely dark brown, the fruit is usually ripe enough to harvest. In jackfruit, peduncle and last leaf of peduncle turned yellow at ripening.

The colour measurement is done on the basis of reflectance or light transmittance by spectrophotometer. Colour perception depends on the type and intensity of light, chemical and physical characteristics of the commodity, and person's ability to characterize colour. Although human eye is used to evaluate colour but results can vary considerably due to human differences in colour perception. Therefore, an instrument (Colour Difference Meter) is used to provide a specific colour value based on the amount of light reflected off the commodity surface or light transmitted through the commodity. This instrument can measure small differences in colour accurately by colorimetric method and can be automated in the packing line. However, colour meters and colour charts have been developed for determining harvest times for apples, tomatoes, peaches, chilli peppers etc. Whereas, some fruits do not exhibit any perceptible colour changes during maturation and thus this parameter cannot be effectively used. Colour changes also differ among different cultivars of the same fruit. For example, the Hayward cultivar of kiwifruit maintains its green flesh during maturation while 'Sanuki Gold' cultivar changes gradually to golden-yellow. Some cultivars of avocado also maintain their green skin colour during maturation.

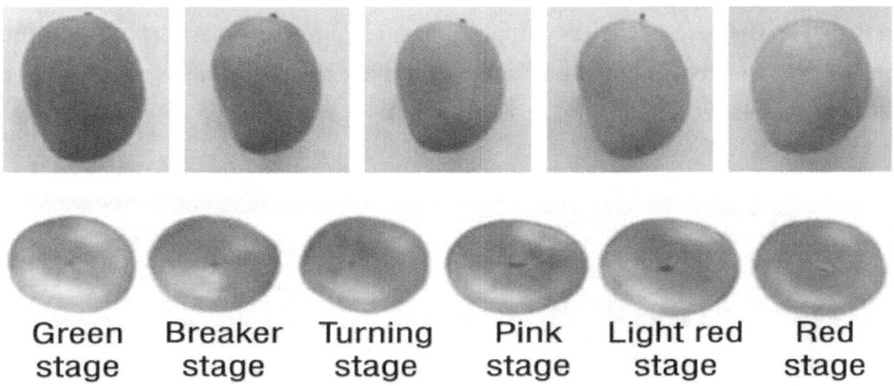

Figure 21. Judging maturity of mango and tomato depends on colour variation

Firmness: Some fruits may change in texture during maturation and these changes can be used to determine the harvest time. Textural changes are detected subjectively by touch or gentle squeezing. The degree of firmness can be estimated subjectively by finger or thumb pressure in peas, beans, okra etc., but more precise objective measurement is possible with pressure tester or penetrometer. It is ideal for determining the best time to harvest fruit or to test its progress to maturity. The firmness of fruit is one of the most widely used parameters in testing its maturity, together with measuring its sugar content using refractometers. In many fruits such as peach, plum, apple, kinnow, pear, guava, nectarines, apricot etc., firmness is the parameter of harvest maturity. The degree of softness or crispiness can be estimated by squeezing produce, or by taking a bite. Objective measurements can be made with inexpensive penetrometers. The most common way to measure firmness is resistance to compression or pounds-force (lbf) or it is also expressed in kg./sq. cm. The effigy fruit penetrometer is a hand held probe with a gauge for pounds-force. To measure firmness, fruits that are uniform in temperature are used since warm fruit are usually softer than cold fruit. Fruits that are uniform in size are used, since large fruits are usually softer than smaller fruits. Two puncture tests per fruit on larger fruits, one on opposite cheeks, midway between stem and blossom ends are made. A disc of skin (larger than the tip to be used) is removed and the appropriate plunger tip is chosen. The fruit is held against a stationary, hard surface and the tip is forced into fruit at a slow, uniform speed (take 2 seconds) to the scribed line on the tip. The reading to the nearest 0.5 lb-force is taken. Electronic penetrometers are available with different diameter penetration points. This allows for fruit of different size and hardness to be tested. It is important that this first part of the measurement process is done correctly if the readings obtained are to be valid in making adequate decisions as to when the fruit should be harvested. Firmness testers should be calibrated by holding the tester vertically and placing the tip on the pan of a scale. Press down until the scale registers a given weight, then read the firmness tester. Repeat 3 to 5 times, if the instrument reads the same as the scale, it is ready to use. The penetrometers can be adjusted by inserting washers in the appropriate locations (follow the instructions that come with the instrument). Firmness tests in mango 1.75 -2 kg at maturity.

Figure 22. Penetrometer measures the firmness or hardness of fruit.

Table 35. Appropriate effigy plunger tip sizes to use when measuring firmness in selected fruits

Tip length	Tip sizes	Fruits
1.5 mm	(1/16 inch)	Olive
3 mm	(1/8 inch)	Cherry, grape, strawberry
8 mm	(5/16 inch)	Apricot, avocado, kiwifruit, pear, mango, nectarine, papaya, peach
11 mm	(7/16 inch)	Apple

Specific gravity: The term specific gravity is used to describe the weight or density of a liquid compared to an equal volume of distilled water at 4° C or 16.7° C. If the liquid comparing will float on this water it has a specific gravity of less than one. If it sinks into the fresh water the specific gravity is more than one. Fresh water at 4° C has been assigned a value of one. Fruits that will float in water (Floaters) has lower TSS, lower specific gravity and hence immature. Fruits that sink in water (Sinkers) have specific gravity more than one, higher total soluble solids and therefore, are mature. This parameter is rarely used in practice to determine when to harvest a crop but it could be where it is possible to develop a suitable sampling technique. To give greater flexibility to the test and make it more precise, a salt or sugar solution can be used in place of water. For example, mango fruits are harvested at the tapka stage (when mature fruits start dropping) having a specific gravity >1.00 are considered to have attained optimum maturity. However, the majority of mature fruits of Baneshan mango (89.4%) harvested at this stage had a specific gravity <1.00. Light stones, low stone: pulp ratio (1:6.04) and a void space between the kernel and internal stone wall in Baneshan mango seemed to be contributory factors. There were more vascular vessels (and these were larger) in the peel of Baneshan fruits than in Bombay Green and Surkha fruits. Quality was slightly higher in Baneshan fruits with a specific gravity >1.00. Example, cherries, watermelon, potato, ber and mango (at 1.015 immature and at 1.02 ready for harvest). This method has been found useful in judging the maturity of guava, cherries, mango and ber fruits.

Fruit retention strength: It is the force required to pull the fruit from the tree which indicates the maturity status of the fruit. Immature fruit required more force to detach from mother plant compared to ripe fruits in cucurbits, citrus fruits, pomegranate and sapota etc.

Acoustic/sound tests: The sound of a fruit as it is tapped sharply with a finger knuckle can change during maturation and ripening. This method of testing fruit is sometimes used by consumers when purchasing fruit. Water melon fruit may be tapped in the field to judge whether they are ready to be harvested, ripe fruit gives dull sound and also in jackfruit.

Aroma/ Orgnoleptic quality: Fruits synthesize volatile chemicals as they may give its characteristic odour and can be used to determine whether it is ripe or not with indication of fruit flies. This method has limited scope in commercial application.

Fruit opening: When the fruit is fully mature on the tree it splits. It is common in fruit of spice tree nutmeg, ackee tree. In vegetable like chow chow distal end of the fruit opens and large single seed emerges and germinates.

Leaf changes: Leaf quality often determines when fruits and vegetables should be harvested. In root crops, the condition of the leaves can likewise indicate the condition of the crop below ground. For example, if potatoes are to be stored, then the optimum harvest time is soon after the leaves and stems have died. If harvested earlier, the skins will be less resistant to harvesting and handling damage and more prone to storage diseases. In banana, potato, onion and garlic, leaf senescence is an indication for harvesting.

Dropping of fruit from the tree: The general criteria for the time of harvesting of mango fruit are when a few ripe fruits begin to fall naturally from the tree. This is called as tapka. The whole crop is considered to be ready for picking.

Abscission: As part of the natural development of a fruit an abscission layer is formed in the pedicel. For example, in cantaloupe melons, harvesting before the abscission layer is fully developed results in inferior flavoured fruit, compared to those left on the vine for the full period.

Chemical Methods

Fruit maturity can also be measured using chemical parameters. The total soluble solids of the fruit can be measured with refractometer, which indicate the harvest maturity of fruits. Acidity is readily determined on a sample of extracted juice by titration with 0.1 N NaOH. The sugar acid or TSS acid ratio is often better related to palatability of fruit than either sugar or acid level alone. Following are methods for chemical measurements:

Total Soluble Solids (TSS): During maturation, starch in non-climacteric fruits is converted to sugars, whereas in climacteric fruits, starch accumulates. Therefore, harvest maturity can be determined by measuring the sugar content or starch content. The sugar content is generally measured in terms of total soluble solids content using a Brix hydrometer or refractometer. Temperature will affect the reading (increasing about 0.5% TSS for every 5° C, so the measurement for the ambient temperature, should be adjusted at 20° C or 68°

F. The refractometer has different scales (0-32° B), (28-62° B) and (56-92° B) which can be read directly. For large size fruits, these should be cut from stem to blossom end and to the centre of the fruit to account for variability in TSS from top to bottom and inside to outside of the fruit. If reading indicates a higher per cent TSS, then produce is better than the minimum standard. Strawberries which are of excellent flavour, for instance, would measure 8 per cent TSS or above.

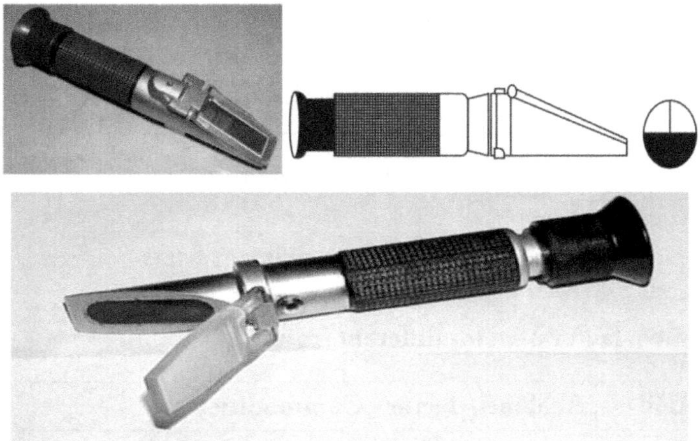

Figure 23. Different types of Refractometer

Table 36. Minimum per cent TSS for selected fruits

Fruits	Minimum %TSS	Fruits	Minimum %TSS	Fruits	Minimum %TSS
Apple	11.5-14.5	Guava	12-14	Pear	13-14
Apricot	10-12	Kiwifruit	6.5 -7.5	Peach	10-12
Ber	15-18	Litchi	17-18	Pineapple	12-13
Blueberry	10 -12	Mandrins	12-14	Plum	12-15
Cherry	14-16	Mango	10-18	Pomegranate	15-17
Citrus	12-14	Muskmelon	10-12	Sweet orange	10-12
Grapes	14-18	Nectarine	10 -12	Strawberry	7-10
Grape fruit	10-12	Papaya	11-12	Watermelon	10-12

Titratable acidity: The total acidity content is arbitrarily based on the principal acid of the fruit or vegetable. The principal acid of citrus, guava, mango, pineapple, pomegranate, strawberry, fig, tomato, watermelons and muskmelons are citric acid. Malic acid is the principal acid of apple, pear, plum, peach, cherry, banana and sweet lime. Very few fruits have tartaric acid for example grapes and tamarind. In many citrus fruits and others acidity progressively reduces on maturation and ripening. Extract the juice from the sample and titrating it against a standard alkaline solution gives a measure which can be related to optimum time of harvest. Titratable acidity (TA) can be determined by titrating a known volume of juice with 0.1N NaOH to end point of pink colour as indicated by phenolphethalin indicator. The milliliters of NaOH needed are used to calculate the TA. The TA expressed as per cent malic, citric or tartaric acid can be calculated as follows:

$$TA = \frac{\text{ml NaOH x N (NaOH) x acid meq. factor*}}{\text{Juice titrated}} \times 100$$

*the following acid meq. factor may be used for different fruits

Table 37. Acid meq. factor used for different fruits

Acid	Acid meq. factor	Commodities
Citric	0.0064	Berries, citrus fruits, pineapple
Malic	0.0067	Apple, pear, peach, tomato
Tartaric	0.0075	Grape

Grapes contain significant amounts of organic acids like tartaric, malic and citric. During ripening, the tartrate and malate content of the fruit decrease and a steady increase in pH. Due to variation in buffer capacity, there is no direct relationship between titratable acidity and pH. The organic acids in wine are primarily derived from grapes. The major acids produced during and after the alcoholic fermentation are acetic, lactic, and succinic acid. The acid content of the wine is determined by titrating a sample (a given volume) with a base such as sodium hydroxide solution to a phenolphthalein end point or alternatively, to a pH of 8.2. The titratable acidity is expressed as grams of tartaric acid per 100 ml.

Figure 24. Automatic tritrator

Materials required for analysis acidity

- 500 ml wide mouth Erlenmeyer flask and one 250 ml beaker
- 25 ml burette and burette stand
- 1 ml & 5 ml pipette
- Stir plate and magnetic stirrer
- pH meter with buffer solutions
- Distilled water
- Normal NaOH (sodium hydroxide)
- Phenolphthalein indicator
- Thermometer
- Pipette bulb or pump

Procedure: Titration without using pH Meter

- Place 200 ml of boiled and cooled distilled water into a 500 ml Erlenmeyer flask and add 1 ml of phenolphthalein indicator
- Titrate the water with 0.1 N, NaOH to a definite pink end point
- Add 5 ml of fruit juice/must/wine sample to the flask

- Titrate the sample with 0.1 NaOH to the same distinct end point
- Note the volume of NaOH used in the titration

TA as tartaric acid (g/100 ml) = (V) (N) (75) (100)_ (1000) (v)

V	=	ml of sodium hydroxide solution used for titration
N	=	Normality of sodium hydroxide solution
v	=	sample volume (ml)

If the above procedure is followed, the total titratable acidity can be calculated by multiplying the volume of 0.1 N NaOH used by a factor of 15.

The consumer acceptability is closely associated to the acid content of fruits. Attempt to assess the stage of ripeness of fruit in terms of titrable acidity and to correlate it with the time of harvest has not found general acceptance because of inconsistent results. The titrable acidity varies more with growing seasons and with the amount of nitrogen fertilizer applied to the crop, between harvest dates. Therefore, the eating quality after storage of a low titrable acidity first at harvest is unacceptable, it is suggested that a minimum level of titrable acidity below which fruit should not be stored, be specified for each cultivar.

TSS/acid ratio: It is an expression of the relationship between the sugar and acid content of fruits. It gives a much more dependable and complete measurement of palatability than sugar content of acidity alone. The ratio can be obtained by dividing the degree brix or TSS (after correction for temperature) by the total acidity. This is index for many fruits such as grape, mango, papaya, kiwifruit, melons etc.

Sugar (Free State or as Derivatives): The sugar formed in the leaves is transferred to the fruit where it is stored as starch. As the fruit matures and ripens the starch changes to sugar. Therefore, determination of starch content in fruit will help to decide the optimum date of picking. The conversion of starch to sugars during maturation is test for the maturity of some apple cultivars. It is based on the reaction between starch and iodine to produce a blue or purple colour. The intensity of the colour indicates the amount of starch remaining in the fruit. This parameter plays an important role in imparting attractive flavour, appearance and texture to the fruits. Sugars are the major soluble solids in fruit juices and therefore, soluble solids can be used as an estimate of sweetness. The sugar concentration increases during onset of ripening of pear fruits in free space of cell. It is may be due to the enhancement of permeability of sugars across the tonoplast thus, giving the fruit a sweet and juicy taste. This method is used in determining the maturity of pear cultivars whereby the fruit is cut into two and dipped into a solution containing potassium iodide and iodine.

Method: For starch iodine test, 1 g of potassium iodide is dissolved in 1/8 cup of water. When it is dissolved, 2.5 g of iodine crystals are added. After dissolving, it is diluted to make 1.14 litre solutions and stored in a brown bottle. The prepared iodine-iodide solution is poured into a shallow pan or glass dish to a depth of 50 to 75 mm. The fruits are cut in half across the core and the surface is immersed in the iodine solution. After one minute, fruits are removed and the treated surface is rinsed with running tap water and starch iodine patterns are read and photographed immediately. The reaction of iodine with the starch on the cut surface of fruit gives dark bluish black colour and used as an indication of starch content thus, maturity. The changes in intensity and pattern of the dark colour as the season progresses has been charted and found reliable index of maturity for several cultivars of apple and pear. Most of the cultivars have a characteristic pattern of disappearance, so starch charts are often cultivar specific, hence the starch iodine value at any optimum harvest date can vary between cultivars.

Hydrometer sugar test: The hydrometer measures specific gravity (the weight per unit volume as compares with pure water). Sugar is the chief constituents affecting specific gravity in ripe fruits, therefore estimate the quality of sugar present by measuring the specific gravity. The term brix hydrometer is used to designate the instrument and degree brix is used for the values. Most brix hydrometers are standard at 20° C. The per cent soluble solids and refractometer reading is the same value as the degree Brix.

Refractometer sugar test: The refractometer measures the refractive index of a solution (light bending property as compared with pure water). The refractive index of a sucrose solution is an accurate measure of the concentration of dissolved substances. This instrument again measures total soluble solids as percent sugar from representative sample of fruit and extracted juice.

Oil content: Oil concentration in fruits, vegetables and nuts is varied on different factors, e.g., cultivars and stage of ripening. It ranges from a low of 57.9 per cent in Hall's Giant to 65.7 per cent in Tombul. Macadamias were the highest in oil content (76.9%) and chestnuts were the lowest (2.8%). Oleic acid and linoleic acid comprised more than 90 per cent of the fatty acid composition in all nuts. Oil content level increases with the onset of maturity. It can be used as a reliable index for the maturity determination of avocado. It should not be less than 8 per cent by weight excluding the skin and seed.

Juice content in fruits: During maturity in citrus fruits, there is an increase in the juice content of the fruit. By taking representative samples of the fruit, extracting the juice in a standard and specified way and then relating the juice volume to the original mass of the fruit, it is possible to specify its maturity. The minimum juice content has been specified for navel oranges (30%), grapefruit, other oranges (35%), lemons (25%), mandarins (33%) and clementines (40%) below which they should not be harvested.

Dry matter content: Rate of dry matter accumulation is used to predict optimum harvest time by using instrument hydrometer. Dry matter is also being used to as the maturity standard in processing varieties of potato. Potato dry matter content at the time harvesting should be in the range of 18 – 24 per cent.

Physiological methods

The physiological method is based on the evolution of CO_2 and ethylene. This evaluation rates have successfully been used as the most reliable method to determine the optimum physiological state for the maturity of fruits.

Respiration rate: Best picking date can be decided by measuring respiration at different picking dates. This can be corrected with position in the respiration curve. It accurately expresses the stage of ripening. In climacteric fruits at physiological maturity and ripening processes, there is an increase in their respiratory activity and increased evolution of CO_2. The preclimacteric minimum (PCM), when CO_2 production is at its lowest level, is, therefore, a useful indicator of physiological state of the fruit. It was initially thought to provide a good method of determining the suitability of fruit for storage, the optimum time of harvest being at the PCM, before ripening commenced. The variability among fruit samples and patterns of evolution CO_2 means it may take several days to ascertain whether the climacteric has been initiated, by what time fruit should be harvested at the optimum date. Thus, measurement of respiratory activity is a useful test in retrospect to determine, for example, what was physiological state in fruit with the best storage when harvested, but not as a practical maturity index.

Ethylene concentration: Maturity of climacteric fruits such as mango, banana, apple, pears, etc. can be judged by level of ethylene. In ripe fruits concentration varies from 100 to 700 ppm. During the climacteric, the temporary but marked increase in CO_2 production is not the only phenomenon occurring. Production of ethylene, a ripening hormone in climacteric fruit, follows a similar pattern concurrently and hence, can be a useful indicator of the physiological maturity of the fruit. The ethylene production was found to be the most useful of all the indices examined for Royal delicious apple. It has been suggested that fruit should be harvested just before the commencement of the climacteric irrespective of internal ethylene concentration (low or higher) is still low, for best storage. The concentration of ethylene (low or higher) production does not mean that fruit will have good eating quality after storage. Ethylene measurements are related with other maturity indices and help to determine the storage period, and which would be more suited for immediate consumption or other end uses. The method for assessing maturity based on ethylene production is not necessarily superior to other methods (soluble solids, acidity, starch index, firmness and colour) in determining the optimal picking date for a good storage. The other indices are like water into free water in tissue, changes in optical density, skin colour, etc are used.

Factors affecting maturity

Temperature: Higher temperature gives early maturity than lower. e.g. Gulabi (Pink) grapes mature in 100 days in Western India but only 82 days are enough in the warmer Northern India. Lemon and guava takes less time to mature in summer than in winter. Sun-scorched portions of fruits are characterized by chlorophyll loss, yellowing, disappearance of starch and other alcohol insoluble material, increase in TSS content, decrease in acidity and softening.

Soil: Soil on which the fruit tree is grown affects the time of maturity. e.g. Grapes are harvested earlier on light sandy soils than on heavy clays.

Size of planting material: This factor in propagated fruits affects fruit maturity. In pineapple, the number of days taken from flowering to fruit maturity was more by planting large suckers and slips than by smaller ones.

Closer spacing: Close spacing of hill bananas hastened maturity.

Pruning intensity: It enhanced the maturity of flordasun and sharbati peaches.

Girdling: Process of constricting the periphery of a stem which blocks the downward translocation of CHO, hormones etc. Beyond the constriction which rather accumulates above it. In grape vines it accelerates maturity, reduces the green berries in unevenly maturity cultivar and lowers the number of short berries.

Table 38. Maturity indices for different fruits

Fruits	Maturity Indices		
	Visual	Physical	Chemical
Almonds	Fruit colour change from light green to brown and light yellow	Splitting of hull, separation of hull from shell, development of abscission zone	Increase sweetness in seed
Apple	Develop yellow colour in Golden delicious and deep red colour in Red delicious varieties	Deep brown colour of seed coat, Golden Delicious & Red delicious 18 lb firmness	TSS 11.5-14.5%, starch/acid ratio: 4, Golden Delicious 12% SSC
Banana	Fullness of finger, brittleness of floral ends, disappearance of angularity in a cross section of the finger	Pulp/peel ratio: 1.2-1.4, Dwarf banana mature within 11-14 months and tall cultivars takes about 14-16 months to harvest.	Acid content 0.25%
Ber	Golden yellow colour	Seed /stone ratio: 12 to 18, specific gravity less than 1	TSS 15-18%
Cherry	Light red colour	Increase softening	TSS 14-15%,
Guava	Colour break stage (when skin colour changes from dark green to light green and light yellow)	Specific gravity 1.01 to 1.02	TSS 12-14%, TSS acid ratio for Allahabad safeda - 35.81, Apple colour guava - 26.39, Chittidar guava - 28.13, Lucknow 49 - 34.25
Grape fruit	Yellow or pink colour	Juice 40-45%, easy separation of berries	TSS 10-12%
Grapes	Appearance of bloom on surface	Sweet taste of berries	TSS/acid ratio about 25, TSS 14-18%, TSS 12-18% Thompson seedless, 12-14% for Bangalore Blue, 14-16% for Anab-e-shahi
Kinnow	Bright yellow or orange colour	Juice 35-40% by volume	TSS 10-12%, TSS/acid ratio 12:1 to 14:1
Kiwi fruit	Bright brown colour	Firmness 14 lbs	TSS – 6.5%,
Lemon	Colour break 40%	Juice 25-40% by volume	TSS 10-12%
Lime	Colour break 50%	Juice 25-40%	TSS 10-12%

Fruits	Maturity Indices		
	Visual	Physical	Chemical
Litchi	Pink colour and flatness of tubercles, bright red in colour	Pulp/seed ratio: 4.3 to 6.3 (seeded) 12 to 28 (seedless)	TSS 17-18%, TSS: total acid ratio of 30-40
Mandarins	Colour break 40-50%	Juice 50%	TSS 12-14%
Mango	Olive colour, fullness of cheeks or bulge of shoulder, growth of seed hair	Specific gravity 1.01 to 1.02, flesh colour yellow to yellowish-orange	TSS 10-18%
Orange	Colour break 35%	Juice 40-50%	TSS 10-12%
Papaya	Trace of yellow colour on apex or between ridges	No milk on tapering	TSS 11-12%
Pear	Greenish yellow colour	Cream to light yellow colour	TSS 13-14%
Peach	Ground colour change from green to light yellow or red strip	Pulp/stone ratio: 12 to 16	TSS 10-12%
Pineapple	Colour break 25% or eyes changed from light to dark green	Specific gravity: 0.98 to 1.02	TSS 12-13%, TSS/acid ratio 21-27
Plum	Dark blue or purple colour	Pulp/stone ratio: 12 to 25	TSS 12-15%
Pomegranate	Light green to red, pink and yellow colour develop, seed colour becomes pink	Juice 40-50%, red juice colour	Sugar content 12-16% and acidity 1.5—2.5%, sugar acid ratio 19.5-20.7%.
Sapota	Light brown colour	No milk on tapering	-
Strawberries	2/3 of berry surface showing pink or red colour		
Watermelon	Flesh colour 75% red, Ground colour turning creamy yellow	Juice content 80-96%, Dull hollow sound when thumped	TSS 10-11%

5

Harvesting of Fruits and Vegetables

Harvesting is the gathering and separating from the vital sources of water, nutrients and growth regulators of plant parts that are of commercial interest. It is one of the important operations, that decide the quality as well as storage life of produce and helps in preventing huge losses of fruits. Harvesting of fruits should be done at optimum stage of maturity. During harvesting operation, a high standard of field hygiene should be maintained. It should be done carefully at proper time without damaging the fruits. Harvesting also bring out wound responses like ethylene production and increased respiration in the tissue. Mature tissue generally shows only small responses to harvesting because it stored carbohydrates reserves and relatively low respiration and transpiration rates, and it's destined for natural separation by abscission. Rapidly metabolizing tissue such as leafy vegetables/immature fruits and vegetables exhibits larger responses to harvesting. Harvest the produce when the heat load is low, however around- the -clock harvesting is done when machinery are used to meet the cost of the machine and factory processing schedule. In harvesting operations are include, identification and judging of fruits, vegetables and flowers for harvesting, detaching or separating of the produce from tree, and collection of harvested produce. Harvested produce should be kept under shaded tree or using tarpaulins/shade nets.

HARVESTING METHOD

The harvesting methods and the time of harvest play a vital role in maintaining the best quality of crop during the course of postharvest handling and storage. Any physical or mechanical injury during harvest leads to shortening of shelf life due to increase water loss, respiration and ethylene bio synthesis, add to microorganism infection through damaged areas and it also possible to help in develop physiological disorder, loss in quality and causes spoilage. Therefore, gentle picking, digging and handling help to reduce losses. Manual pickers should be well trained to harvest the fruit and vegetable crops to minimize damage and waste and should be able to recognize the proper stage of maturity. Pickers should be trained to empty their picking bags or baskets with care, never dumping or throwing produce into the field containers. Harvesting can be performed by hand or mechanically. Fruits and vegetables are harvested by hands for this purpose scateurs, clippers or diggers are used. Mechanical harvesting is used for fresh market crops (roots, tubers, rhizomes and nuts). The fruits like wine grapes, prunes, peaches etc. are harvested with machines because harvest damage does not significantly affect the quality of processed product as the commodities are processed quickly. However, for some crops eg. onions, potatoes, carrots and others it is possible to use a combination of both systems. In such cases, the mechanical loosening of soil facilitates hand harvesting. Fruits and vegetables for the fresh market are hand harvested while vegetables for processing or other crops grown on a large scale are mainly harvested mechanically (peas, beans, potato etc.). In some fruit crops, a natural break point forms at the junction of the stem and the stalk when produce is mature. Harvesters should grasp the product firmly but gently and pull upward. The nut crops (pistachio nuts) should not be knocked to the ground during harvest because of their open shells and relatively high moisture content. They should be harvested putting plastic sheeting or canvas tarpaulins, which is spreaded below the tree during harvesting.

Manual harvesting: Harvesting by one's own hand is called manual harvesting. It is economic with low damage. It is done in several ways:

» Ladder / bag picking method

» Poles/ clippers method

» Harvesting by means of cutting knives

» Harvesting by means of digging tools.

It predominates for the fresh market and particularly the produce which is more susceptible to physical injury and soft fruit like grapes, litchis, jamum, strawberry and others berries which are borne on low growing plants. The main benefit of hand harvesting over mechanized harvesting is that humans are able to select the produce at its correct stage of ripening and handle it carefully. The result is a higher quality product with minimum damage. This method is also usefull reducing incidence of fungal infection in papaya and grapefruit. When fruit are cut from the tree using clipper shows less infection then the harvesting by twisting and pulling. But harvesting small fruits and from thorny plants are major obstacle.

Mechanical harvesting: In this method numbers of mechanical devices are used for harvesting the produce on commercial scale. In region where labour cost is high machine harvest is popular for processing crops because it could damage the produce and subsequent faster deterioration eg. Peas for freezing, peaches for canning and grapes for wine making. It needs mechanical assistance to hand pickers by ladders and worker positioners (tree towers and platforms) and self-propelled carts (to reduce non harvest time by assisting with materials) handling and movement of the workers and to increase productivity during harvesting time by providing optimum working conditions and by making detachment of the fruits easier. Mechanical harvesting devices employ direct contact methods such as combing, cutting, pulling, snapping, twisting, stripping and compacting. Likewise machine harvest is used for robust, low unit value ground crop such as potatoes and onions. The main advantages of mechanized harvesting are speed and the reduced costs per ton harvested. However, because of the risk of mechanical damage, it can only be used on crops that require a single harvest. It is done in several ways:

» Picking pole fitted with cutter device for fruits high on trees like mango, avocados

» The shake and catch machine used in apple and citrus to harvest and collect the fruit by shaking the trunk and collection the fallen fruit on the canvas which spread under the tree.

» Use of vibrating digger for harvest underground roots, tuber and rhizomes.

» Use of robotics to harvest mushroom by method of sucker end-effecter.

The primary advantages of mechanical harvesting are the produce can be harvested at a faster rate, and there is less manpower is requirement as compared to hand harvesting. However the disadvantages are there is a possibilities of crop damage and is not suitable for marketing of fresh commodities.

Table 40. Harvesting techniques for fruits and vegetables

Fruits	Technique
Mango	Hand picking or mango harvester using a cutting tool attached to a long pole and picking with 8-10 mm stalk
Banana	Bunches are cut with stalk retaining 30 cm length
Citrus, guava, grapes, litchi	Stalks clipped with the help of clipper
Papaya	Fruit is harvested by twisting the fruit till it snaps off
Pine apple	Harvested by cutting the stalk with 2 cm

Fruits	**Technique**
Aonla	Hand picking and with local harvester using long pole attached with netting bag
Ber	Hand and hook with pole picking of individual fruits
Date	Harvesting by cutting the whole cluster
Loquat	By clipping the cluster with stalk
Jack fruit	By cutting the stalk with 8-10 cm
Pomegranate	By clipping the stalk
Apple	Just lift the fruit from the bottom with the help of palm and press the stalk gently with the fore finger. The mature fruit will be separated from the mother tree without much force
Peach, pear and apricot	Picked by hand with precaution as mentioned for apple
Dry fruits almond and walnut	By shaking the branches
Vegetable crops	
Leafy vegetables	By number of cuttings
Root, tuber and bulb crops	By digging
Tomato and chilies	Hand picking or pulling the fruit with hand
Brinjal	Clipping the stalk
Cauliflower, cabbage	By cutting the stem
Green peas, okra	Picked by hand
Bitter gourd, pointed gourd and sponge gourd	Hand picking
Bottle gourd, pumpkin, ash gourd and melons	By clipping the stem/stalk

Harvesting time: Exposure to the sun should be avoided as much as possible during and after harvest as produce left out in the sun, gain heat and may become sunburned. Therefore, time of harvesting has also an important role. Night or early morning hours harvest when temperatures are relatively low helps to enhance shelf life of fruits and vegetables. Latex flow is often lower in the morning than it is at dawn for some crops like mango.

Produce Management at Harvesting

The harvest marks the end of the growing season, or the growing cycle for a particular crop, and this is the focus of seasonal celebrations of many religions. On smaller farms with minimal mechanization, harvesting is the most labour-intensive activity of the growing season. Harvesting in general usage includes an immediate post-harvest handling, all of the actions taken immediately after removing the crop like cooling, sorting, cleaning, packing, etc. and up to the point of further on-farm processing, or shipping to the wholesale or consumer market. Use of improper harvesting method may result in the damage of crop. Bruising which is caused by compression (due to overfilling of boxes or in bulky stores), impact (due to dropping of crop or from something hitting the crop) or vibration (due to loose packing during transportation). Harvest timing is a critical decision that balances the likely weather conditions with the degree of crop maturity. Weather conditions such as frost, rain (resulting in a wet harvest), and unseasonably warm or cold periods can affect yield and quality. An earlier harvest date may avoid damaging conditions, but result in poorer yield and quality. Delaying harvest may result in a better harvest, but increases the risk of weather problems. Appropriate production practices, careful harvesting, proper packaging, storage, and transport all contribute to good produce quality. Produce that has been stressed by too much or too little water, high rates of nitrogen, or mechanical injury (scrapes, bruises, abrasions) is particularly susceptible to postharvest diseases. Mold and decay on winter squash, caused by the fungus *Rhizoctonia,* result from the fruits lying on the ground, and can be alleviated by using mulch. The process needs large scale operations, costly handling machinery and high-tech postharvest treatments. Instead, simple, low cost technologies often can be more appropriate for small volume, limited resource commercial operations, farmers involved in direct marketing, as well as for suppliers to exporters in developing countries. Quality cannot be improved after harvest, only maintained, therefore it is important to harvest fruits at the proper stage, size and at peak quality. Immature or over mature produce may not last as long in storage as that picked at proper maturity. Harvest should be completed during the coolest time of the day, which is usually in the early morning, and produce should be kept shaded in the field. Crops destined for storage should be as free as possible from skin breaks, bruises, spots, rots, decay, and other deterioration. Bruises and other mechanical damage not only affect appearance, but provide entrance to decay organisms and also increase moisture loss. Postharvest rots are more prevalent in fruits and vegetables that are bruised or otherwise damaged. The rate of moisture loss may be increased by as much as 400 per cent by a single bad bruise on an apple, and skinned potatoes may lose three to four times as much weight as non-skinned potatoes. Damage can be prevented by training the harvest labour to handle the crop gently, harvesting at proper maturity, harvesting dry whenever possible, handling each fruit or vegetable no more than necessary (field pack if possible) installing padding inside bulk bins and avoiding over or under-packing of containers.

Necessary care during harvesting

- Harvesting should be made by skilled labour during the cool hours of the day at early morning and late afternoon.
- Harvested produce should not be kept on the soil and avoid falling of produce on earth.
- Hand gloves should be used for harvesting on spiny plants.
- Ladders should be used to harvest produce in case of tall trees.
- Produce selected for harvesting should be of right maturity.
- Harvesting should be done gently, without jerks to protect the produce from damage.
- While harvesting underground crops like potato, onion, radish, carrot and beet root etc. care should be taken that produce should not get damaged by digging implements.

Maturity Signs and Harvesting Methods of Important Fruits and Vegetables

In a physiological sense, maturity refers to the attainment of final stage of biological function by a part of plant as a whole. It is a stage of fruit and vegetable at which consumer wants the fruit and vegetable eat. The harvest maturity of vegetable depends upon the purposes for which it is harvested. The french bean and okra pod is matured when it is tender with maximum size, as per horticultural maturity., Fully coloured tomato fruits are harvested for local market and processing, while the produce is shipped to distant markets, or stored, it should be picked in the mature but unripe stage. Poor quality and uneven ripening are caused by early harvesting whereas, late harvesting result in extremely poor shelf life. Here lies the difficulty, because unlike the ripening stage, the boundary between pre maturation and maturation stage is hard to detect. Maturity signs and harvesting of fruits and vegetables as under:

Banana

Maturity signs of banana

- The dwarf bananas are ready for harvest within 11 to 14 months after planting while tall varieties take about 14 to 16 months to harvest.
- Fruits by and large mature in 120 to 140 days after flowering.

- The fruit bunch is harvested when the ridges on their surface changes from angular to round.
- The dried parts of flowers at the top of fruit drop off easily.
- The top most leaf starts drying as the bunch matures.
- Colour of fruits or fingers changes from dark green to pale green.

Harvesting of banana

- The trunk is lopped with a sickle or hatchet over half-way through the stem.
- The bunch will not fall to the ground but hang on, and injury is avoided. The bunch is held and its peduncle is served.
- About 30 cm of the stalk must be left to make handling easy.

Mango

Maturity signs of mango

- Fruits normally require 95 to 115 days to mature after flowering.
- Building of shoulders and a slight depression near the stalk end indicates the maturity.
- Colour of fruit changes from dark green to pale green and red blush develops on the fruit shoulders.
- One or two or three ripe fruits (tapka) fall from the plant naturally and fruit specific gravity between 1.00 to 1.02 indicating maturity.

Harvesting of mango

- Harvesting should be done by hand as much as possible. The fruit is twisted sharply sidewise or upward.
- The pedicel is trimmed short enough to keep it from puncturing other fruits. Mango harvester harvests fruit borne on high branches.

Sapota

Maturity sings of sapota

- Fruits normally mature in about 240 to 270 days after flowering.

» Fruits at full maturity develop a dull orange or brown colour.

» A mature fruit when scratched slightly with nail shows a yellow streak instead of green streak and milky latex content is reduced.

» Brown scaly material disappears from the fruit surface and dried spine like style at the top of the fruit falls or drops off easily when touched as the fruits approach full maturity.

Harvesting of sapota

» The fully matured fruits are harvested with the stalk intact, individually by giving a twist and collected without brushing.

» Sapota harvester (developed by Konkan Krishi Vidyapeeth, Dapoli) should be used to avoid post-harvest lose.

Papaya

Maturity signs of papaya

» Fruits require 125 to 140 days from flowering to maturity.

» Colour of fruit changes from green to pale green or yellowish.

» Portion of fruit exposed to sunlight becomes dark yellow in colour.

» The latex of fruits becomes watery and TSS at harvest should be minimum 6 per cent.

Harvesting of papaya

» It is harvested by twisting the fruit on dwarf trees, the operation can be done without mechanical aids. The peduncle should be trimmed before panicking.

» Ladders are often used to harvest tall trees.

Jackfruit

Maturity signs of jackfruit

» The fruit mature in about 180 to 270 days after flowering.

» Fruit colour changes from green to pale green.

» A dull, hollow sound is produced when the fruit is tapped by the finger.

» Fruit spines become well developed and wide spaced.

» The last leaf of the peduncle yellows and an aromatic odour develops.

Harvesting of jackfruit

» If the fruit is within the reach, it can be cut with a sickle or twisted unit it snaps off. In tall trees, a sack is placed on the fruit with a rope tied on the peduncle.

» After cutting, the fruit is gradually lowered to the ground.

» It is laid on its peduncle against railing for some time to allow the latex to flow and coagulate.

Ber

Maturity signs of ber

» About 125 to 150 days are required for harvesting of fruits after flowering.

» Colour changes from green to pale green / red / brownish, yellow, depending upon variety and fruit become soft.

» In variety umran, specific gravity should be less than 1 with golden yellowish colour.

Harvesting of ber

» It is harvested from November to March in different parts of the country which is previous in South and delayed in North India. The peak season of harvest in northern India is February to April. Individual ripe fruit harvest by the help of long hook and collect on impact proof net.

Guava

Maturity signs of guava

» Fruit mature in 140 to 160 days after flowering.

» Colour changes from green to pale green or yellow.

» Fruit becomes soft, and a slight depression develops near the stalk end.

» Ridges on the fruit disappears and it becomes round.

Harvesting of guava

- Hand-pulling is employed.
- Higher branches can be reached with a ladder. The fruit are collected in basket.
- Picking is done 2 to 3 times a week during harvest season of 8 to 10 weeks.

Grapes

Maturity signs of grapes

- A fruit mature in 120 to 135 days, after flowering.
- Fruit colour changes from green to yellow on golden yellow or black or pink, depending upon the variety
- Development of characteristics like flavours, aroma and easy separation of barriers.
- TSS of anabe shahi fruits 15 to 16 per cent and thompson seedless 18 to 20 per cent.

Harvesting of grapes

- Grapes are non – climacteric fruits harvested when they are fully ripped. For harvesting of the clusters should be removed from the vine by cutting with a sharp knife near its attachment to the cane.

Pomegranate

Maturity signs of pomegranate

- Fruits are ready for harvesting in about 150 to 170 days after flower initiation.
- Colour changes from green to yellow or yellowish brown or bright red depends upon varieties and fruit becomes soft.
- Floral part dry out and ridges develop on the fruits and fruit becomes flat.
- Pressing fruit with finger give cracking or metallic sound.

Harvesting of pomegranate

- The fruits are harvested manually with help of sharp carving knife and clipper. Higher branches can be reached with the help of a ladder. The fruits are collected in basket.

Mandarin and sweet Orange

Maturity sings of mandarin and sweet orange

» Fruit mature in about 210 to 240 days after flower initiation.

» Colour changes from green to pale green or pale yellow or orange. Fruit become soft and easy in detachment from fruit stack.

» Outer rind looks shiny and one can see oil glands on the fruit.

» In mandarin fruits 0.4 per cent acidity and 12 to 14 per cent TSS, in orange 0.3 per cent acidity and 12 per cent TSS is sings of maturity.

Harvesting of mandarin and sweet orange

» In India the fruit picker collect the fruits manually by climbing on a ladder when the tree is tall, with a collecting bag on his shoulders. In U.S.A. mechanical harvesting is done particularly for processing industry.

Kagzi Lime

Maturity signs of kagzi lime

» Fruits mature in 120 to 150 days after flower initiation. Colour changes from green to pale green or pale yellow and start natural fruit dropping.

Harvesting of kagzi lime

» In India the fruit picker collect the fruits manually by climbing on a ladder when the tree is tall, with a collecting bag on his shoulders. In U.S.A. mechanical harvesting is done particularly for processing industry.

Pineapple

Maturity sings of pineapple

» Flowers require 105 to 130 days for maturity.

» Colour of lower portion of fruit turns green yellow.

» Eyes on the fruit becomes smooth or flat and they buldge on the sides.

» Eye bracts become loose and turn brown in colour.

» Mature pineapple should have 12 per cent TSS 0.5 to 0.6 per cent acidity and specific gravity of fruits is 0.98 to 1.02.

Harvesting of pineapple

» Harvesting should be done with a sharp knife severing the fruit stalk with a clean cut and retaining 5-7 cm of the stalk.

Custard Apple

Maturity sings of custard apple

» Flowers require 100 to 120 days for full maturity.

» Scales on fruits becomes prominent, plummy and well space.

» Fruit become scales or segments turn creamy white with softness.

Harvesting of custard apple

» Harvesting may extend from September to November depending on the period of flowering. Fruits are harvested manually by hand with help of sharp knife and scatear.

Kokum

Maturity sings of kokum

» The fruits generally require 94 days to mature after flowering.

» The colour of fruit changes from green to light green and red and purplish red when ripe.

» At mature stage at average weight of fruit 34.01 gm and specific gravity is 1.04.

» Chemical parameters at mature stage are TSS 13.68 per cent, acidity 3.61 per cent and total sugar 6.35 per cent.

Harvesting of kokum

» Generally ripped red to purplish red colour fruits are harvested manually by picking. The Atul Sapota harvester developed by Konkan Krishi Vidyappeth, Dapoli can be used for harvesting kokum fruits.

Jamun

Maturity sings of jamun

» The fruits generally require 120-130 days to mature after flowering. Colour changes from green to deep purple or black with full size and fruit becomes soft and juice.

Harvesting of jamun

» The fruits of jamun are generally harvested daily and sent to the market on the same day, because shelf life of fruit is 2-3 days. The ripe fruits are picked singly by hand. The picker climbs on the trees, with bag of cotton slung on the shoulder and harvests the ripped fruits.

Karonda

Maturity sings of karonda

» Fruit generally require 87 to 90 days to mature after flowering.

» Colour changes from light green to dark green and develop red blush spots on the fruits.

» At maturity specific gravity is 1.012, TSS 13.5 per cent, total sugar 2.35 per cent and acidity 2.62 per cent of fruits.

Harvesting of karonda

» The fully mature fruits are harvested by hand and with the stalk in act.

Coconut

Maturity signs of coconut

» Coconut usually matures in about 350 to 375 days after appearance of the inflorescence.

» Colour of fruit changes from green to yellowish or brownish green.

» The fruit produces peculiar metabolic sound on thumping.

» All nuts in a bunch mature uniformly at same time.

Harvesting of coconut

» Different coconut producing country adopting various method of harvesting like in Sri Lanka, harvesting is done from the ground with the help of a knife attached to a long bamboo pole, in Malaysia and Thailand trained monkeys are also used for harvesting, in Africa, nuts are allowed to fall from bunches.

» In India climbing on the trees for perform harvesting. The climber often uses a small ladder, after climbing top of the ladder uses a rope ring round the feet for climbing the rest of the tree.

Arecanut

Maturity signs of arecanut

» Fruits are ready for harvesting in about 180 to 200 days from flowering.

» Fruits colour changes from green to yellowish or orange.

Harvesting of arecanut

» Harvesting is done by climbing the tree with the help of ladder and rope ring around the feet. On reaching the top the climber harvest the matured bunch with sharp knife or sickle.

Cashew

Maturity sings of cashew

» Fruit matures in about 95 to 120 days from flowering.

» Depending upon variety apple colour changes from green to yellow or red.

» Nut becomes hard and ash in colour.

Harvesting of cashew

» Harvesting and collection of nuts are done over a period of 10-12 weeks. Where apple is not collected, the fruits may be allowed to fall on the ground and nuts collected periodically during first 4 to 6 weeks. Later most of the nuts are collected in the trees. Towards the end of season nuts can again collected from ground.

Cabbage

Maturity signs of cabbage

» At low elevations cabbage varieties in about 62 to 110 days and at high elevations about 81 to 125 days required for mature from field transplanting.

- Solidity and firmness of head are the usual maturity characteristics used.
- Head colour turns a lighter shade of green when full development is an attained.
- They have tendency to burst or loosen the leaves beyond the marketable stage.

Harvesting of cabbage

- The heads are cut with knife frequently attached with some wrapper leaves.

Cauliflower

Maturity signs of cauliflower

- The best stage of maturity is determined by curd size and condition, usually harvest the head upon the desired size and before the curds become discolored, loose or otherwise blemished.
- The head should be compact and not to be broken into segments.
- Over mature head, becomes too large, flower stalks elongates resulting in loose, leafy condition and possess poor market value.

Harvesting of cauliflower

- The curds are cut off the stalk with large sharp knife. The large leaves are trimmed away until sufficient jacket leaves remain to protect the curd.

Knol-khol

Maturity signs of knol-khol

- Harvesting is done when it attains its marketable size *i.e.* 5 to 7 cm in diameters and bright colour depending upon variety.
- At this stage the edible portion is tender and non-fibrous.

Harvesting of knol-khol

- The knobs are harvested with sharp knife, when they are mature. Foliage is kept while harvesting.

Tomato

Maturity signs of tomato: Harvesting depends upon the purpose for which they are used.

Four maturity stages are generally recognized.

- **Green stage:** The mature green fruits are generally harvested to send them to the distant market.
- **Pink stage:** At this stage colour turns to pink or red at the blossom end. They are picked for local market.
- **Ripe stage:** At this stage most of the fruits is red and the soften of fruits begins.
- **Fully ripe stage:** At this stage fruits have approached maximum colour development and are soft. They are generally consumed or used for canning and processing.

Harvesting of tomato

- Depending upon the purpose for which they are used and market distance, tomatoes are harvested manually by plucking the fruits at different maturity stages. Mechanical harvesting also used only for processing purpose of fruits.

Chilli

Maturity sings of chilli: Chillies are harvested at two stages, one for green vegetables and the other as dry chillies.

- Green chillies are harvested when they are fully mature and before they change from green to red.
- Chillies for drying should be harvested when colour changes from green to red.

Harvesting of chilli

- Depending upon the purpose for which chillies are to be used, fruits are picked either green or fully red ripe.

Bell pepper

Maturity signs of bell pepper

- Usually bell peppers are harvested in the mature green, yellow, red and orange colour stage. They should be of good shape, waxy, firm and shiny.
- Peppers for canning are allowed to attain full size and develop a red colour before harvesting.

Harvesting of bell pepper

» The sweet peppers are picked with an upward twist which leaves a piece of stem attached.

Brinjal

Maturity signs of brinjal

» The fruits must be harvested as they have attained the desired size, colour and earlier than they lose their bright lustrous appearances and become dull.

» The fruits are edible from the time they are quarter growth until they are near ripe.

Harvesting of brinjal

» The fruits are harvested from the plant by cutting with a small pruning shears or a sharp knife or hand clipper. The fleshy calyx and a short piece of stalk are left attached to the fruit.

Carrot

Maturity sings of carrot

» Depending upon variety carrot become ready for harvesting within 100 to 120 days from seed sowing.

» Desired size is the primary consideration in harvesting the roots. Generally small to medium sized, soft core and crisp carrots are preferred. Oversized and mature hard fibrous roots are not acceptable.

» At marketable stage carrots should have at least 2.5 to 4.0 cm diameter at the upper end.

Harvesting of carrot

» In India, harvesting is done manually with help of iron shovel. Early carrots for the market are pulled out when they partly developed. They are normally dug out, when the soil is sufficient moist with spade or a khurpi. The roots are trimmed and washed before sending them to the market. In advanced countries, mechanical harvesters are also available for root crop harvesting.

Radish

Maturity sings of radish

» European type is ready for harvesting within 30 days and Asiatic type within 45 to 60 days from seed sowing. At this stage roots are mild, tender and crisp and usually of the proper marketable size. Roots must be harvested before they become pithy, bitter and fibrous.

Harvesting of radish

» In India, harvesting is done manually with help of iron shovel. A light irrigation may be given before harvesting to facilitate lifting of roots. The roots are washed, graded and tied in bunches of 5 to 10 roots. In advanced countries, commercial radish growers use a single row harvester that pulls the plants from the soil.

Onion

Maturity signs of onion: Harvesting of onion based on the purpose for which the crop is planted and maturity will depend upon the variety being used.

» **Green bunch onion:** Green onion is best when they are of lead pencil unit a small bulb is formed.

» **Ripe bulbs:** Ripe bulb crop is ready for harvest in 90 to 120 days after transplanting. Neck fall is the indication of maturity. The best time to harvest onion is when 60 to 70 per cent of tops have broken over. Development of red pigment and the characteristics pregnancy of the variety are also important harvest indices.

Harvesting of onion

» If the bulbs are mature, they are pulled out easily by hand if the soil is not hard or compact, otherwise, they may be dug out with a shovel or khurpi.

Garlic

Maturity signs of garlic

» Crop gets ready for harvest in about 100 to 140 days after planting.

» The crop may be harvested when the top turns yellowish or brownish and shows signs of drying up and cloves are well developed.

Harvesting of garlic

» The bulbs lifted manually with help of shovel, cleaned and the leaves are tied at the top with 10 to 15 bulbs for drying. The bulbs are dried for a week or so under shade.

Okra

Maturity sings of okra

» Fruits are harvested when the pods are still mature, young and tender. They are harvested at an interval of 2 days. Frequent picking promote fruit development and enhances yield and quality of fruits.

» The suitable time of okra fruit harvest being 5 to 6 days after opening of flower.

» Mature pods are fibrous, hard and unfit for human consumption.

Harvesting of okra

» The tender young dark green fruits 7 to 10 cm long should be harvested every alternate day with help of bhindi/okra pullker.

Pea

Maturity signs of pea

» Early varieties require 45 to 60 days and late varieties require 70 to 100 days from seed sowing.

» By means of increasing maturity and size, sugar content and tenderness declines rapidly, with a correspondence increase in starch and protein.

» Harvest the pods when colour of pods changes from dark green to light green with good size.

Harvesting of pea

» The pods should be harvested at proper maturity by hand picking at an interval of 12 to 15 days. Mechanical harvesting also adopted in developed counties.

Potato

Maturity sings of potato

- Potatoes are harvested when they attain desired size as per varieties characteristics. Early varieties are harvested at 60 to 100 days and late varieties are 120 - 160 days after tuber planting.
- Skins slipping from the tuber, starch content and leaf senescence or top drying are the harvest indices.

Harvesting of potato

- Care should be taken to save the tuber from injury and immediately after harvest they should be left in the scorching sun, otherwise they develop sunscalds. Digging is done with spade or other hand tool or by tractor operated potato digger. There should be optimum moisture in the soil at the time of harvest.

Cucumber

Maturity sings of cucumber

- Cucumber fruits are harvest at immature, soft and dark green colour stage. Fruits are ready for harvest at 45 to 55 days after seed sowing.
- Cucumber fruits for slicing should be picked when they are 15 to 35 cm long, whereas for pickling 6 to 15 cm long. In case of slices at marketable stage, spines on fruit become soft and fall down.
- In general cucumber may be picked at any stage of fruit growth, provided yellowish has not started.

Harvesting of cucumber

- The fruit should be picked at frequent intervals in order to avoid losses due to oversized or over mature fruits. The fruits should be harvested at proper maturity by hand picking at an interval of 2 to 4 days.

Water melon

Maturity signs of water melon

- Fruits are ready for harvest in 90 to 120 days from sowing.

- The tendril accompanying the fruit withers as fruit ripeness.
- Ripe fruits when thumped with finger give out heavy dull sound, whereas the immature fruits give metallic sound.
- The portion of fruit, which rests on the ground, turned yellowing at maturity.
- Ripe fruits produce a crisp, cracking noise on being pressed with the flat of the hand.

Harvesting of water melon

- The fully ripe fruits are harvested manually with the help of sharp knife or by twisting.

Muskmelon

Maturity signs of muskmelon

- Fruits will be ready for picking in about 100 to 110 days depending upon variety. There are two groups of cultivars which behave distinctly. In one group, the fruits when mature slips out easily from the vine with little pressure or jerk or if not it will remain separated from the vine next day. This is called full slip stage. In some Indian cultivars, green stripes on the skin begin to turn yellow during maturity.

Harvesting of muskmelon

- Muskmelon is a climacteric fruit which ripens during transit and storage and hence, it is harvested before it is fully ripe. Fruits are harvested manually with the help of sharp knife or by twisting.

Red pumpkin

Maturity sings of red pumpkin

- Fruits require 110 to 120 days for maturity from seed sowing.
- Fruits should be harvested at full stage of maturity when the rind has hardened.
- Colour changes from green to yellow or orange. Some varieties remain green only after full maturity.

Harvesting of red pumpkin

- The fruits are harvested manually by keeping a portion of stem attached. The fruits are placed in heap and allowed to cure for a few days.

Gourds

Maturity signs of gourds

- In most cases fruits require 12 to 15 days after pollination to attain marketable size.
- Usually tender light green colour fruits are ready for harvesting. Proper stage of maturity in bottle gourd, sponge gourd and ridge gourd is judged by its fruit size is as per variety.
- The gentle press of finger nails which easily penetrate the epidermis.
- Plugging the fruits which show fine tender flesh and quite immature seeds. However, this test makes the fruit unfit for market.

Harvesting of gourds

- Harvesting is done at 5 to 7 days interval as per maturation sign of fruits. The fruits are harvested manually by help of sharp knife or cutter with keeping a portion of stem attached.

Sweet potato

Maturity sings of sweet potato

- The roots are harvested at 3.5 to 4 months after planting, when leaves start yellowing.
- After cutting the tuber if white gum remains white, then they are ready for harvest, it turns blackish on green then tubers are said to be immature.

Harvesting of sweet potato

- The crop is harvested by digging with spade or kudali or tractor operated potato digger. At the time of harvesting it is necessary the content optimum moisture level.

Leafy vegetables

Maturity signs of leafy vegetables (Spinach, Amaranthus and Fenugreek)

- Leafy vegetables are harvested 3 to 4 weeks after sowing.
- Harvesting should be done, when the leaves are immature, light green and tender, but large enough. Old leaves are bitter and unfit for consumption.

Harvesting of spinach

» First cutting 3 to 4 weeks after sowing. Leaves are cut manually by help of sickle from the base. It gives 3 to 4 pickings at 15 days interval.

Harvesting of amaranthus

» Uprooting is done 3-4 weeks after sowing. If clippings are to be taken then first cutting 25 to 30 days after sowing and other at 20 to 25 days interval upto total 2-4 clippings.

Harvesting of fenugreek

» When common methi is taken, the younger shoots are nipped off in about 3 weeks after sowing. Later on the whole plant is often pulled out, bunched and marketed. The nagouri/kasuri methi gives many cuttings. First cutting 25-30 days after sowing and succeeding cuttings taken at 12 to 15 days intermission. After 2-3 cuttings plants are left for seed production.

Harvesting containers

The containers used by pickers in the field should be clean and hygienic, have smooth inside surfaces and be free of rough edges. In practice picking baskets, bags, buckets, plastic crates and bins come in many sizes and shapes are used as harvesting containers. Harvesting containers can be made by fitting fabric over opened bottom of ready-made baskets, fitting readymade canvas bags. All of these will decrease mechanical injure to produce. Buckets are better than baskets in protecting produce, since they do not collapse and squeeze produce. Harvesting containers must be easy to handle for workers for picking and cutting the produce in the field. Harvesting bags with shoulder or waist slings (as they are easy to carry and leave both hands free) can be used for fruits with firm skins, like mango, citrus, avocados etc. For high risk products, woven baskets and sacks are not recommended because of the risk of contamination, e.g. strawberry and jamun. The contents of the bag are emptied through the bottom into a field container without tipping the bag. These containers are made from a variety of materials such as paper, polyethylene film, sisal, hessian or woven polyethylene and are relatively cheap but give little protection to the crop against handling and transport damage. Sacks are commonly used for crops such as potatoes, onions, cassava, and pumpkins. Following containers are used for harvesting.

Plastic buckets: Plastic buckets are suitable containers for harvesting fruits that are easily crushed, such as tomatoes, jamun and strawberry. These containers should be smooth without any sharp edges that could damage the produce.

Bulk bins: Commercial growers use bulk bins with a capacity of 250-500 kg, in which crops such as apples and cabbages are placed, and sent to large-scale packinghouses for selection, grading and packing.

Stackable plastic crates: Plastic crates are relatively expensive to purchase, but are durable, reusable and easy to clean. These have required features like stacking strength, ventilation holes and long life. These can be used for harvest, storage, cooling, transport and even for display in retail markets.

Tools for Harvesting

Depending on the type of fruits, vegetables and flowers several devices are employed to harvest produce. Commonly used tools for fruit and vegetable harvesting are secateurs, knives, pruning shears and hand held or pole mounted picking shears that cut and hold onto the stem of the cut product. This feature allows the picker to harvest without a catching bag and without dropping fruits. Using a cutting tool attached to a long pole can aid picking of crops such as mangoes and avocados when the fruit is difficult to reach. When fruits or vegetables are difficult to catch, such as mangoes or avocados, a cushioning material is placed around the tree to prevent damage to the fruit when dropping from high trees. Some fruits need to be sharpened clippers or knives for cut from the parent plant. Penduncles, woody stems or spurs should be trimmed as close as possible to prevent fruit from damaging neighboring fruits during transport. Cutting edges should be kept sharpened and the catching bag should be relatively small. Clean harvesting tools and containers should be used to protect the produce quality. Disinfect tools with one part chlorine bleach and one part clean water solution before use. The use of containers that can be easily filled and carried by workers minimizes damage to produce. The containers that are smooth on the inside, or provide clean, disposable liners made from paper or cardboard should be used. Following tools are generally used for harvesting and collection of fruit and vegetables.

Picking poles and catching sacks: These tools are handmade, a long pole attached to a collection bag, allow the harvester to cut catch produce growing on a tree without climbing on tree. The collection bags can be hand woven from strong cord or sewn from canvas. The hoop used as the collection bag rim and sharp cutting edges can be made from sheet metal, steel tubing or recycled scrap metal.

Clippers and knives: The fruits like mangoes, citrus, grapes are clipped or cut from the plant. Clippers or knives should be kept well sharpened and clean. Peduncles, woody stems or spurs should be trimmed as close as possible to prevent fruit from damaging neighboring fruits during transport.

Tripod ladders: The ladder has three legs and is more convenient and more stable than a common ladder. It helps in harvesting crops such as mango, kinnow, pears, peaches, plums without damaging tree branches.

Collection bags/ Picking bag: These are cloth bag with openings on both ends can be easily worn over the shoulders with adjustable harnesses. In case metallic buckets are to be used for harvesting, fitting cloth over the opened bottom can reduce damage to crop. Fitting canvas bags with adjustable harnesses or by simply adding some carrying straps to baskets also helps to reduce handling losses. The collection bags illustrated below were hand woven from strong cord or sewn from canvas. The hoop used as the basket rim and cutting edges can be fashioned from sheet metal, steel tubing or recycled scrap metal.

Fruit catching sack: Fruit trees are sometimes quite tall and letting fruit fall to the ground when it is cut from the tree will cause severe bruising. If two pickers work together, one can clip or cut the fruit from the tree and the other can use a sack to break its fall. The catcher supports the bag with his hands and one foot, catches the falling fruit, then lowers the far end of the bag to allow the fruit to roll safely to the ground.

Packing in the field and transport to packinghouse

It is the procedure of selection, sorting, trimming and picking the produce in the field. It has great potential to reduce mechanical damage by reducing the number of handling steps between field and consumer. Careful harvesting, handling, and transporting of fruits and vegetables to packinghouses are necessary to preserve product quality and elongation of shelf life.

Polyethylene bags: Clear polyethylene bags are used to pack banana bunches in the field, which are then transported to the packinghouse by means of mechanical cableways running through the banana plantation. This technique of packaging and transporting bananas reduces damage to the fruit caused by improper handling.

Plastic field boxes: These types of boxes are usually made of polyvinyl chloride, polypropylene, or polyethylene. They are durable and can use several years. Many are designed in such a way that they can nest inside each other when empty to facilitate transport, and can stack one on top of the other without crushing the fruit when full.

Wooden field boxes: These boxes are made of thin pieces of wood bound together with wire. They come in two sizes: the bushel box with a volume of 36052 cm^3 and the half-bushel box. They are advantageous because they can be packed flat and are inexpensive, and thus could be nonreturnable. They have the disadvantage of providing little protection from mechanical damage to the produce during transport. Rigid wooden boxes of different capacities are commonly used to transport produce to the packinghouse or to market.

Bulk bins: Bulk bins of 200-500 kg capacity are used for harvesting fresh fruits and vegetables. These bins are much more economical than the field boxes, both in terms of fruit carried per unit volume and durability, as well as in providing better protection to the product during transport to the packinghouse. They are made of wood and plastic materials. Dimensions for these bins are 48 x 40 inches (120 x 100 cm) and approximate depth of bulk bins depends on the type of fruit or vegetable being transported like for citrus 70 cm, pears, apples 50 cm, stone fruits 50 cm, and tomatoes 40 cm.

Management practices for minimum changes in commodities after harvest

In order to minimize changes that occur in horticultural produce after harvest it is imperative that the following practices be accepted:

Harvest at optimum maturity for best eating quality: Immature and over mature produce are more susceptible to decay and cause quality problems. Immaturity increases water loss and shriveling whereas, over maturity reduce in quality and shelf life of fresh produce. When harvested too immature, some fruits such as tomatoes may never ripen satisfactorily others such as watermelons and sweet corn may be low in sugars. When harvested at over mature most crops such as beans, sweet corn and celery become tough and sweet corn will be low in sugars and higher in starch.

Harvest during the coolest part of the day: To minimize the spread of certain diseases and water losses the produce harvest should begin as early morning and late evening for minimize field heat lode in produce. This practice is most important for highly perishable products, because high temperatures lead to rapid deterioration. Harvesting continuously during the day decreases the importance of this factor.

Keep harvested products in the shade: This simple practice will minimize wilting, sunburn damage and prevent unnecessary heating of the produce. On a sunny hot day, tomato fruit in the sun for an hour can be as must 14° C hotter than fruit in the shade.

Wash harvest containers daily: Use water containing about 70 ppm chlorine to thoroughly clean containers. Chlorine kills decay-causing organisms on the container surface and washing removes sand and other waste that may puncture or injure the produce.

Handle all produce gently: Many fruits and vegetables have a natural protective surface. Careful handling helps maintain this surface and results in a more attractive, better quality product. Watermelons that have been handled roughly may appear undamaged but internal bruising may have occurred. Bruises, punctures, and other wounds increase susceptibility to decay and water loss.

Avoid rough roads: When transporting produce from the field to the market, avoid rough roads. Many operators forget that vibration during transit can cause considerable damage to produce. Tie or wedge the load securely to help reduce damage.

Bruising in fruit and vegetable at harvesting and transportation

Bruising is an ever-present problem and fresh-market fruit growers have long been worried. Processing-fruit growers also have grown concerned, because un-brushed fruit commands the best prices. The vast majority of bruising in the harvest process falls into two categories:

» Picking bruises associated with rough handling and detrimental impacts.

» Compression bruises associated with significant vibrations during transport.

Bruising of fruit after harvest ranged from 0.6 to 13 per cent, with an average of 7.1 per cent and it caused 2.7 to 8.1 per cent of the culls. At the retail level in supermarkets, bruising was found to range from 29 to 78 percent, averaging 61 percent. While bruising is a concern, it must be regarded as a defect that can be controlled through basic management principles. The damage inflicted on fruit is related to the energy available for bruising and the characteristics of the product. The energy available for bruising is in turn related to:

» The suspension characteristics of the vehicle transporting the fruit.

» The energy input to the system (a function of roughness of the road and vehicle speed).

» Engineering factor involving both the properties and the packaging of fruit.

The damage suffered by fruits and vegetables is dependent on the number of individual shocks and their severity, and is directly related to the energy absorbed by the fruit. For example the packers of Golden Delicious have learned that packing apples directly on removal from storage may produce more bruised fruit than if fruit is packed after being held at a relatively low humidity for a few days to create an outer layer of bruise-resistant cells. Reducing the amount of bruising in fruit appears to be attainable by reducing the amount of energy that fruit receives in handling. In practical terms, bruising can occur during any of six operations in which fruits and vegetables is removed from the tree and moved into storage. Listed below are seven locations of produce and the six steps involved in moving the fruit and vegetable from tree to storage:

- Fruit on the tree Step 1 Harvesting.
- Fruit in the bins Step 2 Moving bins out of the orchard.
- Bins at the edge of the orchard Step 3 Moving bins to loading area.
- Bins in loading area Step 4 Loading bins on truck.
- Bins on truck in loading area Step 5 Trucking bins to storage.
- Bins on truck at storage Step 6 Forklift hauling bins to storage.

6 Initial Handling of Fruits and Vegetables after Harvesting

After harvest, fruits and vegetables require to be prepared for sale on the farm or at the level of retail, wholesale or supermarket chain. Regardless of the destination, preparation for the fresh market comprises four basic key operations like removal of unmarketable material, sorting by maturity and/or size, grading and packaging. Post harvest activities include cooling, curing, handling, storage, processing, packaging, transport and the market phase. However, this is only really possible with tender or perishable products or small volumes for nearby markets. Products need to be transported to a packinghouse or packing shed for numbers of operations like washing, brushing, waxing, controlled ripening, refrigeration, storage or any specific type of treatment or packaging for distent or demanding markets. In many cases partial field preparation is completed later in the packing shed. Because it is a waste of time and money to handle unmarketable units, primary selection of fruits and vegetables is always carried out in the field where products with severe defects, injuries or diseases are removed. Field preparation of lettuce is an example where a team of three workers cut, prepare and pack. For distant markets, boxes prepared in the field are delivered to packinghouses for palletizing, pre- cooling, and sometimes cold storage before shipping. Mobile packing sheds provide an alternative for handling large volumes in limited time. On completion of loading, the consignment is shipped to the destination market. In mechanized harvesting, the product is transported to the packinghouse where it is prepared for the market. In many cases, harvest crews make use of an inspection line for primary selection in the field. Preparation and packing operations should be designed to minimize the time between harvest and delivery of the packaged product. Delays frequently occur in the reception area, therefore the produce should be protected from the sun as much as possible. Produce is normally weighed or counted before entering the plant and, in some cases, samples for quality analysis are taken. Preparation for the fresh market starts with dumping onto packinghouse feeding lines. Dumping may be dry or in water, but it is important to have drop decelerators to minimize injury as well as to control the flow of produce. Water dipping produces less bruising and can be used to move free-floating fruits however, not all products tolerate wetting. Water dipping through washing helps to remove most dirt from the field. Water rinsing allows produce to maintain cleanliness and be free of soil, pesticides, plant debris and rotting parts. However, in some cases this is not possible because of insufficient water. Chlorination of dumping and washing waters with a concentration of 50–200 ppm active chlorine eliminates fungi spores and bacteria on the surface of diseased fruits, which prevents the contamination of healthy fruit. Bruising should be avoided because it creates the entry for infection by decay organisms. At depths greater than 30 cm and for periods of immersion longer than 3 minute, water tends to penetrate inside fruits, particularly those that are hollow, for example, peppers. It is recommended that the temperature of fruit is at least 5° C lower than that of water. Post harvest management is about maintaining quality from production in the enclosure to the fruit and vegetables being placed on a plate for consumption. Post harvest losses mean that production resources such as land, water, energy, fertilizers, labour and effort are wasted, and ultimately, profitability for growers is reduced.

Quality Deterioration of Fresh Produce after Harvesting

Once detached from the mother plant, the plant organ is deprived of its source of water, nutrients and antisenescent hormones. As a consequence normal factors such as transpiration and respiration lead ultimately to water loss and senescence of the product. Being living organs, fruits and vegetables continue to respire even after harvesting when they have a limited source of food reserves. In addition to degradation of respiratory substrates, a number of changes in taste, colour, flavour, texture and appearance take place in the harvested commodities which make them unacceptable for consumption, if these are not handled properly. Post harvest technology of fresh fruits and vegetables combines the biological and environmental factors in the process of value addition of a commodity. Many factors such as respiration, ethylene, senescence and breaking of dormancy can lead to loss of quality in fresh produce, hence the common description of these products as perishable. Some of these factors are part of the life cycle of living produce, that is, over-ripening of fruits or sprouting in root and bulb crops. The growth of pathogens or physical damage will cause direct loss of product quality through their visual impact but both also stimulate senescence. Furthermore, the storage environment will play a highly significant role in determining the speed of all quality changes.

Respiration of produce: It has been normally observed that the greater the respiration rates of a product, the shorter the shelf-life. Immature products such as peas and beans tend to have much higher respiration rates and short shelf-lives caused by natural senescence whereas the opposite is true for mature storage organs such as potatoes and onions. Respiration is the metabolic process by which cells convert energy from one type of chemical structure into another form, for driving metabolic reactions. Under normal circumstances, fresh produce undergoes aerobic respiration, during which oxygen and glucose is consumed while carbon dioxide, water and heat are produced. In non-storage tissues, for example in leafy crops such as lettuce or spinach or immature flower crops such as broccoli, there is little by way of energy reserves and hence excessive respiration will eventually lead to metabolic collapse. Cell membranes will break down and allow the contents to leak out. Saprophytic bacteria may grow in these tissues and give rise to off-odours. Visible symptoms of tissue collapse and yellowing caused by senescence breakdown of chlorophyll in the chloroplasts may appear. Without adequate cooling, respiratory heat will further stimulate respiration leading to even more rapid deterioration. Certain types of fruits (known as climacteric) can be harvested unripe and ripened artificially at a later stage (e.g. avocados, bananas, mangoes, tomatoes). Without careful temperature control, the fruit will rapidly over-ripen and senesce leading to internal tissue breakdown and the production of volatiles compounds. Failure to control respiratory heat also can increase water loss from the produce. Furthermore, the increased warmth and moisture levels, which can develop in storage, are highly conducive for the development of bacterial and fungal infections.

Ethylene production: Ethylene is a plant hormone that plays a key role in the ripening and senescence of fruits and vegetables. All plant cells produce low levels of ethylene however, anything that causes stress to the plant tissues will stimulate ethylene synthesis. Stressors may include excessive water loss, physical damage or pathogenic attack. Climacteric fruits produce high levels of ethylene during initiation of ripening and the hormone is believed to stimulate and coordinate the physiological and biochemical changes which occur during ripening. Exposure to exogenous ethylene can lead to an acceleration of maturation and senescence, for example, green vegetables lose their chlorophyll more rapidly, thickened fibres can develop in asparagus, premature ripening can occur in unripe fruits and cabbages and cauliflowers can lose their leaves through accelerated leaf abscission.

Senescence of tissues: Senescence is the natural ageing of the plant tissues and is stimulated by the presence of ethylene and anything else that speeds up respiration rates. Senescence ultimately affects all aspects of quality, ending in the death of the product. Although in fresh produce, texture is highly dependent on cell turgor, the integrity of the cell wall is important to the texture of some processed products. In some fruits and vegetables (e.g. apples and tomatoes), the breakdown of intercellular adhesion between cells leads to a condition known as mealiness which is generally perceived as a loss in textural quality. In potatoes, so-called senescence sweetening is where, over time, storage starch is gradually converted to sugars.

Concentrations of reducing sugars of greater than 0.1 per cent in potato tissues being processed into chips and crisps can lead to browning or blackening of the product during the cooking process.

Breaking dormancy of produce: Root, tuber and bulb crops have a natural dormancy period that can be considerably extended under suitable storage conditions. Storage and shelf-life is often limited by the breaking of dormancy. Most commonly this is seen as the growth of sprouts, for example, in onions and potatoes. Under high moisture conditions, the development of roots may also occur. Neither sprouts nor roots are acceptable in marketed produce. Although roots and shoots can be trimmed off during processing, the internal quality of the produce generally deteriorates during the breaking of dormancy owing, for example, to the conversion of stored starch into sugars that are transported to the growing points.

Water loss from fresh produce: Plant tissues are covered with protective tissues, which serve to protect the plant from insect and pathogen attack, physical injury and excessive water loss. The primary protective layer is the epidermis but if the plant organ undergoes secondary growth, a multi-layered periderm may develop, for example, on apples and potatoes. The epidermis is covered with a waxy cuticle of cutin while the cell walls of periderm tissues generally become impregnated with suberin. Both cutin and suberin can reduce water losses from plant surfaces, however some water loss is inevitable. Water vapour can permeate the cuticle and is also lost through lenticels, which are gaps in the periderm which form to enable gas exchange for respiration. If the epidermis or periderm is damaged, water loss can be massively exacerbated. Mature plant organs such as stems, roots and some fruits develop strengthening tissues such as collenchyma or lignified schlerenchyma to maintain their structure. The presence of tough fibrous components is not desirable in fresh produce, so many vegetable crops are harvested immature. Structure and thus textural properties of fresh produce are almost entirely dependent on the maintenance of adequate cell turgor pressure, that is, the force generated when the solute filled vacuole presses against the relatively inelastic cell wall. If too much water is lost from the tissues, turgor pressure will fall, leading to wilting or shrivelling of the product. The speed of post-harvest water loss is dependent primarily on the external vapour pressure deficit; however, other factors will influence the situation. Products with a large surface to volume ratio such as leaf crops will lose a greater percentage of their water far quicker than large spherical fruits. The specific structure of the cuticle and the extent of suberisation in the periderm appear to be more important than thickness in improving resistance to the movement of water vapour. Produce varies in the percentage of water which can be lost before quality is markedly reduced. Fruits with thick peels can lose a considerable amount of moisture from the skin without compromising edible quality, for example citrus species, bananas. The appearance of the fruit will, however, deteriorate steadily with increasing water loss. Other thin-skinned fruits are more susceptible to water loss, for example, table grapes.

Fungal and bacterial pathogens infestation: The most important microorganisms causing post-harvest wastage of fresh produce are fungi. This is particularly true for fruits, where the relatively acid conditions tend to suppress bacterial growth. Vegetables with a higher pH can, however, suffer high losses from bacterial infections. The majority of pathogens rely on damaged tissues to obtain entry into fresh produce (wounds or sites of physiological injury). For example, the Penicillium species which cause blue and green mould infections of citrus and other fruit crops are classic wound pathogens, incapable of invading an undamaged fruit. A whole, fresh commodity is resistant to the majority of potential pathogens. The physical barrier of the skin and the presence of antimicrobial compounds in the skin and flesh are sufficient protection. Some pathogens can gain entry through natural openings such as stomata and lenticels. Bacteria may use this penetration route. The most common group of bacteria causing significant reductions in shelf-life is the soft rotting species of the genus Erwinia. Under suitable conditions of warmness and the presence of free water, the bacteria can readily colonise produce such as potatoes through the lenticels. They produce large quantities of extracellular enzymes which rapidly crush the tissues. Sometimes, soft rots are accompanied by the growth of saprophytic bacteria which give rise to highly unpleasant off-odours. Only a small number of fungal pathogens are capable of direct penetration of the undamaged skin of the produce. On the whole, these latter pathogens are particularly problematic owing to the fact that they may infect produce before harvest but remain quiescent in the tissues until conditions become favourable for growth. This phenomenon is largely seen in fruits, where initial pathogen development and subsequent quiescence occurs in the unripe fruit. As the fruit ripens, quiescence is broken and the pathogen colonises the fruit tissues. *Colletotrichum gloeosporioides* is a common pathogen showing this behavior on a number of tropical fruits such as mango and papaya. Typical symptoms on ripe fruits are sunken, lens-shaped lesions, which may develop salmon-coloured sporing structures. *Colletotrichum musae* causes similar symptoms on bananas. *Botrytis cinerea* may also show quiescent behaviour on certain fruits, for example, in strawberries, fungal spores contaminate the flowers, germinate and the hyphae grow into the developing fruit where they remain symptomless until the fruit is fully ripe. The subsequent disease development can be extremely rapid and the whole fruit is completely colonised and covered with a grey, sporulating mycelium within a few days at 20° C. Skin diseases may remain superficial but cause large market losses owing to the blemished appearance of the produce. The potato industry has a major problem with a number of skin diseases, such as black scurf (*Rhizoctonia solani*), black dot (*Colletotrichum* *coccodes*), silver scurf (*Helminthosporium solani*) and common scab (*Streptomyces scabies*) which can spread rapidly on the tubers after the temperature rises in retail outlets.

Most of the pathogens of fruits and vegetables are variable with respect to their ability to grow and reproduce between 6° C to 35° C. Some will survive and even grow slowly at temperatures as low as 1° C, for example, *Botrytis cinerea*. So, it is particularly important on produce grown in cool temperate climates, whereas infections caused by *Botryodiplodia theobromae* and *Aspergillus niger* tend to cause serious losses in warm regions for example, in citrus fruits infected with Alternaria rot in a consignment can result in off-flavoured juice. The presence of certain cell wall degrading enzymes from infecting pathogens, for example Rhizopus spp., can cause continuing softening of canned products.

Physiological disorders of fruits and vegetables: Physiological disorders are unfavorable quality changes that happen in fresh produce because of metabolic disturbances. These disorders can be caused by internal factors such as mineral imbalances or may be due to non-optimal environmental factors such as unsuitable storage temperatures or atmosphere composition. The symptoms may be unique to a particular condition on a specific produce type however, in many cases the symptoms are similar in a range of conditions with differing fundamental causes. Gentle symptoms are often confined to superficial tissues which may not be too significant if the produce is to be processed, but can strongly decrease marketability of the fresh product owing to visual disfigurement. The onset of disorders may be determined by pre-harvest conditions, the cultivar, maturity or stage of ripeness. One of the most important nutrients in this respect is calcium which plays an important role in maintaining cell wall stability. A typical example is 'bitter pit' in apples in which hard, sunken brown pits develop both on the skin and internally. Excessive exposure to the sun or inappropriate post-harvest heat treatments, may cause skin damage and uneven fruit ripening. During low temperature ice crystals form inside the cells leading to membrane rupture, and the tissue collapses upon defrosting leading to unacceptable textural changes. These changes are less obvious to the consumer in produce with a relatively low water content and/or which will be cooked before consumption, for example, peas, sweet corn, parsnips, potatoes, carrots, broccoli and spin. Symptoms include water soaking, surface pitting, internal discoloration, failure to ripen, accelerated senescence and increased susceptibility to decay. Symptoms may not become obvious until the produce temperature has been raised to non-chilling levels. Irish potatoes are susceptible to reversible low temperature sweetening at 2° C store. If produce is stored in an atmosphere with insufficient oxygen or excessive carbon dioxide, for example in poorly ventilated stores, respiratory disorders like potatoes may develop a black centre whereas lettuces may have pale midribs can develop. Some apple cultivars suffer external injury and others develop internal browning owing to excessive carbon dioxide in the tissues. Very low oxygen levels can lead to alcoholic fermentation with accompanying off-odours. Tolerance levels are variable, for example, some apple cultivars tolerate levels less than 1 per cent O_2, whereas sweet potatoes are highly sensitive and fermentation may set in if O_2 levels fall below 8 per cent. Anaerobic conditions will also encourage the growth of soft-rotting bacteria in potatoes. A range of specific symptoms in stored fruits and vegetables have been attributed to exposure to ethylene. Some examples include russet spotting of lettuce (at concentrations >0.1 ppm) which is associated with increased activity of phenylalanine ammonium lyase (PAL) and phenolic content, formation of the toxin pisatin in peas, and production of phenolics in sweet potatoes and in carrots. In carrots and beetroot, the phenolic isocoumarin gives a bitter flavour.

Physical injury of fruits and vegetables: Physical injury is possibly the significant cause of loss in fresh produce. This is not due to the direct losses, although these can be significant in some crops but rather to the indirect effect of creating a wound in the surface of the produce. This wound is an ideal entry point for many post-harvest pathogens, though cause losses in fruits and vegetables. Injury also allows water loss which compromises the quality of the produce. Furthermore, physical injury stimulates ethylene production in plant tissues, which can lead to premature yellowing or ripening of commodities. Physical injury can arise at any stage of the life of the crop, from insect injuries in the field to poor post-harvest handling. Many fungi invade through the stem end where the produce was severed from the mother plant. Poor packaging can create problems from cuts caused by sharp edges or hard parts of adjacent produce, for example pineapple crowns, to grazes caused by lack of padding or under filling of cartons allowing movement of produce within the pack during transport and handling. Bruising can occur from dropping or compression bruising can occur if produce is stacked too high or packs are overfilled. Significant levels of wastage occur in the potato industry owing to internal bruising of potato tubers during storage and handling. The shelf-life of many fresh products is considerably reduced by physical damage caused by rough handling at the retail level, particularly where the produce is loose and can be picked over by the potential customer.

Table 41. Causes for losses of fruit and vegetables and how to prevent them.

Issue	Where it occurs	Result	Prevention
Misshapen, split product	Production in field	Low pack out and potential rejections from market	Good soil, irrigation and nutrition management. Variety selection to suit growing conditions and careful harvesting.
Wrong harvest maturity	Harvest	Poor flavour and storage life. Mixed qualities in the same packaging and associated quality problems.	Maturity assessments before harvest and no mixing of different maturities in the same packaging.
Water loss after harvest	Throughout the supply chain	Shrinkage, loss of weight, increased susceptibility to diseases.	Maintain adequate humidity during storage. Avoid stresses due to handling, damage, o incorrect storage temperatures.
Mechanical damage	Harvesting, grading, re-packing, stacking of boxes, shipping, retail handling	Careless handling of vegetables and containers causing splitting, internal bruising, superficial grazing, and crushing of soft produce leading to entry points for diseases, increased water loss and increased respiration.	Adequate harvest method, no over or under trimming. Grading line design with minimum drop heights, no sharp corners or points. Adhering to recommended pallet and container stacking patterns.
Increased respiration	Throughout the supply chain	Increased use of stored energy resources (starch, sugars) leading to ageing and death when reserves are exhausted. Stress due to mechanical injury, diseases or temperature fluctuations can increase respiration rates and cause faster use of stored energy and increased water loss.	Adhere to temperature requirements for each vegetable type throughout the supply chain. Take field heat out as soon as possible after harvest
Wrong temperatures, temperature fluctuations	Throughout the supply chain	Cold damage (browning, mushiness). Freezing at temperatures between 0 and -2° C. Rots following low temperature injuries.	Check and maintain cooling equipment for stores, shipping containers and trucks. Do not store vegetables that need higher temperatures e.g. beans, cucumbers, tomatoes or eggplants with those requiring low temperatures.

Low oxygen, High carbon dioxide in storage atmosphere	In store rooms, containers and packaging	Poorly ventilated, controlled atmosphere/ modified atmosphere storage reduces oxygen supply and can lead to accumulation of damaging carbon dioxide, 'suffocating' vegetables. This leads to stress that increases respiration and water loss. Stressed vegetables may release more ethylene and carbon dioxide into the storage atmosphere, which promotes ageing.	Good atmosphere control in controlled atmosphere rooms if used. Good ventilation in normal storage (including leaving gaps between pallets and walls for sufficient ventilation/head space in packaging). Correct use of modified atmosphere packaging and maintenance of the cool chain.
Diseases	Throughout the supply chain	Rotting and contamination of surrounding produce	Do not use dirty, contaminated field boxes or insufficiently sanitised washing water. Do not pack damaged, over mature or diseased produce in the same packaging as healthy produce.
Ethylene (a ripening agent) produced by fruits or from fuel exhausts	Packing sheds, stores, transport containers, depots	Pre-mature ageing (senescence), loss of colour and firmness, vegetables susceptible to decay.	Do not store or transport fruit emitting ethylene with vegetables that are sensitive to it. Do not use gas or diesel forklifts and vehicles in or near packing and storage facilities.

Postharvest Handling of Fresh Fruits and Vegetables

The most important goals of postharvest handling are keeping the product cool, to avoid moisture loss and slow down undesirable chemical changes, and avoiding physical damage such as bruising, to delay spoilage. Sanitation is also an important factor, to reduce the possibility of pathogens that could be carried by fresh produce, for example, as residue from contaminated washing water. After the field, postharvest processing is usually continued in a packing house. Produce handling plays an important role at different stages of supply chain likes at the time of harvest, at the field, time of loading and unloading, transportation, whole sale market, retail market and at customer end. Handling methods of fruits and vegetables directly affects freshness and quality. For most produce, maintaining cool temperatures (to slow deterioration) and high humidity (to prevent moisture loss) are the most effective means of preserving quality. However, there are several things producers, handlers, and retailers can do to assure that fruit and vegetables going to the market or into storage are of high quality.

Harvesting and handling

It has been observed that improper handling of fresh fruit and vegetables is a major cause of deterioration and post harvest losses. Handling at each stage plays an important role in protecting the quality and enhancing the shelf life of produce. Provide gentle harvesting and handling to avoid cuts, abrasions, and bruising damage that allow decay-causing micro organisms to enter the tissue. Harvest produce at the peak of quality, at that point produce will have the greatest shelf life. If possible, harvest during the cool part of the day (usually just after sunrise) will extend their quality. If storage facilities are not available, harvest only as much produce at one time as can pack or sell before the quality deteriorates. The harvested produce put in shaded area during perform sorting and packing operations. Particularly dark colour vegetables, such as zucchini squash, eggplants, peppers, watermelons, green beans, and tomatoes, which are often harvested during sun will absorb solar energy and become warmer. At farm markets, display only good-quality vegetables for sale. Frequent sorting to remove poor quality material will present the best display possible to customers.

Handling at the time of harvest: The throwing of produce during hand harvesting or handling should not be allowed. When crops are harvested at little distance from the packinghouse, the produce must be transported promptly for packing. Avoid the use of dirty containers, contaminated with soil, crop residues and the remains of decayed produce. Growers should make certain that harvesting labours are fully conversant with the quality control strategy employed on the farm. Mechanical damage during harvesting and subsequent handling operations can result in defects on the produce and expose to disease-causing microorganisms. The inclusion of dirt from the field further aggravates the process. Every effort should be made to harvest produce at its optimum maturity, as storage life is reduced proportionate to the immaturity and/or over maturity of a vegetable crop.

Handling at the field: As all fruit and vegetables are tender and have soft texture/skin should be handled gently to minimize bruising and breaking/rupturing of the skin. After harvesting, produce is handled at the field for three main activities before dispatch to the market for sale.

Washing, cleaning and dressing/trimming: After harvest all underground vegetables and most of the leafy vegetables require washing and cleaning before sorting-grading and packing. While washing and cleaning care should be taken that produce does not get damaged while rubbing to clean the outer surface and only clean water should be used for washing to protect the produce from contamination. Washing with 100 ppm chlorine solution is better to control microbial growth. Removal of extra water is a must before packing to avoid rotting. Vegetables like cauliflower, cabbage, radish and other leafy items require to be dressed by removing unwanted leaves and stalk before sending them for marketing.

Sorting/grading: All defective produce such as bruised, cut, decayed and insect infested pieces should be discarded while sorting-grading. This will help to control further deterioration of the produce while in transit. Care should be taken that produce is picked gently and should not be thrown.

Weighment and packing: Packing material and package itself play a protective role against mechanical damage, dust and infection. They also diminish the rate of water loss, or hinder gaseous exchange and thus modify the composition of the atmosphere around the produce. Pack should not have inside sharp edges. Proper cushion in the pack helps the produce to sustain jerks/vibrations during transportation. As far as possible, uniformly graded produce should be packed in one type of pack. After packing, each pack has to be weighed before sending to the market for sale. Each pack should have some extra quantity to take care about the moisture loss during transit.

Handling during loading and unloading: During entire supply chain loading and unloading mainly takes place at field, pack house, wholesale market and at retail market. During loading and unloading of produce, care should be taken (packs should not be thrown) that all the packs should be gently placed on the transport vehicle. Hooks should be avoided for picking the bags and crates. Torn bags and broken crates should not be used. Different grade packages should be kept separately.

Handling during transportation: Post-harvest handling is the ultimate stage in the process of producing quality fresh fruits/vegetables/flowers for market or storage. Getting these unique packages of water (fresh produce) to the point of retail or safely into store with the minimum of damage and exposure to disease risk must be a priority for all growers. Much damage is done to fresh produce during transport and growers should take all necessary remedial measures to ensure that produce leaving the field/firm for markets /storage arrives in the same condition as it left the field or the firm. The vehicle should be well cleaned before loading. Walls, floors, doors, and ceilings should be in good condition. No sharp object should be there inside the vehicle. For refrigerated transport temperature, humidity and air circulation should be checked before loading. Containers, bulk bins or sacks should be loaded onto transport carefully and in such a way as to avoid shifting or collapse of the load during transportation. Bulk loads or open top containers traveling long distances should be covered with hessian/shade net/plastic to prevent excessive dehydration. Transport loaded with vegetables should not halted for long periods, as this causes excessive heat buildup and will accelerate the onset of breakdown, cause condensation and make produce more vulnerable to diseases. In wet weather however loads of vegetables destined for storage should be covered to protect the produce from getting wet.

Handling of produce at market: It should be make sured that produce is unloaded at the shop with proper care. After unloading only required quantity of produce should be displayed in the shop for sale. Display of large quantity produce not only reduce the shelf life of produce, it gives customer a opportunity to pick and choose from a large lot, which results in loss of freshness/luster and damage of the produce by customers touching and mis-handling. Normally the harvested produce stays at retail market for longer duration. Therefore special care is required to store and protect the produce from mishandling by the customers. Produce which is not kept on display should be stored only in polythene bags or in wet gunny sacks to maintain its freshness. Periodic water spray on leafy items helps in maintain the quality for longer period or centralized air conditioned shop. Customers should not be allowed to break/put pressure, squeeze or damage the produce during sale.

Handling at customer end: When customer purchases the fruit, vegetables and flowers, lot of damage is already taken place, sometimes this damage is not visible at the time of purchase, but develop within few hours of purchase. Visibility of any damage to the produce itself is an indication that produce should be consumed as early as possible to avoid further deterioration. At low temperature green leafy vegetables wrap in wet cloth and store. Beans, brinjal, cauliflower, cabbage, radish, carrot, chilli, capsicum and rooty vegetables keep in polythene bag and store. Before keeping in bag extra moisture should be removed. Apple, guava, onion and garlic should be avoided along with other vegetables due to their typical flavour/aroma. Pick all the fruits and vegetables very gently with thumb and middle finger only. Do not pick banana from the body and leafy vegetables by the leaf end. Never press citrus fruits, because it damage oil cells present on skin and turn brown after some time and ripe fruits like sapota, banana and mango etc. to check the ripening. Postharvest handling is the final stage in the process of producing high quality fresh produce. Being able to maintain a level of freshness from the field to the dinner table presents many challenges.

Packing house operations

The packing house can be a simple shed, providing shade and running water, or a large-scale, sophisticated, mechanized facility, with conveyor belts, automated sorting and packing stations, walk-in coolers and the like. There are different operations in packing house to minimize the post harvest losses. If there is proper handling of crop one can minimize the mechanical damage by avoiding drops, rough handling and bruising during the different steps of pack house operations. The operation of packing should be carried out in shaded area which can be created using locally available materials like, shade cloth, woven mats, plastic tarps or a canvas sheet hung from temporary poles. Shade alone can reduce air temperatures surrounding the produce by 8-17° C. Following steps should be taken in packing house:

» **Dumping:** After removal of fruit crop from the field bin or harvesting container, transfer to packinghouse. Dumping is the first step and should be done softly. For this purpose water assisted methods or dry dumping can be used. Wet dumping can decrease bruising and abrasions by using moving, chlorinated (100-150 ppm) water to carry delicate produce. When using dry dumping, padded, sloped ramps or moving conveyor belts can decrease injuries to produce.

» **Pre-sorting**: Pre-sorting of fruits is usually done to eliminate injured, decayed, misshapen or otherwise defective produce (culls) before cooling or additional handling. Pre-sorting saves energy because damaged fruits will not be cooled, packed and transported. The removal of decay produce will limit the spread of infection to other healthy fruits during handling, especially if postharvest pesticides are not being used.

» **Curing in tuber crops**: Curing is an effective operation to reduce the water loss during storage from hardy vegetables viz, onion, garlic, sweet potato, yam and other tropical root vegetables where cuticle are poorly developed. They are relatively susceptible to mechanical wounding during harvesting and handling. The curing of root and tuber crops develops periderms over cut, broken or skinned surfaces wound restoration. It helps in the healing of harvest injuries, reduces loss of water and prevents the infection by decay pathogens. Onions and garlic are cured to dry the necks and outer scales. For the curing of onion and garlic, the bulbs are left in the field after harvesting under shade for a few days until the green tops, outer skins and roots are fully dried.

Method of curing: It is one of the most effective and simple means of reducing water loss and decay during subsequent storage of root, tuber, and bulb crops. Curing is normally undertaken in the field, but in some case curing structure is employed. Produce can be cured in the field by piling them in a partially shaded area. Cut grass or straw can serve as insulating material, while, covering the pile with canvas, burlap, or woven grass matting. This covering will provide sufficient heat to reach high temperature and high relative humidity. The stack can be left in this state for up to four days. Curing in potato starts with deposition of suberin in parenchymatous cell just below the damaged area of the tuber. Suberin (a waxy waterproof substance found in the cell walls of many plants, especially corky in nature) is a group of fatty acids which provides initial protection to the tuber against water loss and infection. Subsequently, below the suberized cells a meristematic layer of cells is formed which is the periderm, also called as cork cambium. This produces new cells which seal off the damaged area. But these processes are temperature and humidity dependent eg. Curing of potato takes place in 1 day at 21° C; 2 days at 15° C; 3 days at 10° C; 5-8 days 5° C.

Table 42. Optimum condition for curing of vegetables

Crops	Temperature (°C)	Relative humidity (%)	Duration (days)
Cassava root	30-40	90-95	2-5
Onion and garlic bulbs	30-45	60-75	1-4
Potato tubers	15-20	85-90	5-10
Sweet potato	30-32	85-90	4-7
Yam tubers	32-40	90-100	1-4

Curing of citrus fruits: Reduce decay through lignifications. Increase in the antifungal chemicals in the fruit peel in orange (30° C), pomelo, lemons and grapefruit at 34° C and 90-100 per cent RH for 2-3 days, not later than 48 hours after harvest.

Drying: Drying is carried out to preserve the fruits and vegetables by reducing the water activities, below that which support the growth of microorganisms and action of enzymes. It also provides a surface barrier to water loss and microbial infection. Sometimes this process is referred to as curing, but since no cell regeneration or wound healing occurs it is clear to refer to it as drying.

Windrowing: Drying of onion and garlic has been carried out in the field called windrowing. This involves the pulling the bulb from the ground and laying them in ground in small heaps for 1-2 weeks, allowing the external layers of skin and neck tissues dry out prior to handling and storage. The dried tops of the plant can be arranged to cover and shade the bulb during the curing process for protecting the produce from excess heat and sunburn. If conditions such as rain do not permit field curing and facilities are not available then a temporary tent can be used for curing. Curing can be assisted by the use of ventilated sheds where relative humidity and solar radiation is high and natural air movement is low. When ground is wet they can cured in pack house for 7-10 days with condition of 30° C and 70 per cent RH. Drying is judged to be complete when necks of bulbs have dried out and tight and the skin rustles when held in the hand.

Washing and cleaning: Before fresh fruits and vegetables are marketed various amounts of cleaning are necessary which typically involves the removal of soil dust, adhering debris, insects and spray residues. Washing with chlorine solution (100–150 ppm) can also be used to control inoculums build up during pack house operations. For best results, the pH of wash solution should be between 6.5 and 7.5. Some fungicides like Difenconuzole (0.10 - 0.25%) or carbendazim (0.2 - 0.5%) for 5 minutes may be used as post harvest dip to control the disorders eg. apple superficial scald. For cleaning of some fruit type vegetables (melons, brinjals, tomatoes, cucumber) they should be wiped with damp cloth. Mangoes and bananas need washing to remove latex, whereas kiwifruits and avocadoes need dry brushing. Washing and cleaning (sanitation) operations are essential to control the spread of disease from one item to another and to limit spore buildup in wash water or in the packinghouse air. Walls, floors and packing equipment can also be cleaned using quaternary ammonium compounds labeled as safe for food processing equipment.

Dressing of horticultural produce: Removal, trimming and cutting of all undesirable leaves/ stem/ stalks/ roots/ other non edible or unmarketable parts is called dressing. Dressing makes fruits, vegetables and flowers attractive and marketable. Trimming is done especially in vegetables and flowers to remove unwanted, discoloured, rotting and insect damaged parts (e.g., cabbage, cauliflower, spinach, lettuce, rose, chyrysanthemum, gladiolus, tuberose etc.) or parts that may favour deterioration or damage during later handling. In case of grapes, trimming of bunches is done to remove the undersize, immature, dried, split and damaged berries. Trimming and removal of decaying parts are preferably done prior to washing. Trimming enhances visual quality, minimizes water loss and other deteriorative processes. Trimming reduces the likelihood of diseases or their spread, facilitates packaging and handling, and reduces damage for other produce.

Water spray: Produce starts loosing water as soon it is detached from the plant. Water spray helps in compensating that water loss and maintaining the quality for longer period. Produce can also be covered with gunny sack soaked in cold water, if it has to store for longer period before sale. Example: Green leafy vegetables.

Sizing/grading: Sorting or grading is, necessary to obtain better returns as per quality of the produce. Sorting not only helps to remove diseased fruits or vegetables and prevent decay loss but also helps to differentiate fruits or vegetables for different purposes. Off grade, injured and diseased produce should be eliminated as soon as possible. Size, shape, colour, weight, soundness, firmness, clearness or appearance, maturity, freedom from disease, insect pest damage and mechanical injury are the basis for grading. Sorting is done by hand to remove the fruits which are unsuitable to market or storage due to damage by insects, diseases or mechanical injuries. The remainder crop product is separated into two or more grades on the basis of the surface colour, shape or visible defects. Grading can be done manually or by automatic grading lines. Size grading can be done subjectively (visually) with the use of standard size gauges. Machine sizers work on weight and diameter of produce. Sizing on the basis of fruit shape and size are most effective for spherical (oranges, tomato, certain apple cultivars) and elongated (delicious apples and European pears or of non-uniform shape) commodities, respectively. Round produce units can be easily graded by using sizing rings. In an apple packing house in India 3 grades viz. Extra fancy, fancy and standard may be packed for marketing. The fourth "cull" grade is meant for processing.

In general there are three categories of grade.

i. **Class A** – Superior quality fruits or vegetables posses uniform colour, size and shape of the variety and without internal defects with 5 per cent tolerance of errors by number or weight.

ii. **Class B** – Medium size produce of same quality as in class A with 10 per cent tolerance of errors.

iii. **Class C** – Fruit and vegetable of some external and internal defects provided they are fit for consumption while fresh. This class is suitable for local or short distance market and suited to those for whom price is more important than quality.

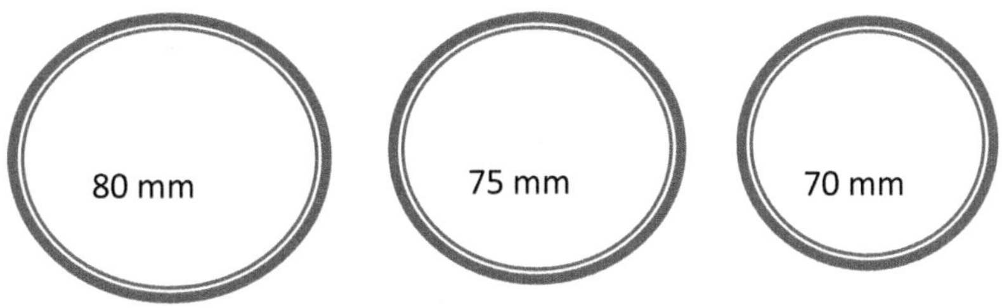

Several types of mechanical sizes are available for small scale operations. One type is composed of a long slanted tray with a series of opening which coverage (largest at the top, smallest at the bottom). This type of sizes works best with round commodities.

The machine used to grade the apples by size and separates them into six different groups (sizes).

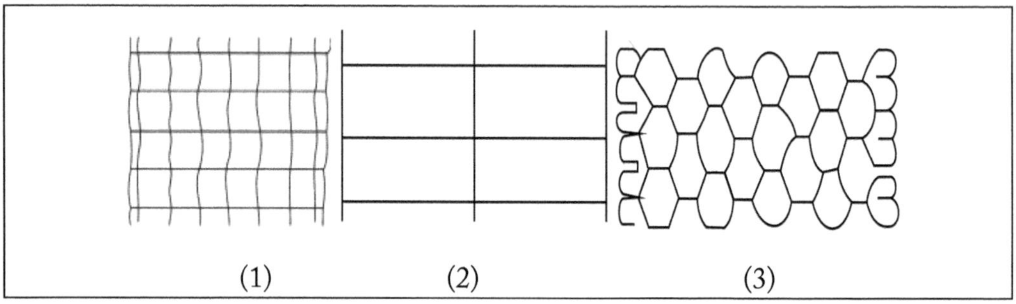

Figure 25. Various types of sizers: (1) Square type sizing chain, (2) Rectangular type sizing chain, (3) Hexagonal type sizing chain.

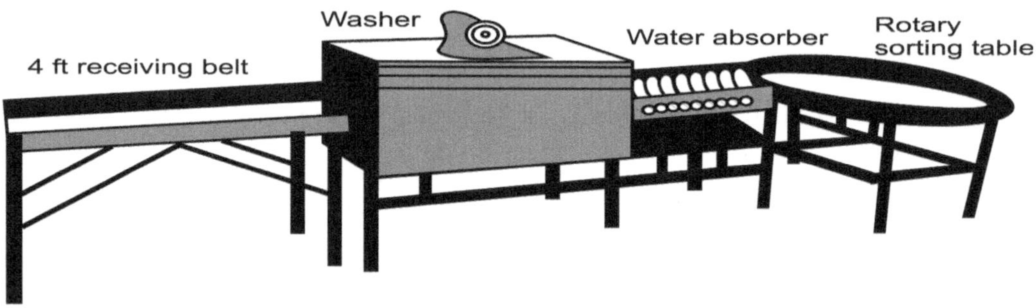

Figure 26. Fruit simple packing line. The Optional equipment includes eliminator, dry brusher, waxer and a variety of sizers, rolling inspection belts, conveyor belts and side tables. (Source: Catalog of TEW Manufacturing Corporation, 2001).

Onion: Diseased, thick necked, damaged, bolted and double bulbs should be removed. Medium sized bulbs of 4-6 cm size of big onions are good. Grades recommended for export quality are-

i. **Extra class** – Superior quality, characteristics of variety and bulb shall be firm and compact, unspotted, properly cleaned, free from swelling caused by abnormal development and free of root tufts or root tufts are allowed if harvested before maturity. Grade tolerance 5 per cent by number or weight.

ii. **Class I** – Good quality, characteristics of variety and bulb should be firm and compact, unspotted, free from swelling and free of root tufts or root tufts are allowed if harvested before complete maturity. The slight defects in shape, colouring and light staining (does not affect the last dried skin protecting the flesh) may be allowed with 10 per cent tolerance by number or weight of the onion.

iii. **Class II** – Do not quality for higher grades but satisfy the minimum requirement. Reasonably firm and defects in shape, colour early signs of shoot growth, traces of rubbing, small healed cracks, slight bruising unlikely to impair keeping qualities, root tufts, stains and cracks in the outer skin with 10 per cent tolerance by number or weight of the onion.

Garlic: Thick necked injured, split, diseased bulbs are sorted out. Size grading is done manually or by a grader. Grades for export qualities are:

i. **Extra class** – Superior quality bulb, characteristic of variety and commercial type. Bulbs shall be intact, regular in shape, cleaned, free of defects, cloves must be compact and roots of dry garlic must be cut off flush with the bulb. Five per cent tolerance by weight of the garlic.

ii. **Class I** – Good quality, characteristic of variety and commercial type, Bulb shall be intact, fairly regular shape, slight defects but not affect the appearance, slight tears in the outer skin of the bulb, cloves reasonably compact 10 per cent tolerance by weight.

iii. **Class II** – Do not quality for higher grades but satisfy the minimum requirements such as sound bulb, fresh in appearance clean, firm, free from defects of damage, sprouts and foreign smell and taste, the defects allowed area healed injuries, slight bruises and irregular shape. Ten per cent tolerance by weight of the garlic not satisfying the requirements of the grade but meeting the minimum requirements as mentioned above. According to size the minimum diameter of the bulb should be 45 mm for extra class and 30 mm for garlic in Class I & II.

Table 44. Weight grading standards of fruits for export purpose

Crop	A grade	B grade	C grade
Mango	200-350 g	351-550 g	551-800 g
Grape (Bunch)	300 (Extra class)	250 (Class I)	150 (Class II)
Pomegranate	350 g & above	250-349 g	<200 g
Figs	50 g & above	40-49 g	30-39 g
Papaya	200-700 g	700-1300 g	1300-1700 g
Guava	>450 g	351-450 g	251-350 g
Pineapple (with crown)	2750 g	2300 g	1900 g
Litchi (diameter)	33 mm	20 mm	-
Lime	Minimum wt: 75 g and minimum diameter: 4 cm		

Table 45. Quality or grading standards for vegetables crop

Vegetables	Specific requirement
Okra	Green, tender, 6-9 cm long
Chilies	Green, 6-7 cm long
Cluster bean	Green, tender, 7-10 cm long
Bitter gourd	Green, 20-25 cm long having short neck
Bottle gourd	Light green, straight, cylindrical, 25-30 cm long
Tomato	Round, medium size in middle east, cherry tomatoes in European countries
French bean	10-12 cm long, straight, round green pods in bush beans flat beans with 12-13 cm & straight are also demand in European markets
Big onion	4-6 cm, light to dark red, round, strong pungency for gulf & SEA markets Yellow/brown colour, 7-10 cm, round or spindle shape for European & Japan markets
Small onions	2-3 cm dark red and round
Garlic	White, round, 5 cm & above, bigger cloves of 10-12 cm & above with 10-15 in number. For Bangladesh and Sri Lanka 4-5 cm size bulbs also acceptable
Potato	White, oval, 4.5 to 6 cm. Bangladesh demands red type and Iran & Iraq demands potatoes with yellow flesh

Pre-cooling of horticulture produce

Pre-cooling refers to removal of field heat (quick cooling) after harvest; if not, its deterioration is faster at higher temperature of 1 hour at 32° C, 1 day at 10° C and 1 week 0° C. The entire products must be pre-cooled as early as possible to the recommended storage temperature and relative humidity. Pre-cooling is done just above chilling and freezing temperature. Cooling condition for a crop vary with the air temperature during harvesting, stage of maturity and nature of crop. Precooling reduces the loss of moisture, inhibits the growth of spoilage organisms and bruise damage from vibration during transit. It also decreases the rate of respiration, production of ethylene, and extends ripening and product shelf life.

Postharvest treatments

Fresh fruits are living tissues subject to continuous change after harvest. Some changes are desirable from consumer point of view but the majorities are not. Postharvest changes in fresh fruit cannot be stopped, but these can be slowed down within certain limits to enhance the shelf life of fruits. The post-harvest treatments play an important role in extending the storage and marketable life of horticultural perishables. The most important postharvest treatments include:

Washing with chlorine solution: Chlorine treatment (100–150 ppm available chlorine) can be used in wash water to help control inoculums build up during packing operations. Maintain pH of wash water between 6.5 and 7.5 for best results.

Ethylene inhibitors/Growth regulator/fungicide treatments: 1-MCP (1-methyl cyclopropene), AVG (Amenoethoxyvinyl gycine), silver nitrate, silver thiosulfate, cycloheximide, benzothiadiazole etc. are some of the chemicals which inhibit ethylene production and/or action during ripening and storage of fruits. The growth regulators or fungicidal application such as GA_3 can be effectively used to extend/enhance the shelf life of fruits.

Mineral application: Deficiencies of certain minerals result in physiological disorders, loss of storage life or quality. Calcium is the nutrient most commonly associated with post harvest disorders and can be overcome by external application of calcium chloride (4-6%) on fruits after harvest by spraying or dipping, vacuum infiltration and pressure infiltration. The post-harvest application of $CaCl_2$ or $Ca(NO_3)^2$ enhancing the storage and marketable life of fruits by maintaining their firmness and quality. Calcium application delays aging or ripening (tomatoes, avocados, pear and mango), reduces postharvest decay, controls the development of many physiological disorders (internal break down, bitter pit, storage scald etc.) and increases the calcium content, thus improving their nutritional value. Vacuum infiltration (250 mmHg) of calcium chloride at concentration of 1- 4 per cent is beneficial. In guava fruit dipped in 1 per cent solution of calcium nitrate reduces weight loss, respiration, disease occurrence and increased the shelf life for more than 6 days as compared to 3 days in the control at room temperature. The post-harvest application of $CaCl_2$ (2–4 %) or $Ca(NO_3)^2$ for 5–10 minute dip extend the storage life of pear up to 2 months, plum up to 4 weeks and apple up to 6 months at 0–2° C with excellent colour and quality. Calcium infiltration reduces chilling injury and increase disease resistance in stored fruit.

Thermal treatments: Thermal treatments included:

Hot water treatments: Dipping of fruits in hot water of specific temperature (51-55° C) for specified periods (5-30 minute) for the purpose of disease control, disinfestations of insect or uniform ripening. Hot water treatment is an approved quarantine treatment for export of many fruits and vegetables against pests. However, for insect disinfestations a longer treatment is necessary than for fungal control. It also adds other additional advantages such as removal of surface residues, removal of sap fallen on the fruit surface during harvesting and facilitates washing.

Table 46. Hot water treatments for different fruits

Commodity	Pathogens	Temperature (°C)	Time (min)
Apple	*Gloeosporium* sp.	45	10
	Penicillium expansum		
Blue berries	Fungi	46-55	3
Grapefruit	*Phytophthora citrophthora*	48	3
Lemon	*Penicillium digitatum*	52	5-10
	Phytophthora sp.		
Mango	*Collectotrichum gloeosporioides*	52	5
Orange	*Diplodia sp.*	53	5
	Phomopsis sp.		
	Phytophthora sp.		
Papaya	Fungi	48	20
Peach	*Monolinia fructicola*	52	2.5
	Rhizopus stolonifer		
Sweet potato	Fungi	40	2

Vapour heat treatment: Vapor heat treatment is a method of heating fruit with air saturated with water vapor (humidified by injection of steam) at temperatures of 40–50° C to kill insect eggs and larvae as a quarantine treatment before fresh market shipment. This treatment proved very effective in controlling infection of fruit flies in fruits after harvest and mandatory for export of mangoes. The treatment consists of a period of warming (approach time) which can be faster or slower depending on a commodity's sensitivity to high temperatures. Then there is a holding period when the interior temperature of the produce reaches the desired temperature for the length of time required to kill the insect. The last part is the cooling down period which can be air cooling (slow) or hydrocooling (fast). A recommended treatment of citrus, mango, papaya and pineapples at 43° C in saturated air for 8 hours and then holding the temperature for a further 6 hours. For control of papaya fruit fly, fruit should be exposed to 43° C and 40 per cent RH for 11 hours, followed by 43° C and 100 per cent for 8 hours. The temperature and exposure time are adjusted to kill all stages of insects (egg, larva, pupa and adult), but fruit should not be damaged.

Hot water and vapour heat treatment may causes both internal and external damage to produce if not properly done -

» **External damage:** Includes peel browning, pitting, or yellowing of green vegetables. Tissue damage caused by heat will also result in increased decay development.

» **Internal damage:** Causes poor pulp colour development, abnormal softening, the lack of starch breakdown and the development of internal cavities as in case of mango and papaya. In addition, the fruit can soften quickly or show abnormal softening where some areas of the flesh remain hard while others soften.

Fumigation: The fumigation of SO_2 is successfully used for controlling post- harvest diseases of grapes. This is accomplished by placing the boxes of fruit in a gas tight room and introducing the gas from a cylinder to the proper concentration. However, special sodium metabisulphite pads are also available which can be packed into individual boxes of a fruit to give a slow release of SO_2. The primary function of treatment is to control the *Botrytis cinerea*. The SO_2 fumigation is also used to prevent discolouration of skin of litchis.

Astringency of fruits and vegetables: Astringent is a Latin word meaning 'to bind fast'. Astringency tastes unpleasant to many which tend to avoid eating astringent fruits. Astringency is the dry, puckering (to gather something around the lips) mouth feel caused by tannins found in numerous fruits such as red grapes, blackthorn, quince, persimmon fruits, beal and immature banana. Varieties of fruits high in phenolics are more astringent than varieties with low in phenolics. Red grapes (var. Cabernet Sauvignon) have high astringency than white grape. The tannins denature the salivary proteins, causing a rough sand papery feeling in the mouth. During maturation, the condensation of phenolics continuously increases and at the same time the astringency decreases, perhaps because highly condensed flavans are less soluble and tightly bound to other cell components. In banana, tannins polymerise as the fruit ripen and lose their astringency. Treatment with high levels of CO_2 and spraying with alcohol (ethanol) will help in reducing the astringency. Storage of persimmon in 4 per cent CO_2 at -1° C for about 2 weeks before removal from the storage followed by 6-18 hours in 90 per cent CO_2 at 17° C removes the astringency. Spraying persimmon fruit with 35 – 40 per cent ethanol @ 150-200 ml per 15 kg of fruit is known to remove the astringency 10 days later in the ambient storage. This treatment takes much longer time than CO_2 treatment, but fruits quality is better.

Irradiation: Radiation can be applied to fresh fruits and vegetables to control micro organism/ insects/parasites and inhibit or prevent cell reproduction and some chemical changes. It can be applied by exposing the crop to radioisotope in the form of gamma-rays but X-rays can also be used from the machine which produces a high energy electron beam. Radiation helps in breaking the chemical bonds in the produce or micro organism. Ionizing radiation involves damage to DNA, the basic genetic information for life. Microorganisms can no longer reproduce and continue their harmfull or pathogenic activities. Irradiation reduces the spoilage, slowing down the rate of metabolism, delay ripening, senescence and extend shelf life of fresh produce. It also controls sprouting in potato, onion, garlic and yams (0.05-0.3 kGy). Most insects are not survived, sterilized at doses of 0.1 -1.0kGy and survived adults progeny are sterile. Egg phase is most sensitive followed by larval, pupal and adult stages. Irradiation being used in Australia to produce sterile male Queensland fruits flies and in Hawaii it is being used in papaya for papaya fruit fly.

Radiation doses are measured in Grays (Gy) and one Gray equal 100 radons. One Grays dose of radiation is the same to one joule of energy absorbed per kg of food material. In radiation processing of foods, the doses are generally measured in kGy (1,000 Gy). Cobalt 60 (Co^{60}) is commonly used as a source of gamma-rays in food irradiation. In general, most vegetables can withstand irradiation dosages up to a maximum of 2.25 kGy, higher doses can, however, interfere with the organoleptic properties of food products. The highest absorbed dose delivered to a food should not exceed 10 kGy. Combining irradiation + temperature control + gaseous environment + adequate processing conditions is one of the most effective approaches to fruit and vegetable preservation. Radioisotopes cannot be switched on or off so they are immersed in a pool of water to permit operators to enter the processing area. When food is to be irradiated the radioisotopes is raised out of the water and material to be irradiated is usually passed through radiation field on the conveyer belts. The whole processing area is surrounded by thick concrete to prevent the radiation out.

Table 47. Desirable effects and dose of irradiation on fresh produce

Commodities	Dosage (kGy)	Effect
Apple	1.5	Control scald and brown core
Apricot and peach	2.0	Inhibit brown rot
Asparagus	0.05 - 0.1	Inhibit growth
Avocado, banana, mango and papaya	0.025 - 0.75	Delay ripening
Avocado, banana, mango and papaya	1-3	Inhibit mould
Mushroom	2.0	Inhibit stem growth and cap opening

Papaya	0.25	Disinfect fruit fly
Potato, onion, yams	0.05 - 0.3	Controlling sprouting
Strawberries, mushrooms, onion etc.	2 - 3	Reducing the decays
Shredded carrots	2.0	Inhibits the growth of aerobic and lactic acid bacteria
Grapes	0.25-0.50	Inhibits grey mould
Tomato	3.0	Inhibits Alternaria rot
Mango, dried fruits and raisins	0.25 - 0.75	Insect disinfestations
Garlic, Ginger	0.03 - 0.15	Sprout inhibition

Waxing: Waxes are esters of higher fatty acid with monohydric alcohols and hydrocarbons and some free fatty acids. Waxing of the produce is an old age art, which was started in the beginning of the 19th century. Food grade waxes are used to replace some of the natural waxes removed during harvesting and sorting operations. It also helps in sealing tiny injuries and scratches on surface of fruits and vegetables. Waxing improves cosmetic appearance, freshness of produces after harvesting and prolongs the storage life of fruits and vegetables by reducing respiration rate. Fruit coating results in the restriction of the gas exchange between the fruit and surrounding atmosphere and builds up of CO_2 and a depletion of O_2 within the fruit, thus causing an effect similar to controlled atmosphere storage. Reduced physiological losses in weight (by 30 to 50 per cent) and retards wilting and shriveling during transportation and storage. Produce prevents from the chilling injury, browning and protect from bruising. Protects fruits from micro-biological infection and increase in the shelf life. Mango fruits treated with wax emulsion containing 8 to 12 per cent solids have one or two weeks longer. It is considered as cost effective substitute in the reduction of spoilage when refrigerated storage is unaffordable. However, it should be remembered that waxing does not improve the quality of any inferior horticulture product. The principal disadvantage of wax coating is the development of off- flavour if not applied properly. Adverse flavour changes have been attributed to inhibition of O_2 and CO_2 exchange thus, resulting in anaerobic respiration and elevated ethanol and acetaldehyde contents. Over waxing also results in abnormal ripening and softening that affects the marketing of such fruits. The fruits which are waxed are apple, avocado, guava, mango, lemon, grape, banana, melon, orange, lime, passion fruit, papaya, peache, pineapple and vegetables are cucumber, brinjal, tomato, potato, pumpkin, carrot, snake gourd, coccinia, capsicum and summer squash.

Skin coating (Protective coating): It is defined as artificial application of a very thin film of wax or oil or other material to the surface of the fruits or vegetables as an addition to or replacement for the natural wax coating. Fruit coatings can be formulated from different materials including lipid, resins, polysaccharides, proteins, and synthetic polymers. Most coatings are a composite of more than one film with the addition of low molecular weight molecules such as polyols that serve as plastisizers (increase the plasticity or fluidity of the material). Otherwise, coatings can be too brittle and will flake or crack on the coated product. Surfactants, antifoaming agents, and emulsifiers are also often used in coatings.

Types of Waxing

Natural waxing: Fruit attains desired stage of maturity on plant, nature provides them with thin coat of whitish substance, which is called bloom or natural waxing. Natural coat is clearly visible on fruits and disappears after harvest due to repeated handling of fruit. Exampl apple, pear, plum, mango and grapes.

Artificial waxing: To prolong the shelf life of produce some of the fruit and vegetables are dipped in a wax emulsion and then dried for few minutes. This process provides thin layer (<1 μ) of artificial wax on skin of the produce by which the small pores present on the skin are fully covered and reduce the transpiration and respiration process resulting in increased shelf life. Artificial waxes like solvent waxes, water waxes and paste or oil waxes are used. Shellac, carnauba wax, bee wax, polyethylene (12% used in Israel for mango), wood resins, paraffin wax are generally used for waxing. As of now, only coating with beeswax and carnauba is allowed in India. Grade name of some extensively used waxes are Citrashine@ from DECCO (India UPL), waxol (12 – oil/ water-emulsion wax containing 12% solids), tal (prolong), semper fresh, frutox (emulsion of different waxes with 12 % solids). Quantity of wax emulsion at 12 per cent concentration required for one metric tonnes orange is 3.6 liter, mosambi 5.4 liter, mango (300-350gm) 3.6 liter, patoto 7.9 liter, apple 5.4 liter, guava 5.7 liter, tomato 5.0 liter, banana 3.0 liter and lime 9.0 liter.

Table 48. Use of wax concentration on the fresh commodity produce

Concentration of wax (%)	Commodity
12	Carrot, brinjal, snake gourd, potato, cucumber, coccinia, capsicum, ribbed gourd, pine apple, guava and papaya
09	Tomato, lime, orange
08	Apple
06	Mango and musk melon

Specifications of a desirable wax

Wax is kneadable at 20° C, easily emulsifiable, resistant to chalking and provide a lasting shine. The wax should not impart undesirable odour (off-flavour) non-sticky or tacky, with low viscosity, translucent to opaque, and should have efficient drying performance. The used wax should never interfere with the quality of fresh fruit and must be manufactured from food grade materials. The commercial available waxes are citrashine, staryfresh, sta-fresh 451, semper fresh, carnauba wax etc.

Table 49. Types of wax used for fruit and vegetable waxing

Animal wax	Bees wax, spermaceti wax, shellac wax, chinese insect wax
Vegetable wax	Carnauba wax, candelilla wax, sugarcane wax, palm wax, esparto wax, japan wax, oricury wax, waxol-12 (is the vegetable wax manufactured by CFTRI, Mysore)
Mineral and synthetic wax	Ozocerite, montan wax, synthetic wax

Colouring waxes - Dyes are sometimes added to waxes for greater consumer appeal, it is being used on red variety of Irish potatoes, sweet potatoes, and other vegetables. They enhance the colour to give the same shade or tint as when the roots were freshly dug. In citrus fruits, dye has been approved for general use. Citrus Red No.2 is 1-2 (2, 5-dimethoxy phenylazo) 2-napththol with an established tolerance of 2 ppm.

Methods of Wax Application

Amount of wax applied and uniformity of application are extremely important. Fruits should be damp dry prior to wax application to prevent dilution. Waxes should never be diluted with water. Coatings may be applied by either by dipping, brushing or spraying on the fruits and vegetables.

Spray waxing: This method is used most commonly. Fruits and vegetables which move on the roller conveyor are sprayed with water-wax emulsion. The waxed produce is dried in a current of air at 55° C. There are two types of spray waxing namely low pressure spraying and high pressure atomizing. The wax is dissolved in a suitable solvent on basis of the pressure employed, volume of wax used, wax temperature, distance of fruit from the spray and number of spray nozzles.

Dipping: Here fruits are dipped in water wax emulsion of required concentration for 30 to

60 seconds. The fruits or vegetables could be waxed by keeping them in wire boxes holding about 100 fruits (30 kg) and dipping in 30 litre capacity tank containing wax emulsion. The fruits are then removed and allowed to dry under electric fan or in the open air or with warm air at 54 to 55° C. The produce should be turned periodically while drying.

Foam waxing: Foaming is a suitable means of application because it leaves a very thin coating of wax on the fruit after the water has evaporated. A foam generator is mounted over a suitable brush head, and water is applied to the fruit or vegetable in the form of foam. Spraying tends to waste wax, but it can be recovered in catch pans.

Flooding: Flooding is similar to dipping and is a safe and convenient method of application.

Liquid paraffin wax method: In this method fruits and vegetables are dipped in hot paraffin. The main disadvantage of this method is too much of coating material is used.

Slab wax method: In this case the wax is pressed against rapidly revolving brushes, but the efficiency is very less.

Lipids and resins: Lipids and resin are added to coating formulations to reduce gas exchange, but mainly to impart hydrophobicity (to reduce water loss) and gloss appearance. Fruits coated with resins will develop whitening of the skin due to condensation that develops when they are brought from cold storage to ambient conditions.

Edible films and coatings for fruits and vegetables

Application of edible films and coatings on fresh fruits and vegetables seems new, but its basic technology is an age old practice. Wax coatings have been applied to reduce dehydration of citrus fruits in China since 12th centuries. In the 19th century, sucrose was initially applied as an edible coating on dry fruits and nuts to prevent oxidative rancidity during storage. The more important application of edible films and coatings extensively started particularly since the 1930s with the use of an emulsion made of waxes spread on fruits to improve their appearance (shininess, colour, softening), onset of mealiness, carriage of fungicides, and to better control their ripening and to delay the dehydration. Edible coatings and films are generally produced from renewable natural and biodegradable polymeric materials such as polysaccharides, proteins, lipids, or the combination of these components (composite and conglomerate coating/film). Though in present days, scientists are interestingly creating a domain for novel application of nanotechnology in various fields of food processing keeping in mind about their probable risks and safety issues, but till now very little research has been done in the area of edible nano-packaging for fruits and vegetables. Observing and analyzing each and every dimension of developments of edible films and coatings author expects that improved edible films and coatings with various active materials including natural extracts, their nanomaterials and safer synthetic nanomaterials will dominate among the existing technologies for preservation and primary packaging of fruits and vegetables.

Edible packagings can also be fabricated and utilized as intelligent packagings because they can be both active, selective and also be incorporated with smart features for infinite potential use. A new generation of edible coatings for climacteric and perishable fruits and vegetables is under development, with the aim of allowing the incorporation and/or controlled release of functional compounds using micro-level and nanotechnological solutions such as nanoencapsulation and multilayered coatings, thus enhancing their viability and stability. Alginate is the most widely used biopolymer for encapsulation, but materials from different other biological sources can also be used. Enzymes, antioxidants, antimicrobials, colorants, flavours, probiotic, prebiotic, essential oils (omega-3-fatty acids) are among the most applied functional substances in encapsulations. Now days, nanotechnologies are being used to enhance the nutritional aspects of food by means of nanosized additives and nutrients and nanolevel delivery systems for functional compounds. The concept of nanocomposites further brought a stimulating route for creating new and innovative materials, also with natural polymers. Coating foods with nanolaminates involves entirely either food-grade ingredients (proteins, polysaccharides, lipids) or could also include various functional materials such as anti- microbials, antioxidants, antibrowning agents, enzymes, colorants and flavourings. Further, the electrodeposition technique has also demonstrated potential that could be used to coat highly hydrophilic food including fruits and vegetables surfaces such as fresh-cut fruits and vegetables.

Edible films and coatings

Any type of material used for enrobing (i.e., wrapping or coating) various food to extend shelf life of the product and/or value addition that may be eaten together with food with or without further removal is considered as edible film or coating. Both edible film and coatings are similar in composition but basic difference among them is edible coating remain adhered with the surface of the product, whereas edible film do not adhere with the product rather it fulfills the utility of wrapping or packaging purposes. An ideal coating is defined as one that can extend storage life of fresh fruits or vegetables without causing anaerobiosis and reduces decay without affecting the quality of the coated fruits or vegetables. When edible coating or film contains any functional or active material (antioxidant, anti- microbial or nutrients etc.) then it is termed as functional edible coating or film, which seems to have tremendous scope for value addition and storage. Edible coatings in foods is driven due to the high demand of consumers for longer shelf–life and better quality of fresh foods as well as of environmentally and health friendly packaging. Indeed, edible coatings can be used as a vehicle for incorporating natural or chemical antimicrobial agents, antioxidants, enzymes or functional ingredients such as probiotics, minerals and vitamins. Edible coatings provide replacement and/or fortification of natural layers to prevent moisture losses, while selectively allowing for controlled exchange of important gases, such as oxygen, carbon dioxide, and ethylene, which are involved in normal respiration processes and ripening of fruits and vegetables. A coating can also provide surface sterility and prevent loss of other important components. The main advantage of edible coatings over traditional synthetics is that they can be consumed with the products and are environmentally safe. The coatings are produced exclusively from renewable and edible ingredients and therefore are anticipated to degrade more readily than polymeric materials. Edible coatings can enhance the organoleptic properties of coated foods provided they contain various components (flavourings, colorings, sweeteners). The coatings/films can be produced with a variety of natural products such as polysaccharides, proteins and lipids, with the addition of plasticizers and surfactants. The functionality and performance of edible coatings mainly depend on their barrier, mechanical and optical properties, which in turn depend on coatings composition and manufacturing process. The choice of an edible coating mainly depends on the specific characteristics of the food product that requires protection and its storage conditions. The new generation of edible coating is being especially designed to increase their functionalities by incorporating natural or chemical antimicrobial agents, antioxidants, enzymes or functional ingredients such as probiotics, minerals and vitamins. Antimicrobial and antioxidant coatings have advantages over direct applications of the anti-microbial or antioxidant agents because they can be designed to slow down the diffusion of the active compounds from the surface of the coating. The sensory quality of coated products can be also improved if flavour and pigment compounds are added in the matrix. Incorporation of antimicrobial agents such as benzoic acid, sorbic acid, propionic acid, lactic acid, nisin, and lysozyme in coating have shown retarded surface growth of bacteria, yeasts, and molds on a wide range of products, including various fruits and vegetables. Various types of polysaccharide based (cellulose, chitosan, alginate, starch, pectin, and dextrin), protein-based (wheat gluten, collagen, corn zein, soy, casein, and whey protein), lipid-based (waxes, acylglycerols, and fatty acids) edible films, as well as their composites are already developed and at present these are a commercial reality. Wide ranges of functional agents are being successfully incorporated into such coatings during manufacture and several are yet to explore to enhance the safety and shelf life of fresh and ready-to-eat foods including fruits and vegetables.

Edible coating materials

Hydrocolloids (i.e., proteins and carbohydrates) and lipids are the base matrix materials that generally being used for the production of edible films and coatings. Combinations of different materials at various ratios in the form of composite films or coatings are the improved type with higher functionality to address the limitations of the individual one.

Hydrocolloid materials: Hydrocolloids are hydrophilic polymers that generally contain many hydroxyl (-OH) groups. Now-a-days they are widely used as the base material for film-forming solution to execute and control the texture, flavour, colour and shelf-life of fruits and vegetables. They are fully or partially soluble in water and are used principally to increase the viscosity with or without the external application of heat to the continuous phase of the film forming solution i.e., as gelling agent or thickener.

Polysaccharide based films and coatings: Polysaccharide films and coatings are principally made from starch, pectin, cellulose, chitosan and alginate etc. Due to their ability to form gel, thickening quality, viscosity and adhesiveness during the application of coating, as well as their efficacy to impart hardness, crispiness and compactness to the final product, they are extensively utilized in the practical field. Because of the makeup of the polymer chains, these coatings exhibit excellent gas permeability properties, resulting in desired modified atmospheres that enhance the shelf-life of the product, especially fresh products like – fruits and vegetables, without creating anaerobic conditions. Additionally, polysaccharide films and coatings can be used to extend the shelf-life of various fruits and vegetables by preventing dehydration, controlling oil and aroma transmission, oxidative rancidity and surface browning but their hydrophilic nature makes them poor barriers for water vapour. The primary advantages of polysaccharide films or coatings are their structural stability and ability to slow down oxygen transmission due to their tightly packed, ordered hydrogen bonded network and low solubility.

Protein based films and coatings

Corn zein, wheat gluten, milk proteins, whey proteins, soy proteins and gelatin are being extensively studied for their film and coating forming ability, their features and functional efficacy in terms of quality retention and shelf life of coated products. Protein based films and coatings exhibit better resistance to gas transmission and mechanical and structural properties than polysaccharides. As like polysaccharide films, protein based films also possess excellent oxygen, aroma, and oil barrier properties. But, it also shows poor water vapor barrier property due to their hydrophilic nature.

Lipid based films and coatings

Lipid substances are utilized as protective coating to improve the gloss and most importantly to prevent moisture transmission, due to its hydrophobicity. Moisture transmission rate decreases with the increase in hydrophobic phase. Lipids offer limited oxygen barrier properties due to the presence of microscopic pores and elevated solubility and diffusivity. Lipid films or coatings are usually opaque, lack structural rigidity, mechanical strength and relatively inflexible. Due to their low polarity, films from lipid become thicker and brittle and thus they must be combined with different film forming hydrocolloids or their derivatives to provide mechanical and structural strength. Several lipid materials including paraffin wax, bee wax, acetoglyceride, shellac resins etc are extensively used to provide edible surface coating to retain the quality attributes of various fruits and vegetables.

Composite films

To overcome the drawbacks of single component edible films and coatings, a concept of combining the unique features of different individual coating materials in desired combination has been developed in the name of composite coating. It is constructed either by making successive layers of different film forming material, or in the form of emulsion, suspension or dispersion in the immiscible layers at a time, or by making solution of different materials in the same solvent and subsequent film or coating application. Improvement of mechanical, structural and permeation properties of the film or coating is the main motto of producing composite films.

Properties of Edible Films and Coatings

Thickness: Thickness of edible films and coatings is an important parameter since it directly affects the biological properties and shelf life of the coated produce by interfering permeability, transparency etc. The effectiveness of edible films and coatings for protection of fruits and vegetables depends primarily on controlling the spreading of the coating solutions, which affect the thickness of the film and thus leading to its affect into porosity, transparency, gas and water vapor permeability etc.

Mechanical properties: Tensile strength, Young's modulus and elongation at break are the most important key parameters to study the mechanical properties of edible film, whereas for edible coating these properties are not directly determined. Still, these properties are indirectly associated with various other properties of edible coating that significantly affect the product quality. Researchers have determined these parameters from the stress–strain curves according to the norm of the American Society for Testing and Materials (ASTM). These properties of films are very important for their application as a packaging material. These properties indicate the strength, elasticity and protect-ability of the film.

Barrier properties: Protecting features of edible films or coatings against the permeation of gases, water vapor, oil, aroma compounds and light are very important key factors for maintaining the quality of the products during storage and extension of the shelf-life, which determine the efficacy of the film or coating. Chemical composition and structure of the film/coating-forming polymers, com position of film/coating solution, interactions between the sorbent and solvent, processing and drying parameters, the characteristics of the product, and the storage conditions are mainly responsible for the barrier properties of the film or coating.

Optical properties: Optical properties are characteristics of surfaces which are detected by human vision affecting some crucial aspects including esthetic properties of produce quality. Gloss, transparency and colour are the main optical properties of edible coatings and films. The internal and surface film microstructure together plays an important role in overall optical properties of the film and coated product. Indeed, the intensity of light reflected by the coated produce is determined both by the light directly reflected at the interface between air and the coated produce surface, which is responsible for the gloss of the product (specular reflection) and by both the light re-emitted out of the surface in all directions after penetrating into the coating of the produce and scattering internally (indirect reflection) depending upon the transparency of the coating. Specular reflection is related to the gloss of the coating or film. The gloss and indirectly specular reflection property of edible coatings as well as film is affected by coating microstructure in particular, by the type and level of surfactant, the distribution and the size, particle diameter of dispersed phase, relative humidity, storage time and the surface roughness. The transparency of films was measured through the surface reflectance spectra in a spectrocolorimeter, which is inversely proportional to the absorbance at a particular wavelength.

Thermodynamic parameters associated to sorption phenomenon

The interactions of between the sorbant and solvent can define the efficiency of the edible film and coating to control the gas permeability and therefore also controls the respiration and ripening of fresh fruits and vegetables, as well as the stability of the coated produce. The relationship between the total moisture content and the water activity, over a range of values, at constant temperature, yields the sorption isotherm when expressed graphically. When obtained at different temperatures, the energy associated to moisture interactions at low water vapor pressures can be estimated. Various mathematical models have been developed to describe sorption isotherms, which are also used to standardize the edible films and coatings to achieve better efficacy.

Advantages of edible films and coatings

Edible coatings and films are used to improve produce appearance and conservation due to their various advantages over usual synthetic packaging materials. These are produced from both animal and agricultural products, hence possess environment friendly, biodegradable and eatable nature. Edible packaging also reduces the wastage of packaging materials in compared to synthetic packaging. They act as barriers to moisture, oxygen and other gases during processing, handling, and storage, and not only retard produce deterioration but also enhance its safety due to either their natural biocidal activity or the incorporation of antimicrobial compounds, that also leads to the advancement of the concept of active packaging. Thus, the use of an edible coatings and films, such as chitosan and yam starch were found to be a viable alternative for controlling microbial growth in minimally processed carrots, strawberries, and raspberries, as well as in table grapes. Advantages of edible films and coatings also include, improved retention of moisture, firmness, colour, acids, sugars, and flavour components, reduced weight loss, reduction of storage decay, improved consumer appeal and extended shelf-life. Edible films and coatings have also found to retard respiration rates, reduce ethylene production, reduce metabolism and oxidation rates, hinder solute movement, retard loss of chlorophyll and other colouring compounds thus retaining the natural colour of produce.

Packaging of Produce

Packaging is the act of putting the produce inside a container along with packing materials to prevent movement and to cushion the produce such as plastic or moulded pulp trays, inserts, cushioning pads and to protect it specifically by plastic films, waxed liners. Packaging must convince to contain products and facilitate handling and marketing by standardizing the number of units or weight inside the package. It protect product from injuries (impact, compression, abrasion and wounds) and adverse environmental conditions (extermily high/ low temperature and relative humidity) during transport, storage and marketing. Packaging provide information to buyers, such as variety, weight, number of units, selection or quality grade, nutritional value, producer's name, country and area of origin. A well-designed package needs to be adapted to the conditions or specific treatments required for the product. If hydro-cooling or ice-cooling is required, the package must tolerate wetting without losing strength. For a product with a high respiratory rate, the packaging should have sufficiently large openings to allow good gas exchange. When produce dehydrates easily, the packaging should be designed to provide a good barrier against water loss. Semi-permeable materials make it possible to generate special atmospheres inside packages and helps in maintaining produce freshness. There are three types of packaging, for instance consumer units or prepackaging, transport packaging and unit load packaging or pallets. Fresh fruits and vegetables are generally packed in bamboo baskets, plastic crates, plastic bags, or nylon sacks for transportation, in many developing countries. Often, they are transported in an unpackaged form. After harvest, fresh fruits and vegetables are generally transported from the farm to either a packing house or distribution center. At the retail level, fresh produce is sold in an unpackaged form, or is tied in bundles. This type of market handling of fresh produce greatly reduces its shelf life if it is not sold quickly.

Previous functions of packaging

Fruits and vegetables are subjected to preliminary treatments designed to improve appearance and maintain quality. These preparatory treatments include cleaning, disinfection, waxing, and adding of colour.

Cleaning: Most produce receives various chemical treatments such as spraying of insecticides and pesticides in the field. Most of these chemicals are poisonous to humans, even in small concentrations. Therefore, all traces of chemicals must be removed from produce before packing. The fruit or vegetable passes over rotary brushes where it is rotated and transported to the washing machine and exposed to the cleaning process from all sides. From the washing machine, the fruit passes onto a set of rotary sponge rollers. The rotary sponges remove most of the water on the fruit as it is rotated and transported through the sponger.

Disinfection: After washing fruits and vegetables, disinfectant agents are added to the soaking tank to avoid propagation of diseases among consecutive batches of produce. In a soaking tank, a typical solution for citrus fruit includes a mixture of various chemicals at specific concentration, pH, and temperature, as well as detergents and water softeners. Sodiumortho- phenyl-phenate (SOPP) is an effective citrus disinfectant, but requires precise control of conditions in the tank. Concentrations must be kept between 0.05 and 0.15 per cent, 11.8 pH and temperature in the range of 43-48° C with soaking time is 3-5 minutes. Low concentrations of chlorine solution (100-200 ppm) are also used as disinfectant for many vegetables. The advantage of this solution is that it does not leave a chemical residue on the product.

Storage of produce

A number of storage techniques loke ground storage, ambient storage, refrigerated storage, air cooled storage, zero energy storage, modified atmospheric storage, hypobaric storage and controlled atmosphere storage are being used for fruits and vegetables depending upon the nature of the commodity and the storage period intended.

Post-harvest treatments to minimize produce contamination

Reduce microbial contamination: Avoidance of microbial contamination of fresh produce is favoured over dependence on corrective actions once contamination has occurred produce get losses. In order to minimize microbial food safety hazards in fresh produce, growers, packers, or shippers should use good agricultural and management practices in those areas over which they have control. The use of animal manure or municipal biosolid wastes as manure should be closely managed in order to minimize the possiblity for microbial contamination of fresh produce. Worker hygiene and sanitation practices during production, harvest, sorting, packing, and transport play a critical role in minimizing the potential for microbial contamination of fresh produce. Clean water containing an appropriate concentration of sanitizers is required in order to minimize the potential transmission of pathogens from water to produce, from healthy to infected produce within a single lot, and from one lot of produce to another, over time. Natural plant surface contours, natural openings, harvest and trimming wounds and abrasion can provide points of entry as well as safe harbor for microbes. When located in these protected sites, microbes are largely unaffected by common or permitted doses of post-harvest water sanitizing treatments. It is therefore essential that the sanitizer concentration is sufficient to kill microbes before they attach or become internalized in produce. The concentration of sanitizer is important in some pre-harvest water uses (such as spraying pesticides or growth regulators) and in all post-harvest procedures involving water, including washing, cooling, water-mediated transport (flumes), and post-harvest drenching with calcium chloride or other chemicals.

Table 50. Sanitizing chemicals used in produce handling

Sanitizing chemicals	Advantages	Disadvantages
Chlorine compounds Calcium hypochlorite Sodium hypochlorite Chlorine gas Chlorine dioxide	Low cost	Corrosive, irritating, trihalomethanes are a byproduct
Iodine compounds	Low cost, non irritating	Slightly corrosive, staining
Ozone	Faster action on microorganisms, fewer disinfection by-products than chlorine	Higher cost than chlorine
Peroxyacetic acid Hydrogen peroxide	More effective in removing and controlling microbial biofilms	Higher cost than chlorine

Minimize water loss: Transpiration of water from the plant tissues is one of the major causes of deterioration in fresh horticultural crops after harvest. Water loss through transpiration not only results in direct quantitative losses (loss of saleable weight), but also causes losses in appearance (wilting, shriveling), textural quality (softening, flaccidity, limpness, loss of crispness and juiciness), and nutritional quality. Transpiration can be controlled either through the direct application of post-harvest treatments to the produce (surface coatings and moisture barriers), packaging in polymeric films or through manipulation of the environment (maintenance of high relative humidity). Treatments that can be applied to minimize water loss in fruits and vegetables such as curing of certain root vegetables (garlic, onion, potato and sweet potato), waxing and the use of surface coatings on fruits (apple, citrus fruits, nectarine, peach, plum, pomegranate and tomato) and misting with water on leafy vegetables. Careful handling to avoid physical injuries, which increase water loss from produce.

Reduce ethylene damage: The promotion of senescence in harvested horticultural crops by ethylene (1 ppm or higher) results in acceleration of deterioration and reduced post-harvest life. Ethylene accelerates chlorophyll degradation and induces yellowing of green tissues (eg. yellowing of cucumbers either by 1 ppm ethylene over 2 days or to 5 ppm ethylene over half day at 10° C), thus reducing the quality of leafy, floral, and immature fruit-vegetables and foliage ornamentals. Ethylene induces abscission of leaves and flowers, softening of fruits and persuades several physiological disorders. Ethylene may increase decay development of some fruits by accelerating their senescence and as a result of inhibiting the formation of antifungal compounds in the host tissue. In some cases, ethylene may stimulate the growth of fungi, such as *Botrytis cineria* on strawberries and *Penicillium italicum* on oranges. Treatment of fruits, vegetables and ornamental crops with 1-methylcylopropene (1-MCP) by 1 ppm, which is an ethylene action inhibitor, provides protection against ethylene damage and significantly extend the shelf life.

Decay control: Post-harvest rots frequently arise as a result of rough handling during the marketing process and are caused by a wide range of microorganisms. The grey mold *Botrytis cineria* is a very important cause of loss in many commodities (such as grapes, kiwifruit, pomegranates, raspberries, and strawberries), and it is a destructive pathogen, even at low temperatures. Virus infection frequently lowers the quality of perishable commodities, usually as a result of visual deterioration, although viruses may also affect flavour and composition. Curing is a post-harvest treatment that facilitates certain anatomical and physiological changes that can prolong the storage life of some root crops. The addition of sanitizers to water dumps and spray or dip washes, reduces inoculums levels of decay-causing organisms from fruit surfaces, inactivates spores brought into solution from fruit or soil and prevents the secondary spread of inoculum in water. Fruits decay can control by heat treatments, such as dipping mangoes in water at a temperature of 50° C, for 5 minutes in order to reduce subsequent development of anthracnose, use of post-harvest fungicides, such as bavstin and/or difenconuzole on citrus fruits, use of biological control agents, such as Bio-Save (*Pseudomonas syringae*) and Aspire (*Candida oleophila*) alone or in combination with fungicides at lower concentrations on citrus fruits, use of growth regulators such as gibberellic acid or 2, 4-D to delay senescence of citrus fruits, use of 15-20 percent CO_2 in air or 5 per cent O_2 on strawberries, cane berries, figs, and pomegranates and use of SO_2 fumigation (100 ppm for one hour) on grapes.

Insect control: Fresh fruits, vegetables and flowers may anchorage a large number of insects during postharvest handling. Many of these insect species, in particular fruit flies of the family Tephritidae (e.g. Mediterranean fruit fly, Oriental fruit fly, Mexican fruit fly, Caribbean fruit fly), can seriously disturb trade among countries. Currently approved quarantine treatments, other than irradiation (ultraviolet radiation), include certification of insect-free areas, use of chemicals (e.g. methyl bromide, phosphine, hydrogen cyanide), cold treatments, heat treatments, and combinations of these treatments, such as methyl bromide fumigation in conjunction with cold treatment. The use of alternative treatments, such as fumigants (carbonyl sulphide, methyl iodide, sulphuryl fluoride) and insecticidal atmospheres (oxygen concentrations of less than 0.5 per cent and/or carbon dioxide concentrations ranging between 40 and 60 per cent and/or ozone at 600ppm) alone or in combination with heat treatments. Most insects become sterile when subjected to irradiation doses ranging between 50 and 750 Gy. The actual dosage required to produce sterility in insects varies in accordance with the species concerned and its stage of development. An irradiation dose of 250 Gy was approved by the Indian quarantine authorities for application to fresh commodities, such as litchi, mangoes, and papayas in light of the efficacy of that dose in preventing the reproduction of tropical fruit flies. Irradiation doses of 250 Gy can be tolerated by most fresh fruits and vegetables with minimal detrimental effects on quality. Doses ranging between 250 and 1000 Gy (the maximum irradiation dose allowed as of 2002), can, however, be damaging to some commodities. Fruits, in general, are more tolerant to the expected dose range (250 to 500 Gy absorbed by fruits on the inside vs. those on the outside of the pallet) than non-fruit vegetables and cut flowers.

Postharvest Management under Hot and Dry Condition

Environmental factors encountered during handling of produce that can affect final quality include extremely low or high temperatures, rain and extremely low relative humidity. Postharvest losses encountered in hot areas with a high incidence of rainfall are likely the highest in the world. Losses in developing countries along the tropical belt are estimated to be about 50 per cent. Fruit that were exposed on the day of harvest, or during the 3 days before harvest, to minimum temperatures higher than 22.4° C showed improved tolerance to quarantine heat treatments. This induced tolerance to postharvest heat treatments has been associated with heat shock proteins. High RH and temperatures around 30° C can enhance wound healing of citrus fruits. This is a major reason for the recommendation that degreening treatments occur in high RH and controlled temperatures. This beneficial effect is associated with a fast wound healing rate, triggered by the formation of lignin and the induction of antifungal compounds, such as scoparone and scopoletin. The proliferation and/or penetration in the fruit by pathogens, such as *Penicillium digitatum*, *Penicillium italicum* and *Geotrichum candidum* is reduced when curing is induced under high RH and temperatures around 30° C. High temperatures during postharvest storage are commonly associated with high transpiration rates and subsequent degradation of quality traits. Chlorophyll degradation significantly increases in leafy vegetables, such as valeriana lettuce, fresh cut rocket and spinach, when storage temperature increases by only 6° C. Diurnal changes in metabolism may explain the prolonged shelf life obtained with harvest during the late hours of the day. Xyloglucan endotransglusylase hydrolase, a cell wall modifying protein, is more active during the late hours of the day, suggesting a role in preventing cell degradation during postharvest. Turgidity of the produce also prolongs shelf life, which may be improved with postharvest calcium treatments. Rapid cooling at sunset inhibits the export of sugars from the leaves to the fruit, which results in increased osmotic and turgor pressure. Extreme rainy conditions can have direct implications on plant disease and food safety. Clinical pathogens that invade fruit and vegetables generally benefit from high RH encountered in the field and during postharvest storage. Hepatitis A virus, which is quite stable in the environment, with the longest survival having been reported at low RH. In contrast, high RH has been found to afford Salmonella longer survival on the surface of tomato plants it is probably wise to schedule harvest for periods when the rain stops and there has been enough time to dry water from the surface of the tissue to be harvested.

Low RH with high temperature is of particular concern, since fresh-harvested commodities tend to lose weight rapidly. Most fruits and vegetables become unmarketable with a loss of 5–10 per cent of their initial weight. Water activity, the differential between vapor pressure of water in produce and vapor pressure in pure water, affects the water loss rate of a fresh commodity. Higher water activity (or less negative water potential) implies higher amount of water that could be easily released to the environment in the form of water vapor pressure. Lettuce with extremely high water activity showed lower shelf life, particularly if accompanied by extremely low RH during postharvest handling. The loss of membrane integrity associated with desiccations, leads to loss of cellular compartmentation, which consequently allows polyphenol oxidase and polyphenol substrates to mix in the damaged cells, resulting in tissue browning. Colour retention and quality of beans was found to be higher when the product had lower moisture content. While initial moisture content of near 14 per cent yielded shelf life of lower than 10 days, initial moisture content of near 11 per cent produced shelf life of above 100 days. Herbs in particular suffer great weight loss when conditions at harvest include high temperatures. The rate of transpiration in herbs is commonly 100 per cent higher than that of vegetable species showing intensive respiration. However, for herbs that are dried, the rapid loss of water did not commonly affect the key active compounds. The quality of asparagus spears clearly varies depending on the weather conditions prevalent during the growth cycle, but postharvest conditions may be more critical in influencing final quality. Vegetable crops such as broccoli, beans, peas and asparagus harvested at an over-mature stage will be prone to excessive accumulation of fiber content, a factor that will become more pronounced if weather conditions include high temperatures. Extremely high temperatures at harvest impact hormonal levels in the marketed, edible portion of harticultural perishables. Abscisic acid seems to peak before the maximum light intensity or prior to the onset of the maximum daily temperatures. This could be explained by the high rate of photo-oxidation during late morning and afternoon hours, which exceeds that rate of synthesis of ABA. It has been shown that higher ABA levels at harvest result in a shorter vase-life of cut roses. In arid condition water stress is one of the most common stresses that plants encounter due to excess of transpiration during certain periods of plant growth. Changes in water availability during late phases of growth are so dramatic that a single stress event can trigger alterations to the anatomy of a leaf. If plants are under severe stress at the moment of harvest, due to extreme weather conditions or abuse through agricultural practices, the result on the postharvest quality is commonly negative. Potatoes and tomato grown under stress have shown undesirable texture and structure of fruit and tubers. However, in some cases in lettuce, mild water stress at harvest under desert conditions has resulted in clear benefits for the quality of intact and fresh-cut produce. When a mist was induced during the hours prior to harvest, the low temperature breakdown of kiwifruit occurring during postharvest storage was reduced or eliminated, depending on the maturity of the fruit at harvest. Increase of dry and fatty matter, polyphenol and antioxidant composition in cauliflower was obtained in a year when plants were subjected to stress conditions. Early harvested product was more suitable for storage due to the higher ration of unsaturated as compared to saturated fatty acids. Light intensity during postharvest storage is an issue that has been overlooked for a long time; however, it could have a significant effect on final quality, particularly when excess intensity reaches the product. Solanine content in potatoes exposed to sunlight was increased as high as eight times that found in potatoes stored in the dark. The concentration of solanine obtained with varieties grown in the desert and exposed to sunlight during harvest reached levels of 150 and 90 mg 100 g^{-1}, far above what has been found in other markets. Levels of solanine above 20 mg 100 g^{-1} of fresh tuber are considered toxic.

Effect of drastic changes occurring during postharvest handling

Adaptation of fruit and vegetables to extreme postharvest handling conditions may be affected by previous conditions occurring during preharvest or early stages of handling and storage. Cucumbers developed no symptoms of chilling injury during postharvest handling when preharvest temperatures in the greenhouse were 5° C higher, i.e. fruits grown at 27° C developed symptoms, cucumber grown at 32° C did not show any symptoms. Temperature sensitivity and ethylene production was reduced in the fruits from plants grown at the higher temperature. However, weight loss was significantly higher in cucumber grown at the higher temperature. When the cucumber, fruit stored at 36–40° C for 24 hours before refrigeration storage significantly reduced subsequent respiration and the appearance of pitting. Postharvest responses of fruit and vegetables to chilling stress are often greatly influenced by pre harvest field temperature. These factors may be associated with biological agents, physiological mechanisms, mechanical damage, cultural practices, genetic variation or environmental factors. Treatment with air temperature at 36–40° C reduced chilling injury of avocado fruit. Tolerance of postharvest heat treatment is affected by air temperatures experienced in the field 3 days before harvest. Retention of nutrients may also be improved with higher preharvest temperature. Fruit grown in the shady inner areas of the canopy have a greater incidence of chilling injury than fruit grown in the outer areas of the canopy. Extremely high temperatures in destination markets during summer months may require adjustments in earlier steps in the postharvest chain to reduce deterioration of quality. Produce in transoceanic shipments often suffers from abrupt changes in temperature when arriving at the destination port. Cassava root for export markets stored at 15° C kept for a longer time than that stored at 10° C when the product was shipped during the summer season to the end market. This was likely due to a higher transpiration rate and subsequent condensation of water between the skin and paraffin wax in the product stored at lower temperatures, due to a drastic change with the extremely high ambient temperature on the patios of the brokers. Oppositely during the winter season at low storage temperatures transportation extended the shelf life of cassava root. The activities of antioxidant enzymes, such as catalase and ascorbate peroxidase, decreased under high RH (85–90 %), whereas superoxide reductase increased and glutathione reductase decreased at low RH (55–60 %), an indication of their possible role in lowering rind staining in Navelina oranges. The shelf life of sweet basil increased by almost 100 per cent when harvest was done late in the day compared to harvest done in the morning, a reaction that could be due partially to changes in carbohydrate composition. Accumulation of carbohydrates increases during the day as a product of photosynthesis. Improved shelf life due to increased extensible cell walls has been correlated to harvesting product during the late hours of the day. Time of day of harvest has also been shown to affect the shelf life of leafy greens. Shelf life was extended by as much as 6 days when the harvest was done at the end of the day.

Table 51. General effects of storage on contents of health related constituents of fruits and vegetables

Constituents	Optimal temp.	Suboptimal temp.	Incident light	Reduced O_2	Elevated CO_2	Elevated O_2	Dehydration
Vitamin C	Decrease	Decrease	Decrease	Slower decrease	-	-	Decrease
Phenolics, fruits and vegetables	Stable	Decrease	Increase	Stable or increase	Stable or increase	-	Variable
Phenolics, berries	Increase	Increase	Variable	Stable or decrease	Stable or decrease	Increase	Increase
Carotenoids, fruits	Variable	Variable	-	-	-	-	-
Carotenoids, carrots	Stable	-	Stable	-	-	-	Decrease
Carotenoids, green vegetables	Variable	Variable	Variable	-	-	-	-
Glucosinolates	Stable or decrease	Decrease	-	Variable	Increase	-	Decrease
Dietary fibre	Stable	Variable	-		-	-	

Table 52. Extreme environmental conditions encountered during different postharvest steps and factors that influence the impact of those conditions on quality of the final product

Extreme environmental conditions	Potential negative impact	Factor that may enhance or reduce the impact
Extremely low relative humidity at harvest	Rapid water loss and early development of dehydration symptoms.	Water activity/potential at harvest affects the weight loss. Product with extremely high water activity shows symptoms of dehydration more quickly.
Extremely high relative humidity at harvest	Excess moisture on surface of product can result eventually in large loss of water and oxidation of tissue. It can also produce condensation of water vapor in packages, affording ideal conditions for proliferation of pathogens	Depending on the product, a time dedicated to release excess water may aid in prolonging shelf life.
Extremely low temperature at harvest / early day hours	Freezing damage. Early day hours produce higher incidence of leaf abscission of vegetative cuttings.	Materials used to cover crops or equipment to elevate the air temperature is currently used commercially.
Extremely high temperature /sunlight intensity at harvest/ late day hours	Accumulation of undesired compounds in certain crops such as potatoes. Reduced production of ethylene and subsequent loss of leaves during transition.	Use of transpirants and rapid movement of product from field to refrigeration conditions is used to ameliorate potential negative effects.
Extremely low temperature during postharvest handling	Chilling injury	Temperature during the growth of the plant can affect sensitivity. Plants grown in higher temperatures better tolerate the chilling temperatures. Postharvest heat treatments may aid in reducing sensitivity as well.
Extremely high temperature during postharvest handling	Occurrence of proliferation of plant pathogens. Microorganisms in general will be afforded better conditions to grow.	Risk is reduced if a good preharvest program to control diseases is in place, and relative humidity previous to harvest was not high.
Extremely high temperature at destination port	Abrupt raise in respiration and transpiration of the product. Condensation of water in product packages and spaces between wax and cuticles. Potential for microorganisms' growth.	Higher temperature and transpiration during transition may reduce problems by reducing condensation. Packages, films and coatings with sufficient permeability for gas diffusion is an option.

Adapted from Fonseca (2009)

Table 53. Classification of vegetable types by anatomical description, and maximum permissible water loss threshold for saleability of selected vegetables at commercial maturity

Vegetable type	Maximum percent water loss	Common name listing of examples
Root	4	Carrot (bunched), beet (bunched)
	7-8	Carrot (topped), beet (topped), parsnip
Bulb	10	Onion
Tuber	7	Potato (cured and immature)
Stem	8	Asparagus
Leafy	3-5	Lettuce, spinach, rhubarb
	7-10	Celery, cabbage, leek, brussels sprout
Floral	4	Broccoli
	7	Cauliflower
Immature fruit	5	Peas, cucumber, snap beans
	7	Peppers, sweet corn
Mature fruit	7	Tomato

Adapted from Toivonen (2011)

7

Precooling of Horticulture Produce

The rapid cooling of fresh produce (fruits, vegetables and flowers) from field heat (pulp temperature at the time of harvesting) to its best storage temperature is called precooling. The term, precooling, also refers to several practices whereby the temperature of the freshly harvested produce is quickly lowered to shorten the period of initial high respiration rate as well as to reduce the loads on the long-term cold storage facilities. Remove field heat as quickly as possible after harvest and prior to shipment, storage and display is essential for slowing down the rate of deterioration of highly perishable products. Deterioration is faster at higher temperature of 1 hour at 32° C = 1day at 10° C or 1 week 0° C. It is an important postharvest operation recommended in almost all flowers, fruits, and few vegetables. This is increases shelf-life, slow or inhibits moisture loss and the wilting of the produce. Removing field heat can suppress enzymatic degradation, undesirable metabolic changes, biochemical reactions and significantly slowed down respiration processes from newly harvested produce. Every fresh horticultural produce get energy through the convection from the surrounding air mainly from the sun in the form of radiation and from the metabolic and chemical reactions within the crop (respiration). The greater need for postharvest precooling in higher respiration rates of a fruit or vegetable. Precooling also reduces spoilage and slow or inhibit the growth of decay producing bacteria, fungi and yeasts. It also reduce production of ethylene, minimize aging due to ripening, softening, textural and colour changes of fruits and undesirable growth, such as sprouting of potatoes. Apricots, avocadoes, peaches, plums, guavas, mangoes, papayas and pineapples, etc. need immediate precooling, however, fruits like grapes, pears and citrus have longer post-harvest life, but they also need prompt cooling for maintenance of their high quality. Produce should be harvested in the early morning and late evening so that it will be at the coolest possible temperature during harvesting and initial cooling. Exceptions to this recommendation are produce, such as some citrus fruit, that are damaged if they are handled when they are turgid in the morning, or situations in which the produce is harvested in the late afternoon so that it can be transported to a local market during the cool night hours. Produce should be shaded to protect it from solar heat gain. Reduce the time between picking and initial cooling, this is particularly critical because fruits and vegetables transpire and respire at high rates at field temperatures. Precooling facility is usually erected within pack house premises and it is the responsibility of grower, shipper or pack house owner. Fruit and vegetables are precooled when its temperature is reduced from 3 to 6° C (5 to 10° F) and is cool enough for safe transport. Success of efficient precooling depends upon fast removal of field heat from all fruits, preferably within 2 to 3 hours.

Table 54. Important methods of precooling and storage conditions for fruits

Fruit	Temperature F° (C°)	% Relative humidity	Precooling Method	Storage Life Days	Ethylene sensitive
Apples	30 – 40 (-1.1-4.4)	90-95	R, F, H	90-240	Y
Apricots	32 – 34 (0-1.1)	90-95	R, H	7-14	Y
Avocados	40 – 55 (4.4-13.2)	85-90	R, N	14-28	Y
Bananas	56 – 58 (13-14.4)	90-95	R, H	7-28	Y
Beans, lima	37 – 41 (2.7-5.0)	95-98	R, H	7-10	-
Blackberries	31 – 32 (-0.5-0)	90-95	R, F	2-3	-
Blueberries	31 – 32 (-0.5-0)	90-95	R, F	10-18	-
Cantaloupe	36 – 41 (2.2-5.0)	95-98	H, F	10-14	Y
Sweet cherries	30 – 31 (-1.1 to -0.5)	90-95	H, F	14-21	-
Cranberries	36 – 40 (2.2-4.4)	90-95	R	60-120	-
Grapefruit	50 – 60 (10.0-15.0)	85-90	R, H	28-42	-
Grapes	32 – 34 (0.0-1.1)	85-90	F	56-180	-
Kiwifruit	32 – 34 (0.0-1.1)	95-100	R, N	28-84	Y
Lemons	50 – 55 (10-12.5)	85-90	R, H	30-180	-
Nectarines	31 – 32 (-1.1 to 0)	95-98	F, H	14-18	Y
Oranges	32 – 48 (0 – 9.0)	85-90	R, H	21-56	-
Peaches	31 – 32 (-1.1 to 0)	90-95	F, H	14-28	Y
Pears	32 – 34 (0-1.1)	90-95	F, R, H	60-90	Y
Pineapple	45 – 55 (7.2-12.5)	85-90	R, H	14-36	-
Plums	32 – 34 (0-1.1)	90-95	F, H	14-28	Y
Raspberries	32 – 34 (0-1.1)	90-95	R, F	2-3	Y
Strawberries	32 – 34 (0-1.1)	90-95	R, F	5-10	-

F= forced-air cooling, H= hydrocooling, I= package icing, R= room cooling, V= vacuum cooling, N = no precooling needed, Y= ethylene sensitive fruits.

Sources: USDA Agricultural Marketing Service, Kansas State University Cooperative Extension Service.

Table 55. Commodity wise cooling methods

Cooling methods	Commodities
Room cooling	All fruits and vegetables
Forced air cooling	Fruits and fruits type vegetables, tubers and cauliflowers
Hydro cooling	Stems, green leafy vegetables, fruits and fruit type vegetables
Package icing	Roots, stems, cauliflowers, green onion, brussel sprouts
Vacuum cooling	Stems, leafy and flowers type vegetables
Transits cooling -Top iceing and channel icing	All fruits and vegetables Roots, stems green leaf vegetables and cantaloupes

Mechanism of pre-cooling

Conduction and convection are the two main heat transfer mechanisms used for cooling of produce. With conduction, the heat is transferred within a produce to its coldest surface. This is direct movement of heat from one object to another by direct methods (from fresh produce to water or warmer to cooler). With convection, the heat is transferred away from the surface of the produce via a cooling medium such as moving water or air. Produce have bigger mass and lesser surface area like potatoes, apples, cauliflower, orange and other fruits and vegetables require more time to pre-cool than produce which is having smaller mass and large surface area like lettuce, green onion, carrot tops, peas, corn, brussel sprouts. This is because of the heat from the inside of the crop has to move to the surface before it is transferred in bigger produce. The large exposed surfaces (leafy vegetables) cool almost 5 times faster than large fruit such as melons (more volume, less surface).

Factors affecting the rate of pre cooling

- The rate of cooling depends on individual volume and the exposed surface of product.
- The velocity of heat transfer from the produce to the air or water (cooling medium) used *i.e.* cold water has a greater capacity to absorb heat than cold air.
- The difference in temperature between the commodity and the cooling medium, *i.e.* the greater the difference between the two the faster the product cools.
- The nature of the commodity which influences the rate of transfer of heat from the crop to the cooling medium, *i.e.* leafy vegetables have a greater thermal conductivity than root crops and so cool faster.

- Air temperature during harvesting (during summer pre-cooling time is more than winter)
- Use of water proof ventilated boxes for good air circulation in the room is helpful in fast cooling.

Cooling time calculations

Rate of cooling is directly related to the temperature difference between the cooling medium and the product. Initially, when the product is warm, temperature drops quite rapidly and later, the rate slows as product temperature drops. The product is considered "*half cool*" when its temperature drops to half the difference between its initial temperature and the cooling medium temperature. After another half-cooling period, the product is considered "*three-quarters*" cool. Product is usually finished cooling at "*seven-eighths*" or "*fifteen-sixteenths*" cool.

Methods of precooling

There are several methods of precooling viz, cold air (room cooling, forced air cooling), cold water (hydrocooling), direct contact with ice (top icing), evaporation of water from the produce (evaporative cooling, vacuum cooling) and combination of vacuum and hydrocooling. Some chemicals (nutrients/ growth regulators/ fungicides) can also be mixed with the water used in hydrocooling to prolong the shelf life by improving nutrient status of crop and preventing the spread of post harvest diseases. Forced air cooling is the most common and widely used method of precooling. A combination of cooled air and water in the form of a mist called hydroaircooling is an innovation in cooling of vegetables.

Cold air cooling

Room cooling

Room cooling is the most widely used system and is based on the products exposure to cold air inside a refrigerated room, in which the bulk or containerized commodity is placed for several days. In this method, produce is simply loaded into a cool room and cool air is allowed to circulate among the cartons, sacks, bins or bulk load. Room cooling is helpful in cooling fruits and they can be stored at the same room thus saves on handling costs. It is suitable for those crops which are marketed soon after harvest. The method has some drawbacks, like the cooling process is too slow, requires more space and unsuitable for packed produce. There is an excessive water loss from the produce due to slow cooling. It is used for potato, onion, garlic, apple and citrus fruits.

Room cooling requires days for packed product to reach desired temperature, but it can be faster for unpacked products with good exposure to the cold air. For example, bunched flowers in buckets can cool in 15 minutes, but the same flowers packed in boxes and loaded on a pallet take days to cool. For best results, containers should be stacked so that the moving cold air can contact all container surfaces. Total fan airflow should be at least 0.3 m^3 min^{-1} per ton of product storage capacity (100 cfm ton^{-1}) for adequate heat removal. Well vented containers with vent alignment between containers greatly speed up room cooling by allowing air movement through containers. After cooling is complete, airflow can be reduced to 20 to 40 per cent of that needed for initial cooling.

Cooling bays: Cooling bays mean a single large room is divided into bays by installing partitions partway into the room from each side. Air supply channels direct the air into the back of each bay. When a single bay is filled with warm product, supply ducts are opened to direct a large volume of cold air behind the product. Air return occurs down the center forklift aisle. When cooling is completed, the air supply is reduced in that one bay to create the desired storage condition. If each cooling bay has a separate cold air supply, cold product in one bay is not warmed by warm product in other bays.

Forced air cooling or pressure cooling

This is the fastest method of pre-cooling it is 4 to 10 times more rapid than room cooling. Air being forced to flow through a box of fruit, cooling can be accomplished in about 96 minutes that would ordinarily require 12 hours in the conventional cold storage room a time ratio of 1:8. Forced air cooling pulls or pushes air through the vents/ holes in storage containers directly. Cooling time depends on the airflow, the temperature difference between the produce and the cold air and produce diameter. Water losses range from virtually none to 1 to 2 per cent of initial weight. The speed of forced-air cooling is controlled by the volume of cold air passing over the product. Maximum feasible cooling requires about 0.001 to 0.002 m^3 sec^{-1} kg^{-1} of product (1 to 2 cfm/lb). Some products can withstand slower cooling and use air volumes of 0.00025 to 0.0005 m^3 sec^{-1} kg^{-1} of product (0.25 to 0.5 cfm/lb). Air is blown at a high velocity leading to desiccation of the crop, which minimize by the air is blown through cold water sprays. This type of cooling is perfect for small, delicate products such as grapes and strawberries that cannot be handled in bulk and that are packed in cardboard cartons before precooling. In India, this method has been most commonly used for precooling grapes for the export market. This system can be applied to all crops mainly berries, ripe tomatoes, bell peppers and many other fruits and vegetables, like anona, avocado, banana, brussels sprouts, capsicum, citrus, coconut, cucumber, eggplant, fig, ginger, grapefruit, guava, kiwifruit, lima bean, litchi, mango, melons, mushrooms, okra, orange, papaya, green peas, pineapple, pomegranate, pumpkin, sapota, snap beans, summer squash, yam and watermelon. For baby corn 5 – 6 hour cooling at 2 - 4° C, leafy vegetables 1 – 2 hour cooling at 6 - 8° C (too low temperature causes chilling injury).

Important aspects of forced air cooling

Air management: Fans used for cooling are supplied with the refrigeration equipment are used to cool the air by forcing it past the evaporator coils. These fans are not large enough, nor are they properly located in most cases, to force air directly past the produce. It must be mixed with the warmer air inside the room to prevent chilling injury. Therefore, additional fans are required to move the air past the produce.

Moisture loss in forced air cooling: Moisture loss in forced air cooling ranges from very little to amounts significant enough to damage produce. Factors that affect moisture loss include product initial temperature and transpiration co-efficient, humidity, exposure to airflow after cooling and whether waxes or moisture-resistant packaging is used.

Controlling relative humidity: Proper selection of the cooling coil or direct expansion evaporator plays a major role in maintaining a high relative humidity of 90 per cent to 95 per cent in the precooling process and thus minimizing moisture loss in the product.

Sweating: Cold produce exposed to a warm atmosphere usually becomes moist or even wet, which is referred to as sweating and is caused when the warm air loses moisture as it is cooled on contact with the produce.

Refrigeration: Refrigeration is the most widely used method for extending the postharvest life of fruits and vegetables. Low temperature slow product metabolism and the activity of microorganisms responsible for quality deterioration. As a result, reserves are maintained with a lower respiration rate, ripening is retarded and vapor pressure between products and ambient is minimized reducing water loss. Under current practice, it may take 12 to 36 hours for even the field heat to be removed from such table fruits as plums, apricots, peaches and grapes. However, orderly harvest and movement of the fruit to adequate refrigeration facilities will shorten the cooling time. Beyond the initial cooling, the shipper should supply adequate cold storage to insure the fruit being marketed in a fresh appearing condition. Natural movement of refrigerated air is quite slow. Even where fans are used in storage rooms to circulate the air, warm air in the boxes of fruit seems to defy replacement with cold air. Actually, the weight difference between the warm air in a box of fruit and the cold air outside is too small to create even a fair exchange rate.

Types of forced air cooling

The use of forced air cooling types depends on the requirement of the grower.

Tunnel-type forced air cooling: In tunnel type forced air cooling system there is a row of palletized containers or bins. They are placed on either side of an exhaust fan, leaving an aisle between the rows. The aisle and the open end are then covered to create an air plenum tunnel. The exhaust fan creates low air pressure within the tunnel. Cooled air from the room moves through the openings in or between containers toward the low pressure zone, sweeping heat away from the product. The exhaust fan circulates air over the refrigeration coils and returns it to the cold room. The exhaust fan can also be a portable unit that is placed to direct the warm exhaust air toward the air return of the cold room or refrigeration evaporators.

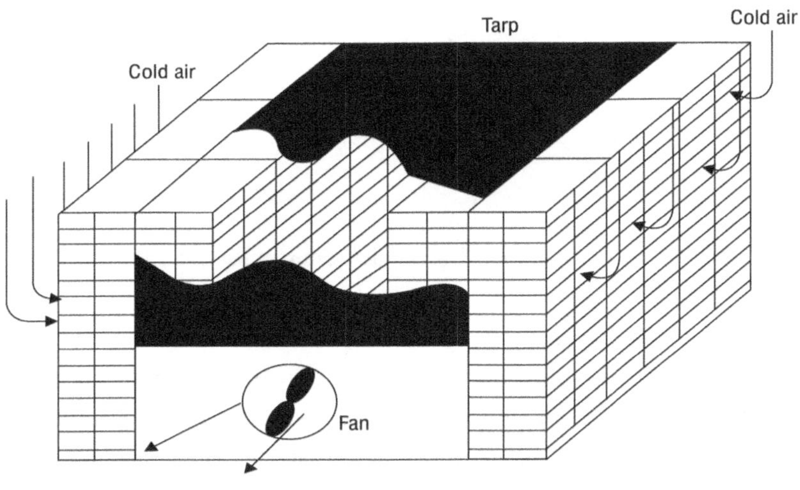

Figure 27. Schematic of a tunnel-type forced-air cooler (Thompson *et. al.*, 2002)

Cold wall cooling: It is permanent air plenum equipment with exhaust fans. The air plenum is situated at one end or side of a cold room, with the exhaust fans so that air can move over the refrigeration coils. The openings are located along the room side of the plenum, against which stacks or pallet loads of containers can be placed. This method is not often used for products in bins. Various damper designs can be used to ensure that airflow is blocked except when a pallet is in place. It is a fast and instant method of cooling.

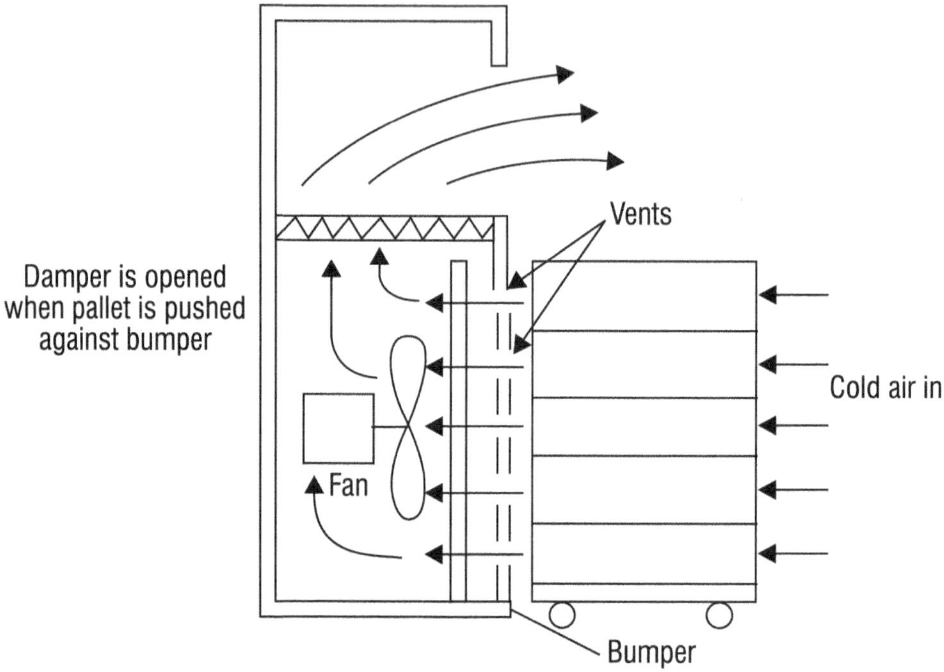

Figure 28. Cold wall forced-air cooler (Source: Rij *et al.*, 1979)

Different packages and even partial pallets can be accommodated by proper design of the damper system. This is a benefit in operations that handle a range of commodities. Each pallet must be independently monitored for temperature and promptly moved from the cooler as soon as it is cold in order to avoid desiccation from continued rapid airflow over the product. The canvas sheets must be well sealed, and pallet opening blocked for the tunnel style cooler to function properly.

Serpentine cooling: A serpentine cooling circuit is formed in a gas turbine blade to cool portions of the tip shroud, primarily the fillet between the airfoil and the tip shroud and the shroud edges. The serpentine system is used for forced-air cooling of produce in bins through bottom ventilation slots. Ventilation on the sidewalls of the bins does not aid cooling, although it probably does not hinder it either. The system requires modification of the cold-wall design to allow the forklift openings between bins to be used as air supply and return plenums. The cold air moves vertically through the product in each bin, causing a slight pressure difference between plenums.

Bins may be stacked up to six rows deep against the cold wall, depending on the cooling speed desired and the available airflow. The airflow capacity of the small forklift opening plenums usually limits airflow to less than 0.0005 $m^3sec^{-1}kg^{-1}$ of product (0.5 cfm lb^{-1}) and cooling is usually slower than tunnel coolers. To achieve the desired airflow pattern, openings are placed in the cold wall to match alternate forklift openings, starting one bin up from the floor. On the room side of the bins, these same openings are then blocked. Air flows into an open slot between bins and passes up or down through one bin of product to reach the return plenum. This system requires no space between rows of bins and bins can be stacked in even numbers as high as the cold store or forklifts can reach.

Forced air evaporative cooling: Cold air is forced to pass through produce by means of a pressure gradient across packages. Cooling is four to ten times more rapid than room cooling and its rate depends on airflow and the individual volume of produce. This system allows for more efficient utilization of space for bulk storage. Air conducts run under a perforated floor and air is forced through the produce. In this method uniform cooling of the produce can be achieved if the stacks of pallet bins are properly aligned.

An evaporative cooler produces effective cooling by combining natural process water evaporation with a simple, reliable air-moving system. Fresh outside air is pulled through moist pads where it is cooled by evaporation and circulated through a house or building by a large blower. As this happens, the temperature of the outside air can be lowered as much as 30° C.

Evaporative coolers can work wonderfully well, provided the outside air they are drawing in is dry and desert like. As the humidity increases, however, the ability for them to cool the air effectively and can be achieve product temperatures up to 16° C to 21° C. In forced air evaporative cooling, the air is cooled with an evaporative cooler instead of with mechanical refrigeration. If designed and operated correctly, an evaporative cooler produces air a few degrees above the outside wet bulb temperature and the cooler air is above 90 per cent RH. This cooling method may be adequate for some products that are best held at moderate temperatures such as tomatoes or for products that are marketed quickly after harvest. Forced air evaporative coolers are much more energy efficient than mechanical refrigeration and if properly designed they provide high humidity cooling air. Cooling time depends on the airflow, the temperature difference between the produce and the cold air and produce diameter.

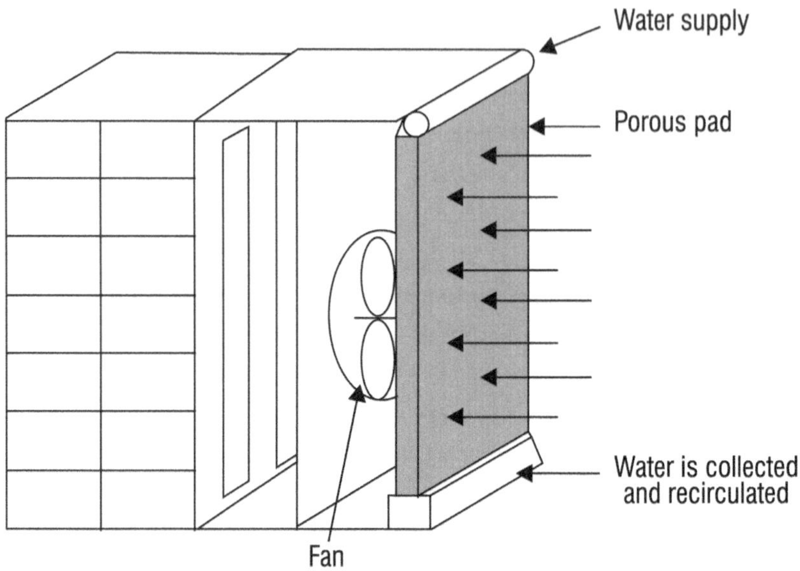

Figure 29. Evaporative forced air cooler (Source: Thompson and Kasmire, 1981; Mitchell in Kader, 1992).

Container venting: Containers are filled with active chemicals or that are subject to changes in pressure, temperature or altitude need vented closures to allow air to maintain equilibrium while retaining liquids safely inside. Failure to vent a container can result in collapsing, swelling or leaking. Effective container venting is essential for forced air cooling to work efficiently. Cold air must be able to pass through all parts of a container. For this to happen, container vents must remain unblocked after stacking. If containers are palletized in register (aligned with one another), container side or end vents will suffice, provided they are properly located in relation to trays, pads and other packaging. If cross-stacking is used, matching side and end vents is essential. For the 400 by 300 mm (or 16 by 12 in) container cross-stacked on the 1200 by 1000 mm (or 48 by 40 in) pallet, vertical vent slots on 100 mm (4 in) centres around the container perimeter should be considered, because they remain matched when cross-stacked.

Too little venting restricts airflow and too much venting weakens the container. A reasonable compromise appears to be about 5 to 6 per cent side or end wall venting. A few large vents (1.3 cm = 0.5 inch wide or more) are more effective than many small vents for speeding the cooling rate. Locating vents midway from top to bottom is adequate unless trays or other packing materials isolate some of the products. Vertical slots at least 12 mm (½ in) wide are better than round vents. Vent design should minimize the effect of product blocking vents. Any type of unvented bag, liner, or vertical divider inside the package may block vents and reduce airflow. If a solid liner is used, slow but acceptable cooling times can be obtained so that cold air can flow over and under each box and boxes are fairly shallow. The illustrations (Figure 30) below show the recommended pattern of vents for cartons used to hold produce that is to be forced-air cooled. Vents should make up 5 per cent of the total surface area and should be located 5 to 7.5 cm (2 to 3 inches) away from the corners. Flower boxes often have closable vents. This allows closing after cooling so that flowers shipped on unrefrigerated transport can maintain temperatures longer than if shipped in a box with open vents.

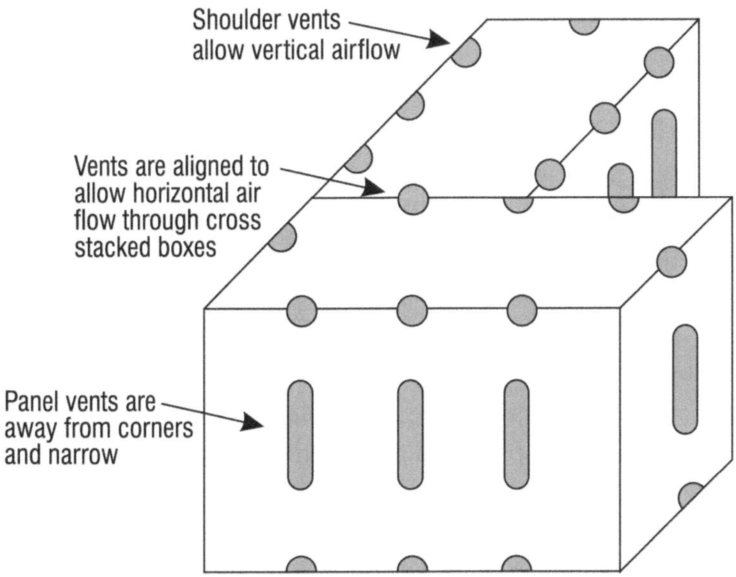

Figure 30. Pattern of vents for cartons (Source: Mitchell, 1972; Thompson, 2002)

Cooling in transport: Marine containers and break bulk ships have enough airflow and refrigeration capacity to achieve slow forced air cooling. Intransit cooling is used for products produced in areas without cooling facilities. But it is usually better to cool the product before it is loaded in the ship or container, since most transport modes do not have the extra refrigeration capacity needed for fast cooling. Refrigerated ships and most marine containers have a bottom delivery air supply system. Air travels from the refrigeration system to a floor plenum. It flows then upward through the boxes and returns to the refrigeration system in the space above the product. To work well, boxes must have vent holes on the top and bottom and the vent holes must align if boxes are cross stacked. The floor must be covered to prevent air from by passing the load. Refrigerated highway trailers with a top delivery air supply system do not have enough airflow to allow transport cooling.

Cold water/ hydrocooling

The use of cold water is an old and effective cooling method used for quickly cooling a wide range of fruits and vegetables before packaging. For the packed commodities it is less used because of difficulty in the movement of water through the containers and because of high cost involved in water tolerant containers. This method of cooling not only avoids water loss but may even add water to the commodity. Chlorination of water (150–200 ppm) is important to prevent the accumulation of pathogens. In most hydrocoolers, a pump moves chilled water into contact with warm produce. The warmed water is then recooled and recycled. For cooling the water, many hydrocoolers have a vapor compression refrigeration system similar to an air conditioner or refrigerator. The capacity of a refrigeration system to move heat is measured in tons. One ton of refrigeration is equivalent to 12,000 BTU per hour. Efficient cooling depends upon adequate water flow over the product surface. Immersion type hydrocoolers usually take longer time to cool produce than shower type cooler. Hydrocooling systems are suitable for precooling of artichokes, asparagus, avocados, green beans, beets, broccoli, brussels sprouts, cantaloupes, carrots, cauliflower, celery, cherries, capsicum, cucumber, eggplant, green onion, kale, leeks, leafy vegetables, nectarines, orange, parsley, peas, peaches, pomegranate, radishes, romaine lettuce, spinach, sweet corn, turnips and other fruits and vegetables.

Principle of hydrocooling: The transmission of heat from a solid to a liquid is faster than the transmission of heat from a solid to a gas because water is better heat conductor than air. In this method cold water is used for quick cooling of a wide range of fruit and vegetables. Under this method, water is usually cooled by mechanical refrigeration, but ice may be used to make process faster. For quick cooling of the produce, cold water must constantly be passed over the crop. Chlorine (150-200ppm), iodine, nutrients, growth regulators and fungicides can be added in water to sanitize or improve nutrient status and prevent post-harvest diseases of the produce. Cooling time depend on nature of produce such as 2 minutes for asparagus (long & narrow) and leafy vegetables (more surface to volume ratio), 10 minutes for small produce like capsicum (large and globular), cherries, tomato and upto 1 hour for large products such as melons.

Types of hydrocoolers

Shower type: In this type of hydrocooler, cold water is pumped to an overhead perforated pan which produces a shower over the produce which may be in bins or boxes or loose on a conveyer belt passing beneath. The water leaving the produce may be filtered to remove debris then passed over refrigeration coil where it is recooled. Efficient cooling depends upon adequate water flow over the product surface. For product in bins or boxes, water flows of 75-100 liter minute^{-1} feet^{-1} (400-600 liter minute^{1}/m^{2}) of surface area are generally used.

Immersion type: It is simplest type of a hydro cooler in which produce is dipped in cold water. Here product are normally in bulk, is in direct contact with the cold water as it moves through a long tank of cold. This method is best suited for products that do no float, because, slow cooling would result if the product simply moved out of the water. Immersion hydro coolers convey product against the direction of water and often have a system for agitating the water. Depth of the water tank should be >30 cm and water tends to penetrate inside fruits, particularly those that are hollow such as peppers. Water temperature also contributes to infiltration. It is recommended that fruit temperature is at least 5° C lower than liquid.

Ice cooling

Ice cooling is probably one of the oldest methods used to reduce field temperature for individual packages. Crushed ice is placed on top of the produce before the package is closed. Ice layers may also be interspersed with produce. As it melts, cold water cools the lower layers of produce. Liquid icing is another system where a mix of water and crushed ice (40 % water + 60 % ice + 0.1 % salt) is injected into open containers so that a big ice block is formed.

Top icing and direct contact with ice: Top icing involves placement of measured amount of crushed ice over the top layer of product in a container prior to closure. Although relatively inexpensive, the cooling rate can be fairly slow since the ice only directly contacts the product on the top layer. This method is sufficient in many instances but can result in uneven cooling because the ice generally remains where it was placed until it has melted. Because each carton must be opened, iced, and reclosed, the process is slow and labour intensive. Individual package icing has been automated to some extent by ice-dispensing devices and the use of package conveyors and roller benches. Top ice is used for certain products during transport to help maintain a high relative humidity. Top ice can be used only with water tolerant, non-chilling sensitive products such as carrots, sweet corn, cantaloupes, escarole, lettuce, spinach, radishes, broccoli, and green onions with water tolerant packages (waxed fibre board or wood). Top-ice on loads should be applied in rows rather than a solid mass. It is important not to block air circulation inside the transport vehicle.

Package icing: Crushed ice distributed within the container is known as package icing. Cooling is faster and more uniform than for top icing, but it can be more labour intensive to apply.

Disadvantages of top and package icing

- Package-icing requires costly, water tolerant packages.
- The packages should be quite tight but should have enough holes to drain melt water.

» The product should be tolerant of prolonged exposure to wet conditions at 0° C temperature.

» Some low-density products have excess space in which to load ice within the package and ice not melted during cooling can remain in the package even after transport. This excess ice can keep the product cold if the cold chain is broken.

Liquid icing (slush icing): Liquid icing is the preferred method when large amounts of produce must be iced in a relatively short period of time. Large operations use liquid ice machines to automatically ice pallet loads of packed cartons. The process, which requires only a few minutes, is used for cooling some field packed vegetables, particularly broccoli. The iced packages should be placed into a cold room after filling to minimize ice melt. The required equipment consists of an ice crusher, a slurry tank with mixer, a pump, and delivery hoses. This method is a significant improvement over hand icing because the water carries the ice into the voids throughout the package, thus bringing it into greater contact with the produce. Packaged icing can be used only with water tolerant, non-chilling sensitive products with water proof packages (waxed fiberboard, plastic or wood). In the liquid icing of broccoli, it has been estimated that as much as 40 per cent of the total product cooling is accomplished by the coold water and balance by the ice as it melts inside the carton. This method is fast and effective, and it does not require that the cartons be opened or removed from the pallet. With the proper equipment, two workers can liquid ice a pallet of 30 cartons in about 5 minutes.

Evaporation of water from produce

Evaporative cooling: This is one of the simple, low cost technologies and effective methods for cooling of fresh produce at farm level. Such type of structure maintains high relative humidity with temperature nearer to wet bulb and thus reduces shriveling, extend storage life and prevent the weight loss considerably in stored fruits and vegetables. Though cold storage is the best yet, due to highly energy extensive, huge capital investment and perennial power shortage, most of the small farmers find it difficult and are forced to sell their produce at distress price soon after harvest. Under these conditions the evaporative cool storage help the farmers to get better returns for their produce. When the heat is removed by way of evaporative cooling then the fresh produce must not be sealed in moisture proof film like polyethylene bags. This method has a low energy cost but cooling efficiency is limited by the capacity of air to absorb humidity.

Vacuum cooling: Vacuum cooling is the fastest and most uniform method of pre-cooling because of the high amount of heat required for evaporating the water (about 1000 BTUs per pound of water evaporated). It is, one of the most expensive systems to set up. In this method, air is pumped out from a larger steel chamber in which the produce is loaded for pre-cooling. Removal of air results in the reduction of pressure of the atmosphere around the produce, which further lowers, the boiling temperature of its water. As the pressure falls, the water boils quickly removing the heat from the produce. Vacuum cooling cause about 1 per cent produces weight loss (mostly water) for each 6° C of cooling. Vacuum cooling takes place by water evaporation from the product at very low air pressure, at normal pressure of 760 mmHg, water evaporates at 100° C, but it does at 1° C if pressure is reduced to 5 mmHg. For example, water at the freezing point boils when the pressure is reduced to 0.08859 psi absolute. Normal atmospheric pressure is about 14 psi absolute. As a result, a small amount of water in the warm produce evaporates causing a corresponding lowering of the produce temperature. Vacuum cooling of fresh vegetables or fruit occurs by rapid evaporation of water from the product works best with those products that have a high ratio of surface area to volume and a high transpiration coefficient e.g. leafy vegetables and cabbage, whereas with low ratios between mass and surface area and effective water barrier like wax on surface is not suitable e.g.tomato. To overcome the more loss of water from the produce, another method of water spray vacuum is used called hydro-vacuum cooling.

Table 56. Comparison of methods used for cooling

Variable	Cooling method					
	Ice	**Hydro**	**Vacuum**	**Water spray**	**Forced-air**	**Room**
Typical cooling times (h)	0.1-0.3	0.1-1.0	0.3-2.0	0.3-2.0	1.0-10.0	20-100
Water contact with the product	Yes	Yes	No	Yes	No	No
Product moisture loss (%)	0-0.5	0-0.5	2.0-4.0	No	0.1-2.0	0.1-2.0
Potential for decay contamination	Low	High	None	High	Low	Low
Capital cost	High	Low	Medium	Medium	Low	Low
Energy efficiency	Low	High	High	Medium	Low	Low
Water-resistant packaging needed	Yes	Yes	No	Yes	No	No
Portable	Common	Rarely done	Common	Common	Sometimes	No
Feasibility of in-line cooling	Rarely done	Yes	No	No	Rarely done	No

(Source: Thompson *et al.* 1998)

Table 57. Classification of crops according to cooling method

Method of cooling	Crops
Room cooling	Artichoke, banana, beans (dry), beet, breadfruit, cabbage, cactus leaves, cassava, coconut, custard apple, garlic, ginger, grapefruit, horse radish, kohlrabi, kumquat, lime, lemon, melons, onion, orange, cucumber, pineapple, potato, pumpkin, radish, summer squash, sweet potato, tomato, turnip, watermelon
Forced-air cooling	Avocado, banana, barbados cherry, berries, brussels sprouts, cactus leaves, cassava, coconut, cucumber, eggplant, fig, ginger, grape, grapefruit, guava, kiwi fruit, kumquat, lima bean, mango, melons, mushrooms, okra, orange, papaya, passion fruit, pepper (bell), pineapple, pomegranate, prickly pear, pumpkin, snap beans, snow peas, strawberry, summer squash, tangerine, tomato
Hydro-cooling	Artichoke, asparagus, beet, broccoli, brussels sprouts, cantaloupe, cauliflower, carrot, cassava, celery, chinese cabbage, cucumber, eggplant, escarole, green onions, horse radish, kiwi fruit, kohlrabi, leek, lima bean, orange, parsley, parsnip, peas, pomegranate, potato (early), radish, snap beans, snow peas, spinach, summer squash, sweet corn, swiss chard, watercress
Ice cooling	Belgian endive, broccoli, brussels sprouts, cantaloupe, chinese cabbage, carrot, escarole, green onions, kohlrabi, leek, parsley, pea/snow peas, spinach, sweet corn, swiss chard, watercress.
Vacuum cooling	Belgian endive, brussels sprouts, carrot, cauliflower, chinese cabbage, celery, escarole, leek, lettuce, lima bean, mushrooms, snap beans, snow peas, spinach, sweet corn, swiss chard, watercress

Specific Product Recommendations for Precooling

Fruits and vegetables have a notable range of pre cooling and storage potentials, which are based on genetically. The potential storage life of different tomato varieties vary from a few days to a few weeks, and these differences reflected in quality factors such as intensity of flavour many newer varieties often store well, but have lesser flavour than heirloom varieties. Important fruits and vegetables are precooled as below.

Apple: For apple fruit room cooling is generally used. Care should be taken to load rooms in a manner that will allow good air circulation around the bins. The effects of slow cooling are magnified as the storage period increases. The recommended storage temperature for most apple varieties is 0° C with 90 to 95 per cent RH. Varieties that are susceptible to low temperature injuries should be stored at 2.2-3.3° C, with 90 to 95 per cent RH.

Apricot: Apricots should be cooled in the field or immediately after that, usually by room cooling. If picked at the firm mature ripening stage and stored at 0° C with 90 to 95 per cent RH, fruit can be stored for one to two weeks or more, depending on the variety. Controlled atmospheres of 2 to 3 per cent oxygen and carbon dioxide can have moderate commercial benefits.

Artichoke: Artichokes should be cooled by room cooling or hydro-cooling to below 5° C within 24 hours of harvest. The recommended storage for artichoke is 0° C with 95 to 100 per cent RH. The storage period under these conditions can be two weeks, this length decreasing with increasing storage temperatures.

Asian pear: Asian pears are usually cooled using room cooling. Forced air cooling is not recommended as it may increase risk of flesh spot decay. The recommended storage temperature is 0° C with 90 to 95 per cent RH. Fruit are susceptible to low oxygen and high carbon dioxide injuries during storage. For retail outlet displays, Asian pears should be kept as cold as possible without contact of fruit with ice.

Asparagus: Hydro-cooling of spears is the most effective cooling method. The storage temperature of asparagus should be 0° C for periods of less than 10 days, but 1.5° C for longer periods because of sensitivity of spears to chilling injury. A high RH of 95 to 99 per cent is important to prevent desiccation and maintain freshness. Controlled atmosphere storage, typically elevated carbon dioxide concentrations of 5 to 10 per cent, retards bacterial soft rot, toughening, water loss and yellowing of the spears.

Beet: Bunched beets should be precooled rapidly after harvest using hydro-cooling, forced air or icing. Topped beets can be cooled using room or forced air cooling. Bunched beets should be stored at 0° C with 98 per cent RH, while topped beets should be stored at 0.5 to 2° C with 98 per cent RH. Red beets can be stored in air ventilated common storage for four to six months and in refrigerated storage for eight to ten months. For retail outlet display, bunched beets should be iced or refrigerated at 3 to 5° C, while beetroot should be kept at less than 10° C.

Blackberry and Raspberry: Rapid cooling within an hour of harvest is critical to maintain quality of berries after harvest, but the rapidity of cooling is more important with longer storage requirements. Room cooling is commonly used, but forced air cooling is highly recommended to maintain the quality of blackberries and raspberries. The berries are not chilling sensitive and they should be stored at 0° C with 90 to 95 per cent RH.

Broccoli: Hydro-cooling and forced air cooling are options to rapidly remove field heat from the broccoli. It can be stored at 0° C with 98 to 100 per cent RH for 2 to 3 weeks. Broccoli is not susceptible to chilling injury and responds well to controlled atmosphere storage.

Cabbage: Cabbage is generally room cooled without pre-cooling. Cabbage should be stored at 0° C with 98 to 100 per cent RH, and even a small increase in storage temperature to 1° C may promote senescence related losses in long term air storage.

Carrot: Hydro-cooling is the best option for pre-cooling of carrots, after they have been pre-washed. Carrot need to be cooled on the day of harvest to reduce wilting, shriveling and decaying, prevent heating due to respiration. Smaller carrots are more susceptible to water loss than larger carrots because of their large surface area to volume ratio. Mature topped carrots can be stored for several months at 0 to 1° C at 98 to 100 per cent RH. Commercial storage can run as high as 4° C with 90 to 95 per cent RH.

Cauliflower: Cauliflower is mostly pre-cooled by hydro-cooling or vacuum cooling, but forced air cooling is also effective. Quality of the heads can be maintained for 3 to 4 weeks if they are stored at 0° C with 95 to 98 per cent RH with good airflow. Storage at warmer temperatures increases riciness development, expressed as loose curds with floral parts protruding. Controlled atmosphere storage can extend storage life, although cauliflower is sensitive to both low oxygen and high carbon dioxide injuries. Cauliflowers produce very low levels of ethylene, but they are highly sensitive and develop curd discoloration, leaf yellowing and abscission to ethylene exposure.

Celery: Celery can be kept for two to three months if cooled rapidly and stored at 0° C with 98 to 100 per cent RH. Precooling options include hydro cooling, forced air and vacuum cooling. Celery is not sensitive to chilling injury. Celery produces low levels of ethylene, and it is less sensitive to the gas at low storage temperatures.

Cherry (Sweet and sour): Sweet cherries should be pre-cooled within 4 hours of harvest by hydro cooling, forced air cooling and room cooling methods. Sweet cherries can be stored at 0° C with >95 per cent RH for 2 to 4 weeks. They produce only very low levels of ethylene and respond to ethylene with increased respiration, quality loss and a shortened storage life.

Chinese cabbage: Precooling can be carried out using hydro-cooling, vacuum cooling or forced air cooling. It can be stored at 0°C with 95 to 100 per cent RH for period of two to three months.

Cranberry: Cranberries are chilling sensitive and should not be stored below 2-3° C. They are low ethylene producers with less sensitivity to ethylene exposure. Controlled atmosphere conditions with 90-95 per cent RH can maintain the fruit in good condition for 1-3 months.

Cucumber: Cucumbers are chilling sensitive, but if exposure to cold temperatures during pre-cooling is short (less than six hours), hydro-cooling and forced air cooling methods can be used. Room cooling is recommended only for those cucumbers harvested in mild weather conditions. Cucumbers should be stored at 10 to 12° C, with 95 per cent upto two weeks. Cucumbers produce low levels of ethylene, but exposure to ethylene concentrations of 1ppm or more accelerate yellowing and decay.

Eggplant: Room cooling is effective, but hydro-cooling and forced air cooling is more effective. Eggplants are sensitive to chilling injury (surface pitting and scald) and should be stored at 10 to 12° C with 90 to 95 per cent RH. Ethylene production of the fruit is low, but exposure to exogenous ethylene (>1ppm) can cause calyx abscission and browning.

Grape: The preferred precooling method is forced air cooling. Moisture loss can result in stem drying, fruit browning, berry shatter and shriveling. Grapes stored at 0° C with 90-95 per cent RH should remain in good condition for 1-2 months. Traditionally sulfur dioxide is used as a fumigant at the time of pre-cooling and periodically during storage. A controlled or modified atmosphere with high carbon dioxide can be used instead of sulphur dioxide to control decay for up to 4 weeks.

Herbs: Herbs may be hydro cooled before packaging and room cooled after packaging. For most herbs the optimum storage temperature is 0° C with 95 to 98 per cent RH. Ethylene production by herbs can vary and is generally higher than that for leafy green vegetables. Sensitivity to ethylene is high, and even low levels of ethylene can cause leaf abscission, epinasty (stem curvature) and yellowing.

Kohlrabi: Kohlrabi can be pre-cooled by hydro-cooling, package icing or forced air cooling. Topped kohlrabi not susceptible to chilling injury and can be stored for two to three months at 0° C with 98 to 100 per cent RH, but only two to four weeks with leaves. Kohlrabi produces very little ethylene and are not sensitive to the gas.

Leafy greens (spinach, swiss chard, collards, kale, rape, mustard, turnip): Leafy greens are highly perishable, especially prone to wilting, and therefore harvested leaves should be shaded and cooled immediately. Precooling can be carried out using hydro-cooling, liquid icing, package icing, and top icing. Leafy greens are not chilling sensitive and they should be kept at 0° C with 95 to 98 per cent RH. Harvest of most leafy greens in the cooler part of the day, rapid cooling, and topping with ice can result in 10 to 14 days of storage. Ethylene production of leafy greens is very low, but exposure to ethylene will cause yellowing of the leaves.

Leek: Hydro-cooling, icing, and vacuum cooling can be used to pre-cool leeks. The leeks should be stored at 0° C with 95 to 100 per cent RH upto five to six weeks and leeks can be stored in crushed ice also. Controlled atmosphere storage can retard yellowing and decay development. Leeks produce very low levels of ethylene, but they are moderately sensitive to exposure ethylene can induce softening, yellowing and decay.

Lettuce: Rapid cooling will improve market quality and shelf life of lrttuce. Advocateed cooling methods are vacuum cooling, forced air cooling and ice made with potable water cooling. Lettuce is not sensitive to chilling injury. Both crisp head and leafy lettuce types should be stored at 0° C and at least 98 to 100 per cent RH. Ethylene production of lettuce is extremely low, but exposure to ethylene can cause russet spotting and yellowing of the leaves.

Melons: Melons are often pre-cooled by forced air or sometimes by hydro-cooling and packed in cartons to be shipped. Full slip melons can be held for 14 days at 2-5° C with 85-95 per cent RH without deteriorating, and even longer for less mature melons. Casaba melons have low ethylene production and sensitivity while crenshaw, honeydew and persian melons are moderate producers with high sensitivity to ethylene. Cantaloupes are high ethylene producers but have moderate sensitivity. Effects of ethylene are loss of firmness, yellowing and an increase in volatiles.

Mushroom: Mushrooms should be precooled to 0° C immediately after harvest by room cooling or forced air cooling, and stored at 0° C with 95 per cent RH. Mushrooms are not sensitive to chilling injury, and they can be stored for seven to nine days at low temperatures. Mushrooms produce especially low levels of ethylene, exposure to ethylene causes browning of the caps.

Onion (dry) and garlic: Curing is carried out using forced-air ventilation that is controlled thermostatically by an air proportioning system. At a weight loss of about 5 per cent, the tight neck and dry scales form an effective barrier to further reduce water loss and susceptibility to pathogens. Pungent, dry onion bulbs should be precooled to 0° C immediately after drying. The optimum storage temperature for onions is 0° C at 65 to 75 per cent RH. High temperatures induce sprouting, high RH induce root growth, while both are increases decay and loss of quality. Mild or sweet onions have a shorter storage period than pungent types and can be kept in refrigerated rooms or common storage.

Onion (Green) and shallot: Both can be precooled with package icing, hydro cooling or water spray vacuum cooling. Both last about 3 weeks when stored at 0 to 2° C with 95-100 per cent RH. Green onion odour is absorbed by figs, grapes, mushrooms, rhubarb and corn so care must be taken when multiple crops are stored together. Green onions can be stored for 6-8 weeks under controlled atmospheres. Both are low ethylene producers but shallots have low sensitivity to ethylene while green onions are highly sensitive.

Parsley: Precooling can be carried out by icing, forced air, hydro-cooling or vacuum cooling. Parsley is not sensitive to chilling injury, it should be stored at 0° C with 95 to 100 per cent RH. Storage life can be one to two months under these conditions. Packing the parsley in perforated polyethylene bags helps maintain freshness. Parsley produces little ethylene, but is highly sensitive to ethylene in the atmosphere. Effects of ethylene on yellowing are reduced by use of the proper storage temperature.

Parsnip: Parsnips, like carrots, should be hydro cooled or ice packaged. Rapid cooling to below 5° C immediately after harvest minimizes decay and moisture loss during long term storage. Parsnips store well for 4 to 6 months at 0 to 1.1° C with 98 per cent RH. They produce very little ethylene but exposure to low levels of the gas causes bitterness.

Pea: Peas should be precooled using forced air cooling, hydro cooling or vacuum cooling. Peas are best stored at 0° C with 92 to 98 per cent RH upto one to two weeks. Garden peas store better unshelled than shelled, possibly because shelling damages the peas, although it has been suggested that pea pods actually act as modified atmosphere packages with lower oxygen and higher carbon dioxide around the intact peas.

Peach and nectarine: Peaches are typically cooled in cold rooms, but forced air and hydro-cooling is more effective. The fruit can be cooled to 5 to 10° C provided that packing is carried out the next day. Peaches and nectarines should be stored at 0° C with 90 to 95 per cent RH.

Pear: Cooling, typically room cooling, should be prompt. Care should be taken to load rooms in a manner that will allow good air circulation around the bins. Negative effects of slow cooling on the storage life are magnified as the storage period increases. The recommended storage temperature for most pear varieties is -1.0 to 0° C with 90 to 95 per cent RH.

Pepper: After harvest, fresh market peppers can be pre cooled to 5-7° C using forced air, hydro-cooling or vacuum cooling. Peppers are susceptible to chilling injury and should be stored at no lower than 7° C with 90 to 95 per cent RH. Peppers are non-climacteric fruits and produce very low levels of ethylene, but they may be affected negatively by exposure to ethylene. Peppers should be stored separately from ethylene producing fruit.

Plums: Plums can be passively cooled in cold rooms, but forced air and hydro-cooling is more effective. Plums should be stored at 0° C with 90 to 95 per cent RH. Controlled atmosphere storage helps maintain flesh firmness and ground colour of some varieties. Plums are susceptible to chilling injury (flesh translucency, flesh internal browning, flesh mealiness, flesh bleeding, failure to ripen and flavour loss).

Potato: Curing is used to stimulate suberization or wound healing, making less susceptible to damage and disease. Suberization is the primary means of reducing water and quality loss and minimizing disease development during storage. Temperatures of 10° C to 15° C with high RH should be provided for 1 to 3 weeks at the beginning of the storage period. After curing, the storage temperature should be lowered 2 to 5° C per day until the desired temperature and RH are reached. Potatoes produce very low levels of ethylene unless they are injured, but they are not very sensitive to the gas.

Radish: Radish store best at 0° C with 95-100 per cent RH upto 1-2 months. It produces extremely low ethylene and has minimum sensitivity, primarily resulting in bunched tops getting yellow after prolonged exposure in storage. Radishe can develop freezing injury where roots appear water soaked and glossy and shoots become water soaked, wilted and turn black if the temperature drops to -1.1° C in storage.

Rhubarb: Precooling can be by room cooling or hydro-cooling. After precooling, whole, packed stalks can be stored for two to four weeks at 0° C with 95 to 100 per cent RH, but cutting stalks to shorter lengths will markedly reduce the potential storage periods achievable. Ethylene production by rhubarb is low, but it is sensitive to the gas.

Rutabaga (Swedes turnips): Roots should be rapidly cooled to 0° C and this is most commonly done through room cooling. The rutabaga can also be hydro cooled, forced air cooled or package iced to speed the cooling, and help control weight loss, decay and the development of storage burn. Rutabagas can be kept at 0° C for 4 to 6 months with 98 to 100 per cent RH. Rutabagas produce scarce amount of ethylene and are not sensitive to it.

Summer squash: Precooling can be done by room, forced air or hydro-cooling methods. Rapid cooling reduces water loss from the fruit. Summer squash are susceptible to chilling injury and the optimum storage regime is 5 to 10° C with 95 per cent RH. Summer squash produce low to moderate levels of ethylene. Exogenous ethylene can cause yellowing of green skinned varieties.

Winter Squash: Squashes are susceptible to chilling injury and they should be stored at 10 to 12° C with 50 to 75 per cent RH. For long term storage, a higher RH of 70-75 per cent helps to reduce water loss. In spite of storage type, good air circulation is desirable to maintain dryness and uniform temperature around the fruit throughout the storage period. Winter squashes produce very low levels of ethylene, but exposure to ethylene can cause skin colour changes to orange yellow and stem abscission.

Strawberry: Room cooling is commonly used, but forced air cooling is highly recommended to maintain berry quality. Strawberries are not chilling sensitive and they should be stored at 0° C with 90 to 95 per cent RH for a week or more, but the storage periods achievable are affected greatly by variety, berry maturity, and pre and post harvest conditions. Controlled atmosphere storage using high carbon dioxide (10 to 15%) maintains berry firmness and inhibits decay. Strawberries produce very low levels of ethylene and do not respond to the gas.

Sweet corn: Precooling to 0° C within an hour of harvest results in optimum quality retention, but is rarely achieved in commercial practice. Package and top icing is a rapid means of cooling the sweetcorn and helps keep the husk fresh. Hydro cooling and vacuum cooling is also highly effective. Sweet corn is not susceptible to chilling injury, it should be stored at 0° C with 90 to 98 per cent RH.

Sweet potato: Immediately after harvest the roots should be cured for 4 to 7 days at 28-30°C temperature and 90 to 97 per cent RH with good ventilation. The roots should then be moved to storage at 13-15° C with 90 per cent RH. Sweet potatoes store well without sprouting for up to a year but will lose sensory quality with extended storage. Storage at 18-20°C or above results in sprouting after a few months, and storage at 10-12° C or lower, results in chilling injury. Sweet potatoes produce very low or no ethylene. Ethylene exposure should be avoided since levels as low as 10 ppm result in increased discoloration and reduced β-amylase activity. So long as the storage is properly ventilated with an air flow of 36 ft^3 min^{-1} ton^{-1}, this should not be a problem.

Tomato: Room cooling is common for ripe tomato, but forced air cooling is more uniform and maintains better quality fruit. Mature green tomatoes should be stored at 12-13° C with

90-95 per cent RH as fruit are susceptible to chilling injury at lower temperatures and will ripen faster at warmer temperatures. Red ripe tomatoes can be stored for a few days at 7 to 10° C temperature with 85-90 per cent RH, but flavour and aroma may be compromised and the fruit susceptible to chilling injury.

Turnip: Turnips can be pre-cooled in wash water, so long as the temperature of the water is not more than 7° C cooler than the temperature of the roots to prevent cracking. They can be stored for 4 to 5 months at 0.5- 2.2° C with 90 to 95 per cent RH. Storage at 0° C may result in freezing damage while storage at > 5° C accelerates soft rot development and weight loss. Turnip does not produce measurable ethylene and are not sensitive to it.

Watermelon: Watermelons are usually not pre-cooled. They develop chilling injury, expressed as rind browning and fading of flesh colour, if stored at less than 10° C for more than a few days, and therefore watermelons should be stored at 10° C to 15° C with approximately 90 per cent RH. Watermelons do not respond to controlled atmosphere storage. Watermelons produce low levels of ethylene, but exposure of fruit to as little as 5ppm ethylene causes softening, rind thinning, flesh colour fading, and over ripening.

8 Transportation and Marketing of Fruits, Vegetables and Flowers

The most important purpose of transportation is to hand out the fruits, vegetables and flowers from production center to consumers as rapidly as possible with proper packaging and minimum change in quality of the produce. Losses in quality and quantity during transportation occur due to physical and mechanical injury. Fresh fruits and vegetables are primarily transported by road and flowers by air from farmer to consumer. A marketing concern is that fresh produce should be of the highest quality and kept in the best condition during transportation and storage. Minimizing losses during transportation, necessitates that special attention be given to vehicles, equipment, infrastructure and handling.

MEANS AND MODES OF TRANSPORT

Different means are used to transport fresh fruits, vegetables and flowers. In earliest the selected fruits, vegetables and flowers are transported only by manually in decorated wooden baskets at nearby markets and specified consumers. After that the bullock cart, camels, donkey, rickshaw, bicycle and tempo are used for transport of small quantity of fruits, vegetables and flowers in local markets. In modern time the motorcycle, pickup, truck, rail, steamer and air transport are used for commercial transportation depending on market distance. Rail transport causes losses due to loading unloading at various stations. Moreover, it is not available at producer's door steps as compared to road transport. Ships and air transport are very expensive but these are used for export of fresh fruits, vegetables and flowers. Among the various transportation means, road transportation is fasted one and easily available. Temperature management is critical during long distant transport, so loads must be staked air circulation to keep away heat from the produce itself as well as incoming from atmosphere. Most fresh produce in India and other countries of the world is transported from farmers field to nearby market or wholesale market and from wholesale market to terminal market up to final retailers shops in open and non-refrigerated vehicles. Only few reputed firms use refrigerated vehicles for transportation and distribution in summer months only, starting from March to May/June in India. It is mainly due to the increased cost of transportation by reefer van. In open truck vehicles (Non-reefer), produce is always susceptible to a loss of quality and spoils the produce. All imported fruits are transported in air condition containers by ships only. In every country, a devoted port is assigned for receiving and dispatch of fresh produce containers. In India, a large number of fresh produce containers are received at Mumbai and Chennai port. Fresh produce is transported using both refrigerated and non-refrigerated vehicles.

Non-refrigerated vehicles: Non-refrigerated vehicles are generally open-sided trucks, with wire mesh frames. This type of transportation is low-cost, convenient and easy. Truck, pickup, loading car and other transport vehicle can be used for the delivery of fresh produce. Layers of produce are not, however, separated in order prevent heat generation. Often the produce is stacked too high. Non-refrigerated vehicles are used for transporting produce over distances of upto 850 km. On arrival the produce must be unloaded or sold as quickly as possible to overcome overheating. Fresh produce must not be watered prior loading, as this will lead to decay, rotting and extensive losses. Improper handling during loading and unloading, over loading, overheating, mechanical injury, rough roads, long distance and lack of ventilation causes of losses during the non refrigerated transportation of fresh produce.

Refrigerated vehicles: Refrigerated vehicles (trucks, trains, ships, airplanes, etc.) contain installed refrigeration units with adequate low temperatures to keep up freshness in fresh produce. These types of vehicles are sealed with insulation material inside the walls of the container, which maintains the inside container temperature at desired level and thus preserves maximum quality. Fruits and vegetables must be classified in order to separate those susceptible to cold temperatures (mango, banana, tomato, etc.) and those not (apple, pear, cauliflower, peppers, etc.). This reduces the possibility of product damage (chilling and freezing injury) when cooling at low temperatures during transport. Refrigeration temperatures can vary from 0° C (32° F) to 13° C (55.4° F) and RH from 70 to 95 per cent.

LOADING AND TRANSPORTATION OF HORTICULTURAL PRODUCE

An open air vehicle can be loaded in such a way that air can pass through the load and provide some cooling to the produce as the vehicle moves. Bulk loads of produce should be carefully loaded so as not to cause mechanical damage. Vehicle can be padded or lined with a thick layer of straw woven mats or sacks can be used in the beds of small vehicles. Other loads should not be placed on top of the bulk commodity. When cartons of various sizes are too loaded together, the larger or heavier containers should be placed on the bottom of the load. Parallel channels should be left for air to move through the length of the load. Traveling during the early morning and night can reduce the heat load on a vehicle which is transporting the produce. The speed of vehicles is not higher than limit and non-reefer transport to move continuously as under sunlight and stop or park vehicle under a tree shade, especially during sunny days.

Pattern of produce loading

Pattern of loading also take part in critical role in maintaining quality of fresh produce. Here pattern of loading means number of packed boxes in one layer (stacking height). In case of fresh produce, stacking height depends on degree of perishable nature of packed commodities and strength of packing materials. If produce are more perishable or box strength is weak, stacking height is kept low and vice versa. For example, height of grape boxes is kept low or it is packed in five ply corrugated boxes or thermocol boxes. This preventative measure must be taken to protect postharvest quality of this highly perishable commodity. Loading and unloading fruits and vegetables directly affect quality of fresh produce. It can be done either by hand or with the aid of a forklift. Generally, fruits and vegetables are stacked on pallets to ease the loading and unloading process and to prevent damage to the product and packages. Export/Import commodities arrive at the docks in containers and are unloaded directly into the vehicle with the aid of conveyor belts connected from the vehicle to the container. In interlocking systems of loading a little space is left in each layer on alternate basis (once in left side and once in right side). This facilitates air movement through the produce and provides strength to the boxes during transportation. Exposure to sun while awaiting loading at local mandis or transport can reduce quality drastically. The exposed portion turns black or brown and starts decaying.

Table 58. Produce load compatibility

Group 1. Temperature 0-2° C, relative humidity 90-95 per cent			
Apple	Figs	Mushrooms	Pomegranates
Apricots	Grapes	Nectarines	Prunes
Asian Pears	Horseradish	Oranges	Quinces
Beets, topped	Leeks	Peaches	Radishes
Cashew apples	Loquats	Pears	Turnips
Coconuts	Lychees	Plums	Rutabagas

Group 2. Temperature 0-2° C, relative humidity 85-90 per cent			
Amaranth	Cauliflower	Kiwifruit	Radishes
Bean sprouts	Celeriac	Kohlrabi	Rhubarb
Beets	Celery	Leeks	Rutabagas
Broccoli	Cherries	Lettuce	Spinach
Brussel sprouts	Corn	Mushrooms	Turnips
Cabbage	Grapes	Parsley	Water chestnuts
Carrots	Green onions	Peas	Watercress
Group 3. Temperature 0-2° C, relative humidity 65-75 per cent			
Garlic	Onions	Shallots	
Group 4. Temperature 4° C, relative humidity 90-95 per cent			
Cactus pears	Clementines	Lychees	Tamarillos
Caimito	Cranberries	Mandarins	Tangelos
Cantaloupe	Kumquats	Oranges	Tangerines
Cassavas	Lemons	Pepino melons	Uniq fruit
Group 5. Temperature 10° C, relative humidity 90-95 per cent			
Beans	Eggplant	Okra	Squash
Calamondin	Cucumbers	Peppers	Tamarindos
Chayotes	Malangas	Potatoes	Pummelos
Group 6. Temperature 10-15° C, relative humidity 85-90 per cent			
Atemoyas	Cherimoyas	Lemons	Pineapple
Avocados	Coconuts	Limes	Potatoes, new
Bananas	Ginger root	Mangoes	Pumpkins
Bitter melons	Grapefruit	Mongosteen	Sugar apples
Breadfruit	Guavas	Papayas	Tomatoes, ripe
Carambolas	Jackfruit	Passion fruit	Winter squash
Group 7. Temperature 15-20° C, relative humidity 85-90 per cent			
Pears (for ripening)	Sweet potato	Tomatoes, mature-green	Watermelon

Factors affecting produce quality during transport

During transport, produce is seldom held under optimum environmental conditions. Approximately 25-30 per cent of the fruit and vegetables never make it to the supermarket shelves due to damage during transit. This produce loss can only be reduced by improved refrigerated transport. Temperature is the main environmental condition influencing produce quality. Excessively low temperature causes chilling or freezing injury and high temperature accelerates produce respiration and water loss, and causes a decrease in internal flesh quality, shriveling and premature softening. Other factors affecting produce quality are initial quality of produce, temperature, environmental humidity and water loss, atmospheric gas concentration, mixed loads, physical injury and stress and transport conditions.

Initial quality: Produce leaving the packing facility must be suited to the handling it will receive as it is transported to market. Locally produced fruits can be moderately mature and ripe because the time to market is short. Produce shipped from great distances is often a little less mature than locally produced and must be free of mechanical damage and other conditions predisposing it to noticeable quality loss in a long postharvest handling period.

Temperature: For fruits, vegetables and flowers respiration increases by a factor of two to five for each 10° C increase in temperature above its recommended holding temperatures. For example, berries like strawberries, raspberries and blackberries, have a shelf life of 7 days at 0° C but only 1 day at 20° C. Longer lived produce such as green beans, mushrooms, green onions, and green peas last only 2 to 3 days at 20° C. Above 30° C, increases in respiration rate, and produce will die or lose quality very quickly. Temperatures below recommended levels cause freeze damage or chilling injury to produce. Chilling injury usually occurs in fruits and vegetables native to tropical or sub-tropical regions. Signs of chilling injury are tissue darkening or drying, surface pitting, failure to ripen normally, off flavours, or increased susceptibility to decay.

Humidity and water loss: At 90 to 95 per cent relative humidity atmosphere is needed for maximal shelf life of most fruits, vegetables and flower. A few produce such as bulbs onions, garlic, winter squashes, and ginger should be kept below 70 per cent RH. Low RH around produce causes wilting or shriveling. This is aesthetically displeasing and generally reduces marketability. Loss of produce weight caused by water loss is a direct marketing loss. Water loss also weakens the plant cells, making them more susceptible to decay. In addition, fungal growth causes increased ethylene production, chlorophyll loss and yellowing. Produce susceptible to wilting should be waxed or packaged in liners, bags or plastic boxes to slow moisture loss. Special packaging capable of slowing moisture loss is particularly important in air freight. Airplanes usually have very low RH, often around 10-15 per cent. Some marine containers are equipped with spray humidification systems. These increase RH but air temperatures have to be slightly above 0° C to prevent ice from clogging the spray nozzle. Fiberboard and wooden boxes absorb moisture from the produce and can cause more than 1 per cent weight loss from the produce. The absorbed moisture also weakens the box. For example, fiberboard held in moisture equilibrium of 90 per cent RH retains only 40 per cent of its original stacking strength after a relatively short exposure period. Under high humidity conditions, waxed fiberboard, plastic and wooden boxes would perform better, as they can better withstand high humidity conditions. Marine containers are much more airtight than highway trailers and they are equipped with an adjustable fresh air exchange vent to control the amount of air entering the container. Excessive air exchange in humid areas results in unnecessary use of refrigeration capacity to condense excess moisture from the air.

Atmospheric composition: As fruits, vegetables and flowers respire, CO_2 and ethylene levels may increase, and oxygen levels decrease around the produce. CO_2 levels should be kept below 0.3 percent in a tightly sealed container. Highway trailers usually have enough air leakage to prevent damage from accidental build up of CO_2. AFAM and AFAM$^+$ system to enhance ventilation management of marine containers. Both systems have an electric controller to set the ventilation rate and to restrict ventilation during initial cooling to reduce the heat load. The AFAM+ system has a gas sensor to adjust the ventilation rate based on CO_2 concentration. Controlled atmosphere systems typically increase the storage life of produce by 30-40 per cent.

Mixed loads: Fresh fruits, vegetables and flowers have varying requirements for temperature and humidity, and have varying sensitivity to absorbing odours or ethylene induced damage from other produce. Many trucks have several types of produce in a single trailer. If some produce is mixed with a commodity with differing storage requirements, the quality of the load can be compromised in the three to four day trip to markets. The problem is even more severe in marine containers travelling several weeks to their destination. If logistically possible, produce should be loaded as mixed loads only if their temperature and humidity requirements and ethylene sensitivity and odour absorption capability are compatible. Incompatibility between ethylene producing produce and ethylene sensitive produce can be dealt with in several ways. Damage in refrigerated containers or semi-trailer load may also be reduced by using a fresh air exchange rate of 0.2 L s^{-1} or by using ethylene scrubbers (1-MCP). Mixed loads cause incompatibility problems in transport also because packaging dimensions are different for different produce and it is not fully stackable. It is therefore not advisable to transport mixed lots for long distance. However, for short distance, there may not be any problem. There is usually little or no humidity control available during transport and marketing. Thus, the packaging must be designed to provide a partial barrier against movement of water vapor from the product. Plastic liners designed with small perforations to allow some gas exchange may be an option. Multi-compartment trucks or semi-trailers have removable partition panels to divide the vehicle into smaller compartments. Mini-containers are usually the size of a standard pallet load. Some are insulated compartments and others are also refrigerated. The mini containers are highly flexible, easy to handle and can be used in different sizes of vehicles. Some produce have a short postharvest life and are not suited for container shipment. This is particularly true if they are held at non-optimal temperatures. Modified atmosphere (MA) packaging or CA can sometimes increase shelf life and allows produce to be shipped to destinations that require several weeks of transportation time. If a MA environment is used to hold different produce all together within a single space, it should, as a minimum requirement, not reduce the postharvest life of any of the mixed commodities. Dried vegetables should not be mixed with other produce when transit times are of a week or more. These vegetables should be held in a 50 per cent to 70 per cent RH environment to prevent decay. Most vegetables in the lowest temperature range (0° C to 2° C) are sensitive to moisture loss and should be held at higher than 90 to 95 per cent RH or packaged to minimize water loss.

Physical injury: Vibration, compression and impact cause physical damage to produce. The damage can be minimized by using proper packaging, good package management, and correct placement in the refrigerated vehicle. Compression damage occurs when the weight of the load is supported by the produce, rather than the produce container. Compression occurs generally when the boxes are overfilled, not properly palletized, and not strong enough or lose strength through moisture absorption or mechanical damage. Reusable and recyclable plastic containers are structurally stronger than corrugated fiberboard boxes, and provide more protection against compression damage. Pallet loads should be unitized and secured so that they do not shift during handling or transportation. Free space at the periphery will allow the load to shift in transport. Most reusable plastic containers are designed with an interlocking system and when they are properly stacked they do not shift in transport. Vibration damage is due to the constant vibrating motion of a vehicle when it is transported over the road. Vibration damage is greatest in locations over steel spring suspended axles. Air ride suspension dramatically reduces vibration damage.

Road condition and duration of transportation: Both road condition and duration of transportation affect quality of fresh produce. In hilly tracks and rough road surface, more touching and bruising take place as compared to smooth surface. Longer duration during transportation also affects quality.

System For Preserving Produce Quality During Transport

Fruit, vegetable and flowers should be kept at optimum temperature, high relative humidity and proper air circulation systems are the most important means to ensure quality preservation of perishables horticultural produce.

Cooling requirements: The refrigeration system used in a transport vehicle must have adequate cooling capacity. With properly cooled produce, most of the heat input is from air infiltration and heat conducted across the walls of the vehicle. During the cooler months of the year, the temperature control system may need to add heat to the vehicle to prevent chilling or freezing damage. The three main sources of heat are internal heat, external heat and residual heat loads. As a basic rule, the total cooling capacity of the refrigeration system is the sum of these three heat loads multiplied by a safety factor. Precise cooling capacity calculations tend to be more complicated as they account for additional effects such as solar heat gain and air speed past the vehicle.

Internal heat loads: The internal heat loads include residual field heat and heat of respiration generated by the produce during transit. Produce should be properly precooled before loading and maintained at the recommended temperature during transit to minimize this type of heat load.

External heat loads: External heat loads depend on the external environmental conditions and enter the transport vehicle by conduction, convection, infiltration and radiation. External heat loads are the main source of heat gain or loss in refrigerated transport. Insulating material is used to reduce the heat gained through conduction. Infiltration is the largest external heat load in refrigerated trailers. Infiltration can be minimized by repairing damage to trailer walls, doors and door seals, and opening doors only when necessary. Radiation is heat transferred through a vacuum or a gaseous medium, like the sun heating the earth. Light coloured objects tend to absorb radiant heat at a lower rate than dark coloured objects and using clean reflective surfaces like polished steel and aluminum, or reflective paints to reflect radiant heat from the sun or the hot road. Frequent cleaning is also required to maintain the reflective properties of exterior surfaces.

Residual heat loads: Residual heat loads include heat initially contained in the transport vehicle and any heat load not included as internal or external loads. The usual sources of residual heat are the uncooled mass of the transport vehicle and any heat in boxes, pallets and devices used to secure the load. The refrigeration systems installed in transport trailers have the capacity to remove some residual heat, but they do not have the capacity to remove the large amount of heat in uncooled produce. Therefore, produce must be cooled to the optimum transport temperature prior to loading.

Refrigeration Systems During Transportation

There are several systems used to control the temperature of perishables during transportation. The most commonly used cooling systems are mechanical refrigeration, ice and cryogenic cooling.

Mechanical refrigeration: Mechanical refrigeration is the most commonly used temperature control method in refrigerated transport vehicles. It is based on a vapour compression cycle common to most refrigeration equipment and has been used for transport vehicles. Chlorofluorocarbon refrigerants were commonly used due to their excellent physical characteristics, however they were found to contribute to the depletion of the earth's protective ozone layer. Two control modes are available in mechanical refrigeration systems, continuous and automatic.

Ice cooling: Ice is one of the simplest and oldest methods of maintaining a low temperature of produce. It is most commonly used with produce shipped in a non-refrigerated transit system. Ice has a cooling capacity of $335 kJ\cdot kg^{-1}$ and one unit weight of ice will drop the temperature of three times its weight of produce by about 28° C. Ice also maintains a high humidity around the produce and minimizes produce moisture loss. Finely crushed ice, flaked ice or liquid ice (a mixture of crushed ice and water) can be blown onto the top of produce boxes; this is called top-icing. The ice remaining after cooling protects the produce from warming and dehydration during transport. The use of ice for produce cooling and transport is declining because of a number of disadvantages. A typical liquid ice machine used for a 9 kg broccoli box requires 14.5 kg of ice although the produce temperature reduction requires the equivalent of only 3 kg of ice. Water resistant containers have to be used with ice. In a mixed load, the melting ice can damage neighbouring containers that are not water resistant. Liquid icing is used with only a few produce items such as broccoli, sweet corn and leafy vegetables. Also, the weight of ice reduces the quantity of produce that can be loaded on trucks and trailers, since there are limits on the weights of vehicles circulating on highways. Gel ice usually has a lower melting temperature than water ice and may have to be separated from the produce to prevent freeze injury to sensitive produce.

Cryogenic cooling: Cryogenic cooling with liquid N_2, liquid CO_2 or solid CO_2 (dry ice) is occasionally used to maintain temperature of fresh fruits and vegetables. It was developed before mechanical refrigeration was widely used in transport. An advantage of dry ice is that it does not produce a liquid, but sublimates to a gas. As with all cryogenic materials, the gaseous N_2 or CO_2 must be vented from the cargo area to prevent an unintended modified atmosphere around the produce. Due to the difficulty in controlling atmospheric conditions and the development of efficient mechanical refrigeration, cryogenic cooling is limited mostly to frozen products, local deliveries, and air transport. Dry ice is solidified carbon dioxide and has 92 per cent more refrigeration capacity per mass (645 kJ kg^{-1}) than water ice (335 kJ kg^{-1}). In air transport, this may make freight costs lower for dry ice than ice if there is a weight restriction. Dry ice has a temperature of -78° C which is too cold to be packed with horticultural produce. Often dry ice is put on a sheet of foam insulation placed above the produce so that the cold CO_2 gas and air mix together before they finally contact the produce. One kg of dry ice produces about 1100 m^3 of CO_2 gas under standard atmospheric conditions.

Heating under low temperature condition

During cold weather conditions, transport vehicles must be heated to avoid chilling or freezing of fresh produce. Mechanically refrigerated trucks, trailers, railcars and containers are usually heated by electric heating or by running the refrigeration unit in reverse cycle. However, the heating capacity of the mechanical refrigeration system operating in reverse cycle is only about a third of the effective refrigeration capacity of the unit when set at 0° C in a 30° C ambient temperature. Thus, supplemental electric heating may be required when transporting produce in very cold conditions.

Air circulation systems during transport

Air circulation plays a fundamental role in maintaining optimal temperatures during transport. Regardless of the capacity or sophistication of the refrigeration system, without an air circulation system the produce will not be protected from temperature extremes. Good air circulation is needed to transfer heat inside the temperature controlled volume to the refrigeration system. The insulation of trailers or containers can only slow down the external heat from penetrating, however it cannot economically stop the heat flow. Air circulation systems must be designed to allow rapid heat removal and uniform air distribution. Rapid heat removal also helps to reduce temperature gradients across the evaporator and prevents frequent defrosting of the refrigeration unit. If air is unevenly distributed throughout the load, parts of the load may over-heat and other parts may be over-cooled, which accelerates produce spoilage.

Top-air delivery: The top-air delivery system is used for air circulation in refrigerated semi-trailers and railcars. The refrigeration unit blows cold, high velocity air along the ceiling above the load from the front to the rear of the trailer. As the air flows toward the rear of the vehicle, some of it moves downward along the sidewalls. When air reaches the rear of the trailer, it then flows down along the rear door and then flows underneath the load along floor, returning to the front of the trailer.

Bottom-air delivery system: The bottom-air delivery system is commonly used in intermodal and marine containers. In sea containers, it generally consists of a T-beam floor, vertically ribbed rear doors, vertically ribbed sidewalls and a solid bulkhead in the front. It requires these extra features to support the pressurized vertical airflow that is used for this delivery system. Like the top-air delivery system, the refrigeration system is in the front and circulates the air from the front to the rear. However, in the bottom-air delivery system, air is blown to the bottom along the T-beam floors and through pallets, rather than along the ceiling. As the air flows from the front to the rear, it is forced upwards through the cargo. When air reaches the rear of the container, it flows up between the load and the rear doors to the ceiling and then returns to the refrigeration unit, at the front of the container, through bulkhead openings. Pallets, boxes and inner packaging should have enough venting and airspaces to allow vertical airflow through the pallet load. If air cannot flow through the packages, it will flow around the pallet loads, causing greater temperature variation in produce throughout the container load.

Physical and Structural Considerations during the Selection of a Transport System

Floor of vehicle:

Top air delivery: In a top-air delivery system, the space between the load and the floor of the trailer acts as a plenum for returning air to the evaporator. If there is insufficient space for return air between the floor and the load, airflow will be inhibited preventing conditioned air from reaching the load. To achieve maximum fan capacity around 0.15 to 0.19 m^2 of return air space is required. The most common types of floors found in refrigerated semi-trailers are flat floors (without any channels), duct board floors, duct-T floors, and T-beam floors. Each design offers a different cross sectional area for return air passage, whereas flat floor provides no return air space. T-beam floors provide more return air passage than other designs, but it has several disadvantages. It is more susceptible to forklift damage during loading and unloading operations compared to other floor designs. Debris can easily accumulate in the deep channels, making the floor very difficult to clean and reducing the effectiveness of the floor design. Furthermore, forklifts are more susceptible to slippage during handling operation on this type of floor due to the reduced contact area between the floor and tires. Placing produce on pallets significantly increases the area available for air circulation and should be used in trailers with duct board floors and is a necessity in flat floor trailers.

Bottom air delivery: A uniform temperature can be maintained in the load only if the floor of the trailer or container is completely covered with produce on pallets or in boxes, or with other solid material from the bulk-head to the end of the floor rails. When the floor is covered, refrigerated air is forced up through and around the packages. If the produce is palletized, the pallet opening for forklift tines (sometimes called pallet pockets) should also be covered to prevent air from traveling horizontally through these openings and escaping into an open vertical channel between pallet loads. Under hot ambient conditions, produce surrounded by poor airflow will tend to warm up during the transit period, especially if it had not been thoroughly cooled prior to transport or if it has a high respiration rate. An open space in the front of the container allows refrigerated air to flow through these open areas, causing produce in the rear of the container to be warm due to very little refrigerated air reaching it. The floor area between pallets can be covered with sections of fiberboard. Covering the floor and any pallet openings with fiberboard held down with boxes of produce is an effective way to cover the space at the rear of the trailer and avoid short circuiting in airflow. Under cold ambient conditions, produce in areas with low airflow is more susceptible to freezing or chilling injury.

Ceiling: The air-delivery duct helps to distribute air from the outlet of the refrigeration unit to the rear and both sides of the load. The duct is usually made of canvas or vinyl and is connected to the blower discharge through an adapter. Quick release fasteners can be used as connectors to the refrigeration unit or bulkhead adapter to ensure accessibility of the evaporator coils for cleaning, inspection and maintenance. The air-delivery duct should be mounted in the middle of the ceiling. The national perishable logistics association/refrigerated transportation foundation (npla/rtf) recommends a minimum cross-sectional area of 0.155 M^2 for the air duct. Rather than having air spill down the side walls along the full length of the air delivery duct, it recommends discharging the air at three different positions along the ceiling: 20 per cent should be discharged near the front of the vehicle, 50 per cent at approximately 1/3 of its length from the front to the rear, and 30 per cent at 3/4 of its length. This configuration ensures air will spill down the side walls along the full length of the vehicle. The size of the air duct is normally matched to the type of refrigeration unit used and the ceiling area of the trailer. One disadvantage of the air duct is that it can obstruct the movement of the forklift during loading or unloading.

Ventilation of marine containers: Ventilation should be specified based on volume of airflow per unit of time ($m^3\ h^{-1}$ or $L\ s^{-1}$). Ventilation settings described as per cent vent opening are not meaningful because performance characteristics and the wide open vent capacity can vary considerably for different container designs. Ventilation rates above the recommended levels should not be used. Excess ventilation increases energy consumption. In hot, humid environments, it may also increase evaporator coil icing, decrease the cooling capacity and reduce the consistency of temperature control.

Doors and walls: Door seals should be kept in good condition to prevent air leakage. Doors should be kept closed whenever the refrigeration system is operating. Opened doors are very detrimental to temperature uniformity maintenance. Vertical channels on the rear doors and side walls have been found to improve air movement and reduce heat conduction through the rear doors and side walls. However ribbed or fluted walls are more likely to be damaged by forklifts during loading and unloading. Most semi-trailers are equipped with smooth doors and flat side walls.

Insulation: Insulation materials are used to reduce the amount of heat load entering the trailer. The amount of heat transferred by conduction through a wall depends on the wall surface area, the type and thickness of insulation, and the temperature difference between the inside and outside of the vehicle. Insulating materials should be of moderate cost, easy to apply, light in weight, and should have low thermal conductivity, low moisture permeability and very low water retention capacity. Insulation used for the vehicle floor should have enough strength to support loading equipment and the load. Insulation effectiveness decreases as the insulation material ages and gains water in normal use.

Return air bulkhead: A return-air bulkhead is a false wall that provides a clear pathway for air to return to the evaporator and isolates the load from the front wall. It also forces air to go around and under the load without short circuiting. The bulkhead can cover the full width and half the height of the front wall. Frame and solid bulkheads are commonly used. A solid or pressure bulkhead generates a pressure difference across the outlet and inlet of the fan. This causes air to circulate through, around and underneath the load before returning to the refrigeration unit rather than short circuit from the ceiling directly to the evaporator. The standard pressure bulkhead is usually made of fiberglass reinforced plywood and aluminum. The bulkhead should have a space of at least 75 mm between it and the front wall and a minimum open space of 150 mm between its bottom edge and the trailer floor. The top of the bulkhead must have an open area of 0.02 to 0.03 m^2 to allow mixing of top and bottom-air and allow some airflow to the thermostat in case of blockage at the bottom of the bulkhead.

Management of Fruit, Vegetable and Flowers Transport Systems

Selection of transport equipment: A semi-trailer for produce transportation should be in good physical condition, equipped with a duct floor, an air delivery duct and a solid bulkhead. It is also important to ask for a trailer equipped with an air ride suspension system, which will absorb the shock and reduce vibration damage to the produce.

Preparation of the load: Produce should have enough postharvest life for the trip and subsequent marketing after it reaches its destination, otherwise it should not be included in the load. Produce stored before shipping will have a shorter shelf life for transportation and at destination. Most transport refrigeration systems do not have the refrigeration or airflow capacity to rapidly cool produce which should be cooled to its optimum storage temperature before stowing. One exception to this rule is cut flowers that can be cooled in about 24 hours, in refrigerated highway trailers, because of their low mass.

Preparation of the trailer

Cleaning: Cleanliness of all refrigeration system components ensures proper air circulation around the load. It also reduces plant pathogen levels, chemical and odour contamination. Refrigerated trailers can be used for transporting a wide variety of commodities including non-food or even toxic material, thus previous loads may be a source of contamination if the cleaning process is not adequate. Floors should be washed or swept thoroughly before loading. Floor ducts should be free of debris to provide clear pathways for air circulation. To ensure proper drainage and prevent disease development, floor drain holes should be free from obstructions. Semi-trailers, containers, or railcars used for transporting fruits and vegetables may also be used to deliver seafood or frozen produce. In such cases, thorough cleaning and airing of the vehicle may be necessary to eliminate odours. This is particularly important when shipping odour absorbing produce such as apples, banana and berries.

Trailer condition: The trailer should have no serious damage. Doors should seal tightly when closed, the air chute should be intact, and the interior wall surfaces should be in good repair. Refrigerated trailers should have a solid front bulkhead. The air delivery duct should be inspected for damage and cleanliness prior to each trip. Dirt accumulation inside the air duct will be blown by the air and cause contamination of the load, so the air duct should be removed and cleaned at regular intervals using cleaning agents.

Precooling or pre-warming of transport vehicle: Transport vehicles should be precooled or pre-warmed before loading. This reduces the initial cooling or heating load on the vehicle's refrigeration or heating system. During summer, precooling the vehicle reduces the chance of the produce warming after loading and causing a larger demand on the refrigeration system. During winter, pre-warming of the vehicle reduces the chance of chilling or freezing injuries for produce. Modern mechanically refrigerated vehicles are equipped with microprocessors that automatically run the precooling cycle. In all cases, sufficient time has to be scheduled ahead of loading, as the precooling process may be long. For all types of vehicles, it is important to turn off the mechanical refrigeration system once the doors are opened for loading. When the refrigeration system is on, open doors allow the fan to draw in outside air. Outside air usually has a higher dew point temperature than the air in the vehicle, and excess moisture condenses on the refrigeration system coils. The ice formed on the evaporator coils will limit or block the air circulation during transport. Similar conditions occur when the vehicle's doors are left open longer than required, even when the refrigeration system is turned off. Thus, reducing the open door period to its minimum is highly recommended.

Loading: During the loading of transport vehicles, the following handling practices should be followed to prevent an excessive loss of quality during the transportation step:

» Avoid bumping pallets during handling

» Avoid puncturing MA plastic bags

» Load together produce that is compatible

» Load produce in centerline loading pattern

» Apply bracing between the pallets and both side walls

» Apply bracing to secure the rear pallets

» Do not load produce sensitive to vibration damage directly above steel-spring, suspended axles.

Additional considerations: Some produce, such as green onions and broccoli, are sometimes shipped with crushed ice in waxed corrugated cartons or plastic container units. They should never be stowed on top of other produce not packaged in ice. Water from the melting ice damages packages that are not water resistant. If ice packed produce must be shipped with produce packed in conventional corrugated boxes, use moisture proof divider sheets to protect corrugated boxes. However, dividers should not extend to the floor where they can block the air flow through pallet openings. Protective wrapping should cover the top, sides and bottom of the pallets or the load. Reflective wrap is more effective than opaque plastics or paper for reducing heat gain when the load is exposed to the sun. Clear plastic films are not use as a pallet cover, because solar heat passes through it and it is then trapped under it as in a greenhouse, resulting in a very rapid temperature increase. Insulated covers add some protection from heat gain, but the benefit is relatively small compared to the use of a simple reflective cover. Covers that incorporate gel ice are also available.

Loading patterns: The loading pattern used for arranging produce in a vehicle depends on the type of transport vehicle (semi-trailer, container, railcar, ship or airplane), the type of load (palletized or unitized), and the type of air delivery system (top or bottom). Horticultural produce can either be handled as palletized or unitized loads. The palletized load is the most commonly used loading arrangement because of its ease of handling and its reduced labour requirement. Nevertheless, unitized loads also allow better air circulation through the cargo when boxes are stacked evenly and air passages are left between rows of boxes. Refrigerated produce should be loaded away from sidewalls when transported on long trips through extreme hot or cold outside temperatures to prevent heat conduction. Also, loading produce away from the walls of the trailer increases air circulation in the space between the produce and the wall. This air circulation allows any heat transferred through the wall of the trailer to be eliminated before it reaches the produce.

Unloading and pre marketing storage of produce: Simple set of stairs can be constructed to ease the work of both loading and unloading produce. The stairs can be folded and pushed back under the back of the truck when the vehicle is in the motion. The step can be made of wood or steel matting and steel bars can be used for support. It is important to avoid rough handling, minimize the number of handling steps and maintain the lowest feasible temperature. If produce is to be stored before sale than whole sale and retail market need clean and well insulted storage room. If produce is held at destination for a short time (2-5 days), the handler can help to maintain quality and reduce losses by storing the produce at their most suitable temperature. Relative humidity is maintained between 85-95 per cent and ethylene level is kept below 1ppm by ventilating or using ethylene absorbent. Most of the fruits and vegetables grouped in to three categories:

1. Low temperature: Store room temperature is 0.2° C with 90-95 per cent relative humidity (Apple, pear, plum, peach, cherry, date, fig, grape, cabbage, carrot, cauliflower, spinach and mushroom, etc),
2. Slightly low temperature: Store room temperature is 7-10° C with 85-90 per cent relative humidity (guava, litchi, mandarin, sweet orange, olive, pine, apple, pomegranate, snap beans, brinjal, okra, pepper, chilies, cucumber, watermelons etc.),

3. Moderate temperature: Store room temperature is 16-18° C with 80-90 per cent relative humidity (mango, lime, lemons, banana, coconut, papaya, potato, ripe tomato, mature green squash, pumpkin, yam, ginger, dry onion, etc.).

Handling at destination and during retailing: When the produce arrives at destination, pallet covers and wraps should be removed and temperature and quality of the produce should be inspected. Prior to sale the produce to the consumes, some handler may desire to sort for quality or at least to discard damaged and decayed produced in order to give the produce better market appeal. If ripeness or maturity is not uniform, sorting at destination can provide the seller with a better price for the better quality. The ethylene releasing compound such as ether/ethephon in to a special environment or ripening room is more effective way to ensure uniform ripening. Produce still within 2° C of its optimum temperature range should be immediately transferred to a refrigerated room. At retail, produce should be refrigerated upon arrival. Since display cabinets or counters do not have the refrigeration capacity to cool produce, fruits and vegetables should be at their optimum temperature before being stacked for display. To prevent an increase in temperature during display, produce should not be stacked above the load line of the display cabinet or counter. The required details of temperature and humidity should be change as on quality of produce and external or internal factors of shipping environment. The shippers instructions should be maintained at all times.

Table 59. The approximate condition and shelf-life of fruits and vegetable during transportation

Fruit & vegetable	Approx. max storage, transit and shelf life (Days)	Optimum transit temperature (°C)	Container temperature set points (°C)	Highest freezing points (°C)	Relative humidity (%)
Apples – non chilling-sensitive	90-240	-1.1 to +1.0	1.1 to 2.2	-1.5	90-95
Apples – chilling sensitive	35-45	1.5 to 4.5	4.4 to 5.6	-1.5	90-95
Apricots	7-14	-0.5 to + 1.0	1.1 to 2.2	-1.1	90-95
Asparagus	14-21	2.2	2.2	-0.6	90-95
Avocados	21-28	5.0 to 8.0	5.0 to 12.8	-0.3	85-90
Bananas – green	14-21	13.0 to 14.0	13.0 to 14.0	-0.7	90-95
Blueberries	10-18	-0.5 to 0.0	1.1 to 2.2	-1.3	90-95
Carrots – topped	30-180	0.0	0.6 to 1.7	-1.4	95
Cherries – sweet	14-21	-1.1	1.1 to 1.7	-1.8	90-95

Fruit & vegetable	Approx. max storage, transit and shelf life (Days)	Optimum transit temperature (°C)	Container temperature set points (°C)	Highest freezing points (°C)	Relative humidity (%)
Clementines	14-28	4.4	3.3 to 4.4	-1.0	90-95
Coconut – flesh	30-60	0.0	1.1 to 2.2	-0.9	80-85
Sweet corn	4-6	0.0	0.6 to 1.7	-0.6	90-95
Courgettes	14-21	7.2	7.2 to 10.0	-0.5	90-95
Cucumbers	10-14	10.0	10.0 to 11.1	-0.5	90-95
Dasheens	42-140	13.3	11.1 to 13.3	-	85-90
Garlic	140-210	0.0	0.6 to 1.7	-0.8	65-70
Ginger rhizomes	90-180	13.3	12.8 to 13.3	-	85-90
Grapefruit	28-42	13.3	14.4 to 15.6	-1.1	85-90
Grapes	56-180	-1.1	1.1 to 2.2	-2.2	90-95
Guavas	14-21	10.0	9 to 10	-	85-90
Kiwi Fruit	28-84	0.0	1.1 to 2.2	-0.9	90-95
Kumquats	14-28	4.4	4.4	-	90-95
Lemons	30-180	12.2	10 to 12.8	-1.4	85-90
Lettuce – Iceberg	10-18	0.0	1.1 to 2.2	-	90-95
Limes	42-56	9.0 to 10.0	9 to 10	-1.6	85-90
Lychees	21-35	1.7	1.7 to 2.2	-	90-95
Mandarins	14-28	7.2	7.2	-1.1	90-95
Mangoes	14-25	13.3	12.8	-0.9	85-90
Melons – Honeydew	21-28	10.0	7.8 to 10	-1	85-90
Mineolas	21-35	3.3	3.9 to 6.7	-1	90-95
Nectarines	14-28	-0.5 to 1.0	-0.6 to +/-1	-1	90-95
Onions – dry	30-180	0.0	0.6 to 1.7	-0.8	65-75
Oranges – Blood	21-56	4.4	4.4 to 6.7	-	90-95
Oranges – California	21-56	6.7	6.7 to 7.8	-0.8	85-95
Parsnips	120-150	0.0	0.6 to 1.7	-0.9	95
Peaches	14-28	-0.5	0.6 to 1.7	-0.9	90-95
Pears	120-180	-1.1	0.6 to 1.7	-1.6	90-94

Fruit & vegetable	Approx. max storage, transit and shelf life (Days)	Optimum transit temperature (°C)	Container temperature set points (°C)	Highest freezing points (°C)	Relative humidity (%)
Pears – Bartlett	70-90	-1.1	0.6 to 1.7	-1.6	90-94
Sweet peppers	12-18	10.0	10.0	-0.7	90-95
Hot peppers	14-21	10.0	10.0	-0.7	90-95
Pineapples	14-36	10.0	10.0	-1.1	85-90
Plantains	10-35	13.0	14	-0.8	85-90
Plums	14-28	-0.5	1.1 to 2.2	-0.8	90-95
Potatoes – seed	84-175	4.40	5	-0.8	90-95
Potatoes – table	56-140	6.0	7	-0.8	90-95
Satsumas	56-84	4.0	4	-	85-90
Sweet potatoes	90-180	14.0	14	-1.3	85-90
Tangerines	14-28	7.0	7	-1.1	85-90
Tomatoes – green	21-28	13.3	13.0 to 14.0	-.5	90-95
Tomatoes – turning	10-14	9.0	10.6	-0.5	90-95
Watermelons	14-21	10.0	8.0 to 10.0	-0.4	85-90
Yams – cured	49-112	16.0	16.0	-1.1	70-80

Marketing of fruit, vegetable and flowers

Marketing of horticultural crops is quite complex and risky due to the perishable nature of the produce, seasonal production and bulkiness. The spectrum of prices from producer to consumer, which is an outcome of demand and supply of transactions between various intermediaries at different levels in the marketing system, is also unique for fruits, vegetable and flowers. Moreover, the marketing arrangements at different stages also play an important role in price levels at various stages viz. from farm gate to the ultimate user. These features make the marketing system of fruits to differ from other agricultural commodities, particularly in providing time, form and space utilities. While the market infrastructure is better developed for food grains, but fruits and vegetables markets are not that well developed and markets are congested and unhygienic. Marketing system in India involves chain of commission agents from producer to consumers and these agent rotate market as per their desire.

Different marketing channels for horticultural produce

The marketing cycle starts from the small farmers – large farmers – commission agent – trader – commission agent – wholesaler and finally the retailer. Due to high perishable nature and long marketing channels in the horticultural produce market the producers have to sell their produce at very low prices, whereas consumer has to purchase it at high prices. Marketing efficiency is depended on length of the marketing channel to a great extent. Those farmers present at dense consumers area are get very good market price through direct marketing, whereas truck gardening farmers face a long chain of commission agents during storage, transportation, marketing and retailing of the produce. There are several marketing channels are present for sell the horticultural produce as:

Long chain marketing

i. Producer-trader-wholesaler-retailer-consumer

ii. Producer-cold store-trader-wholesaler-retailer-consumer

iii. Producer-trader-retailer-consumer

Short chain marketing

i. Producer-trader-consumer

ii. Producer-retailer-consumer

iii. Producer-consumer

Continuation of lengthy marketing channel deprived of the producer of his proper share in the consumer rupee and thereby plummeting the marketing efficiency.

An efficient marketing system can:

- » Reduce post harvest losses
- » Enhances farmers' realization
- » Reduce consumer price
- » Promote grading and food safety practices
- » Induce demand driven production
- » Enable higher value addition
- » Facilitate export

Supply chain management: Supply chain management represents the management of the entire set of production, manufacturing/ transformations, distribution and marketing activities by which a consumer is supplied with a desired product. Supply chain management encompasses the planning and management of all activities involved in sourcing procurement, conversion, and logistics management. It also includes coordination and collaboration with channel partners, which may be suppliers, intermediaries, third-party service providers, or customers. Supply chain management integrates supply and demand management within and across companies. Hence, supply chain management is defined as the design and operation of physical, management information and financial systems needed to transfer goods and services from point of production to point of consumption in an efficient and effective manner. The entire supply chain management process is a value chain where bottlenecks, value adding factors and liability factors are identified and addressed, thus enabling the retail organization to have an efficient supply chain. The supply is the part of retail operations that ensures the right product is in the right place, at the right time and at the right cost. The supply chain perspective can help the retailers identify superior suppliers and distributors and help them to improve productivity, which ultimately brings down customers costs. Supply chain development not only benefits the private sector but also creates spin-offs that stimulate social, economic and environmental sustainable development in the region (employment generation, added value, minimization of product losses etc.). The specific gains are: Reduction of product losses in transportation and storage, increasing of sales, dissemination of technology, capital and knowledge among the chain partners, better information about the flow of products, markets and technologies, transparency of the supply chain, tracking and tracing to the source, better control of product safety and quality, large investments and risks are shared among partners in the chain, productivity improvement, high customer satisfaction, increased profit and on time delivery fruits, vegetable and flowers.

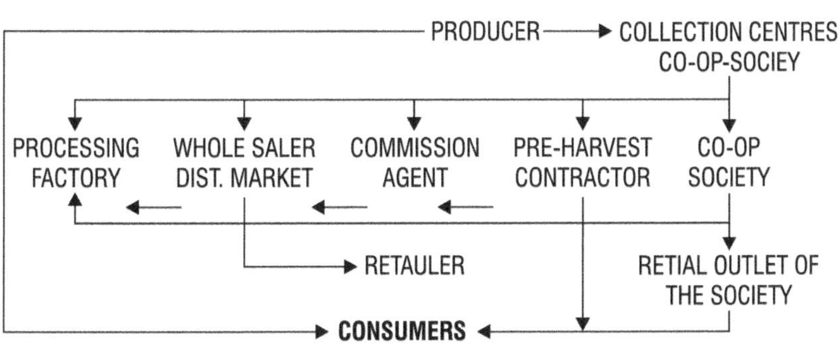

Figure 31. Fruits and vegetables supply chain in India

Input - Production – Procurement and storage - Primary processing- Secondary processing – Retailing

Present scenario of supply chain in India: There is lack of basic as well as specialized infrastructure such as cold storages, refer vans, cool chains, ripening chambers etc. Also there is a missing link between production and research system and consumers demand. The system lacks in capacity building, market information, research and intelligence. India has a huge opportunity to become a leading global food supplier if only it has an agile, adaptive, responsive and efficient supply chain. Some of the problems that are to be mentioned in Indian fresh produce supply chain are the presence of numerous stake holders which are working in isolation and the infrastructure connecting these partners is very weak. There is lack of demand estimation and technology applications such as cold chain, logistic supply chains and product tracking and tracing. Lack system integration and presence of large number of unorganized retailers result in making unorganized supply chain practices further inefficient with several inter mediaries and manual handling. The result is lots of wastages as much as nearly 25-30 per cent and also less remuneration for the farmers.

Table 60. Estimated/assumed waste percentage in each step of the FSC in India

Commodity	Agricultural	Post-harvest	Processing and Packaging	Distribution	Consumption
	Production stage	Handling and storage stage			
Fruits and Vegetables	5-15%	3-9%	10-25%	5-10%	5-7%

Factors affecting India's supply chain management of horticultural produce

Availability of cold storage: Total number of cold storages in India is about 7916, with the total installed capacity of 36.23 million MT. Uttar Pradesh, Maharashtra, West Bengal, Punjab and Gujarat account for more than 60 per cent cold storage capacity followed by Andhra Pradesh, Haryana and Madhya Pradesh. Region wise number of cold stores in India are Central Region 352 (6%), South Region 778 (7%), East Region 947 (33%), North Region 2386 (47%) and West Region 853 (7%).

Government policies: The Governments play a leading role in formulating the policies for food and agriculture which includes farming, production, processing, distribution besides the financing and retailing these activities. There are several multitudes of departments handled between both State and Central Governments. In view of the globalization of this industry, the decisions if are mutually reinforcing and consistent would be extremely productive. Further research initiated in various other fields of food and agriculture may be helpful to further strengthen our economy.

Connectivity: Transferring of fruits and vegetables from one place to other place connectivity plays a vital role in its successful implementation. In India the road infrastructure is not so good so transferring fresh produce such as fruits, vegetables and flowers is a real big challenge. Thus to improve the efficiency of supply chain more attention is needed in this aspect.

Sorting and grading technology: For fruits, vegetables and flowers sorting and grading methodology needs to be updated technically as farmers do it themselves and they have very little knowledge about the processes. To improve the efficiency of the supply chain and improve the skills of the farmers.

Handling and packaging: Most of the fruits and vegetables are in remote village areas and all the places are not connected with proper facilities of handling and packaging. Because of this lot of fruits and vegetables are getting spoiled, wasted and quality deteriorated to be used in supply chain.

Skilled labour: Government should organise training program on regular basis under skill India programme on different post harvest technologies and value addition of fruits and vegetables and may also build some institutions which provide scientific training.

Linkage in marketing channel: Proper marketing and its information on other factors needs to be properly available for providing good efficiency for fruit and vegetable supply chain and thus development of strong economy.

Problems in Horticultural Produce Marketing

Marketing of the fruits, vegetables and flowers from to consumer poses a number of problems which are the bottlenecks for getting remunerative price to the producers on one hand at the reasonable price to the consumers on the other side. India has a huge opportunity to become a leading global fruits and vegetables supplier and consumer if only it has an agile, adaptive, responsive and efficient marketing system. Some of the problems that are to be mentioned in are:

Commission agent's exploitation: The trader is commission agent, does not attend to the small and marginal farmers properly and these farmers have to wait for a long period for auction of their produce.

Presence of large number of unorganized retailers: At present in fruit and vegetable market the unorganized retailers are linked with farmers through wholesalers or commission agents. The commission agents and wholesalers redundant supply chain practices make unorganized further inefficient.

Marketing distortion: These are high and unjust market charges levied on producer- seller, pooling by traders, lack of open bid system and delayed payment. Malpractices are rampant and the rational, operates in a highly segmented manner.

Lack of system integration: The fruits, vegetables and flower marketing chain need to be designed and built as a whole in an integrated manner. The process of product procurement and order to delivery processes should be well designed and well supported with the help of IT tools and software.

Lack of bargaining power: The maximum flowers and vegetables producers are small and marginal and all these growers are unorganized. They are uncapable to group action and have less bargaining power. As a result the traders exploit these farmers. Unless collective efforts are made awareness is created among the farmers.

Numerous producers working in isolation: Marketing of flower and vegetables is complex due to highly perishable nature and numerous small stake holders. In India, the infrastructure connecting these small farmers is very weak.

Lack of transport facilities: Small and marginal farmers cannot afford to transport their produce to the distant mandies/market on account of non-availability of transport facilities, expensive transportation cost and malpractices in the market such as heavy deductions. Hence, these farmers sell their produce at farm level to the village merchants, retailers or pre-harvest contractors on stumpy rate.

Lack of market intelligence: At present the farmers lack knowledge and equipment for grading, treatments and packaging of fruits, vegetables and flowers. Information regarding demand and supply, price, market outlook, consumer's preference, marketing channels and practices are important for marketing of produce which is lacking at the moment. Demand forecasting is totally absent and the farmers try to push whatever they produce into the market.

Lack of technology applications: Cold chain logistic supply chains should take advantage of technology improvements in data capture and processing, product tracking and tracing, synchronized freight transport transmit times for time compression along the supply chain and supply demand matching.

Measures for Improving Horticultural Produce Marketing and its Effectiveness

There has to be structural changes at different levels - farmers, intermediaries and consumer. The government, private, public-private partnership, cooperatives, technology providers, and even media can play a crucial role. Infrastructure like roads, transport, information and communication technology and cold storage are basic requirement for better results in marketing of fruits, vegetables and flowers. The Department of Horticulture and Department of Agricultural Marketing acts as the facilitator for creation of infrastructure facilities for marketing of fruits vegetables and flowers in the states.

Demand forecasting of produce: Demand forecasting is one of the important requirements for improving marketing system of fruits, flowers and vegetables effectiveness. Due to poor forecasting, there is an imbalance between supply and demand. In some months vegetables are either not plucked from the farm due to lack of demand. Where as some time the, produce is not available and a result, prices are boost up.

Market linkage and coordination: Vertical coordination of farmers through cooperatives, contract farming and retail chains would facilitate better delivery of output, reduce market risks, provide better infrastructure, attract more public interest, acquire better extension services, and create awareness regarding the prevailing and new technologies.

Linking production and marketing: There is an urgent need to integrate fruit, vegetable and flower production and marketing. Establishing linkages between farmers to consumers is an important aspect of the agro business. Products should meet the requirements of consumers. Therefore, production of fruits and vegetables must be linked with the buyer's requirement in respect of variety, quantity and application of farm practices.

Development of market infrastructure: Marketing Boards should develop rural markets, create market intelligence system, provide grading and packing facilities in the market and monitor the marketing system. Creation of transport and communication facilities in rural areas is very important. The private sector may be encouraged to invest in such infrastructures. Customized logistics is another important immediate requirement to make logistic effective. This reduces the cost, facilitates the maintenance of quality of the produce and fulfills the requirements of targeted customers.

Better information system and public private coordination: Information system for better coordination among different stakeholders from farmers to consumers is the need of the hour. The internet and mobile communication can also be used to enable information and financial transfer between the stakeholders. Public private partnership is another strategic solution. Supply chain like washing, waxing, grading, sorting, packing, pre-cooling, handling facilities, insurance, finance, transport and processing facilities would add value to marketing of fruits, vegetables and flowers functioning.

Organization of cooperative and self-help group: Small and marginal farmers, growing fruits and vegetables should be encouraged to from their group in the form of cooperatives or self-help groups (SHGs). This will not only to realize economies of scale but will also be able to provide business orientation to their activities, gain bargaining powers, avails quality inputs, higher share in consumer price. In addition, such cooperatives or groups may start processing activities, buy transport means, and build up storage infrastructures and put pressure on the marketing system to improve efficiency.

Contract farming: The concept can establish a perfect linkage between the buyers and processors and thereby they can get an assured business in terms of quantity of take and guaranteed price for the produce.

Training to educate the farmers: Scientific practices are becoming the order of the day; hence, farmers and processor also need suitable training to get equinted with all aspects of production, quality control, handling practices and distribution aspects.

Development of agro export parks: The main objective of establishing export parks it to promote agro and processing industries in cluster in area where there is predominant production of process able horticulture products for export. These parks will also provide the required infrastructural and common facilities which are essential for sustenance of the industries. Quality assurance laboratories, ware housing including cold storages, common effluent treatment plants etc.

9

Ethylene: Fruit Ripening Hormone and Fruit Ripening

HISTORICAL BACKGROUND OF ETHYLENE

One century ago, Anton Nicolyovitch NelJubov resolute that the gaseous olefin ethylene is capable of altering the development of higher plants in different ways. He used the triple response of etiolated pea epicotyls (diageotropism, inhibition of elongation, stimulation of lateral expansion) to assay the various components of coal gas for biological activity. The constituent with the greatest activity was found to be ethylene and Gane positively identified the gas as a natural plant product in 1934. In 1968, plant physiologists recognized it as a true growth regulator or plant hormone.

$$H_2C = CH_2$$

Ethylene

Sources of ethylene in the environment:

Ethylene is ever present in nature and available at low levels in the atmosphere. It is a common pollutant that can only be detected with sensitive instruments. Ethylene can come from many sources both within and outside the package. The major sources of ethylene are combustion, produced by plants and microorganisms, in ripening rooms, in fluorescent ballasts and rubber materials, etc.

Table 61. Physical properties of ethylene gas

Physical properties	Characteristic
Appearance	Colourless, hydrocarbon gas with a faint, sweetish odour that is easily detected in parts per million concentrations
Molecular weight	28.05
Boiling point	
at 760 mm Hg	-103.7°C (-154.6°F)
at 300 mm Hg	-118°C (-180°F)
at 10 mm Hg	-153°C (-243°F)
bp/p at 750 to 770 mm Hg	0.022°C per mm Hg
Freezing point at saturation pressure (triple point)	-169.2°C (-272.6°F)
Surface tension at -103.7°C (-154.6°F)	16.4 dynes^{-1} cm
Flammable limits in air*	
Lower	3.1 % by volume
Upper	32% by volume

Note: * All compositions between the upper and lower limits are flammable and can be explosive.

Overcoming ethylene's undesirable effects: Removing ethylene from the atmosphere around the commodity is the preferred method of preventing deterioration of ethylene sensitive produce.

» It is necessary to eliminate the sources of ethylene around the fruit environment.

» There is should be proper ventilation outside storage and handling areas.

» Ethylene can be removed by using potassium permanganate, ultraviolet lamps, activated or brominated charcoal, catalytic oxidizers, bacterial system and hypobaric storage.

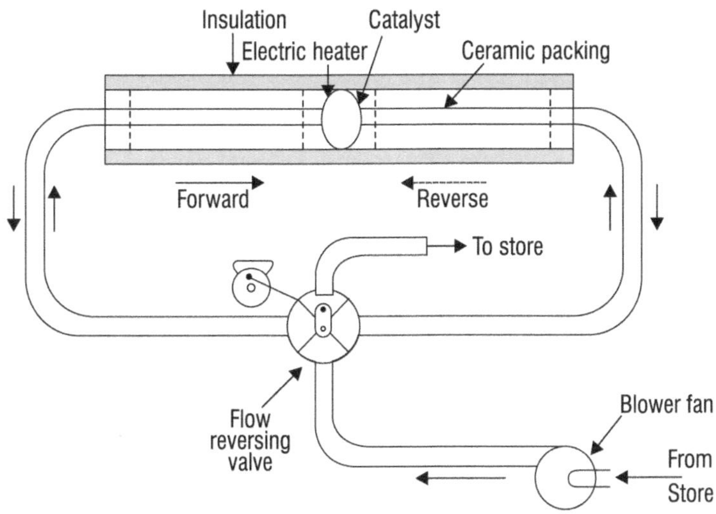

Figure 32. Swingtherm unit removes ethylene from air using a catalytic oxidizer and is equipped with heat exchangers to reduce energy consumption (Blanpied *et al.*, 1985).

FDA status: The use of ethylene gas to promote ripening of fruits and vegetables is sanctioned under FDA Regulation 120, 1016. Ethylene is exempted from the requirement of a residue tolerance when used as a plant regulator either before or after harvest.

Measurement of ethylene: A remarkable feature of the effects of ethylene on plants is the minute concentration required. Fruit ripening, for example, typically occurs at the maximum rate at levels of 1 part of ethylene in 1 million parts of air (1 ppm). Effects on opening of roses can be seen at ethylene concentrations as low as 10 parts per billion parts of air (10 ppb). The development of the gas chromatograph and the flame ionization detector, which made it possible to rapidly measure ethylene at such low concentrations, was crucial to our present understanding of the role of ethylene. New techniques, such as laser-acoustic devices, offer even more sensitivity. For the postharvest technologist, gas-sampling tubes giving a colorimetric reaction can be read reasonably easily down to 1 ppm. These are satisfactory for occasional use by ripening room operators but cannot monitor the low levels of ethylene that may be of concern in storage rooms and marketing outlets.

Safety precautions: Because of the explosion hazard of ethylene mixed with air at concentrations from 3 to 30 per cent, the rules listed below must be followed stringently to prevent buildup of these concentrations and to prevent ignition if they should form.

» Do not permit open flames, spark-producing devices, fire, or smoking in or near a room containing ethylene gas or near the cylinder.

- » Use an approved meter for accurately measuring the gas when discharging ethylene from the cylinder.
- » Ground all piping to eliminate the danger of electrostatic discharge.
- » Store ethylene cylinders in accordance with all instructions and standards of the National Board of Fire Underwriters.
- » All electrical equipment, including lights, fan motors and switches, should fulfill with the National Electric codes for Class 1, Group D equipment and installation.
- » Instruments that detect the concentration of ethylene in air can be set to sound an alarm if the concentration approaches explosive levels.

Biosynthesis of Ethylene

Plants in vegetative growth phase, flowers and immature fruits produce hardly detectable levels of ethylene until they are subjected to stress or undergo maturation events after which ethylene production accelerates in a spatially and temporally specific pattern. All vascular plants analyzed to date synthesize ethylene via the yang cycle wherein s–adenosyl methionine (adomet) is diverted from alternative fates to make the ethylene precursor 1–aminocyclopropane–1–carboxylic acid (ACC). Expression of the enzyme catalyzing this reaction, ACC synthase, is induced by stimuli that lead to increased ethylene production. ACC can be readily converted to ethylene, CO_2 and HCN by ACC oxidase or it can be conjugated to a malonyl or glutamyl group to limit its availability for ethylene production. The gene encoding ACC synthase was first isolated in zucchini and cloning of homologs from other plants quickly followed. A multigene family whose members are differentially expressed in response to developmental, environmental and hormonal factors encodes ACC synthase. For example, by using gene specific probes the differential expression of tomato ACC synthase family members has been investigated. Transcript of one isoform increased during fruit ripening, those of another increased in response to wounding and those of a third form increased in response to treatment with auxin. ACC synthase genes expressed in response to a particular stimulus are more similar to genes controlled by the same stimulus in other species than they are to other ACC genes in the same species.

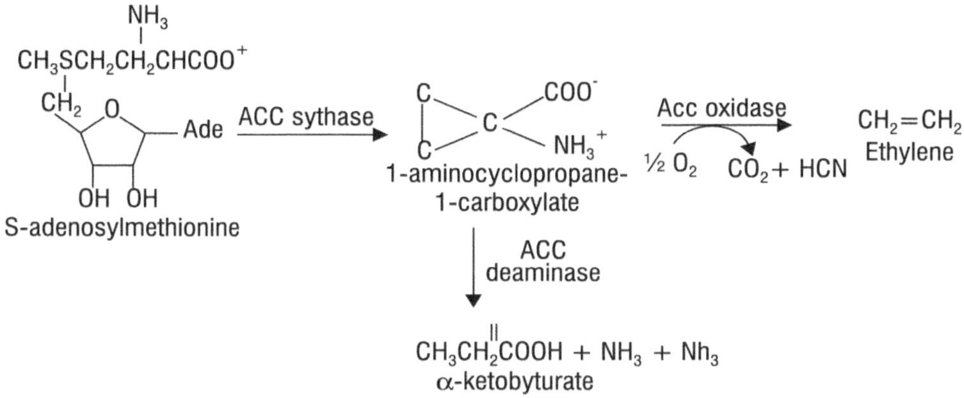

Figure 33. Enzymes involved in ethylene biosynthesis. Ethylene is synthesized from S–adenosylmethionine via the intermediate 1–aminocyclopropane–1–carboxylate (ACC). Genes encoding the ACC synthase and ACC oxidase enzymes have been cloned from several plant species. Synthesis of ethylene can be inhibited by expression of the gene encoding ACC deaminase in transgenic plants. This enzyme shunts ACC to a–ketobutyric acid and ammonia, thus preventing its conversion to ethylene.

The final step in ethylene biosynthesis, the conversion of ACC to ethylene, is catalyzed by ACC oxidase. ACC oxidase was first identified by expressing the tomato cDNA pToM 13 in an antisense orientation, which resulted in greatly reduced ethylene production in tomato. ACC oxidase requires CO_2 and ferrous ion for activity, CO_2 has been postulated to covalently modify an internal lysine side chain as a carbamylate moiety in the ACC oxidase protein. There are seven conserved lysine residues in ACC oxidases and independent substitution of each of them suggests that some of the lysine residues are required for full enzyme activity. It has not yet been demonstrated whether the substitutions directly interfere with carbamylation. Small multigene families encode ACC oxidase. The differential expression patterns of ACC oxidase gene suggest that oxidases contribute to regulation of ethylene production as well. Periods of ACC oxidase induction correlate with ethylene regulated events in several instances including senescence, fruit ripening and wounding. In addition, ACC oxidase and ACC synthase together contribute to the positive feedback loop wherein ethylene treatment leads to increased ethylene production.

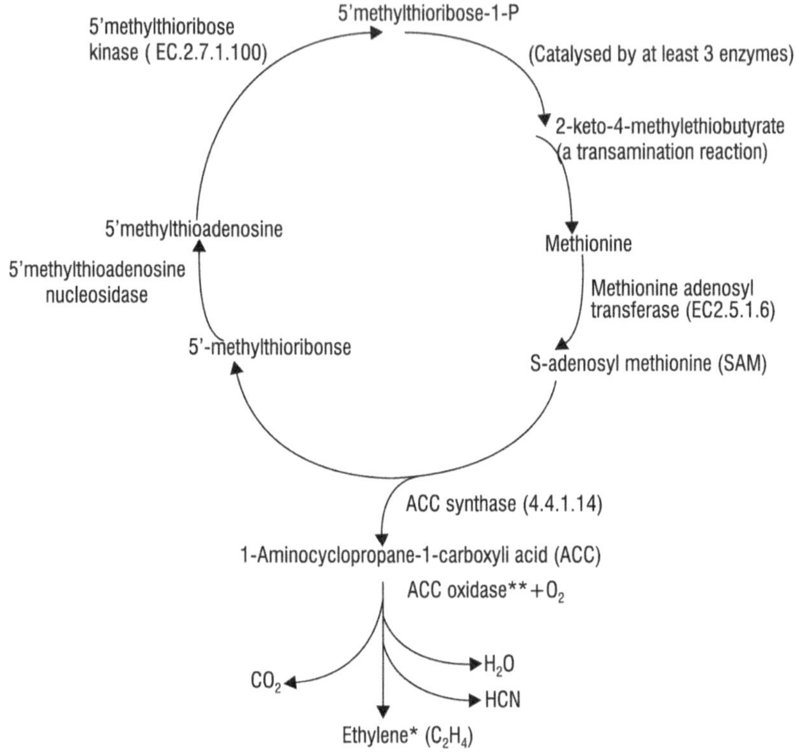

Figure 34. Biosynthetic pathway of ethylene

Ethylene formation in model systems: The Cu^{2+} ascorbate rapidly converts methonine into ethylene and methionol seems to be intermediate, and ethylene is produced from carbons-3 and 4 of methionine. In this reaction Cu^{2+} is reduced to Cu^{2+} by ascorbic acid and reoxidation of Cu^{+} to Cu^{2+} by O_2. Linolenic acid also forms ethylene with Cu-ascorbate system in some situations. The conversion of linolenic acid yields not only ethylene but also ethane. A non-enzymatic conversion of methionine to ethylene is demonstrated by photochemical reaction based on FMN. However, under natural conditions, enzymes play the role of catalysts.

Regulation of ethylene biosynthesis: Rates of ethylene formation vary in different organs and tissues and depend upon growth and developmental stages of an organism. Ethylene production is correlated with several physiological processes like seed germination, vegetative growth, formation of flowers, fruits ripening, leaf senescence, abscission of leaf, flowers and fruit. Different environmental factors like light, temperature, minerals, stress conditions, wounding and chemical factors are also responsible for ethylene biosynthesis.

Ethylene binding: Probably all plants produce ethylene, and all cells are exposed to ethylene at some stage. But all cells are not equally affected and many cells that respond to ethylene do so only at certain stages of development. The affected cells at that particular time possess a particular molecular complement, which recognizes and responds to ethylene. Ethylene is thought to bind to this receptor, enabling a primary reaction that then initiates other reactions, including modification of gene expression.

Ethylene metabolism: Besides being converted to ethylene, ACC can also be irreversibly conjugated to form N–malonyl–ACC. Malonylation of ACC regulates the level of ACC and thus production of ethylene. Plant tissues to ethylene oxide and ethylene glycol can metabolize ethylene but the physiological significance of this metabolism remains to be established. As a gas, ethylene can readily diffuse from plant tissues so metabolism is not essential for its removal.

Biochemical Action of Ethylene

Ethylene has various biochemical actions on plant such as RNA synthesis, protein synthesis, protein loss, enzymes (cellulase, glucanase, pectinase, polygalacturonase, a-amylase, catalase, chitinase, phenylalanine ammonia lyase, malic enzyme, peroxidase, leakage of peroxidase, release of alfa-amylase, stabilization of enzyme activity), modification of ribosomes, mitochondria, ATPase stimulation, swelling, volume change, inducement of alternative path of respiration, chloroplast (chromoplast), chlorophyll synthesis, chlorophyll degradation, degreening, carotenoid synthesis, pigments and formation of anthocyanin. The pigment metabolisms result from the ethylene enhancement of related enzymes. The modification of endoplasmic reticulum indicates a change in protein synthesis or transportation of proteins. Eventually ethylene action results from DNA, RNA and protein synthesis of efficient utilization.

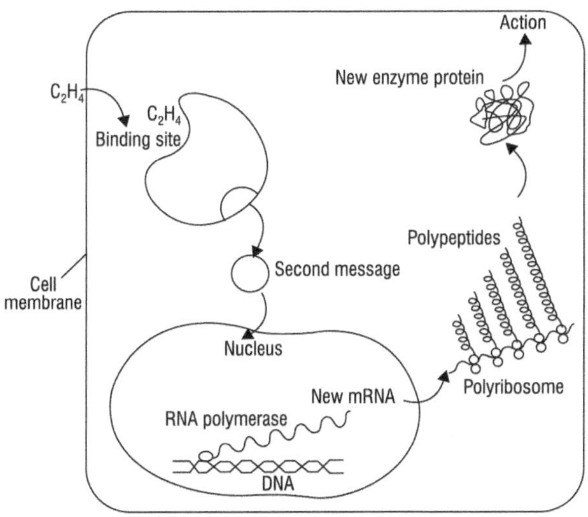

Figure 35. Mechanism of ethylene action

Biological effects: Ethylene is a powerful elicitor of morphological changes during all stages of the plant life cycle. At germination, ethylene causes the hypocotyl to swell and broaden, girding it as it penetrates through soil. As the plant matures, ethylene influences sex determination and promote fruit ripening. Ethylene also protects plants from the onslaughts of nature, strengthening trunks of wind baltered trees, assisting flooded plants to survive submergence and fending off pathogenic attack. Finally, ethylene plays a role in senescing flowers and leaves.

Ethylene and defense responses: Plants have a diverse arsenal of weapons with which to fend off pathogenic organisms. The initial response to bacterial or fungal pathogens called the hypersensitive response (HR) is the formation of necrotic lesions at the site of infection. A secondary, long lasting response, called systemic acquired resistance (SAR) develops throughout the plant in infected and uninfected tissues. Plant tissues displaying signs of pathogenic infection or wounding also produce elevated levels of ethylene. The ethylene insensitive mutants ETR^1 and EIN^2 and the ethylene over producer ETO^1 has indicated that ethylene is not essential. None of the mutants affect the resistance response in arabidopsis and both the hypersensitive response and inhibition of bacterial growth occur when the mutants are infected with avirulent bacteria. Interestingly, all mutants behave as wild type plants when infected with virulent bacteria except for the EIN^2 mutant. EIN^2 shows far fewer necrotic lesions than its wild type counterpart and no restriction in bacterial proliferation. The ethylene over producer ETO^1 neither confer greater disease resistance with avirulent bacteria nor greater disease susceptibility with virulent bacteria. These results indicate that EIN^2 plays an essential role in causing necrotic lesions during pathogen attack. During SAR response, numerous PR proteins are induced, one of which is known to confer increased tolerance to pathogen infection. Using the ethylene insensitive mutants ETR^1 and EIN^2, it was shown that the development of SAR response does not require ethylene.

Senescence, fruit ripening and abscission: As a plant develops eventually, organs that once were dangerous for early stages of growth become obsolete. In a prime example, newer, large leaves that shade them, rendering the old leaves ineffective for photosynthesis, supersede early leaves. To avoid squandering nutrients on nonfunctional structures, plants reclaim nutrients from the dying structure and throw out it by the ethylene regulated processes of senescence and abscission. A progressive yellowing of leaves, followed by necrosis and death, marks its decline. Several senescence associated genes have been identified whose expressions increase as leaves become yellow. Several catabolic enzymes or proteins have been implicated in catabolic events including cysteine and aspartic proteases, components of the ubiquitin proteolytic pathway and ribonucleases. Senescence is also characterized by chloroplast breakdown and remobilization of lipids and nutrients from the dying organ to the rest of the plant. The effect of ethylene is modulatory rather than an absolute requirement. Application of ethylene causes premature expression of senescence markers only in mature leaves. Ethylene mutants such as ETR[1] and EIN[2] show a significant delay in senescence of older leaves but eventually they do yellow and die. These data suggest that there are mandatory developmental events that must transpire before ethylene can potentiate the senescence process.

Leaf senescence is closely linked to fruit maturation so that energy can be shunted from the drying leaves to produce more vigorous fruits. Fruit ripening in the climacteric tomato fruit has been one of the most intensely studied ethylene mediated developmental processes. The tomatoes produced by the "Never–ripe" mutant fail to mature into red fruit and the mutation has been shown to map to a homolog of the ERS, ethylene receptor gene. A third phenomenon tied to senescence and fruit ripening is the process of abscission. It is complex series of molecular and morphological changes that result in the release of leaves, flower petals or fruits. The hydrolytic enzymes that act on cell wall components are expressed in the cells of abscission zone. The cells of abscission zone form a fracture plane and then develop into scar tissue once the organ has been released. Like senescence, ethylene promotes abscission.

Chlorophyll degradation: Degreening does not occur during exposure to ethylene in the peel of satsuma mandarin fruits, but does by exposing the ethylene treated fruits to air. Chlorophyllase enzyme is responsible for degradation of chlorophyll in plants. Its activity is low during exposure to ethylene and during the first 6 h of the subsequent exposure to air. These observations indicate that a lag phase is a prerequisite for the activation of chlorophyllase also that oxygen is required for ethylene enhanced chlorophyll degradation and chlorophyllase activity. The constructing effect of these two inhibitors of protein synthesis indicates that de novo synthesis of cytoplasmic chlorophyllase is essential for ethylene enhanced chlorophyll degradation. Ethylene degreening of citrus fruits is an ordinary commercial practice in USA, Japan, Israel, and Australia to assure product quality in the market.

Amount of ethylene gas needed

The recommended treatment concentration is 10 to 100 ppm (1 m^3 of C_2H_4 in 10,000 m^3 of room space, or 1 ft^3 of C_2H_4 in 10,000 ft^3 of room space). Lower concentrations are used in well scaled rooms that will maintain the ethylene concentration or in rooms where the trickle system is being used. Higher concentrations are used in leaky rooms to compensate for the fall in concentration during treatment. Concentrations higher than 100 ppm do not speed up the ripening process.

Sources of ethylene

Ethylene is an explosive gas therefore, it is advisable to use explosion proof ethylene mixtures in the ripening rooms. The oversupply of ethylene to a ripening room is eliminated by using mixtures of ethylene with inert gases. The proportion of the inert gas should be such that at high concentrations of ethylene not enough oxygen remains in the ripening space to provide an explosive mixture. For example, one commercial formulation, ripegas, contains 6 per cent C_2H_4 in CO_2 by weight. When using these mixtures, the calculation of volumes of gas required must be modified to reflect the composition of the mixture. In the case of ripe gas, for example, the weights, volumes or flow rates calculated using the formulas above would be increased by a factor of 100/6 or 17, to give the required concentration.

Ethylene generators: Ethylene generators, in which a liquid produces ethylene when heated in the presence of a catalyst, are now widely used for supplying ethylene in ripening rooms. The liquid, a proprietary product, comprises ethanol and agents that assist in catalyzing its dehydration:

$$C_2H_5OH - H_2O \text{ à } C_2H_4$$

The generator combines a heated platinized asbestos catalyst with a system for attaching a bottle of the generator liquid. The liquid comes in 1pint and 1quart bottles and is used up by the generator at the rate of 1 pint every 8 hours. The generator delivers about 14 l (0.5 ft^3) of ethylene gas per hour, adjustable on newer models. These figures can be used to determine the number of generators to be used in a given size of room, knowing the air leakage or ventilation rate (gas exchanges per hour). For example, in a 5,000 ft^3 ripening room where there is one air exchange per hour, a 1quart bottle will generate 100 ppm for 16 hours.

Ethylene related plant growth regulators: Ethylene as a gas can only be applied in more or less closed systems like glass houses or incubation chambers, limiting the usefulness of the hormone.

Table 62. Common and chemical names of some important ethylene releasing chemicals

Common name	Chemical Name
Ethephon	(2-chloroethyl) phosphonic acid
Silaid	(2-chloroethyl) methylbis (Phenylmethoxy) silane
Alsol	(2-chloroethyl) tris (2-methoxyethoxy) silane
ACC	1-aminocyclopropane-1-carboxylic acid

Ethephon: 2-chloroethylphosphonic acid (CEPA, Ethephon) is the best known compound of this group. It releases ethylene above pH 4.0 in aqueous solutions but is relatively stable below pH 4.

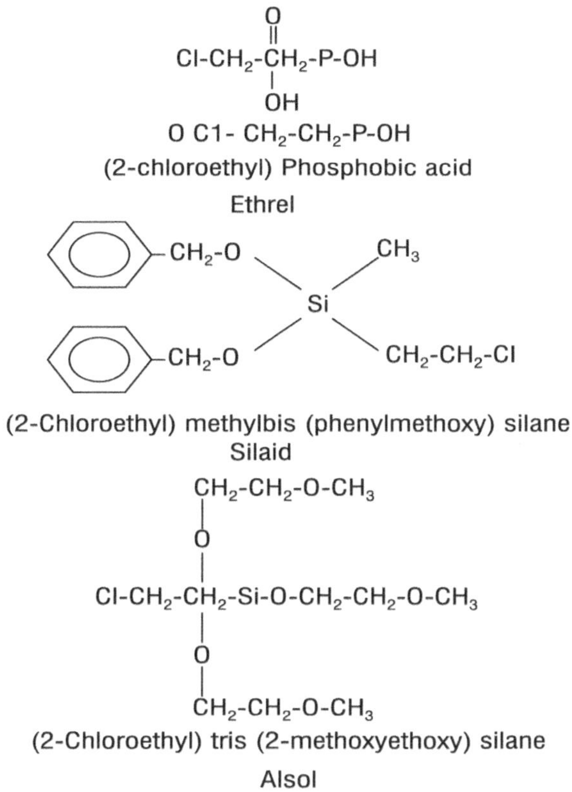

Figure 36. Molecules of ethephon, silaid and alsol

Table 63. Approved uses of ethephon (Kays and Beaudry, 1987)

Use	Approved crops
Postharvest fruit ripening	Bananas, tomatoes
Preharvest fruit ripening	Peppers, tomatoes
Fruit removal	Apples, crabapples, olive
Defoliation	Apple, pomegranate, lasoda, buckthorn, roses
Fruit loosening	Apples, blackberries, cantaloupes, cherries, tangerines
Maturity or colour development	Apples, cranberries, figs, filberts, grapes, peppers, pineapple, tomatoes
Degreening (preharvest)	Tangerines

Use	Approved crops
Degreening (postharvest)	Lemons
Dehiscence	Walnuts
Flower induction	Pineapple and other bromeliads
Sex expression	Cucumber, squash
Flower bud development	Apple
Stimulate lateral branching	Azaleas, geraniums

Caution: Caustic soda and ethephon are corrosive. Wear safety glasses and rubber gloves.

Etacelasil: 2-chloroethyl-tris (2-methoxyethoxy) silan (Etacelasil) belongs to the group of 2-chlorosilanes, also releasing ethylene in aqueous solutions. The chemistry of ethylene evolution is comparable to that of ethephon.

Haloethylsulfinic acids: A third group of ethylene releasing compounds is the 2-haloethylsulfinic acids and derivatives. These compounds also have a similar mechanism of ethylene release as compared with ethephon.

2-Hydroxyethylhydrazine: The growth regulator 2-hydroxyethylhydrazine causes typical ethylene reactions in plants.

Others ethylene releasing compounds: Three chemically unrelated growth regulators should be mentioned; cycloheximide (trade name Act-Aid), glyoxal-dioxim (trade name Pik-off), and 5-chloro-3-methyl-4-nitro-1H-pyrazol (trade name Release). These compounds are useful only for a very limited field of application, namely citrus fruit abscission.

Table 64. Comparative effectiveness of ethylene and related compounds

Compound	Relative Activity (moles/unit)	Compound	Relative Activity (moles/unit)
Ethylene	1	Carbon monoxide	2900
Propylene	130	Acetylene	12500
Vinyl chloride	2370	1- butene	140000

Ethylene management

Management of ethylene in the storage environment is obviously significant. However, it is also important to understand that as with storage temperature, the degree of concern about ethylene will be affected by the size of the operation involved. Detrimental effects of ethylene depend on a number of variables including its concentration, duration of exposure and temperature. Therefore, ethylene management in a small retail operation with rapid turnover of product is less critical that one that involves prolonged storage in a consolidated facility.

Prevention from sources of ethylene: Avoidance of exposure to ethylene begins with careful harvesting, grading and packing to minimize damage to the commodities. In the case of climacteric products it is difficult to reduce the internal levels of ethylene once autocatalytic production has started. Products should be cooled rapidly to their lowest safe temperature to reduce naturally occurring ethylene production and to decrease sensitivity to ethylene. Use of internal combustion engines around ethylene sensitive commodities should be avoided, by using electric forklifts, or isolating vehicles from handling and storage areas. Natural sources of ethylene such as overripe and decaying produce should be removed from storage and handling areas. Ethylene producing and ethylene sensitive commodities should not be stored together for long periods of time. Retail displays should avoid placement of ethylene producing fruits such as apples and tomatoes close to commodities such as lettuce and cucumbers, although good ventilation in such areas probably reduces the severity of contamination.

Ventilation: Ethylene concentrations in the storage environment can be reduced by ventilation with clean fresh air. However, the fresh air has to be cooled and increasing ventilation is therefore energy intensive. Higher ventilation rates will also reduce the ability to maintain high RH in the cold room. Ventilation is also not suitable for CA storages or even packaged produce within normal storages because atmospheres are tightly controlled.

Ethylene adsorbers, oxidation and catalysis: Ethylene in storage rooms can be lowered by adsorption or oxidation. Adsorbers such as activated carbon and zeolites (microporous aluminosilicate minerals) have been available in market. Zeolites incorporated into plastic films can maintain sensory quality and reduce microbial growth during storage. Ethylene can be oxidized using a number of strategies. Potassium permanganate ($KMnO_4$) is available in sachets, films and filters direct contact with products must be avoided because of its toxicity. Ethylene is absorbed by the potassium permanganate its effectiveness is based on the presence of a large surface area, although systems have been developed where room air is drawn though the scrubber to increase efficiency. Ozone also oxidizes ethylene, and its use in slowing down ripening, as well as a disinfectant and sanitizer, that lowers mold and bacterial contamination. Ozone is a toxic gas and OSHA recommends limited exposure of workers of 1.5 minutes for a 0.3ppm concentration. Most work on ethylene removal has centered on Pd (palladium) and TiO_2 (titanium dioxide) using activated carbon as the catalyst support. For delayed ripening of tomatoes and avocadoes has been demonstrated using Pd activated carbon.

Inhibitors of ethylene action: Inhibitors of ethylene action such as SmartFresh™ (1-methylcyclopropene; 1-MCP) are not permitted by organic growers and storage operators. Non-chemical means that are available include the use of controlled atmosphere (CA) storage. CA storage involves the modification of the atmosphere around the commodity, usually by decreasing the oxygen concentration from 21 per cent in ambient air to levels in the range of 1-5 per cent, and increasing the carbon dioxide concentrations from 0.03 per cent in the ambient air to as high as 20 per cent. Modification of the storage atmosphere has the effect of inhibiting ethylene action and/or rates of ethylene production.

Fruit Ripening Process

The processes of ripening generally occur in a relatively coordinated manner in most fruits under natural conditions. The nature of the desirable changes in colour, flavour and texture varies to some extent from fruit to fruit, and the desirable properties of the ripe fruit are characteristic of a particular fruit. The point during the sequence of ripening at which a fruit is ripe is, of course, to some extent a subjective evaluation.

Maturation: The stage of fruit development during which fruit attain full growth is known as maturation. Ripening is a terminal period of maturation during which fruit attain maximum aesthetic and edible quality.

Ripe: It is derived from Saxon word 'Ripi', which means gather or reap. This is the condition of maximum edible quality attained by the fruit following harvest. Only fruit which becomes mature before harvest can become ripe.

Senescence: Senescence is defined as the period when synthetic (anabolic) biochemical process gives way to degradative (catabolic) process, leading to ageing and finally death of the tissue.

Ripening of fruits: Ripening is a dramatic event in the life of a fruit during which structure and composition of unripe fruit is so altered that it becomes acceptable to eat by which fruits attain their desirable flavour, quality, softening, colouring, sweetening and increases in aroma compounds, palatability and loss of chlorophyll. Ripening marks the completion of development of a fruit and the commencement of senescence and it is normally an irreversible event with change in composition *i.e.* conversion of starch to sugar. On the basis of ability to ripen after harvest, sharp rise in respiration rate during onset of ripening and production and/or response to ethylene, the fruits are divided into two categories.

Climacteric fruits: Climacteric fruits are defined as fruits that enter climacteric phase after harvest *i.e.* they continue to ripen. During the ripening process the fruits emit ethylene along with increased rate of respiration. Ripe fruits are soft, delicate and generally cannot withstand rigours of transport and repeated handling. These fruits are harvested hard and green, but fully mature and are ripened near consumption areas. Small dose of ethylene is used to induce ripening process under controlled conditions of temperature and humidity.

Non-Climacteric fruits: It became clear in the 1940s that not all fruits had the distinctive and well coordinated pattern of ripening of the typical climacteric fruit. Non-climacteric fruits once harvested do not ripen further. Non-climacteric fruits produce little amount of ethylene and do not respond to ethylene treatment (except in terms of degreening), and these fruits should be picked when fully ripe to ensure good flavour and quality. There is no characteristic increased rate of respiration or production of carbon dioxide. In order to improve external skin colour and market acceptance, citrus like orange, lemon, mousambi and kinnow can be treated with ethylene, as a degreening agent. Ethylene treatment breaks down the green chlorophyll pigment in the exterior part of the peel and allows the yellow or orange carotenoid pigments to be expressed.

Table 65. Difference between climacteric and non-climacteric fruits

Climacteric Fruit	Non-climacteric Fruit
Normally they ripen after harvest	Fruit that does not ripen after harvest. Ripen on the plant itself.
The quality of fruit changes drastically after harvest characterized by softening, change in colour and sweetness	The quality do not change significantly after harvest except little softening. Do not change to improve their eating characteristics
Exhibits a peak in respiration	Does not exhibit a peak
More ethylene is produced during ripening	Little / No ethylene production
Significant increase in CO_2 production	No increase in CO_2 production
Decrease in internal oxygen concentration	Increase internal oxygen concentration

Low concentration of ethylene 0.1-1.0μ L/L/day is sufficient to hasten ripening	Not much response is seen to exogenous application of ethylene.
Eg. Apple, apricot, banana, ber, date palm, fig, guava, kiwi, mango, papaya, passion fruit, pear, plum, stone fruits and sapota	Eg. Bell pepper, cucumber, eggplant, okra, pea, summer squash, tree tomato and rin & nor tomato, blackberry, blueberry, cacao, cashew apple, cherry, citrus fruits, carambola, grape, grapefruit, kinnow, litchi, loquat, mousambi, olives, orange, pineapple, pomegranate, pumpkin, raspberry, rambutan, strawberry, tart cherries, watermelon

Fruit respiration: Respiration is a process in which stored organic materials (carbohydrates, protein and fat) are broken down into simple end products with release of energy. Oxygen is used in this process and carbon dioxide is produced. The respiration rate of produce is an excellent indicator of the metabolic activity of the tissue and thus is a useful guide to the potential storage life of the produce. If the respiration rate of a fruit or vegetable is measured as their O_2 consumed or CO_2 evolved. A characteristic respiratory pattern is observed during the course of the development, maturation, ripening and senescent period. Oxidation of glucose generates an equal amount of CO_2 for the O_2 consumed, whereas oxidation of malate generates more CO_2 then the O_2 consumed. This ratio between the oxygen consumed and carbon dioxide produced is called respiratory quotient. The O_2 concentration at which anaerobic respiration commences varies between tissues and is usually below 1 per cent V/V and off flavour may results from fermentation. Respiration influences the product as reduced food value (energy value), flavour, sweetness and weight of the fruits. Factors responsible for the respiration (external and internal) are temperature, RH, gas composition in the ambient and within the cell, moisture content of the tissue, wounding or injury, type of the plant parts, stage of development of tissue, surface area to volume of the produce, pre-harvest treatments and post-harvest methods employed, chemical composition of tissue, size of the produce and presence of natural coating on the surface.

Table 66. Classification of horticultural commodities according to their respiration rate

Classes	Range at 5°C ($mg\ CO_2\ Kg^{-1}\ hr^{-1}$)	Commodities
Very low	< 5	Dates, nuts, dried fruit and vegetables etc.
Low	5 - 10	Apple, beet, celery, cranberry, citrus fruits, grapes, honey dew melon, kiwi fruit, onion, papaya, pineapple, potato (mature), sweet potato, watermelon etc.
Moderate	10 - 20	Apricot, banana, blueberry, cabbage, cantaloupe, carrot (topped), cherry, cucumber, fig, gooseberry lettuce (head), mango, peach, pepper, pear, plum, potato (immature), radish (topped), tomato, summer squash
High	20 - 40	Avocado, blackberry, carrot (with tops), cauliflower, leeks, lettuce (leaf), lima bean, radish (with tops), raspberry, strawberry
Very high	40 - 60	Artichoke, bean sprouts, broccoli, brussels sprouts, endive, cut flowers, green onion, okra
Extremely high	> 60	Asparagus, mushroom, parsley, peas, spinach, sweet corn

Respiration rate is proportional to temperature, approximately doubling every 10° C. Due to high respiration, heat buildup will be more, which in turn increases the temperature and respiration of produce. The heat produced may be calculated by 1 mg CO_2/kg hr^{-1} = 61.2 k.cal ton-1 day-1 = 220 BTU ton-1day^{-1}.

Fruits Ripening Facilities

Ripening room: Fruits are ripened in specially built rooms that must be gas tight, have systems for controlling humidity and concentrations of carbon dioxide and ethylene, and have equipment to control product temperature. Ripening rooms are usually insulated but they typically operate at temperature 15-21° C. The ripening process is always done at relative humidity above 85 per cent.

Temperature: Ripening is controlled on the basis of fruit pulp temperature. It should be measured during each cycle with a calibrated pulp thermometer. Simultaneously, room air temperature must also be regularly monitored with calibrated thermometer.

Relative humidity: The refrigeration system must be designed to contain 85-95 per cent RH. Humidity below this range causes excessive product weight loss. Humidifiers are needed to add moisture to the air in rooms. Air humidity should be periodically monitored with a wet and dry bulb psychrometer.

Air flow: Air flow is needed to distribute ethylene gas to the product and to add or remove heat from the product during ripening cycle. Boxes must be stacked with space between them to allow good air flow around each box. Boxes or pallet bins should be vented to allow air flow. If packaging materials are placed in the boxes they should not block vents. Poor venting will cause high fruit temperatures and non-uniform ripening.

After product has reached ripening temperature ethylene gas is added. The gas can be added with three systems.

1. A patented ethylene generator produces ethylene from ethyl alcohol. It produces a relatively constant flow of gas for at least 24 hours.
2. Pure ethylene can be purchased in small pressurized cylinders holding three cubic feet of gas. The cylinder is opened in the room providing a rapid release of gas, which will provide an adequate level during the process, if the room is fairly gas tight.
3. For larger operations use ethylene gas available in larger cylinders and use flow meters to measure a prescribed amount into the room.

Methods of Fruit Ripening

India is facing a typical problem of using synthetic chemicals in accelerating ripening of climacteric fruits like mango, banana, papaya, tomato and jackfruit. Some nonclimacteric fruit like pineapple are also being exposed to ripening chemicals. In the developed country, ethylene gas is used to accelerate ripening of climacteric fruits. This is healthy, and it does not pose any threat to human health. Both conventional and chemical ripening methods are used in India.

Conventional methods of fruit ripening

Conventional methods of fruit ripening include:

i. Application of heat (banana and papaya)

ii. Polyethylene covering and application of heat by lighting candles (Banana)

iii. Fruit piercing through enlarged fleshy peduncle (Jackfruit)

iv. Home fruit ripening by incorporating high concentration ethylene liberating fruits like banana and apples in enclosed container of climacteric fruits (mango, banana and papaya) to be ripened.

» A simple technology practiced in households to trigger ripening is to keep unripened and ripened fruits together inside an air tight container. Since the already ripened fruits release ethylene, ripening will be faster.

» Another method is to place the fruits intended for ripening inside an air tight room and induce ripening through smoking inside smoke chambers. Smoke emanates acetylene gas. Several fruit traders follow this technique to achieve uniform ripening especially in edible fruits like banana and mango. But the major drawback of this method is that the fruits do not attain uniform colour and flavour. In addition, the persistence of smoke odour on the product impairs its quality

» Spreading unripe fruits as layers over paddy husk or wheat straw for a week to ripen is an alternative.

» Another practice is that some farmers dip unripe mature fruits in 0.1 per cent ethrel solution (1 ml of ethrel solution in 1 litre of water) and wipe it dry. The fruits are then spread over a newspaper without touching each other and a thin cotton cloth is covered over this. In this method, the fruits will ripen within two days.

» In one of the simple and harmless techniques, 10 ml of ethrel and 2 gm of sodium hydroxide pellets are mixed in five litres of water taken in a wide mouthed vessel. This vessel is placed inside the ripening chamber near the fruits and the room is sealed air tight. About one third of the room is filled with fruits leaving the remaining area for air circulation. Ripening of fruits takes place in about 12 to 24 hours. In order to reduce the cost of chemical, some ethylene releasing fruits such as papaya and banana can also kept in the same room.

» Ethylene gas filled in pressurized cans promote fruit ripening in 24-48 hours.

Chemical methods of fruit ripening

Lack of easier and rapid methods for uniform ripening poses a major problem in the fruit industry. Almost all methods of ripening, either conventional or the modern chemical methods, come with their own merits and demerits. Ethylene and ethylene releasing compounds, abscisin, ascorbic acid and hydroxyethyl hydrozine (BOH), acetylene and calcium carbide, alcohol, fatty acids all are used for fruit ripening. Normally, the number of days taken for edible ripening varies for different fruits and prevailing climatic conditions. For instance, it takes about 5 to 6 days for mangoes and 6 to 7 days for sapota.

Use of ethylene for fruit ripening: Ethylene (CH_2=CH_2) being a natural gaseous hormone is used for fruit ripening and does not pose any health hazard for consumers of the fruits. Application of ethylene gas at a recommended concentration of 0.1-1.0 $\mu L\ L^{-1}$ (10-100ppm) for 24 h is sufficient to hasten full ripening of climacteric fruits. It is a degreening agent, which can turn the peel from green to perfect yellow (in the case of bananas) and maintain the sweetness and aroma of the fruit, thus value addition in the fruit is possible as it looks more appealing. It has been known for a long time that treatment of unripe fruits with ethylene would merely stimulate natural ripening until the fruit itself starts producing ethylene in large quantities.

Methods of applying ethrel: Use of diluted ethylene gas mixtures is safer than using pure ethylene, which is explosive and flammable at concentrations of 3 per cent or higher. There are two methods of exposing fruit to ethylene. Trickle method involves trickling ethylene gas into room so as to maintain a concentration of 10 ul litre^{-1}, usually for a period of 24 hours. During this time, relative single initial charge of ethylene at a concentration of 20 to 200 ul litre^{-1}. Room is then ventilated after 24 hours to prevent carbon dioxide exceeding 1 per cent concentration, which would retard ripening. Rooms that are poorly sealed are packed in vented cartons stacked on pallets, and fruit temperature is controlled by forced air circulation as in a cooling facility. A small fan can be used to ensure a uniform continuous flow of ethylene into and through the room. Fruit to be ripened ideally is placed in an airtight ripening room maintained at a constant temperature (18-21° C for most fruits, but 29-31° C in mango). Significance of use of ethylene for fruit ripening.

» Ethephon (2-choloroethyl-phosphonic acid) is commercially available and is registered for pre-harvest use on a variety of crops for controlling developmental processes or inducing ripening. This chemical is approved for post-harvest use on fruits crops for enhancing uniform ripening.

» Ethylene is available in large steel cylinders where it is stored under pressure. Typical mixtures are 95 per cent nitrogen and 5 per cent ethylene or 95.5 per cent nitrogen and 4.5 per cent ethylene. The measured quantities of ethylene are introduced in ripening room at regular intervals or continuously and the flow is regulated through metering devise or flow meter.

» On small scale, commodities can be treated using shot method with ethylene liberated from ethephon. A calculated amount of ethephon in stainless steel bowl is placed around the room. The fruits are stacked in the room and sodium hydroxide is added to ethephon and all ventilation to the room is then blocked. When sodium hydroxide reacts with ethephon, ethylene gas is released that ripens the fruits.

» Ethylene generator is a device that is portable and placed inside the ripening room. A liquid (ethyl alcohol) is filled into the tank fitted with ethylene generator and it is connected to an electric power source. The ethyl alcohol gets heated in a controlled manner in the presence of a catalyst that produces ethylene gas. Gas is maintained inside the ripening room until colour break occurs in the fruits.

Table 67. Optimum storage and ripening temperatures for fruits.

Commodity	Ethylene conc. (ppm)	Ethylene exposure time (hr.)	Ripening temp. (°C)	Storage temp. (°C)
Apple	10-20	125-150	20-25	0-4
Avocado	10-100	12-48	15-18	4.4-13
Banana	100-150	24-48	14-18	13-14
Honey dew melon	100-150	18-24	20-25	7-10
Kiwifruit	10-100	12-24	0-20	-0.5-0
Mango	100-150	12-24	15-25	12-14
Orange degreening	1-10	24-72	20-22	5-9
Stone fruit	10-100	12-72	13-25	-0.5-0
Tomato	100-150	24-48	20-25	12-16

Table 68. Internal ethylene concentrations in several climacteric and non-climacteric fruits

Fruits (Climacteric)	Ethylene ($\mu l\ L^{-1}$)	Fruits (Non-Climacteric)	Ethylene ($\mu l\ L^{-1}$)
Apple	25-2500	Lemon	0.11-0.17
Pear	80-100	Lime	0.30-1.96
Peach	0.9-20.7	Orange	0.13-0.32
Avocado	28.9-74.2	Pineapple	0.16-0.40
Banana	0.05-2.1		
Tomato	3.6-29.8		

Table 69. Optimal ripening conditions for fruit ripening

Components of ripening	Optimal ripening conditions
Temperature	18 to 25° C (<18 delay ripening, > 25 microbes)
Relative humidity	90 to 95 per cent
Ethylene concentration	10 to 100 ppm
Duration of treatment	24 to 74 hours depending on fruit type and stage of maturity
Air circulation	Sufficient to ensure distribution of ethylene within ripening room
Ventilation	Require adequate air exchange in order to prevent accumulation of CO_2 (more than 1%) which reduces effectiveness of C_2H_4.

Mode of Action: On the whole two systems exist for the regulation of ethylene biosynthesis, first system is initiated or perhaps controlled by an unknown factor, probably involved in the regulation of senescence. System first then triggers system second which is responsible, during ripening of climacteric fruits, for the production of the large amounts of ethylene that are necessary for the full integration of ripening. In Non-climacteric fruits system second do not have an active and the treatment of climacteric fruits with ethylene circumvents system first.

Chemical properties of ethylene

Active ingredient: Ethephon, Other names: Chlorethephon, Chemical name: 2-Chloroethylphosphonic Acid.

Potential health hazard: Ethylene cause eye and skin irritation.

Ingestion: Gastrointestinal irritation with nausea, vomiting and diarrhea, cardiac, disturbances, central nervous system depression.

Inhalation: cardiac abnormalities, high concentration causes CNS depression. Oral lethal dose for rat is: LD50 = 3400 mg kg^{-1}.

Ethylene biosynthesis: Ethylene has been shown to be produced from methionine via an intermediate S-adenosyl-methionine (SAM) and 1-aminocyclopropane-1-carboxylic acid (ACC). The conversion of SAM to ACC by the enzymes ACC Synthase (ACS). In higher plants, ACC can be removed by conjugation to form malonyl ACC (MACC) or glutamyl ACC (GACC). Ethylene forming enzyme (EFE) or ACC oxidase (ACO) is required to convert ACC to ethylene. ACO is a liable enzyme and sensitive to oxygen and attached to outer layer of the plasmalemma. Factor that effect the activity of the ACS includes fruit ripening, senescence, auxin, physical injuries and chilling injury. This enzyme (ACS) is strongly inhibited by aminooxyacetic acid (AOA), rhizobitoxine and amino ethoxy vinayl glycine (AVG). ACO is inhibited by anaerobiosis, temperature above 35° C and cobalt ions. Among various chemical used for extension of shelf life fruits 1-MCP has been found to be very effective. The 1-methyl cyclo propene (1-MCP) has been shown to be highly effective inhibitors of C_2H_4 action. 1–MCP binds irreversibly to the C_2H_4 receptors in sensitive plants tissues and a single treatment with low concentration for a few hours at ambient temperatures confers protection against C_2H_4 for several days. Many fruits respond to 1-MCP in extension of storage life by retarding the process of ripening. The pattern of C_2H_4 production in tomato is it rises before the onset of ripening, where as in, apple and mango it does not rise before increase in reparation. Immature tomato fruit has high rate of C_2H_4 production and it extremely tolerance to C_2H_4 but banana and melons can readily ripen with C_2H_4 even when immature.

Table 70. Classification of fruits and vegetables on based of ethylene biosynthesis

Class	Range at 20° C (µl C_2H_4 release $kg^{-1}hr^{-1}$)	Commodities
Very low	< 0.1	Artichoke, asparagus, cauliflower, cherry, citrus fruits, grape, jujube cut flowers, leafy vegetables, pomegranate, potato, root vegetables, strawberry
Low	0.1-1.0	Brinjal, blackberry, brinjal, blueberry, casaba melon, cranberry, chilli, cucumber, green capsicum, okra, pepper, persimmon, pine apple, pumpkin, raspberry, tamarillo, water melon
Moderate	1.0 -10	Banana, guava, fig, litchi, melon, mango, tomato
High	10-100	Apple, apricot, avocado, cantaloupe, kiwi fruit (ripe), nectarine, papaya, peach, plum, pear
Very high	> 100	Cherimoya, sapota, passion fruit

Table 71. Practical use of ethephon in some of the fruit and vegetable crops for ripening

Fruits	Stage of application	Concentration (ppm)	Days Hasten/ Application
Mango	Mature green	500	6/ Pre or post-harvest
Banana	Mature green	500	6/Post-harvest
Guava	Mature green	600	7/ Pre or post-harvest
Citrus(degreening)	Colour break	100	7/ Pre or post-harvest

Fruits	Stage of application	Concentration (ppm)	Days Hasten/ Application
Papaya	Mature green	1000	6/Pre-harvest
Grape	Colour break	500	10/Pre-harvest
Litchi	Colour break	500	7/Pre-harvest
Date	Turning stage	1000	15/Pre-harvest
Phalsa	Colour break	250	7/Pre-harvest
Fig	Drupelets colour break	500	21/Pre-harvest
Pine apple	Two weeks before harvest	1000	15/Pre-harvest
Bael	Mature fruits	1000	45/ Post-harvest
Apple	Two months before harvest	500	21/Pre-harvest
Pear	Mature green	250	7/ Post-harvest
Plum	Colour break	250	12/Pre-harvest
Prune	Mature fruits	80	14/Pre-harvest
Tomato	Turning	200	10/ Pre or post-harvest
Musk melon	Mature fruits	1000	7/ Pre or post-harvest
Watermelon	Mature fruits	1000	7/Pre-harvest
Pepper	Chocolate brown	500	7/ Pre or post-harvest
Chilies	Turning stage	250	8/Pre-harvest

(Sourse: Prof I.S.Singh (2009) post harvest handling and processing of fruits and vegetables pp.36-37)

Fruit ripening using calcium carbide

In India and many other developing economies, the banana, mango and most climacteric fruits are ripened with the use of calcium carbide, which releases acetylene and ethylene on interaction with moisture relise from fruits. However, calcium carbide contains chemical impurities such as arsenic hydride and phosphorus hydride that are highly carcinogenic compounds. Acetylene is believed to affect the nervous system by reducing oxygen supply to brain. Further fruits ripened with calcium carbide though develop attractive surface colour, but inferior in taste, flavour and spoil faster. Arsenic and phosphorus are toxic and exposure may cause severe health hazards. In India too, use of calcium carbide is strictly banned as per PoFA (Prevention of Food Adulteration) Act 1954 [Section 44AA].

Chemical properties of calcium carbide

Calcium carbide produces acetylene gas after absorbing moisture. Acetylene is an analogue of ethylene, the ripening hormone. Calcium carbide poses dangers of explosion and carries toxic materials to the consumers.

$$CaC_2 + H_2O > CaCO_3 + CH{\equiv}CH$$

Chemical name: Acetylenogen/Calcium acetylide

Physical properties: Grey-brown solid (rock-like) with garlic odour

Hazard of calcium carbide

Calcium carbide is extremely hazardous compound due to presence of arsenic and phosphorus. Contact can severely irritate or burn the eyes and skin causing permanent eye damage and ulcers on skin. Exposure can severely irritate the mouth, nose and throat causing sores cough and wheezing or shortness of breadth. Higher exposure may cause a build up of fluid in the lungs (pulmonary oedema), a medical emergency, with severe shortness of breath. Early symptoms of arsenic or phosphorus poisoning include vomiting, diarrhoea with or without blood, burning sensation of the chest and abdomen, thirst, weakness and difficulty in swallowing and speech and garlic like odour of breath. Other effects include numbness in the legs and hands, general weakness, cold and damp skin and low blood pressure. While most cases of arsenic and phosphorus poisoning are detected before they become fatal. Pregnant women are particularly vulnerable, the chemical residue in the fruit could lead to miscarriage. The major hazard from the consumption of calcium carbide treated fruit is the residual effect, so using the carbide in closed and separate atmosphere to release acetylene and use the gas in separate ripening room seems to be safe. Carbide wrapped in newspaper can be used as the generator in separate section. Water vapour from the fruit releases sufficient amount of gases from carbide to cause ripening, 100gm of carbide can be used per 50 kg of fruit.

Variations in the pattern of ripening within climacteric and non-climacteric groups

There are some exceptions in typical climacteric fruits such as the mango, in which even in initiated fruit the rise in the internal ethylene concentration is relatively slow, so it is possible to promote ripening by exogenous ethylene treatment even during the climacteric. In honeydew melons, both the stature of the respiration and the rate of ethylene production at peak are very dependent on the stage of maturity at which the fruit is harvested. Very immature fruit fail to give a respiratory response, but as the fruit matures on the vine the extent of the respiration rise and its capacity to produce ethylene increases. Immature melons respond to ethylene treatment in much the same way as do non-climacteric fruit, in that the height of the respiratory response appears to be related to the ethylene concentration applied and there is no autocatalytic increase in ethylene production. On the other hand, with fruit held on the vine beyond a critical time, the response to ethylene is characteristic of the climacteric fruits, ethylene shortening the time taken for initiation of the climacteric rise and inducing its own production autocatalytically.

The fig is one of the fruits that were originally classified as non-climacteric, but which, on more detailed study have been placed in the climacteric category. The fig is a compound fruit made from the ovaries of different flowers, which shows a double sigmoidal pattern of growth. There is a relatively low concentration of ethylene (less than 0.1 $\mu l\ l^{-1}$) in the internal atmosphere of the fruit during growth period II. At the onset of growth period III, there is a rise in respiration reaching a peak (roughly double the rate at the end of period II) at about halfway through period III. During this rise in respiration, the internal ethylene concentration rises from less than 0.1 $\mu l\ l^{-1}$ to a peak of from 4 to 5 $\mu l\ l^{-1}$. The onset of the increase in the internal ethylene content of the fruit precedes the increase in respiration during growth period III.

In grape as in the fig, cell expansion and growth occur in parallel with some aspects of ripening, since the grape also shows a double sigmodial pattern of growth. The onset of period III is associated with changes in colour, texture, loss of acid, accumulation of sugar, and with the onset of renewed growth. This period is also accompanied with a substantial rise in respiration, measured either by CO_2 production or oxygen uptake. The rise in respiration is associated with an increase in RQ from 1.0 to 1.3 or more, and this is the change that would be predicted if there was a change in respiratory substrate at the time of version from carbohydrates towards organic acids. The basis for the classification of grapes as non-climacteric was originally that detached grapes showed a slow decline in respiration during storage. However, the significance of this slow downward drift is not clear, since grapes do not continue to ripen properly when detached from the vine. Ethylene, applied either as gas or as 2-chloroethyl-phosphonic acid (a compound that breaks down on contact with plant tissues to release ethylene at the site of its action), will delay ripening of grapes if applied in the early part of the slow growth period (II). The effect when applied beyond the mid-point of period II is to shorten the delay before the onset of period III, so that both ripening and renewed growth are advanced. While there is an increase in respiration at version there is no corresponding increase in ethylene production. Thus, the grapes have in common with citrus fruits the ability to respond to ethylene, but lack the ability to produce endogenous ethylene in response to exogenous treatment with the gas, and in these characteristics behaves like the typical non-climacteric fruits. However, in its response to ethylene at different stages of growth, in its rise in respiration at the start of period II, and in the fact that growth and senescence phenomena occur simultaneously, it resembles the behavior of the fig, a climacteric fruit. The fact that grapes do not produce ethylene auto-catalytically seems a good reason for classifying them as non-climacteric. In cherries, the ethylene content of the fruit is low about 0.1 to 0.2 $\mu l\ l^{-1}$ throughout the ripening period from the beginning of period III to the fully ripe stage at II. The rate of CO_2 production of fruit on the tree declines from the end of pit hardening (period II) to the end of ripening. This confirmed that cherries did not show the climacteric pattern of behaviour. In general, it is fairly easy to place the cherry in the non-climacteric category, since ripening is achieved in the absence of increases in either respiration or ethylene production. The ripening does occur following the period of renewed growth, and it is interesting that the increased energy demands required for renewed cell expansion do not, in this case, lead to an increase in the rate of respiration.

The strawberry produces only very small amounts of ethylene and ripens in the absence of a rise in respiration and exogenous ethylene has no effect on the ripening process. However, the raspberry contains significant amounts of ethylene in its internal atmosphere at the onset of ripening and this concentration increases as the fruit changes colour. Both blueberry and cranberry have been considered non-climacteric, their respiration did not rise during ripening. In the cranberry, exogenous ethylene treatment has no effect on the respiration rate, on the sugar or on acid content of the fruit, but it does stimulate degreening.

Physiology of Fruit Ripening

Several metabolic changes are initiated after the harvest of fruits and vegetables. In the case of vegetables, harvesting induces stress responses through reduced availability of water and nutrients, wounding and exposure to shelf life. In fruits, an increase in the biosynthesis of the gaseous hormone ethylene serves as the physiological signal for the initiation of the ripening process. During the ripening process, some fruits evolve large amounts of ethylene, sometimes referred to as an autocatalytic increase in ethylene production, which occurs in conjunction with an increase in respiration referred to as the respiratory climacteric. In climacteric fruits such as apple, pear, banana, tomato, and avocado, ethylene evolution can reach 30–500 ppm kg h^{-1} (microliter L^{-1}), whereas in nonclimacteric fruits such as orange, lemon, strawberry, and pineapple, ethylene levels usually range from 0.1 to 0.5 ppm kg h^{-1} during ripening. Climacteric fruits respond to external ethylene treatment by an early induction of the respiratory climacteric and accelerated ripening in a concentration-dependent manner. Nonclimacteric fruits, on the other hand, show increased respiration in response to increased levels of ethylene concentration without showing acceleration in the time required for ripening. Vegetables produce very low amounts of ethylene most of them with less than 0.1 μL kg h^{-1}, with slightly higher levels as in cassava (1.7 μL kg^{-1} h^{-1}), breadfruit (1.2 μL kg^{-1} h^{-1}), and cucumber (0.6 μL kg^{-1} h^{-1}) when measured at 20–25° C. During ripening, activation of several metabolic pathways often leads to drastic changes in the biochemical composition of fruits. Fruits such as banana store starch during development and hydrolyze the starch to sugars during ripening that also results in fruit softening. Most fruits are capable of photosynthesis, store starch, and convert them to sugars during ripening. Fruits such as apple, tomato and grape have a high percentage of organic acids, which decreases during ripening. Fruits contain large amounts of fibrous materials such as cellulose and pectin. The degradation of these polymers into smaller water-soluble units during ripening leads to fruit softening as exemplified by the breakdown of pectin in tomato and cellulose in avocado. Anthocyanins are the major colour components in grapes, blueberries, apples, and plums, whereas carotenoids, specifically lycopene and carotene, are the major components that impart colour in tomatoes. Aroma is derived from several types of compounds that include monoterpenes (as in lime, orange), ester volatiles (ethyl, methyl butyrate in apple, isoamyl acetate in banana), simple organic acids such as citric and malic acids (citrus fruits, apple), and small chain aldehydes such as hexenal and hexanal (cucumber). Fruit ripening is a dynamic transitional period during which many easily perceived changes, such as alterations in pigmentation, firmness, sweetness, and acidity take place. These changes make fruit desirable for human consumption and capable of seed dispersal by birds, animals, and environments. Fruit firmness is associated with several attributes including crispness, mealiness, grittiness, chewiness, succulence and juiciness, fibrousness, toughness and oiliness. Additionally, development of various organoleptic components such as sweetness, sourness, astringency, bitterness, and production of volatile compounds leading to characteristic aroma is connected with fruit textural changes. Although most of these changes impart desirable traits to various fruits and vegetables, some of the fruit softening associated changes make them unacceptable for marketing. These include development of off-flavours and off-odors with excessive softening of tissues. Textural softening can also increase susceptibility to phytopathogens due to their proneness to solute leakage that provide rich media for their growth, and resulting in severe losses during postharvest storage and marketing.

Stimulation of ripening: There are several aspects regarding the usefulness of stimulation of ripening fruits. The most motivating consequence is the possibility of an once-over mechanical or manual harvest in crops that produce ripe fruits over a large time period, speed up the time of cultivation of a crop, and increase possibility to plan harvest and marketing in times and earlier in the year. Ethephon is used for stimulation of tomato fruit ripening. The fruit quality is not negatively affected compared with normally ripened tomatoes. The application of the growth regulator about 8 days before harvest the 20 per cent coloured fruits in the field allows a once over mechanical harvest without losing a substantial amount of unripe fruits. Tomato fruit ripening cannot be enhanced by ACC sufficiently. However, a derivative of ACC, N-Formyl-ACC, is very effective in stimulation of tomato fruit ripening.

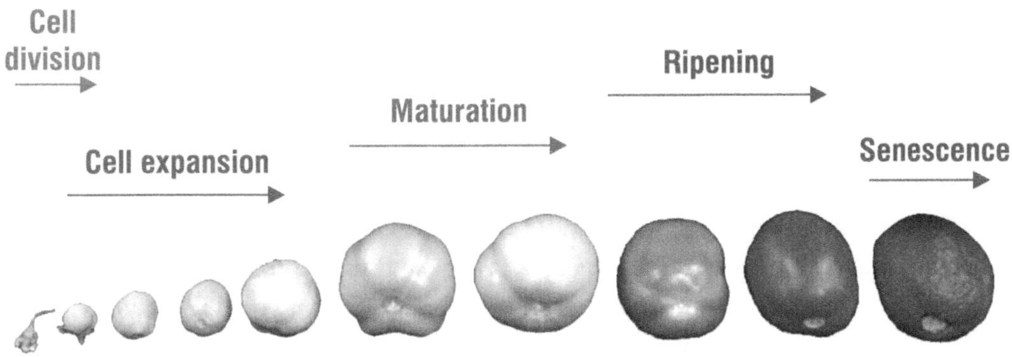

Figure 37. Process of fruit development, ripening and senescence of tomato (Photograph courtesy D.A. Brummell)

Fruit development can be divided into a series of stages, as shown here for tomato. Early in development fruit are enlarging rapidly and are small, hard, green and accumulating organic acids. The seeds become mature prior to ripening. During ripening fruit become soft textured, and accumulate soluble sugars, pigments and aroma volatiles. Eventually fruit will become over-ripe, cell structures will deteriorate and the fruit will become susceptible to pathogens.

Changes during ripening

During development and ripening process fruits and vegetables undergo different chemical and biochemical changes. Some of these changes are very important due to nutrition and health values. The stage of maturity to ripening at which fruits are harvested has the greatest effect on flavour, storage life and marketability. Ripening process includes ethylene synthesis, solubilization of insoluble cellular compounds, changes in cellular membrane permeability, mixing of substrates and their enzymes and initiation of metabolic process.

Changes in structure and composition: The visible signs of ripening are those changes in the colour, texture, taste, and aroma of the fruit tissue that lead to the state in which the fruit is fit to eat, and these are determined by changes in the structure and metabolism of the fruit tissue. Ripening involves complex changes in metabolism of the fruit in which both anabolic and catabolic processes are involved. New pathways of metabolism are brought into play, which may lead, for instance, to the production of new pigments of new volatile compounds.

In addition, degradative processes such as conversion of chloroplasts to chromoplasts, and reduction in adhesion between cells leading to tissue softening also occur. These metabolic changes must reflect changes in the in vivo activity of the enzymes involved in the synthesis, degradation, release, activation, or inactivation. These changes, at least in the climacteric fruit, have to be considered in relationship to the changes in respiration and in ethylene production. The underlying structural and biochemical changes responsible for the visible changes of ripening will now be considered.

Changes in cell structure during ripening: The edible tissue of fleshy fruits is composed usually of parenchymatous cells, frequently of very large size. The increase in ploidy may be a factor determining the large size of fruit cells. These cells have a high water content and accumulate organic substances such as sugars and organic acids. Mature green tomatoes contain chloroplasts with relatively small grana and frequently with one or two vacuoles bearing starch. The first signs of the onset of ripening changes are the appearance of electron dense regions in the granal and intergranal membranes which are thought to represent the initial stages in lycopene deposition, and an increase in the number of plastoglobules. Later the inter-granal region swells and the granal diameter increases markedly. By the time the orange stage is reached, new structures called thylakoid plexes appear; these resemble the loose type of prolamellar body found in plastids in etiolated leaves exposed to light for short periods. Lycopene crystalloids of up to 18 in length appear by the orange-red stage, and crystalloids composed of concentrically arranged lamellae are frequently found. In the later stages of ripening there is a reduction in the amount of membranous material in the stroma and a large increase in the size and number of the plastoglobules. All the nonplastid cytoplasmic structures present in the cytoplasm of immature tomatoes are observed in the cytoplasm of ripe fruit also. During ripening cellulose fibre becomes soft and starts breaking. Middle lamella dissolves or solubilized, cell membrane becomes permeable and intercellular space or porosity reduced. Chloroplast is changed into chromoplast, number of mitochondria initial increase and break down in over riped fruit. In grape and pear deposits of cuticle and wax, in rambuton number of epidermal hair reduced or disappered. Number of lenticels increased as a result fruits and vegetables shriveled. Stomata activities increased, sclerenchyma fibre, vascular bundles, endocaryp and sclereid become lignified.

Textural changes during ripening: Changes in the texture of fruit during ripening result from changes in the structure and composition of their cell wall. The cells of fruit are frequently large parenchymatous cells with large air spaces in between them. Changes in the degree of contact between cells, as well as in the structure of the cell wall affect texture. The texture of the fruit softens with ripening. This is because of the action of enzymes like hydrolases (poly galacturonase, pectin methyl esterase and cellulases) which breakdown the pectins, cellulose and hemicellulose. Propectin is insoluble form of pectic substances binds to calcium and sugars in the cell wall. On maturation and ripening, propectin gradually broken down to lower molecular weight fraction which are more soluble in water. The rate of degradation of pectic substances is directly correlated with rate of softening of the fruit. Moisture content of the pulp increased as the fruit ripen. The additional water derives from the carbohydrate utilized in respiration. Similarly total dry matter content increased, hence, fruit weight also increased. Fruit firmness decreased at ripening, hence, fruit becomes soft. Flavour, aroma and specific gravity increased during ripening. During ripening in the apple there was a decrease in the galactose content of cell wall hydrolysates. In addition there is an increase in water-soluble polyuronide, which has been correlated with cell separation and the softening of the tissue. An increased solubility of the cell wall polyuronides is the characteristic cell wall change occurring during softening in a whole range of ripening fruits. A number of enzymes attacking polyuronides have been detected in ripening fruit. Frequently the activity of such enzymes is low or absent in unripe fruit and increases during the ripening process.

Change in taste of fruits: The primary change in taste is the development of sweetness in fruits after ripening. During ripening the starch break down into simple sugars like glucose, fructose and sucrose which are responsible for sweetness. This change is also mediated through the action of various enzymes like amylase, invertase, phosphorylase, etc.

Colour development in fruit: Change in colour development is common except avocado, kiwi fruit and Granny Smith Apple. The skin colour change from green to yellow or red depending upon the types of fruit. Flesh colour also changes from white to yellow or creamy as the case may be. Seed colour also changes from light brown to dark brown e.g. papaya, apple, watermelon, etc. The change in colour is either due to synthesis of plant pigments and unmasking of already existing colour. These are the water soluble phenolic pigments, the anthocyanins, which are responsible for the red and blue colours of many fruits. They exist as complex conjugates of the parent aglycones, the anthocyanidins, and are thought to be stored in the vacuole of the cell. The six principal anthocyanidins that are found in fruit are usually present as the 3-glycosides. The organic acids involved are commonly cinnamic acid derivatives such as p-coumaric, cafeic, and ferulic acids. Interaction between anthocyanins and other colourless phenolic pigments (the co-pigmentation effect) and between anthocyanins and metal ions may be important in the blue colours of fruits with an acidic sap. The principle agents responsible for the degradation is change in pH, oxidation systems or enzymes chlorophyllase. Carotenoids are stable pigments and remain there till senescence. They are either synthesized during developmental process or they are masked by the presence of chlorophyll. This kind of change is seen in case of banana.

While in tomato, the colour pigment lycopene is developed simultaneously with degradation of chlorophyll. Pigment anthocyanines are red-purple or blue water soluble phenolic glucosides that are found in vacuoles like in beet root and epidermal cell of apple and grape. They produce strong colour, which often mask carotenoids and chlorophyll. In acidic pH levels the anthocyanins are red in colour and in alkaline pH they tend to become blue. This gives rise to phenomena in roses known as blueing, whereas shift from red to blue coloration occur with aging. This is due to depletion of CHO and release of free amino acids resulting in more alkaline pH in the cell sap. The enzyme, phenylalanine ammonia lyase, may play an important role in the regulation of anthocyanin biosynthesis in the apple.

Biochemical Changes During Ripening

Hormones: The ripening hormone ethylene increased during ripening. In fruits the most likely precursor is L-methionine. Transminase, glucoseoxidase and peroxidase activity has been shown to be involved in the conversion of methionine to ethylene in fruits. The ethylene might act by producing small changers in enzymatic activity and cell membrance permeability. Besides ethylene, abscisic acid (ABA) is known to enhancae fruit ripening and endogenous concentration of ABA increased during ripening of fruits such as grape, pear, peach, etc. Auxins decreased with a concomitant increase in IAA oxidase and peroxidase activity during ripening. Endogenous level of gibberellic acid and cytokinin also decreased at ripening.

Respiration: The respiration value is used for evaluateing the physiological age and condition of a fruit or vegetables. High rate short life and low rate, long life. The increase in respiration associated with ripening has been termed as respiratory climacteric. Based on this the ripening period has been divided in to four stages like preclimactric, climacteric rise, climacteric peak and post-climacteric. It is during ripening, most of the changes (colour, texture, flavour, sugar, acid, etc.,) take place in the fruit leading to an edible ripe in the post-climacteric period. On the basis of climacteric pattern fruits and vegetables have been classified into two groups:

1. **Climacteric fruits**: They exhibit climacteric peak (rise) at ripening. They have higher content of ethylene and starch. Respiratory response did not change as function of ethylene concentration. Such fruits ripen after harvest e.g. mango, banana, papaya, guava, apple, pear, stone fruits, tomato, muskmelon, chilies, etc.

2. **Non-climacteric fruits**: These fruits do on exhibit climacteric peak. They have lower content of ethylene and starch. Respiratory response increases as function of ethylene concentration. Such fruits do not ripe after harvest e.g. citrus, grape, pineapple, litchi, jamun, straw berries, watermelon, etc.

Carbohydrates: Carbohydrate serves as the storage material in most of the fruits. Starch which is present almost in all fruits decreased or disappeared during ripening of fruits like mango, banana, apple, tomato, muskmelon, etc., leading to the buildup of sugars.

Pectic substances: They are derivatives of poly galacturonic acids. Parts of the carbohydrates are present as cellulose and pectic substances in the cell walls. Pectins together with cellulose serve as an integral part of the cell wall. Structure of pectin is related with texture of fruit. During ripening of fruits water insoluble protopectin decreases with a simultaneously increase in soluble pectin thus leading to softening of the fruit tissues. (Protopectin → (Protopectinase) → Pectinic acid → (Pectin methyle esteras) → Pectic acid). Total pectic substances decreased (melon, banana, citrus, and mango) during ripening due to depolymerization and deestirification.

Sugars: Sugars are responsible for sweetness or quality of the fruits. The most commonly occurring sugars are: glucose (date, grape, fig, mulberry, straw berry, etc.), fructose (date, grape, pomegranate, apple, pear, etc.) and sucrose (mango, banana, pineapple, peach, apricot, melons, etc.). Fruit also contain minor sugars like galactose, arabinose, xylose, lactose, maltose, raffinose and stachyose. Sugar acids present in fruits are galacturonic, gluconic and galactoric acids. Sugar alcaohols like sorbitol and myoinositol are also present in fruits. Sugar phosphates (G-6-P, F-6-P, F-1, 6 diphosphate) area also present at low concentration in fruits, e.g., banana, apple, etc. Sugars are known to increase during ripening of fruits.

Acids: The pH of the cell sap of fruit cell is frequently below 7.0 and can be as low as 3.0 in lemon cells. Organic acids are an important source of respiratory energy in the fruit tissues. The organic acids in fruits are stored in vacuoles. Fruit contain several organic acids, most of them are in traces, some of them present in large amounts (citrate and malate) and usually one acid pre-dominates. For example, citric acid in mango, citrus fruits, guava, goose berry, pineapple, pomegranate and fig, malic acid in apple, pear, apricot, plum, peach, banana, sweet lime and cherry and tartaric acid in grape and tamarind. Generally organic acid content decreases during fruit ripening. There area few exceptions to this like banana in which the pH falls during ripening from 5.4 to 4.5 and the total acidity increases during ripening due to accumulation of malic and citric acids. Despite, the total organic acid content increases as malic acid in mango and tomato and citric acid in apple and tomato during ripening.

Pigments: Pigments are responsible for colour of fruits. The chief pigments of fruits can be classified into chlorophyll (green), carotenoids (yellow to orange), anthoxanthins (yellow) and anthocyanins (red, blue and purple). Most of the pigments occur in plastids and they develop as the fruit ripens. Changes in colour development during fruit ripening are due to both degradation of chlorophyll (most of the fruits) and synthesis of carotene (mango, papaya, muskmelon, carrot), xanthophylls (citrus fruits), lycopene (tomato, watermelon, pink grapefruit), anthocyanin (red delicious apple, plum, coloured grapes, phalsa, beet root, chillies) and leuco anthocyanin (white grape).

Lipids: Lipids play important role in aroma or flavour of the fruits. Fruits contain little amount of lipids with a very exceptions like avocado and olive, which are very rich in lipids. There are four parts of lipids- fattyacids (saturated – Palmitic, stearic and unsaturated – Palmitoleic, oleic, linoleic), reserve lipids (triglycerides), natural coating (wax and cutin) and living membrane lipids (Phospholipids and glycolipids). During ripening oleic and palmitic in avocado, stearic in banana and triglyceride in mango increases. Total lipids and fatty acids increase during ripening of mango. The ratio of palmitic and palmitoleic determines the aroma or flavour of mango varieties. For example, if ratio is less than 1 (0.8-0.9) varieties (Alphonso, Langra) have strong aroma and if more than 1 (1.1-1.5) varieties (Neelam, Banglora) have very little aroma or flavour. Linolenic acids increased in tomato and decreased in pumpkin during ripening.

Protein and nucleic acids: Fruits in general contain little amount of protein but play an important role in maintaining cellular organization and complement of enzymes in metabolism during ripening and post-harvest life of fruits. Blocking of protein synthesis by cycloheximide at early climacteric stage can inhibit the fruit ripening and ethylene synthesis. Protein and total RNA content is known to increase during ripening.

Amino acids: Fruits contain fewer amounts of free amino acids but they are of fundamental importance in controlling post-harvest behaviour of fruits and vegetables. Amino acids predominate during ripening are alanine (mango, gooseberry, date), histidine (banana), proline (orange), aspartic acid (apple, lemon, banana). Total free amino acids increase during ripening of grape berrics, arginine and proline being the major amino acids of the ripe berries. Glutamic acid and aspartic acid concentration increased during ripening of tomato. Alanine and glycine increased while serine and asapartic acid demised during ripening of muskmelon fruits.

Enzymes: Enzymes are responsible for physical and chemical changes in fruit/ vegetable during ripening and storage. Pectolytic enzymes such as poly galacturonase in tomato and pectin methyl esterase in mango, banana and oranges increase during ripening. Cellulase activity increased during ripening of tomato and peaches. These enzymes are involved in softening of tissue. Starch degrading enzyme amylase increases during ripening of fruits such as in mango, banana, pear, etc., thus, leading to the accumulation of sugar. Sucrose hydrolyzing enzyme invertase and phenoloxidising enzyme poly phenoloxidase increase in banana and grapes. Lipase activity increases during ripening of avocado with parallel increase in fatty acid. Enzyme synthesizes citric acid (citrate synthetase) in citrus and mango and degrades malic acaaida (malic enzyme) in apple, mango, grape, etc., increased during ripening. Acid phosphatase involved in carotenoid biosynthesis increases during ripening of mango and banana. Peroxidase and polyphenoloxidase are involved in ethylene biosynthesis.

Vitamins: They are organic compounds that have enzymatic function. Vitamin A or carotene content of the fruits increased during ripening of mango, date, papaya, and jackfruit, Vitamin B such as pyridoxine (avocado, banana), thiamin (grape, apricot, etc.), riboflavian (papaya, pineapple, etc.), niacin (pineapple, apricot, etc.), pentothenic acid (apricot, grape, etc.), increased at ripening. Normally ascorbic acid content decreased (ascorbic acid → Dehydro ascorbic acid → 2, 3 diketogulonic acid) at ripening of many fruits such as in aonla, guava, citrus, ber, etc., with exception of papaya and strawberry where it increases during ripening. The polygalacturonase (PG) involved in the solubilization of pectin, which acts on de-esterified pectin molecules and attacks linkages between galacturonic acid groups in the polyuronide. Two types of polygalacturonase have been found in fruit; one, an Exo- polygalacturonase, which acts by cleaving free galacturonic acid groups from the non-reducing end of the molecule and the other, an endo- polygalacturonase, which acts by cleaving the chains of polyuronides in a random fashion. The pectin methyl esterase increase in the water solubility of pectins during ripening, and then it is likely that another enzyme, pectin methyl esterase (PE) may also play a role. The polyuronides of the middle lamella are highly methylated, while polygalacturonase is relatively inactive towards methylated polyuronides and requires free carboxyl groups for its action. Pectin methyl esterase is a wall-bound enzyme that can be readily desorbed from cell walls in vitro by addition of neutral salts at slightly alkaline pH, and its role may be to modify pectin structure prior to polygalacturonase action. The role of cellulases which catalyse the breakdown of cellulose in ripening fruit is controversial. In avocado, cellulase activity rises during the climacteric to very high values and cellulases may be synthesized or activated very early in the climacteric and play an important role in the early stages of softening. In other fruits, however, cellulases appear to play only a minor role in softening. It is clear that changes occurring in the softening of fruit tissues during ripening are extremely complex, so it is perhaps not surprising that there is still doubt about the role of the various enzymes that they play in the softening reaction of ripening fruit.

Minerals: Fruits are rich source of minerals such as calcium (almond, citrus, litchi, karonda), potassium (custard apple, jackfruit, banana, apricot, grape, etc.) phosphorus (almond, walnut, litchi, etc.), iron (karonda, date, cashew nut, mango, etc.), magnesium (banana, pineapple, etc.), copper (lemon, pear, etc.), sodium (mango, lemon, etc.), and sulphur (custard apple, banana, etc.). Calcium is one of the important constituents of cell wall, which provides firmness to the fruit and vegetables. There are little change in mineral contents of fruits during ripening. In grape, level of calcium and ash decreased and potassium and sodium increased at ripening. Total amount of minerals in apple fruit increases during ripening.

Phenols: They are mostly byproducts of aromatic amino acids which are important for disease resistance and determining colour and flavour. Polyphenols present in fruits are cinnamic acid derivatives such as chlorogenic acid (apple, peach, carrot, etc.), p-coumaric acid (grape, pineapple, etc.) and cafeic acid (grape) and flavonoids such as catechin (apple, grape, etc.) epicatechin (date), quercetin (mango, apple, strawberry etc.), tangeritin (loose skin orange), hesperidin and linnolin (sweet orange) and naringenin (grapefruit) decrease in ripe fruit and reduces the astringency. Phenols are oxidized into quinones by polyphenol oxidase and quinines polymerized into tannins. Anthocynidin such as cyanidin (apple, grape, plum, fig, etc.), pelargonidin (strawberry), delphinidin (pomegranate), peonidin (plum), petunidin (grape) and malvidin (grape) which are also phenolics increase during ripening of fruits.

Volatile compounds: Most of the compounds responsible for fruit aroma are volatiles. Fruit volatile includes ester, aldehyde, ketones, hydrocarbon, acetate, alcohol, etc. They present in small amount and they are characteristics of each fruit. For example, ethyle-2 methyle butyrate (apple), eugenol (banana), limonene (citrus) geraniol and linalool (vinifera grape), ethyl acetate (pineapple), methyle and ethyle ester (pear) and linalool (papaya) increase during ripening which gives aroma of the fruits.

Changes in flavour of fruit: The changes in sugar balance, an important factor the flavour for quality of fruits is their aroma, which arises from the production of volatile compounds in the fruit during ripening. Even compounds of low volatility can contribute to the aroma of fruits on account of the extraordinary sensitivity of the olfactory epithelium tissue of the nose to some compounds. These volatile compounds include many different classes of organic molecules such as organic acids, alcohols, esters, carbonyl compounds, hydrocarbons, terpenoids. There are wide differences in the contributions that individual compounds make to the resulting aroma, which are not simply related to their rates of production. The production of most volatile odorous compounds is generally very low in unripe pre-climacteric fruit, but it rises many-fold during ripening. The formations of the volatile compounds, trans-2-nonenal, trans-2-cis-6-nonadienal and trnas-2-hexenal, which are important constituents of cucumber flavour. These important flavour constituents are not present in vivo, but are formed during chewing.

After harvest changes in vegetables: Eating quality of vegetables is determined by flavour and texture, not by physiological age. Generally seeds are sweeter and tender at an immature stage. With advancing maturity, the sugars are converted to starch, with a result of loss of sweetness, water content also decreases and amount of fiber material increases. In edible flower, buds, stems, leaves textures is an often dominant character that determines the both harvest date and quality, as loss of turgor through water loss causes a loss of texture. The natural flavour is often less important than texture, as many of these vegetables are cooked and seasoned with salt and spices. Bulbs, roots and tuber in these crops using appropriate storage condition their storage shelf life can be prolonged.

Interrelationships of the different aspects of the ripening process

Two views of interrelationships of different aspects of ripening process exist. One considers the rise in respiration as a response to the increased energy demands made on the fruit by the various ripening changes. There is now, however, considerable evidence that the many aspects of ripening are not dependent on the respiration rise. For instance, the application of inhibitors of protein synthesis such as cycloheximide to unripe pears and bananas inhibited the production of ethylene, softening, and loss of chlorophyll during subsequent storage without inhibiting the rise in respiration. In bananas, in which ripening had been initiated by a short exposure to ethylene, and subsequent by stored in a low oxygen (0.5 to 1%) atmosphere, ethylene accelerated the change in fruit colour without rise in respiration. These results show that it is possible to isolate many aspects of ripening from the rise in respiration, and they clearly throw doubt on the view of ripening as a highly integrated process. The view has been expressed that the respiration climacteric is neither dependent upon nor integrated with other aspects of ripening.

An alternative view of the role of the respiration climacteric has been proposed according to which the activity of the mitochondrial system as operating relatively independently of the other ripening changes. The rise in respiration should be regarded, in Romani's view, as just another independent ripening change and this be related to the quasi-autonomous status of mitochondria in cells. Studies on the properties of mitochondria isolated from fruit at various stages of ripening suggest that, although there may be compositional changes in the mitochondria (such as changes in the ability to metabolize oxaloacetic acid during the climacteric), mitochondria remain fully functional throughout the period of the respiratory rise and beyond. Isolated mitochondria oxidize substrates at increasing rates during ripening, and retain the ability to phosphorylate ADP with a high respiratory control ratio throughout ripening. Indeed the evidence appears to suggest that there is an increased capacity for phosphorylation as measured by ADP/O ratios as the fruits ripen.

Biochemical control of fruit ripening

In the climacteric fruits, ripening is seen as the final phase in development of the organ and, as such is under genetic control. There is a sharp transition in metabolism at the time of the onset of the climacteric and this involves the promotion of both anabolic and catabolic processes. Two theories have been advanced to explain the underlying biochemical control of the levels of enzyme activity responsible for the promotion of these processes. In one, the important feature is the de novo synthesis of a small group of specific enzymes, which catalyze critical steps in the changes leading to ripening, but which are not present before the climacteric. Inherent in this theory is the synthesis of new messenger RNA species for these enzyme proteins. The alternative theory suggests that the activation or release of pre-existing enzymes is the dominant feature and this is linked with changes in the compartmentation of the cell, which could lead to the release of hydrolytic enzymes from lipoprotein-bound cell organelles and changes in the permeability of cell membranes, leading to increased availability of substrates, cofactors or activators to the active site of the enzymes. However, for a number of climacteric fruits such as apples, avocados, pears, and tomatoes there are evidences for a period of enhanced incorporation of amino acids into proteins during the early stages of the climacteric. As in the case of proteins, so with RNA there are evidences for a period of enhanced incorporation of RNA precursors such as uridine into RNA at the early stages of the climacteric in apples, avocados, pears, and tomatoes. There are good evidences for a period early in ripening in which both RNA and protein synthesis is active. Further, evidence for the importance of protein synthesis in ripening has come from the use of inhibitors of protein synthesis. There are many reports of increases in the in vitro activity of enzymes extracted from fruit at various stages of ripening, but often it is unclear whether these results from changes in the extractability of enzymes as the structure and composition of fruit changes as it ripens, due to changes in the form and activity of the isolated enzymes or from genuine changes in concentration of the enzyme in the tissue. Although, an increased rate of amino acid incorporation is a feature of the early stages in ripening in a number of climacteric fruits. The incorporation of amino acids into proteins during the early stages of ripening represent the turnover of a substantial proportion of the proteins of the pulp cells, and that the incorporation was into a wide range of soluble proteins rather than incorporation into few novel proteins specifically involved in ripening. On the other hand, it was found that there was a peak in the incorporation of tritiated uridine into RNA just before the onset of colour changes in ripening tomatoes. At the time of onset of ripening there was a change in the composition of the m-RNA fraction which could be involved in directing the synthesis of a few novel enzymes responsible for ripening.

Changes in membrane permeability have been estimated by measurements of changes in apparent free space (AFS) to organic molecules and to inorganic ions, and by measurements of changes in the leakage of solutes from tissue slices. The measurements of changes in AFS and of solute leakage are relatively crude measures of changes in the cell permeability characteristics. They give no information on changes in compartmentation within the cytoplasm, but gives measurement of tonoplast and/or plasmalemma membrane permeability. It is of interest to know how such permeability changes, is they involved in active or passive transport, or affect the activity of enzymes involved in ripening. An example of this has been proposed in a study of the activity of the enzyme phosphofructokinase (PFK). The activity of phosphofructokinase measured in vitro does not increase during the ripening of either bananas or tomatoes, yet there is good evidence from the accumulation of its product, fructose 1:6 diphosphate, during ripening of bananas and tomatoes that glycolysis is activated during the climacteric rise in respiration, and that the in vivo activity of PFK increases markedly during this period. Ripening is a differentiation process that is ultimately determined at the genome level. The process by which the control is expressed ultimately in changed activities of enzymes and pathways is unknown. The extent to which the increased enzyme activities responsible for these changes are due either to de novo synthesis, activation, or release of enzymes from membrane-bound structures cannot be assessed, it seems that all three processes may be involved. It is clear that increased turnover of proteins occurs, at least in bananas, as a direct response to ethylene and this may obscure the formation of new enzymic proteins at this stage. In some cases, at least, membrane permeability changes occur during ripening, and changes of the very little of the particular membranes that are affected is known and it can only be speculated on the possible metabolic significance of such changes.

External control of fruit ripening

The use of external factors that to control the ripening process has important implications in the storage of fresh commodities prior to marketing and this control of the ripening process must be achieved while minimizing the deleterious effects of water loss, microbial infection, and the development of physiological storage disorders.

Effect of temperature: In maximum cases fruit development and maturation require a relatively high range of temperatures, the changes in pigmentation often favour low temperatures. Loss of chlorophyll in orange fruits maturing on the tree is dependent upon day temperatures (below 20° C). Citrus fruits from the tropics remain green long after internal maturation has been attained, because of the high temperatures prevailing in round the year. Degreening of detached citrus fruits is performed commercially with ethylene under high humidity at 25 to 30° C. The distinct temperature (and light) requirements indicate that lycopene and β -carotene are produced via separate biochemical routes. Accumulation of anthocyanins is favoured by low temperatures, this has been demonstrated in grapes, blood oranges, apples, and other fruits. In apple, more anthocyanins were produced in cool temperatures in the range of 21 to 10° C. At temperature below a certain critical level, many fruits of tropical and sub-tropical origin develop a disorder known as chilling injury. Colour changes associated with ripening in the tomato are inhibited in storage at temperatures above 30° C. Many fruits fail to ripen properly beyond 30 to 35° C and that in most cases this inhibition is reversible. For example bananas fail to ripen normally, the pulp becoming soft and watery at the temperature above 30° C.

Gaseous composition of storage atmosphere: The raising the CO_2 level and decreasing the O_2 in the storage atmosphere delayed ripening. Such a modified atmosphere, coupled with the use of temperatures just above the limit for the onset of low temperature disorders, is the basis for the widely used method of controlled atmosphere storage that has proved so successful with a wide range of fruits.

Effect of O_2 tension: Low levels of O_2 (between 1 and 5%) delay the onset of ripening in a number of different fruits. The inclusion of low levels of ethylene in the low oxygen atmosphere reversed the inhibition of these ripening changes and ultimately normal ripe fruits are produced, showing that the ripening as such was not inhibited by low O_2. Thus, oxygen can act in system by affecting ethylene biosynthesis, ethylene action, and the sensitivity of the tissue to ethylene. One is due to the inhibitory effect of low O_2 tensions on respiratory metabolism and the other is due to its effect on ethylene action and the initiation of ripening processes. Oxygen tensions greater than atmospheric tend to promote ethylene biosynthesis and consequently in promote ripening in a number of fruits. This effect is significant in citrus fruits, which, if placed in atmospheres of 34 to 99 per cent O_2 will have a normal climacteric rise, and their production of ethylene is markedly increased.

Effect of CO_2 tension: Increasing CO_2 concentrations in the atmosphere around fruits delay the onset of the climacteric. Concentrations of CO_2 from 3 to 10 per cent are generally effective in this, although higher CO_2 levels can, in some fruits (such as apples) lead to physiological disorders. The effect of CO_2 in delaying ripening has been related to its effect on ethylene action. CO_2 acts as a competitive inhibitor for ethylene action, presumably by competing with ethylene for the attachment site and also by inhibiting respiration in bulky organs. In contrast to this, CO_2 stimulates the respiration of lemons, but this is due to a stimulation of dark fixation of CO_2 and the resultant accumulation in the tissue of organic acids, which act as respiratory substrates.

The effect of atmospheric pressure: The ripening of bananas was completely inhibited if the fruit was exposed at one fifth atmospheric pressure in pure oxygen, so that the atmospheric tension of oxygen was maintained. However, the inclusion of a small amount of ethylene in the atmosphere overcame the inhibitory effect of the reduction in pressure. The fact that added ethylene reversed this inhibition showed that the removal of ethylene, rather than reduced atmospheric pressure was the factor responsible for the delay in ripening. It was shown that, with mature green tomatoes, stored at 19 cm Hg pressure at a partial pressure of oxygen of 0.20 did not delay ripening compared to similar fruit stored under normal atmospheric conditions. However, storage under reduced pressures at lower O_2 tensions did delay ripening.

Light: Light is required for optimum fruit colouration. The loss of chlorophyll was enhanced by Red light in detached tomatoes. The accumulation of lycopene in tomato could also be induced by Red light and reversed with Far Red light. β-carotene, on the other hand, attained similar levels with Red, Far Red, and dark treatments. Phytochrome appears, therefore, to be involved in the control of the biosynthetic route leading to lycopene, but not of the one leading to β-carotene. The light intensities required to saturate the lycopene system in tomato are quite low. Exposure to light is necessary also for anthocyanin biosynthesis in apple, grape, and other fruits. For full colour development in grapes, 54 per cent of sunlight intensity was sufficient, but further reduction caused proportional decrease in anthocyanin levels down to an absence of anthocyanins in darkened clusters. The need for high light intensities points to mechanisms other than phytochrome in at least some fruits.

Senescence in Fruits

The process of senescence refers to endogenously controlled deteriorative changes, which are natural causes of death of cells, tissues, organs or organisms. Senescence may be thought of as a natural developmental process, namely, terminal differentiation. The concept that death may be actively induced by endogenous factors has been gaining acceptance.

Changes during senescence: Components which have been used as measures of senescence include chlorophyll, total protein, nucleic acids, photosynthesis (reduction in photosynthetic enzymes such as ribulose bisphosphate carboxylase or phosphoenolpyruvate carboxylase), plant growth substances, membrane permeability and abscission. In fruits that show a single sigmodial grown pattern, ripening normally occurs during the final phase of slow growth of the fruit. In fruit ripening on the tree, there is probably no absolute separation in time between ripening and other growth processes and probably some small degree of cell expansion is occuring at least early in ripening. However, in those fruits that store large quantities of reserve materials, and which can ripen normally after detachment from the parent plant, there can be an almost complete separation of senescence and growth phenomena.

In addition to those fruits showing a single sigmodial growth curve, there are fruits showing more complex growth patterns with double and even triple sigmodial patterns. In these there are two or three rapid growth phases interrupted by one or two periods of little or no growth. The double sigmodial pattern is shown by many fruits including the stone fruits, figs, and grapes. It is thus clear that in fruits showing double sigmoidal growth patterns active cell expansion accompanies some aspects of the ripening process. In the grape, the onset of ripening is marked by the start of sugar accumulation, loss of acidity, softening, and skin colouring these characteristic ripening changes are associated with renewed cell expansion. In stone fruits, the two periods of rapid growth are associated with the development of different tissues. It is tissues in which the ripening changes occur these rapidly expanding during the period just prior to, and possibly at ripening. It is clear that in at least some fruits of this type it is not possible to separate senescence phenomena from those of growth, and this can lead to complications in the interpretation of the observed biochemical changes when both processes occur together.

Table 72. List of fruits showing different patterns of growth

Growth pattern	Produce
Single sigmodial	Apple, avocado, banana, date palm, mango, melon, orange, pear, pineapple, strawberry, tomato
Double sigmodial	Apricot, black currant, blueberry, fig, grape, olive, peach, plum, raspberry, sour cherry, sweet cherry
Triple sigmodial	Chinese gooseberry

Acceleration of Ripening

A technique to induce early ripening would permit, staggering harvest, early crop in the season to obtain premium price and an opportunity to avoid early rains which is a limiting factor in commercial cultivation of some fruits such as grape, date, litchi, etc. Agro techniques such as growth regulators are very useful in order to regulate ripening and improve fruit quality. Ethephon or ethrel (2-chloroethyl-phosphonic acid) an ethylene releasing chemical has dramatic influence on ripening. It is stable in aqueous solution below pH 4, when pH rises above 4; disintegration of the chemical takes place and release ethylene along with chloride and phosphate. Accelerating the ripening changes with ethephon suggests the possibility of increasing early yield (better price), programming harvest schedule, uniform ripening (reduce sorting cost), fast ripening rate (reduce weight loss), facilitating once over harvest (reduce the number of picking) and improving colour (market appearance). In mangoes where in anthracnose is a problem, ethephon treatment shortened the ripening period, giving no time for the development of anthracnose, in some citrus hastening chlorophyll degradation. Pre-harvest application of ethephon has been found more effective to concentrate fruit ripening than post-harvest treatments in many fruits.

Chemicals for Accelerate Ripening (Hasten Ripening)

Ethylene and ethylene releasing compounds: A more important break through in C_2H_4 effects and physiology is the chemical synthesis of Ethephon because C_2H_4 is released from this compound. Thus all biological effects attribute to C_2H_4 gas are duplicated by aqueous solution of ethephon. 2, 4 – Chlorophenythio triethyl amine hydrochloride (CEPTA) are also a ethylene releasing early in the season to obtain marketable fruit with premium price.

Abscisin: Application of abscisin accelerates senescence in detached fruits and leaves. Abscisic acid treatment of oranges resulted in accelerated breakdown of chlorophyll and increased synthesis of cartoenoids. Ripening of banana slice was also accelerated by abscisic acid.

Ascorbic acid and Hydroxyethyl hydrazine (BOH): Ascorbic acid, Cu-EDTA (Cupric ethylene diamine tetracetate) and BOH were found to include C_2H_4 formation, when sprayed before harvest.

Acetylene and Calcium carbide: Calcium carbide (CaC_2) treatment to generate acetylene is also used to, hasten fruit ripening in banana and citrus. Gas chromatographic analysis of the smoke from any burning material such as leaves, twig or showed that the active ingredients were C_2H_4 and acetylene is needed to obtain the same rate of degreening by C_2H_4.

Alcohol: It is reported that among the several alcohols, such as ethanol, hexanol, octanol, heptanol etc, used in dip treatment of green tomatoes prior to ripening. Hexanol immersion seemed to accelerate the ripening of the tomato fruit.

Fatty acids: Olerification has been known as a practice to hasten fig fruit maturation. Since the 3rd century BC studies were conducted to elucidate the physiological mechanism of olerification and applied various kinds of oils and related chemicals to the eye of fig fruit in a day or two, fruits treated with several vegetable oil began to increase in size reaching full colour and maturity within few days. By fatty acid oxidation in the cells of fruits acetaldehyde or C_2H_4 are derived which are volatile substance and hence induces enzymatic activities for early ripening.

Chemicals for delay in ripening

A technique to delay ripening may extend the period of availability and extend shelf life of fruits and vegetables. Chemicals which delay the ripening are: Potassium permanganate which absorbs ethylene during transport or storage and delay the ripening in many fruits such as mango, banana, apple etc. Calcium nitrate as pre-harvest spray of 1 per cent delay ripening in guava, grape etc. Hypochlorite as post-harvest treatment with 2 per cent solution maintain colour and extend shelf life of litchi fruits both at ambient and low temperature storage. CIPC commercial preparation both as dust or liquid retard sprouting in potato. Growth regulator such as BOA (Auxin) 100-200 ppm delays ripening in grapes, gibbrelic acid as pre-harvest spray (50-100 ppm) delays ripening in citrus fruits and post-harvest treatment retard ripens in banana, guava, mango and tomatoa. BA (cytokinin) 10 ppm as post-harvest treatment retard senescence in spinacha, lettuce, beans, peppers, green onions, cabbage and cauliflower. Maleic hydrazide (inhibitor) at 2000 ppm as pre-harvest spray retard sprouting of onion and as post-harvest dip delayed ripening in mango. Use of 1-MCP (Smart fresh) delayed ripning in mango, guava and apple.

Kinins and Kinetins: These chemicals were shown to delay chlorophyll degradation of leafy vegetables, spinach, peppers, beans, cucumber and others. The growth effect is a retardation of yellowing by maintaining high protein level on the applied tissue.

Gibberellins: Post harvest treatment of GA markedly retards ripening of tomatoes, guava and banana then effect on ripening was indicated by lowered respiratory rate retarded climacteric and delayed colour change. Pre-harvest sprays of GA to have a striking effect in decreasing the rate of development, maturation ripening of kaki fruits, lemons and navel organs because fruit will remain on the tree beyond normal maturity. Some of the effects feel firmness delayed accumulation of cartoenoids on 'Navala orange' higher TSS and ascorbic acid in lemons. GA probably increase peroxidise and catalase activities.

Auxins (CIPA and NOA): Pre-harvest foliar application of CIPA at 25 ppm and NOA at 25-30 ppm delayed the physio-chemical deterioration of coorg mandarins in storage.

Growth retardants (MH and CCC): The effect of maleic hydrazide on the ripening process varies with different type of fruit and may depend on the application time and amount of chemical absorbed. Harvested mangoes dipped in 1000 and 2000 ppm, maleic hydrazide showed delayed ripening. Maleic hydrazide has inhibitory action on storage sprouting of onions, radish, sugar beet, turnips, carrot and potatoes. Shelf life lettuce was duffled and deterioration in quality of broccoli heads and aspragous spare was retarded by immersion of the cut stem bases of dipping the vegetables in solution of CCC.

Metabolic inhibitors (Cycloheximide and Actimomycin –D): Cylocheximids treatment severely inhibited flesh softening, chlorophyll degradation and C_2H_4 synthesis. When cycloheximide was applied to fruits that had begun to ripen, ripening proceeded at a reduced rate, similar effects on ripening were also observed on pre-climacteric fruit treated with actinomycin-D, C_2H_4, did not reverse the complete inhibition of ripening imposed by actinomycin-D.

Vitamin-K: Vitamin K and Vitamin K_3 to inhibit ripening in banana at temperature as high as 14°F (0-2° C).

Maleic acid: Maleic acid in any acetate retarded to the ripening of jarnaica, banana, pineapples, citrus, and others but it is of transient effect.

Ethylene oxide: Mangoes treated with ethylene oxide also show a definite delay in ripening. Ethylene oxide may be an endogenous metabolic antagonist of C_2H_4.

Na-DHA (Sodium Dehydro Acetic Acid): In strawberry, it is found to delay the ripening and reduced rate of respiration.

Carbon monoxide: In mushroom used to extend storage life upto four folds.

Ethylene absorbents: Waxing, low O_2, high CO_2, and ripening inhibitors are sometimes combined to prolong storage life. Banana is packed in film bag containing $KMnO_4$ (to absorb C_2H_4). A commercial preparation of this absorbent called "Purafil". Alkaline $KMnO_4$ on a silicate barrier proved effective in the complete absorption of C_2H_4 from bananas held in sealed polythene bags.

1-MCP (1-methylcyclopropene): The 1- methylcyclopropene (1-MCP or C_4H_6) is an ethylene action inhibitor. It binds with ethylene receptors and thereby prevents ethylene dependent responses in many horticultural commodities. 1-MCP has been formulated into a powder that releases its active ingredient when mixed with water. This nontoxic compound can be used at very low concentrations (nL L^{-1}). The beneficial effects of 1-MCP in fresh produce include the inhibition of respiration and ethylene production, delayed fruit softening, restricted skin colour changes, prolonged cold storage life and alleviation of certain ethylene-induced post-harvest physiological disorders. 1-MCP treatment is also useful in reducing chilling injury symptoms and decay in tropical fruit during cold storage.

Degreening of fruits and vegetables: Post harvest treatment of citrus fruits with ethylene under controlled conditions hastens the loss of chlorophyll, a process known as degreening. Degreening consists of chlorophyll degradation to allow the expression of natural pigments masked by the green colour e.g. yellowing of citrus fruits. Degreening process can be hastened by applying ethylene and done mainly in non-climacteric fruits like citrus. On a small scale dip treatment in ethrel solution also bring about degreening. The concentration of ethylene required and time of exposure is significantly high in case of degreening when compared to ripening. The main causes for greening are climatic conditions before harvest. For example, citrus often reaches commercial maturity with traces of green colour on the epidermis (flavedo). Although not different from fruits with colour, consumers sense that they are not ripe enough and have not reached their full flavour. Exposure to low temperature during maturation is necessary for an orange-coloured peel to develop. Hence fruit grown in low altitude tropics fail to degreen completely.

Methods of degreening

Exposing to ethylene: Degreening is done at 25-30° C and 85-95 per cent RH with ethylene gas trickled into room to achieve 20-30 ppm or 10 $\mu L\ L^{-1}$ for 24 - 72 hours with regular ventilation of the chamber to prevent CO_2 build up and injury (1%). In batch process it is at 20-200 μL ethylene L^{-1}. Trickle process is faster than batch process; since degreening condition accelerate deterioration and decay of citrus fruits. This is most popular methods for citrus fruits in which citrus fruits are exposed from 1-3 days (depending on degree of greening) to an atmosphere containing ethylene (5-10 ppm) under controlled ventilation, 20-26° C and 90-95 per cent RH. The ceiling of room is relatively high, allowing boxes to be stacked at least four boxes high. In another method the false ceiling is added to provide for adequate air. Degreening is carried out in ripening rooms, with 10 to 100 ppm ethylene concentrations. This process takes 2 to 3 days, and it is again necessary to ventilate daily to ensure that carbon dioxide levels do not exceed 1 per cent. The most rapid degreening occurs at temperatures of 25 to 30° C but the best colour (concentration of peel carotenoids) occurs at 15 to 25° C.

Artificial colouring: When weather is not conducive for the development for colour in orange; legally permitted dye can be used to colour the peel of the fruits like orange, with Citrus Red No.2 (1-2 (2, 5-dimethoxy phenylazo) 2-napththol) this process is called as Colour Add. It is used on mature fruit which are not intended to processing. Dye is applied to fruit by dip at 49°C for 4 minute for oranges. Rinsed enough to prevent bleeding and residue tolerance is 2ppm (2 mg^{-1} kg^{-1} of fruit).

Banana ripening process: Majority of world banana is ripened under controlled condition by following methods.

Batch/shot process: In which the chamber is charged with ethylene gas at once to a concentration of 20-200μL^{-1}. The chamber has to be ventilated after 24 hours to prevent the accumulation of CO_2. Concentration of CO_2 should not exceed 5000μL^{-1} (0.5%) to allow personal to enter to inspect the fruits. If the chamber is poorly sealed, it may be necessary to recharge the chamber with C_2H_4 after 12 hours.

Trickle/flow process: Ethylene is introduced into the room slowly in thin stream continuously into the chamber at a rate just sufficient to maintain the required concentration. The ripening chambers should be ventilated at the rate of about one room volume each 6 hours, to prevent the accumulation of CO_2. In practice it not necessary to install a ventilation system in rooms < 60 m^3 because they have natural air leakage rates higher than the required minimum rates. When banana is placed in chamber so as to expose at least two faces to the circulating air, ensuring that fruits temperature is even. Fruit is packed in vented cartoon, unitized on pallets, and fruit temperature is controlled by forced air temperature. A minimum air flow of about 0.34 L sec^{-1} kg of banana is required. Regulating RH during the course of ripening is important for banana. A RH range of 85-90 per cent has been recommended at stage 2 (green, trace of yellow), but this should be reduced to 70-75 per cent during the later colouring stages to avoid the skin splitting. If RH is high, fruits will become too soft and may split and if it is low it may cause weight loss, poorer colour and more blemishes on fruit. Regular cleaning with chlorine is requiring avoiding mould growth due to high RH during storage. It is important to harvest at the correct stage of maturity otherwise quality will be inferior after ripening. At full maturity it is only necessary to hold fruit at desired temperature and RH and ethylene is not always necessary to ripen fruits, some fully developed fruits produce sufficient C_2H_4 to ripen itself and adjacent fruit (triggering effect).

10

Packaging of Horticulture Produce

Packaging fresh fruits, vegetables and flowers is one of the important steps in the long and complicated journey from grower to consumer. The main function of packaging fruits, vegetables and flowers is to assemble the produce into convenient units for better handling and to protect them. Packaging is the act of putting the produce inside a container along with packing materials to prevent movement and to cushion the produce such as plastic or moulded pulp trays, inserts, cushioning pads, etc. and to protect it i.e., plastic films, waxed liners, etc.

Packaging must satisfy three basic objectives:

i. Contain product and facilitate handling and marketing by standardizing the number of units or weight inside the package.

ii. Protect product from physiological injuries (impact, compression, abrasion and wounds), pathological deterioration, adverse environmental conditions (temperature, relative humidity) and most prominently against excess moisture loss during transport, storage and marketing.

iii. Provide information to buyers, such as variety, weight, number of units, selection or quality grade, producer's name, country and area of origin.

Frequently included are recipes, nutritional value, bar codes or any other relevant information on traceability. A well designed package needs to be adapted to the conditions or specific treatments required for the product. If hydro-cooling or ice-cooling is required, the package must tolerate wetting without losing strength. For a product with a high respiratory rate, the packaging should have sufficiently large openings to allow good gas exchange. When produce dehydrates easily, the packaging should be designed to provide a good barrier against water loss, etc. Too much of a moisture barrier will cause an excessively high relative humidity in the package and result in accelerated spoilage due to micro organisms or in skin splitting on some fruits. Fresh fruits and vegetables are generally packed in bamboo baskets, plastic crates, cartons, plastic bags, nylon sacks, bulk bins and palletized containers for handling, transporting, and marketing in many developing countries.

Levels of Packaging

Primary package: It is the one which is in direct contact with the contained product. It provides the initial, and usually the major protective barrier. Example: Metal cans, paperboard cartons, glass bottles and plastic pouches, beverage can, cushioning envelopes, plastic bottles, skin pack.

Secondary package: It contains a number of primary packages. It is outside the primary packaging perhaps used to group primary packages together. It is the physical distribution carrier and is sometimes designed so that it can be used in retail outlets for the display of primary packages like corrugated case and boxes.

Tertiary package: It is made up of a number of secondary packages. It is used for bulk handling. Example being a stretch-wrapped pallet of corrugated cases.

Quaternary package: It is frequently used to facilitate the handling of tertiary packages for transportation of fresh fruits, vegetables and flowers. This is generally a metal container up to 40 meter in length which can be transferred to or from ships, trains, and flatbed trucks by giant cranes. Certain containers are also able to have their temperature, humidity and gas atmosphere controlled.

PACKAGING ASPECTS OF HORTICULTURAL PRODUCE

Fruits and vegetables have high moisture content, ranging from 75-95 per cent and their equilibrium relative humidities are as high as 98 per cent. Under any normal atmospheric condition, they will dry rapidly. This causes wilting and shriveling due to loss of rigidity and shrinkage of the cells. Proper packaging can prolong the storage life of fresh fruits, vegetables and flowers by preventing moisture loss and thereby wilting. The rate of moisture loss varies with the product and water vapour permeability of the packaging film. The use of small perforations for oxygen permeation has an insignificant effect upon moisture loss. A type of spoilage much prevalent in fruits and vegetables is that caused by microorganisms such as yeasts, molds and bacteria. These organisms can cause destruction by growing on the exterior of the product or they may invade the interior through a surface bruise or cut and cause internal decay. Therefore careful handling and packaging are very important in preservation of freshness and quality. Normal ripening of fruits and vegetables causes alterations in colour, texture, odour and flavour. The product becomes overripe and quality deteriorates. The primary goal of fresh produce merchandising is to deliver the product to the consumer at such a point in the ripening scale, that it will achieve perfect ripeness at time of eating. In practice, the produce is delivered somewhat underripe at the time of purchase and the consumer delays consumption until ripening is completed. Since all these processes are highly sensitive to temperature, they can be slowed down by storing the produce under refrigeration. Each fruit or vegetable has an ideal temperature for storage. If this temperature is not used, injurious result occurs for example, tomatoes will not ripen if chilled below 4° C, bananas will turn black below 11° C and potatoes develop a sweet flavour below 4° C. An additional reason for the necessity of refrigeration for fresh produce is the heat generated due to metabolism or respiration. For example, green beans, sweet corn, broccoli, green peas, spinach, and strawberries generate from 15000 to 50000 BTU of energy per ton per 24 hours of storage at 15° C. Even when chilled to 0° -4° C, they still evolve 2500-17000 BTU per ton per 24 hours. This heat must be taken into consideration in the design of refrigeration equipment. Some types of fresh produce give off volatile compounds during ripening which will impart unacceptable odour and flavour if not allowed to escape, or they may prematurely ripen the fruit. Different types of packages are used for produce in the India and the number continues to increase as the industry introduces new packaging materials and concepts. Although the industry generally agrees that container standardization is one way to reduce cost, the trend in recent years has moved toward a wider range of package sizes to accommodate the diverse needs of wholesalers, consumers, food service buyers, and processing operations.

Primary Functions of Packaging

Containment: It is a basic requirement for movement of a product from one point to another. The package type and size will be a function of the product and market requirements. The produce should fit well inside the container, with little wasted space. For example, a shallow container will be used for a product such as a peach or strawberry to avoid compression damage compared with a harder fruit such as an apple. Small produce items that are spherical or oblong (such as potatoes, onions and apples) may be packaged efficiently utilizing a variety of different package shapes and sizes. However, many produce items such as asparagus, berries, or soft fruit may require containers specially designed for that item. Packages of produce commonly handled by hand are usually limited to 25 kg.

Protection: Packages provide protection for the product against environmental factors such as dust and water, as well as impact, compression, bruising, and friction injuries that can occur during handling and transport. It provides protection from microbial contamination from the surrounding environment and from consumer contact. Produce containers must be sturdy enough to resist damage during packaging, storage, and transportation to market. Cartons must have stacking strength and durability to prevent collapse or crushing while they are on pallets, especially under high relative humidity conditions. Trays, cups, or other protective measures may be contained within the package to prevent movement and/or contact between product items.

Convenience: Products are packaged in sizes that are convenient for the consumer, but may be larger for transport because of economies of scale. The product may be removed from the container and placed in display as single units or repackaged. Smaller, primary packages may be placed in larger secondary containers. Package size is affected by regulations that ensure that package weights are safe and manageable for everyone.

Identification: The package must identify and provide useful information about the produce. It is customary to provide information such as the produce name, brand, size, grade, variety, net weight, count, grower, shipper, and country of origin. It is also becoming more common to find included on the package, nutritional information, recipes and other useful information directed specifically at the consumer. Universal Product Codes (UPC or bar codes) may be included as part of the labeling. The UPCs used in the food industry consist of a ten digit machine readable code. The first five digits are a number assigned to the specific producer (packer or shipper) and the second five digits represent specific product information such as type of produce and size of package. Although no price information is included, UPCs are used more and more by packers, shippers, buyers, and retailers as a fast and convenient method of inventory control and cost accounting. Efficient use of UPCs requires coordination with everyone who handles the package.

Communication: In addition to advertising the type and source of the produce, information is provided about gross and net package weight, unit size of product, as well as any declaration of the use of any postharvest treatments that may be required by state or national regulations.

Sales appeal: High quality graphics are increasingly being used to boost sales appeal of fresh produce. Multi-colour printing, distinctive lettering, and logos are now common.

Shelf life: Modern produce packaging can be custom engineered for each commodity to extend postharvest shelf life and reduce waste. It helps to maintain the freshness of produce.

Requirements of a Good Package

The following are among the more important general requirements and functions of fruit and vegetable packaging materials/containers.

» Package should be environment friendly, light protection, cost effective, non-toxic and compatible with the automatic packing/filling, handling machines and facilitate special treatments like pre-cooling, easly printable and simplicity of opening and disposal.

» It should have sufficient strength against compression, impact and vibrations.

» Package should have consumer appeal, sanitary protection, sizable, easy in handling and packaging material should be recyclable or biodegradable, or both specially, for export purpose.

» It should be easily transported when empty and occupy less space than when full.

» It must be easy to assemble, fill and closes either by hand or by use of a simple machine.

» It must provide adequate ventilation for the content during transport and storage.

» Its capacity should be suitable for market demands and dimensions and design must be suited to the available transport in order to load neatly and firmly.

Produce containers should be produce friendly, helping to maintain an optimum environment for the longest shelf life. This may include special materials to slow the loss of water from the produce, insulation materials to keep out the heat, or engineered plastic liners that maintain a favorable mix of oxygen and carbon dioxide. A properly designed produce container should contain, protect, and identify the produce, satisfying everyone from grower to consumer. Because almost all produce packages are palletized, produce containers should have sufficient stacking strength to resist crushing in a low temperature, high humidity environment.

Specific applications of package for fruits and vegetables

The particular type of package used depends upon the shape and perishability of the product. There are five main classifications of horticultural produce like soft fruit, hard fruit, stem products, root vegetables and green vegetables. Soft fruits are highly perishable, bruise and crush easily and easily subject to anaerobic spoilage. They are packaged in semi-rigid containers with a cover of cellophane, cellulose acetate, polystyrene or other suitable film cover. Polyethylene bags with ventilation holes are used because shelf-life is limited due to individual damage and decay. Typical soft fruits are cherries, grapes, blueberries, strawberries, raspberries and plums retain top quality for 2 or 3 days under ideal conditions. Hard fruits are less perishable, better able to resist damage from handling and have lower respiratory rates. The most common package is an open tray, a plastic film overwrap or sleeve. Hard fruits may also be bagged in perforated polyethylene film or in nets. Examples of hard fruits are apples, bananas, citrus fruits, peaches, pears and tomatoes. Stem products (celery, rhubarb, potato) and asparagus roots are highly perishable as they rapidly lose moisture for long periods storage it is desirable to protect them against moisture losses. They should be bagged or wrapped in moisture proof cellophane or polyethylene with ventilation, or they should be banded or sleeved with shrink film. Typical root vegetables include carrots, turnips, radishes, onions, beets and yams are relatively less parishable. They are washed, graded and sized prior to packaging, which is usually in durable polyethylene bags with perforations.

Package Environments

The packaging has to perform its functions in three different environments. Failure to consider all three environments during package development will result in poorly designed packages, increased costs, consumer complaints and even avoidance or rejection of the product by the customer.

Physical environment: This is the environment in which physical damage can be caused to the product. It includes shocks from drops, falls and bumps, damage from vibrations arising from transportation modes including road, rail, sea and air and compression and crushing damage arising from stacking during transportation or storage in warehouses, retail outlets and at home.

Ambient environment: This is the environment which surrounds the package. Damage to the product can be caused as a result of gases (particularly O_2), water and water vapour, light (particularly UV radiation) and temperature, as well as micro-organisms (bacteria, fungi, molds, yeasts and viruses) and macro organisms (rodents, insects, mites and birds). Contaminants in the ambient environment such as exhaust fumes from automobiles, dust and dirt can also find their way into the product unless the package acts as an effective barrier.

Human environment: This is the environment in which the package interacts with people and designing packages for this environment requires knowledge of the variability of consumer's capabilities including vision, strength, weakness, dexterity, memory and cognitive behaviour. Since one of the functions of the package is to communicate, it is important that the messages are clearly received by consumers. In addition, the package must contain information required by law such as, nutritional content and net weight.

Materials for Packaging

For fruit and vegetables packaging, various kind of materials are in use, like wood (boxes, bins, trays, barrels, pallets), jute/canvas sacks, paper and card board (liners, boxes, trays), plastic (rigid crates, pallets, trays), flexible (films - single and multi layered), polystyrene boxes / trays, combined materials (CFB and plastic). CFB has almost replaced wood and jute and is considered as most important package material to be used in combination with other materials. Wood and expanded plastic packages are inherently strong are resistant to high humidity, condensation and rain compared to fiberboard packages. Rigid expanded polystyrene is light weight yet strong but require much space (collapsible i.e. foldable crates require less space on return journey) and costly. In comparison, fiberboard is attractive and can be made stronger by using two or three thickness, such as the bottom and lid of fully telescopic cartons. The strength of the fiberboard lies in the fluting between the inner and outer liners. Fiberboard comprises of two layers of fluting sandwiched between three layers is stronger than the single layer of fluting. Normal fiberboard carton rapidly absorb moisture under storage can be protected if fully impregnated with wax but wax impregnation is expensive and not fit for recycle. Some materials used for fresh horticultural produce include:

» Wooden box having CFB liners.

» CFB box with plastic film wraps

» CFB trays with wooden corner supports

» CFB laminated or waxed containers.

» CFB box with plastic retailer packs (strawberry boxes)

» CFB or polystyrene trays/boxes with plastic film wraps.

Apart from these, bags and sacks of paper, polyethylene film, woven polypropylene are also used. These give little protection to the crop from handling and transport damage.

Types of packages material: Packaging material for fresh produce is of several types:

Natural material: Baskets and other traditional containers are made from raw material locally available such as bamboo, rattan, straw and palm leaves. Both raw material and labour costs for the manufacture of the containers are normally low and if the containers are well made, they can be reused for several times. They are mostly used for handling the produce in the field, at farm level and seldom for the local market.

Disadvantages are:

- They are contaminate and difficult to clean.
- They lack rigidity and easily bend out of shape when stacked.
- They cause pressure damage if tightly filled.
- They often have sharp edges or splinters and cause damage unless they internally lined with sacks, cotton or other materials.

Wood

Sawn wood is used to manufacture reusable boxes or crates in small quantity due to high cost. Veneers of various thicknesses are used to make various boxes or trays. Wooden boxes are rigid and reusable and if made to a standard size, it will stack well on trucks and in storage. They are used for produce handling at farm and packing house levels.

Disadvantages

- Difficult to clean adequately for multiple uses.
- Heavy and costly to transport for the reutilization.
- They have sharp edges, requiring some form of liner to protect the content.

Cardboard and fiberboard

Containers are made from solid or corrugated cardboard. The types closing with either fold-over or telescopic tops are called boxes or cases. Shallower and open-topped ones are called trays. Boxes are supplied in a collapsible form and are set up by the users. The setting up and closing often requires taping, gluing or stapling. Cardboard boxes are lightweight and clean and in some cases are reusable because they may be easily collapsible when empty.

Disadvantages

- They may, if used only once, prove an expensive recurring cost.
- They are easily damaged by careless handling and stacking.
- They are seriously weakened if exposed to moisture, unless they are wax coated.
- They can be ordered economically only in large quantities.

Molded plastics

Reusable boxes molded from high-density polythene are widely used for transporting produce in many countries they can be made to almost any specifications. They are strong, rigid, smooth, easily cleaned and made to stack and nest when empty in order to conserve space. In spite of the cost, however, their capacity for reuse can make them an economical investment.

Disadvantages

- They can be produced economically only in large quantities, but are still expensive.
- They are attractive and have many alternative uses and are subject to high pilferage.
- They require a very tight organization and control for use in a regular go-and-return service.
- They deteriorate rapidly when exposed to the sunlight, unless treated with ultraviolet inhibitor.

Natural and synthetic fibers

Sacks or bags for fresh produce can be made from natural fibers like jute or sisal or from synthetic polypropylene. Bags usually refers to small containers of up to about 5 to 10 kg capacity, woven to a close texture or made in a net form of about 15 kg bags or sacks are mostly used for less easily damaged product such as potatoes, onions and roots avoiding in any case to damage the produce with a careful handling.

Disadvantages

- Lack of rigidity and rough handling may damage the content.
- They are often too large for careful handling and for the weight they are easily dropped with result of damage.
- The reduced ventilation when stacked if they are finely woven.
- They are difficult to stack and may be unstable and collapse.

Paper or plastic films

They are often used to line packing boxes in order to reduce water losses or to prevent friction damage. Paper sacks may have a capacity of about 25 kg and are mostly used for produce of low value. Closure is done by machine stitching or in the field by twisting wires ties around the top.

Disadvantages

» Walls of papers are permeable by water, vapor or gases.

» Heat can be slow to disperse from stacks of sacked produce.

» Mishandling provokes damage to the content.

Classification of packaging: Packaging can be classified in a number of ways the most important one is by stages of distribution system for which it is primarily intended. Packaging materials, storage containers or bins must be free of synthetic fungicides, preservatives or fumigants. They also must be free of any residue from cleaning and sanitizing. Containers that are to be reused must be of the type that can be thoroughly cleaned prior to use.Packages can be classified as follows:

» Flexible sacks; made of plastic, jute, such as bags (small sacks) and nets

» Wooden crates

» Cartons (fiberboard boxes)

» Plastic crates

» Pallet boxes and shipping containers

» Baskets made of woven strips of leaves, bamboo, plastic, etc.

Classification based on packing unit load

Consumer packaging: The package in which consumer receives the produce is called consumer packaging. Consumer packages are small in size and designed to hold half dozen or one dozen fruits or half kg to two kg of vegetables. Packaging depend upon the need of produce for protection, expected transport and storage time, required shelf-life, packaging material costs, packaging, sorting and storage cost and latest knowledge of the market requirements.

Types of consumer package

Nets: Net bags are used to provide desired ventilation and allow free air movement for the produce. Nets are only suitable for hard produce such as coconuts, potatoes, onions, yams and citrus fruits, etc.

Bags: Bags are the most common and favoured retail packs because of their low material and packaging cost. In terms of cost to strength ratio, 25-40 mm low density polyethylene or 12.5 mm high density polyethylene bags are most suited. The bags can be made of paper, perforated polyethylene or polypropylene film, plastic or cotton nets.

Tray with overrape: The trays used are usually made of moulded pulp, card board or plastic material like polystyrene, expanded polystyrene, PVC, LDPE, LLDPE and PP are overwrapped with heat shrinkable or stretch films. The produce is placed in individual cavities so that abrasion and bruising is avoided during transportation. This film can be applied without the application of heat. The films are transparent, semi-permeable and allow exchange of gases for respiration of the product.

Foam sleeve packs: This is a hygienic, nontoxic and odourless plastic tubular film made of polyethylene foam available in different colours, diameters and lengths. It can be easily slipped over the individual to as many as ten fruits in a snug fit form. Sleeve packs provide a cushioning effect and protect the fresh produce against abrasion and scratches during transit.

Flexible plastic films: Different types of flexible plastic films like LDPE (Polyethylene), PVC (Poly Vinyl Chloride), PP (Polypropylene) and cellulose acetate films are used for packaging of horticultural produce. These films are mostly used as pouches with holes punched at regular intervals to allow respiration.

Plastic punnets: These are strong, versatile, clear, bright containers, which offer product visibility and are provided with holes for ventilation, which keeps the produce fresh. These containers are food-grade, odourless, light weight, stackable and recyclable and give good presentation. They are either made of PET, PVC or PP.

Plastic net bags (Extruded & Woven): The plastic net bags have the feature to stretch and accommodate all sizes and shapes of produce. These are generally made of HDPE (High Density Polyethylene) or PA (Polyamide). These bags are available in roll form or in precut lengths with stretch width of 200 mm – 400 mm. By allowing air to circulate in and around the produce, these net bags prolong the freshness and shelf-life of the fresh produce and preventing spoilage and wastage.

Light weight plastic crates: These are lightweight can be made of HDPE or PP, which need not be put into an outer pack for transportation. The perforations provide ventilation and keep the produce fresh. The crates are stackable and have high compression strength and therefore provide adequate protection to the fresh produce packed inside. These crates are hygienic, clean, reusable and can be recycled.

Shrink wrap: One of the newest trends in fresh produce packaging is the shrink wrapping of individual produce. By reducing the transpiration rate and maintaining the fruit firmness the film forms a barrier which increases the resistance to water vapour. The transpiration rate can be reduced 5 to 20 times using selective permeable plastic films. The films most commonly used are LDPE or LLDPE.

Corrugated Boxes/Cartons: Many fruits like mangoes, apples, grapes, etc. are packed in small packs of 2-4 kgs, either in corrugated boxes made of paper board or polymers like polypropylene. These boxes/cartons are light-weight with good compression strength. They can be printed to have a good shelf appeal.

Transport Packaging

Transport packages are designed for long distance transportation in capacities ranging from 15 to 25 kgs for man carrying and 200 – 500 kgs suitable for fork lift handling. These packs must withstand impacts, compression and vibration during transport. The transport packages can be broadly categorized as rigid containers made of wood, corrugated fibre board or plastics and flexible containers such as sacks made of plastic. Along with these materials some traditional materials used are jute (jute sacks), wooden boxes and bamboo baskets.

Types of transport packaging

Wooden boxes: The conventional baskets have been replaced by wooden boxes as they give better protection to the fresh produce against transportation hazards. Includes natural wood and industrially manufactured wood based sheet materials comprise ply wood, hard board and particle board. All wood that is used for the production of the packaging should be well dried in order to prevent cracks and mould growth later. They have high puncture resistance, good tensile strength as well as compression strength but they occupy more space and add on to the tare weight.

Wooden crates: Wooden crates are typically wire bound crates used for citrus fruits and potatoes, or wooden field crates used for softer produce like tomatoes. Wooden crates are resistant to weather and normally have good ventilation. Disadvantages are that rough surfaces and cracks can cause scratch to the produce, they can retain undesirable odours when painted, and raw wood can easily become contaminated with moulds.

Corrugated fibreboard boxes (CFB): Corrugated fibreboard boxes are the most commonly used shipping containers for fruits and vegetables packaging. It has low cost to strength and weight ratio, good cushioning properties, smooth and non-abrasive surface and good printability on the outer surface of the board. It can be provided with ventilation by punching holes. The CFB boxes have poor wet strength, now-a-days they are laminated with plastic film like LDPE, PP or PVC.

Plastic corrugated boxes: Its advantage over CFB is low weight to strength ratio, reusability and excellent printability. But CFB box has an edge over plastic fibreboard boxes when cushioning properties are taken into consideration. The disadvantages are ultraviolet degradation and temperature resistant, which can be taken care by use of additives.

Fibreboard boxes: They are used for tomato, cucumber and ginger transport. They are easy to handle, light weight, come in different sizes, and colours that can make produce more attractive to consumers. They have some disadvantages, such as the effect of high humidity, which can weaken the box, neither are they waterproof, so wet products would need to be dried before packaging.

Plastic crates: These are usually made of HDPE or PP by injection moulding, it's normal capacity varies between 20 – 40 kgs. They are easy to clean due to their smooth surface and are hard in strength, giving protection to products. Polyethylene has higher impact strength and a low degradation by ultra-violet radiation while polypropylene has a better scratch resistance. These crates are either of stackable, stack-nest or collapsible in design. However, plastic crates can damage some soft produce due to their hard surfaces, thus liners are recommended when using such crates.

Sacks: These are flexible shipping containers which are generally used for transport fruits and vegetables from the field. If the weight of content is more than 10 kg then it is called sack otherwise bag. The commonly used materials for sacks are cotton, jute, and flan woven plastics (HDPE, polypropylene). They are very useful because of low cost, high strength, re-useability and require less space for empties. However, they have low protection against puncture, compression, vibration, impact injuries and poor stackability due to low coefficient of friction, which can be overcome by making anti- slip bags.

Bamboo baskets: Bamboo baskets are widely used even today as transport packs in domestic market. They are available in various shapes, sizes and designs but they do not have rigidity and stackability during long distance transport.

Unit load packaging

The sizes of the pallets are of strategic importance since they correspond directly to the sizes of various types of containers, ship cargo compartments, trucks, fork trucks, etc. Most commonly used pallet sizes are 120cm x 80cm (Euro pallet) and 120cm x 100cm (Sea pallet). Reusable plastic pallets are in use, they are made up of HDPE or polypropylene. They are easy to clean and light in weight as compared to wooden pallets. Palletized loads are used in order to reduce handling costs by allowing substitution of mechanical handling for manual methods. It also decreases in sorting operations, reduced labelling requirement and total distribution time. Better utilization of storage space and maintenance of product quality. Two principles are used in the assembly of pallet loads are-

» The modular principle, in which all packages are oriented in the same direction.

» The two-way principle, in which the packages in each tier form a pattern such that some packages are oriented lengthwise and others cross-wise on the pallet.

Pallet boxes: Pallet boxes are very competent for transporting produce from the field to the packinghouse or for handling produce in the packinghouse. Pallet boxes have a standard floor size (120cm x 100cm) and approximate depth of bulk bins depends on the type of fruit or vegetable being transported like 70cm for citrus, 50cm for pears, apple, stone fruits and 40cm for tomatoes. Advantages of the pallet box are that it reduces the labour cost of loading, filling, and unloading, reduces space for storage and increases speed of mechanical harvest. The major disadvantage is that the return volume of most pallet boxes is the same as the full load. Higher investment is also required for the forklift truck, trailer, and handling systems to empty the boxes.

Classification based on condition

Field packing: Field packing requires a skilled labour force and a mobile packing line to supply containers and packing materials. Many very perishable and fragile fruits and vegetables are field packed to reduce the amount of handling.

Shed packing: For shed packing, harvested produce is placed in temporary holding containers until it is moved to the packing shed where it as graded, cleaned and packed. Harvest containers can be specialized to fit the crop and the packing line requirements.

Conventional packing: It is essential to minimize physical damage to fresh produce if it is to have optimal shelf-life. The most common form of packaging in this sector is the use of the fiberboard carton, however, for most produce, additional internal packaging, for example tissue paper wraps, trays, cups or pads, is required to reduce damage from abrasion.

Environment friendly packaging: Polyethylene and polypropylene bags of 100 gauges

(25μ) are normally used for mushroom packaging. The containers (small basket) are made from environment friendly material viz., Sal leaves (*Shorea robusta*) and Arecanut leaf sheath (*Areca catechu*). These containers are in turn wrapped with low density shrink wrap (50 gauge, 12.5μ) commercially called as L-50 cling film.

Modified atmosphere packaging (MAP): Polymeric films have been used to package fresh produce, with a number of benefits, including control of water loss, protection from skin abrasion and reduced contamination of the produce during handling. These films can lead to the development of lowered O_2 and raised CO_2 levels within the package and, as with CA storage, this can reduce the respiration of the produce and potentially extend shelf-life.

Individual seal /shrink packaging technique: Individual seal packaging, which may be considered as the MAP for an individual fruit, involves sealing of a fruit in a plastic film with or without heat shrinking to conform the shape of a fruit. Individual seal packaging would help to reduce the fruit decay by prevention of secondary infection during long term storage or shipment. Seal packaging has been found to extend the shelf life of several fruits like apple, pear, kiwifruit, citrus and pomegranate.

Grape guard: Corrugated fiber board boxes three ply with a capacity of 2 kg and 5 kg having a dimension of 25x20x12 cm is being used for packing. Different packaging materials like kraft paper, butter paper and soft tissue paper for preparation of grape guard are used. Berry decay and berry drop was controlled by the use of grape guards with containing 6 and 9 gram sodium bisulphite. Grape guards extended the shelf life of Thompson seedless, anab-e-shashi, dilkush and sharad seedless grapes upto 14, 8, 8 and 16 days respectively at ambient temperature and upto 75, 60, 75 and 85 days respectively under cold storage conditions.

Classification based on packing materials

Wooden packing material

Pallet: Pallet boxes are most commonly base on the standard size for pallet of 100cm ´ 120cm and about 50 cm high, have capacity of about 200-500 kg. They are usually made from wood but plastic ones are also available. They are used for whole range of crops, which are commonly loaded into them in the field and transported directly to the store.

Slip sheets: They are considerably less expensive than pallets to buy, store, and maintain. They may be re-used many times and they reduce the tare weight of the load. However, they require the use of a special fork-lift attachment at each handling point from packer to retailer.

Pallet bins: Substantial wooden pallet bins of milled lumber or plywood are primarily used to move produce from the field or orchard to the packing house.

Wooden crates and lugs: Wooden crates, once extensively used for apples, mango, stone fruit, and potatoes have been almost totally replaced by other types of containers.
Wooden baskets and hampers: Wire-reinforced wood veneer baskets and hampers of different sizes were once used for a wide variety of crops from strawberries to sweet potatoes. They are durable and may be nested for efficient transport when empty.

Wooden field box: These are made from thin pieces of wood, widely spread so they are light in weight and cheep to make. These can be used for all types of fruits and vegetables.

Woven baskets: These are traditional containers in which crops are placed after harvest. The produce is damaged in these baskets when they are stacked one above the other during transport and distribution.

Wire-bound crates: Although alternatives are available, wooden wire-bound crates are used extensively for snap beans, sweet corn and several other commodities that require hydrocooling. Wire-bound crates are sturdy, rigid and have very high stacking strength that is essentially unaffected by water. Although few are re-used, wire-bound crates may be dissembled after use and shipped back to the packer (flat).

Bulk bins: Bulk bins of 200-500 kg capacity are used for harvesting fresh fruits and vegetables. These bins are much more economical than the field boxes, both in terms of fruit carried per unit volume and durability, as well as in providing better protection to the product during transport to the packinghouse. They are made of wood and plastic materials with dimensions of 48 x 40 inches or 120 x 100 cm.

Paper based packing material

Corrugated fiberboard: They are made from either laminated or more commonly corrugated fiberboard. They may be used for directly field packing of produce and transported to pack house or destination market. Because of its relativity low cost and versatility, it is the dominant produce container material and will probably remain so in the near future. Most corrugated fiberboard is made from three or more layers of paperboard manufactured by the Kraft process. Double-faced corrugated fiberboard is the predominant form used for produce containers. It is produced by sandwiching a layer of corrugated paperboard between an inner and outer liner (facing) of paper-board. The inner and outer liner may be identical, or the outer layer may be preprinted or coated to better accept printing. The inner layer may be given a special coating to resist moisture.

Waxed boxes: Waxed cartons are specifically suited to cool room storage facilities. The benefits of waxed cartons are that they can be packed in the paddock, stored in the cool room, transported in a refrigerated truck and delivered to customer. The carton is able to do this because wax adds strength and gives the carton a high level of water resistance, ensuring product arrives in the same condition it was packed in. Cabbage, melons, potatoes, pumpkins, and citrus have all been shipped successfully in these containers.

Pulp containers: Containers made from biodegradable recycled paper pulp and starch binders are mainly used for small consumer packages of fresh produce. Pulp containers are available in a large variety of shapes and sizes and are relatively inexpensive in standard sizes. Pulp containers can absorb surface moisture from the product, which is a benefit for small fruit and berries that are easily harmed by water.

Paper and mesh bags: Consumer packs of potatoes and onions are about the only produce items now packed in paper bags. The more sturdy mesh bag has much wider use. However, bags of any type have several serious disadvantages. Large bags do not palletize well and small bags do not efficiently fill the space inside corrugated fiberboard containers.

Plastics packing material

Plastic bags: Plastic bags (polyethylene film) are the predominant material for fruit and vegetable consumer packaging. Besides the very low material costs, automated bagging machines further reduce packing costs. Film bags are clear, allowing for easy inspection of the contents, and readily accept high quality graphics. The shelf life of fresh produce is extended considerably by this packaging due to the availability of modified atmosphere codition.

Plastic field boxes: They are strong, durable and made from moulded polyvinyl chloride, poly propylene or polyethylene. They have smooth surface, which does not damage the produce. Initially, they are expensive to buy, but can be used repeatedly.

Polystyrene boxes: Popularly called thermocol boxes, provides enough protection to produce, are light in weight and keeps the produce for longer periods at pack temperature.

Shrink wrap: Shrink wrapping has been used successfully to package potatoes, sweet potatoes, apples, onions, sweet corn, cucumbers and a variety of tropical fruit individualy. Shrink wrapping with an engineered plastic wrap can reduce shrinkage, protect the produce from disease, reduce mechanical damage and provide a good surface for stick-on labels.

Rigid plastic packages: Packages with a top and bottom that are heat formed from one or two pieces of plastic are known as clamshells. Clamshells are gaining in popularity because they are inexpensive, versatile, provide excellent protection to the produce, and present a very pleasing consumer package. Packaging material is starch-based polyethylene substitutes that will break down in a landfill as fast as ordinary paper.

Precautions for packaging

- The packages should be sturdy and capable of standing up to handling, cooling and storage conditions.
- Rough packages such as baskets and wooden crates should be lined with cardboard inserts.
- Packages used for heavy produce should be reinforced with corner supports or folded dividers.
- Avoid using very large packages since the produce suffers more damage during handling.
- For delicate produce such as berries, grapes, summer squash and ripe stone fruits use shallow packages having single layer or double layers.
- Avoid overfilling or under filling packages and for adequate ventilation of packages about 5 per cent of the surface area per side should be vented.
- For immobilizing the produce in a package, use packaging materials such as trays, cups, wraps, liners and pads.
- To decrease the rate of water loss from produce like cherries perforated plastic film liners should be used.
- Take care while using fillers and liners in packages so that the ventilation holes are not blocked.
- Labeling containers with logo or farm name help in creating a brand name for quality produce.
- Consumer packages that can be used to display produce during marketing should be considered.

- Use large packages to pack consumer packages such as gift packs or display trays for providing better protection to the containers during stacking, transport and marketing.
- For reducing produce damage, packages with handles to carry should be considered.

Quality of modern packaging

- The package must have sufficient mechanical strength to protect the contents during handling, transport and stacking.
- The packaging material must be free of chemical substances that could transfer to the produce and become toxic to humen.
- The package must meet handling and marketing requirements in terms of weight, size, and shape.
- The package should allow rapid cooling of the contents. Furthermore, the permeability of plastic films to respiratory gases could also be important. Mechanical strength of the package should be largely unaffected by moisture content (when wet) or high humidity conditions.
- The security of the package or ease of opening and closing might be important in some marketing situations.
- The package must either exclude light or be transparent and appropriate for retail presentations.
- The package should be designed for ease of disposal, re-use, or recycling.
- Cost of the package in relation to value and the extent of contents protection required should be as low as possible.

Damages suffered by packed produce

Fruits, vegetables and roots vary widely in their susceptibility to physical damage and in the type of damage that is likely to be sustained. Four different causes of mechanical injury affect the produce are vibration (transportation light rubbing), impact (dropping), and compression (over stacking) and cut (sharp edges, punctures, nails etc.). Compression resistant produce are water melon, pumpkin, onion, carrot and potato these vegetables are also called as hard vegetables. Important practical requirement for packaging are to avoid under filing (vibration injury) and overfilling (compression and impact bruising). Individual items should be held firmly, but not too tightly, within the package. Susceptibility to the different types of damage will require a different type of protection to be incorporated into the package. The choice of packaging material will also be influenced by factors such as susceptibility to water losses and microbial infections or heat accumulation and may be the primary consideration in the type of package required. The physical damage suffered by the packaged fresh produce is caused mainly from the following:

» Injuries caused by sharp objects piercing the packages. This results in deep puncture, leading to water loss and rapid decay.

» Impact caused by throwing or dropping the packages. This results in bruising the content and / or bursting the package.

» Compression caused by overfilling the container or stacking them too high. This result in bruising or crushing the content.

» Vibration of the vehicle from rough roads, bursting the packages and bruising and crushing the contents.

» Heat damage caused by exposure of packages to direct sunlight or poor ventilation within the package, these results in fruits becoming overripe or soft and decay develops rapidly.

» Freezing damages caused by subzero ambient temperature the result is the damage to chilling sensitive produce. Most produce is affected by break down after thawing.

» Moisture damages caused by exposition to rains, dew or high humidity these results in softening and collapse of the containers, crushing the produce and decay.

Prevention of injuries to the produce

Suitable packages and handling techniques can reduce the amount of damage to which fresh produce is exposed during marketing:

» To keep the packaging itself from damaging the produce during handling and transport, wooden boxes and / or cardboard cartons must be properly assembled nails, staples and splinters are always a danger in wooden boxes and should be used carefully and properly.

» Individual items of produce should be packed to avoid rubbing against each other during handling and transport, loose fill packs are particularly susceptible to vibration damages.

» Much bruising results from overfilling containers or from the collapse of boxes, collapse may be caused by weak walls of boxes, by the softening of cardboards walls because of moisture or by failing to stack boxes so that the side or the walls support those above, stack of boxes should never exceed the height recommended by the manufacturer or the experience.

» Produce in woven jute sacks or nets is especially susceptible to shock damage, sacks of 25 or 50 kg capacity are normally used for relatively low valued produce, such as roots, tuber crops, cabbages, etc. and are often roughly handled on account of their weight. Whenever possible, handling of bagged produce should be minimized by stacking sacks in unit loads on pallets and avoiding over stacking.

Standardization of packaging

The wide variety of package sizes and material combinations is a result of the market reacting to demands from many different segments of the produce industry. For example, many of the large-volume buyers of fresh produce are those most concerned with the environment. They demand less packaging and the use of more recyclable and biodegradable materials, yet would also like to have many different sizes of packages for handiness. Packers want to limit the variety of packages they must carry in stock, yet they have driven the trend toward preprinted, individualized containers. Shippers and trucking companies want to standardize sizes so the packages may be better palletized and handled. Produce managers, want individualized, high quality graphics to entice retail buyers with in-store displays. Selecting the right container for fresh produce is seldom a matter of personal choice for the packer. For each commodity, the market has unofficial, but nevertheless rigid standards for packaging therefore it is very risky to use a nonstandard package. Packaging technology, market suitability and dumping regulations are continuously changing. When choosing a package for fresh fruits and vegetables, packers must consult the market, and in some markets, standard packages may be required by law.

General operations in packing house

A sequence of operations occurs in a packinghouse. Dumping can be done using either dry or water-assisted methods, depending upon the sort of produce being handled. Cleaning, as well, can be by washing with chlorinated water or dry brushing alone. Waxing, if practiced, occurs after washing and removal of surface moisture. Grading, as demonstrated, separates the product into processing and fresh market categories. Sizing further separates the product, with the smallest size going to the local market or to processing. Typically, the best quality produce is packaged and marketed at the regional or national level.

Bulk bins of 200-500 kg capacity are used for big packaging houses but in India, reusable plastic crates are used. Freshly harvested produce is stacked up for temporary storage. Latest damage, sustained during harvesting, will show as visual defects several hours later and can be detected. Storing field-warm produce for few hours in a cool place, such as cold storage corridors, whereby produce temperature is reduced by several degrees before processing will help reducing subsequent spoilage and damage in the packaging line. The most important process in fresh produce is respiration, a biochemical oxidation of all living cells. The produce is subjected to a cleaning process, which begins with soaking tank, where dirt clods and pesticide residues are softened and diluted by warm or cold solution of water detergents and disinfectants. The fruit is thoroughly washed by piece of cloth or soft brushes with water spray. Before processing further, produce is dried by air streams from overhead fans. In the next stage, the undersized produce called as culls is eliminated, which can be sent for converting, called as presizer stage. The grading process follows next to segregate the produce into quality groups, such as ripe fruits which must be marketed immediately, grade A, B, C, export grade or culls. Before or after grading, high quality fresh produce processing may include waxing operation, especially when long shelf-life is desired. Most produce has a natural wax layer on its rind or skin which protects it from excessive moisture loss while allowing free metabolic gas exchange. This wax is largely removed by the cleaning operation. By applying an artificial wax, the keeping quality of produce is restored or even improved, by including chemical additives to inhibit spoilage or to add gloss to colour for sales appeal. Sizing is an additional sorting operation whereby sorting is done according to size. Depending upon produce type and grade, distance to markets, cost and availability of packaging materials, the produce may be packed in a large variety of shipping containers. The packaging operation usually includes setting up of container, before quantitizing of produce, filling and container closure. Quantitizing may be by count or by weight or a combination of the two. Accurate sizing provides proportional link between count and weight whereby only 'check-weighing' is required after filling by count. Container fill may be random or pattern packed. Pattern packing increases protection to produce by minimizing contact pressure by increasing number of contact points to a maximum of 12 and maximizes volume utilization. Sometimes, the produce is pre-packed in packaging house in consumer packs, mostly different types of plastic bags or overwrapped trays. The final packaging operation is container closing which may be performed by gluing, stapling and strapping. Unitization and palletization may be done according to the container requirement and market need.

Layout of packing house: Packing operations that are done in an unsystematic manner can cause delays, add costs or effect produce quality. Save time and money by laying out the packing shed in an organized, step-by-step system.

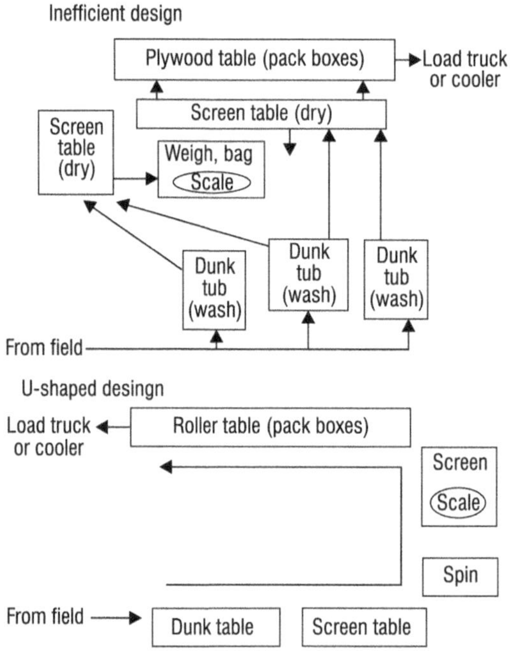

Figure 38. Packing house layout (Source: Meyer, 1999)

Need for a packinghouse

The site for the packing process may mostly be the provision of a portable or temporary shelter in the field adjacent to the harvesting area to protect the produce and workers from the weather during field handling. This is of considerable importance for produce harvested in hot, sunny conditions where exposure to the weather for only a few hours can markedly accelerate senescence or in rainy periods where the probability of microbial infection is significantly enhanced if the produce becomes wet. A packing place constructed with a bamboo frame and grass thatched roof without water or electricity services is satisfactory for a low manual sorting and packing operations. The complexity of packing house operations increase, as market requirements for quality and uniform grading become more demanding. The standards of the required building will also increase along with more sophisticated equipment and a larger and more skilled work force.

Packing house operations

The real operations to be carried out will be determined by the market system and volume of produce passing through the packing place. It may, accept produce from the field and prepare it for the market or act as a preliminary quality control operation to ensure that only quality produce is placed into storage. A packing house may also be used to repack produce after storage for retail marketing. All packinghouse operations can be executed manually, mostly small scale, or with a range of mechanical devices of varying sophistication, mostly for larger operations. For packinghouses with small throughput, economics will favour manual operations with workers performing more or all the tasks. The advantage of human labour is that it provides flexibility of operation since it can perform a variety of tasks with equal efficiency.

Receipt of produce: On arrival to the packinghouse produce is usually unloaded, weighed, sampled shlegt down for quality checking and transferred to a temporary storage area.

Sorting: A preliminary sorting of produce should remove unmarketable pieces and foreign matters, such as plant debris, soil and stones, before the produce passes on to further operations.

Cleaning and washing: The removal of soil and stones mentioned above can be done by hand-picking or by sieving. Washing is required to clean produce which has acquired latex stains from injuries from bad harvesting or handling, such as mango, papaya, breadfruit and most of the root crops. The soft rotating brushes located inside the washer are used to remove surface debris and other unwanted contaminants. It is important to note that washing should be carried out only when absolutely essential. If it is necessary to wash produce, a fungicide should be normally applied immediately afterward. The washing of produce in recirculated water should be avoided because it can quickly become heavily contaminated with decay organisms, leading to rotting of the washed produce. Hypo chlorites or chlorine (100-250ppm) may be added to washing water but its use is not recommended for small-scale operations, because they are quickly inactivated by the organic material present in the dirty water. Washed produced to be treated with fungicide should first be drained after washing or cleaned with a soft cloth to minimise the danger of reducing the concentration of the fungicide. When washing is not followed by chemical treatment, the washed produce should be spread out in a single layer on racks of mesh, in the shade but exposed to good ventilation to aid rapid drying.

Fungicide treatment: Contamination of molds or bacteria may occur before or after harvest, either through injuries or direct penetration of the intact skin of the produce. Post-harvest application of fungicide is usual in crops such as apples, bananas and citrus fruits which have to be stored for a long period or those which undergo long periods of transport. Fungicide is normally applied only after the produce has been washed and drained. In small-scale packinghouse operations, fungicide can be applied by:

» **Dipping:** Treatment is normally carried out by hand operation using a suspension of fungicide agitated using stick. Wire-mesh baskets are used to dip the fruits. After dipping produce should be drained and dried.

» **Spraying:** This can be done using a hand-operated sprayer while the produce is still in trays or racks after washing and drying. It is often done on bananas following the de-handing of bunches.

Quality selection and size grading: Sorting to remove substandard commodities and grading into different classes could be done manually. Sizing according to weight, length or diameter is more often a mechanized procedure which can be done with sizing equipment attached to the packing line.

Waxing: The application of wax or similar coating to enhance appearance and limit water losses from produce such as citrus requires specialized equipment and has little relevance to small scale packing. Waxing is normally preceded by fungicide treatment.

Packaging: Packaging in small-scale operations means the filling of marketing containers by hand, often using a packing stand or a table. There are various methods of packing:

» Loose fill jumble pack is used where there is no advantage to size-grading, weighing is necessary.

» Multilayer pattern pack has size graded produce normally sold by count of the produce such as citrus and apples.

» Multilayer size graded pack mostly used in mechanical packing has separator trays between layers, it is normally sold on per box basis.

» Single layer packs for high value produce may have each piece wrapped in tissue (often chemically treated) or placed in a divider holding it alone.

Storage: After commodities are packaged they should be immediately placed in cold rooms at the recommended temperature and relative humidity.

Dispatch: At the point of dispatch, commodities are handled in the condition that they will reach the buyer. It is therefore essential that rough handling, overloading of trucks, infestation and exposure to extreme weather conditions are kept to a minimum. The dispatch area should be cool, clean and spacious to allow for temporary storage of packed produce and permit unrestricted movement of loading staff and their vehicles.

Under or over filled boxes

The box should not be under filled or over filled. Over filling is generally noticed in India where a farmer fills more in a box beyond the capacity of the box designed by the manufacturer. They do this primarily for two reasons:

» **Demand by the commission agent (Traders):** It was observed that those lots containing more weight in a box priced more. In general, all apple boxes are designed for 20 kg but it contains 22–23 kg and in many cases it is found 25–30 kg.

» **To save the cost of packing and transportation:** Farmers want to save the cost of empty boxes, packing charges, handling and transport charges by saving number of boxes, and to do so they prefer overfilling. The practice results in fruit damage, touching marks (pressure marks) and bruising during handling in almost all fruits.

Cooling the produce in the package

Containers designed for pressure cooling should have holes occupying about 5 per cent of the surface area on each of the air entry and exit ends. Ideally respiratory heat should be able to escape readily from the packages. In case of small and /or tightly packed commodities such as green beans, small fruits, green leafy vegetables and cut flowers, the heat of respiration are removed largely by conduction to the surface of the package. Therefore, the mass of the contents (i.e. minimum dimension of packages from the centre to the surface) becomes important factor. The acceptable mass depends on the respiration rate of the commodity. If the mass of the produce is excessive, that near the centre of the package will heat up because respiratory heat cannot dissipate fast enough. Under dry conditions, produce in containers like wooden boxes, plastic crates may be sprayed with water. Fresh cut flowers and foliage are often transported wet usually in plastic buckets (eg. rose, gerbera etc.) and sometimes individual stem in veil of solution (eg. Anthuriam, orchids etc.).

Need for ventilation in packages

Suitable packaging for any product will consider the need to keep the contents well ventilated to prevent the build-up of heat and carbon dioxide during postharvest stages of transport, storage and marketing. A tight stack pattern is acceptable only if packages are designed to allow air to circulate through each package and throughout the stack. The effectiveness of ventilation during transport also depends upon the air passing through the load via vehicle.

Use of cushioning materials

Cushioning materials are used in many stages during postharvest handling operations for preserve postharvest quality of fresh produce. It is specialy use in putting harvested produce into plastic crates or any rigid container which helps in reducing touch marks. During transportation of produce from field to pack house and from pack house to destination markets which help in reduce vibration, bruising and touching marks. In addition to protecting the package contents, cushioning materials may also be used to adjust the packages to a standard size, in which case they act as adapters between nonstandard package contents and the packaging (modularity of shipping packages).

Cushioning materials: The cushioning material used for packaging fruits/vegetables are newspaper sheets, newspaper cuttings, bubble sheet, specially designed foam nets, moulded trays, gunny bags, leaves, khas-khas, dry grass, paddy straw, leaves, saw dust, paper shreds, thermocol, foam nets (apple, pear, citrus), bud net (rose) etc. Recently, moulded pulp tray, honey comb portion, cell pack etc. are used replacing the cushioning material.

Table 73. The cushioning materials commonly used for fruits, vegetables & flowers packing

Airbags	Bubble films	Rubberized fibre	Plastic foam	Polystyrene
Polyurethane	Foam in place	Polyethylene foam	Loose fill packing	Tissue paper
Buffered tissue	Acid free tissue	Moulded pulp tray	Honey comb portion	Cell pack
Air cellular cushioning - sealed air bubble wrap, bags, jumbo bubble wrap, transit bubbles, and bubble wrap sheets				

Mode of action: Cushioning materials absorb a proportion of the kinetic energy arising when the package suffers impact or is dropped and increase the braking distance of the package contents. Correct selection and sizing of the cushioning material thus ensure that the package content suffers no damage.

Wrapping: Covering the fruits after harvest with any material in order to improve its postharvest life is known as wrapping. The materials commonly employed as wrappers are old newspaper, tissue paper, waxed paper, shrink film, poly film, cellophane paper, aluminium foils and alkathene paper etc.

Advantages of wrapping

» It minimizes shrivelling by the loss of moisture and increases self-life.

» It protects against the spread of diseases from one to the other and reduces rotting loss.

» It reduces bruises damage during transport or in storage.

» It makes the fruit more attractive/appearance and maintains quality.

Characteristics of cushioning materials

Cushioning materials must in particular fulfill the following requirements:

» Cushioning materials itself should be physiologically inactive and free from contamination so that it does not pass on the same to the fruits and vegetables.

» It should have flexible property and able to scatter the heat of respiration and ethylene gas generated by the produce.

» Recovery is one of the most important properties of a cushioning material, it ensures that the package contents continue to be protected even when repeatedly subjected to similar stresses. If recovery is too low, the braking distance declines on constant exposure to stress, such that the resultant kinetic energy can no longer adequately be absorbed and the package contents may be damaged.

» Cushioning materials must be insensitive to climatic conditions, such as moisture due to elevated relative humidity, direct solar radiation and extreme variations in temperature and their action must not be impaired by such exposure.

» Especially in the case of package contents which are at risk of corrosion, it is important that the cushioning materials are not hygroscopic and consequently do not promote corrosion. They should furthermore not contain any aggressive constituents (neutral pH), which could contribute towards corrosion. The cushioning material and package contents should not interact and possibly impair each other's properties.

» Use of the cushioning material should be effective, simple, environmentally compatible and cost-effective.

Various kinds of cushioning materials

Airbags: It consists of an elastic film which is inflated with air. When at rest, only the static load generated by the weight of the package contents bears upon the cushioning. When dynamic loads occur, these are absorbed by compression of the cushion. Airbags are commercially available in various sizes and designs, ranging from spheres, standard cushions to corner and edge cushioning and tubular cushioning. Airbags are mainly used in containers and railroad freight cars and only rarely in trucks.

Bubble films: They consist of two plastic films, one of which is completely flat and the other has small, round indentations, which, once the two films have been heat sealed together, contain the necessary air. Bubble cushioning material provides outstanding protection from damage caused by shock, vibration or abrasion, easy to use.

Rubberized fibre cushioning: Rubberized fibre cushioning provides high quality protection for demanding items. This cushioning is made from animal hair or coconut fibre, which is cleaned, converted into nonwoven mats, coated with rubber and vulcanized to form solidly bonded sheets. Rubberized fibre cushioning is relatively insensitive to the effects of moisture and high or low temperatures and exhibits very good recovery even on long-term exposure to loads.

Polystyrene: Polystyrene is a somewhat soft, elastic foamed plastic with closed cells containing the air required to provide resilience. Polystyrene is not itself hygroscopic and thus remains fully functional on exposure to moisture. Appropriate action must accordingly be taken to protect package contents which are at risk of corrosion. Polystyrene cushioning material is produced either as relatively large mouldings, such as cushioning frames, edge or corner pads and as a loose fill cushioning material, known as polystyrene chips.

Polyurethane: Polyurethanes are produced in flexible, semirigid and rigid forms with an open cell structure. It is primarily flexible and semirigid grades of polyurethane which are used in packaging applications. The shock absorbing properties of polyurethane foams increase with foam hardness, while recovery and elasticity decline.

Air cellular cushioning: Air cellular products are the most versatile and flexible packaging material available. Air cellular cushioning material has numbers of qualities like increased cushioning performance, reduced material costs, labour savings and lower shipping costs.

Foam in place: A simple cushioning or blocking and bracing process for a variety of items in varying shapes and sizes. Instapak foam is dispensed into carton lined with high strength instamate film. The instamate film is folded over and the product is placed on the rising foam. A second sheet of instamate film is placed over the product and more instapak foam is dispensed.

Polyethylene foam: It is closed-cell product, exhibit excellent cushioning characteristics, which are comparable with those of rubberized fibre cushioning. Even when exposed to major loads, they retain their cushioning capability. Polyethylene foams costly, which exclude them from many applications and do not have good weather resistance.

Mic-pac loose fill packaging: Mic-Pac is super light and extremely suitable for void filling and cushioning. They flow quickly and evenly, filling all spaces around different shape and size articles. It reduces freight cost as it is lighter than most conventional packaging and dunnage materials.

Korrvu packaging: It is dynamic packaging product that uses a strong, highly resilient low slip film surround the products, protecting them from shock and vibration and an effective display merchandiser. It may complete packaging solution that holds the product securely in place protecting it from damage due to shock, vibration and impact. Korrvu packaging stores flat, reducing storage and warehouse space requirements, reusable for return shipments so minimizing waste at both ends of the distribution cycle.

Tissue paper, buffered tissue and acid free tissue: Tissue paper is very cheap and is made from a pulp that contains an acidic component lignin, whereas acid free tissue paper is made from higher quality pulp. Acid free tissues can also be buffered, and it can help to scavenge acidic materials as pollutants. It is also soft and is excellent for building up layers on objects to protect smaller areas or to build out areas with protrusions and filling voids three dimensional objects.

Trends in Fruit and Vegetable Packaging

Remaining to perishability, the bulk of fresh fruit and vegetables are still sold unpacked at retail or wholesale level in perforated polyethylene bags, single use printed polyethylene bags or nets. Long term transport of these produce must remain minimal to avoid brushing and consequent decay of the loosely package product. The packaging film should have high gas permeability and anti-fog properties to allow the product to breathe. Green vegetables, in the presence of sufficient light, could consume substantial amounts of CO_2 and produce O_2 through photosynthesis. Shock and vibration leads to damage to produce cells which causes an increase in respiration and may lead to enzymes being released that will cause browning reactions to begin. It is essential to minimize physical damage to fresh produce in order to obtain optimal shelf-life. The most common form of packaging is use of the fiberboard carton, however, for most produce, additional internal packaging, with tissue paper wraps, trays, cups or pads, is required to reduce damage from abrasion. For very delicate fruits, smaller packs with relatively few layers of fruits are used to reduce compression damage. Moulded trays may be used which physically separate the individual piece of produce. Individual fruits may also be wrapped separately in tissue or waxed paper. This improves the physical protection and also reduces the spread of disease organisms within a pack. Providing a uniform size package to protect the produce, there are other requirements for a container:

Quality of containers

» It should be easily transported when empty and occupy less space than when full, e.g. plastic boxes which nest in each other when empty, collapsible cardboard boxes, fibre or paper or plastic sacks.

» It must be easy to assemble, fill and close either by hand or by use of a simple machine.

» It has to provide adequate ventilation for contents during transport and storage.

» It should be uniform in size and capacity suited to market demands.

» Its dimensions and design must be suited to the available transport in order to load neatly and firmly.

» It must be cost-effective in relation to the market value of the commodity for which used.

» It must be readily available, preferably from more than one supplier.

Technological innovations in packaging

Several new technologies offer the packer opportunities to modify the atmosphere inside the shipping package during distribution even though such control traditionally finishes with the sealing of the package. Packaging is termed as active when it performs some desired function other than to provide an inert barrier to the external environment. The goal of developing such packaging is the achievement of a more ideal match of the properties of the package to the requirements of the fruits and vegetables. Most applications of active packaging involve the use of polymers. The role played by the polymer may be that of a conventional packaging material as in the fabrication of sachets that are used commercially to scavenge oxygen or carbon dioxide, or to release ethanol or carbon dioxide. The polymer can play an even more active role when it interacts physically or chemically with the components of the headspace of the package. Such interactions include absorption as in the case of humidity buffering linings for produce cartons, or reactions as in the case of oxygen scavenging plastics.

Permeability regulation via thermally sensitive packaging materials: A significant development is that of side- chain-crystallizable (SCC) polymers with the ability to effectively and reversibly melt as the temperature increases and thus foster increased gas transmission through them. SCC polymers are acrylics with side-chains independent of the main chain. Varying the side-chain length, the melting point can be altered. By making the appropriate copolymers, it is possible to produce any melting point from 0 to 68°C, well within the extreme distribution temperature range of minimally processed produce. When elevated to the switch temperature, SCC polymers become molten fluids which are inherently high in gas permeability. The permeation properties may be modified by inclusion of other polymers to change the carbon dioxide to oxygen permeability ratios. The resulting materials can permit the packaging technologist to achieve the lowest oxygen concentration without going anaerobic within the package. Thus, optimum gas concentration may be employed with minimum concern for elevated distribution temperatures. In addition to the reversible temperature sensitivities, the materials are generally capable of 100 time's greater oxygen permeability than mainstream polyethylene films without compromising the carbon dioxide to oxygen permeability ratio. This is accomplished by coating a porous substrate with a proprietary SCC polymer and applying the membrane as a package label over an aperture on an otherwise reasonably well sealed package.

Ethylene scavengers: Ethylene (C_2H_4) light-activated scavengers, incorporated in the packaging film, are Zero 2 TM developed by CSIRO and marketed by Southcorp Packaging (Australia) and OS1000 developed by Cryovac Sealed Air. It accelerates respiration, leading to maturity and senescence, and also softening and ripening of many kinds of fruits. Furthermore, ethylene accumulation can cause yellowing of green vegetables and may be responsible for a number of specific postharvest disorders in fresh fruits and vegetables. To extend shelf life and maintain an acceptable visual and organoleptic quality, accumulation of ethylene in the packaging should be avoided. Most of these ethylene absorbers are supplied as sachets or integrated into films. Potassium permanganate ($KMnO_4$), which oxidizes ethylene to acetate and ethanol and in this process, colour changes from purple to brown indicating the remaining C_2H_4 scavenging capacity. Rengo Co. (Japan) developed 'Green Pack', a sachet of $KMnO_4$ embedded in silica. The silica adsorbs the ethylene and the permanganate oxidizes it to acetate and ethanol. Another type of C_2H_4 scavenging concept is based on the adsorption and subsequent breakdown of ethylene on activated carbon. Charcoal containing

PdCl as a metal catalyst was effective at 20°C in preventing the accumulation of ethylene, in reducing the rate of softening in minimally processed kiwifruits and bananas and in reducing chlorophyll loss in spinach leaves but not in broccoli. Other C_2H_4 adsorbing technologies are based on inclusion of finely dispersed minerals such as zeolites, clays and Japanese oya into packaging films. Most of these packaging films, however, are opaque and not capable of adsorbing C_2H_4 sufficiently. Although the incorporated minerals may adsorb ethylene, they also alter the permeability of the films: C_2H_4 and CO_2 will diffuse much more rapidly and O_2 will enter more readily than through pure PE. These effects can improve shelf life and reduce concentrations head space C_2H_4 independently of any C_2H_4 adsorption. Commercially available examples of these mineral containing materials are the Orega plastic film (Cho Yang Heung San Co., Korea), Evert- Fresh (Evert-Fresh Co., USA), Peak-fresh TM (Peakfresh Products, Australia), BO film (Odja Shoji Co., Japan) and ProfreshTM (Europe). C_2H_4 scavengers are not yet very successful, probably because of insufficient adsorbing capacity.

CO_2 scavengers: CO_2 is formed in some fruits and vegetables due to deterioration and respiration reactions. The CO_2 produced has to be removed from the package to avoid fruits and vegetables deterioration and/or package destruction. The active compound $Ca(OH)_2$ reacts at sufficiently high humidity with the CO_2 to produce $CaCO_3$. Multiform desiccants have patented (US 5322701) a CO_2 absorbent sachet including a porous envelope containing CaO and a hydrating agent such as silica gel on which water is adsorbed.

Humidity regulators: This approach allows the fruits and vegetables packer to uphold in relative humidity pack. Humidity safeguard is being developed for use in the distribution of horticultural produce which is normally distributed in fiberboard cartons, usually with a polyethylene liner or made from very expensive waxed fiberboard without a liner. A recent development has been the water-barrier coating of the inside of fiberboard cartons to allow moist produce to be placed directly into the carton. Besides the introduction of liquid water with the produce, packing into closed spaces allows the build-up of water vapour. Since temperature cycling is very difficult to avoid during handling there is every likelihood of condensation and with this the growth of microorganisms on fruits and vegetables. Two widely different approaches have been used to buffer the humidity in the cartons in order to prevent condensation while not concurrently causing desiccation of the produce. One is to include micro-porous bags or pads of inorganic salts and the other is to line the carton with a protected layer of a solid polymeric humectant. The main purpose of moisture control is to lower water activity, thereby reducing the growth of moulds, yeast and spoilage bacteria on fruits and vegetables. The shelf life of packaged tomatoes at 20° C was extended from 5 to 15-17 days with a pouch containing NaCl, mainly by retardation of surface mould growth.

Oxygen scavengers can be used alone or in combination with MAP. Their use alone eliminates the need for MAP machinery and can increase packaging speeds. However, it is usually more common in commercial applications to remove most of the atmospheric O_2 by MAP and then use a relatively small and inexpensive scavenger to remove the residual O_2 within the produce package. For an O_2 absorber to be effective, the packaging material needs to have an O_2 barrier of intermediate performance (20 ml/m^2.d.atm), otherwise the scavenger will rapidly become saturated and lose its ability to trap O_2.

Antimicrobial packaging: The use of edible coatings as carriers of antimicrobial compounds is another potential alternative to enhance the safety of fresh fruits and vegetables. Antimicrobial edible coatings may provide increased inhibitory effects against spoilage and pathogenic bacteria by maintaining effective concentrations of the active compounds on the produce surfaces. There are several categories of antimicrobials that can be potentially incorporated into edible coatings, including organic acids (acetic, benzoic, lactic, propionic, sorbic), fatty acid esters (glyceryl monolaurate), polypeptides (lysozyme, peroxidase, lactoferrin, nisin), plant essential oils (cinnamon, oregano, lemongrass), nitrites and sulphites. Currently, organic acids and plant essential oils are the main antimicrobial agents incorporated into edible coatings for fresh fruits and vegetables. The classes of antimicrobials listed range from acid anhydride, alcohol, bacteriocins, chelators, enzymes, organic acids and polysaccharides. Examples of commercial antimicrobial materials in the form of concentrates (e.g. AgIONe, AgION, nisin) and films (Microgarde Rhone-Poulenc, USA) were also presented. Antimicrobial packages have had relatively few commercial successes except in Japan where Ag-substituted zeolite is the most common antimicrobial agent incorporated into plastics. Edible coatings and films prepared from polysaccharides, proteins and lipids have a variety of advantages such as biodegradability, edibility, biocompatibility, aesthetic appearance and barrier properties against oxygen and physical stress. Antimicrobial packaging can play an important role in reducing the risk of pathogen contamination, as well as extending the shelf-life of fruits and vegetables it should never substitute for good quality raw materials. It should be considered as a hurdle technology that in addition with other non-thermal processes such as pulsed light, high pressure and irradiation could reduce the risk of pathogen contamination and extend the shelf-life of perishable fruits and vegetables. Several other compounds have been proposed and/or tested for antimicrobial activity in fruit and vegetable packaging including organic acids such as sorbate, propionate and benzoate or their respective acid anhydrides, bacteriocins e.g. nisin and pediocin, enzymes such as lysozyme, metals and fungicides such as benomyl and imazalil. A film contains a natural antimicrobial compound derived from grape fruit seed. The choice of the antimicrobial is often limited by the incompatibility of the component with the packaging material or by the heat unstability of the component during extrusion. Two commercial biocidal films chlorinated phenoxy compound and chlorine dioxide are currently marketed. A commercial antifungal coating containing chitosan is also sold as a shelf-life extender for fresh fruit. CSIRO (Australia) is developing systems which gradually release SO_2 to control mould growth in some fruits. However, excessive release of SO_2 from pads of sodium metabisulphite incorporated micro-porous material has been shown to cause partial bleaching problems in grapes.

Table 74. Properties of packaging films

Material	Properties
Paper	Strength, rigidity, opacity, printability.
Aluminum foil	Negligible permeability to water vapor, gases and odours, grease proof, opacity and brilliant appearance, dimensional stability, dead folding characteristics.
Cellulose film (coated)	Strength, attractive appearance, low permeability to water vapor (depending on the type of coating used), gases, odours and greases, printability.
Polythene	Durability, heat sealability, low permeability to water vapor, good chemical resistance, good low-temperature performance.
Rubber hydrochloride	Heat sealability, low permeability to water vapor, gases, odours and greases, chemical resistance.
Cellulose acetate	Strength, rigidity, glossy appearance, printability, dimensional stability.
Vinylidene chloride	Low permeability to water vapor, gases, copolymer odors and greases, chemical resistance, heat sealability.
Polyvinyl chloride	Resistance to chemicals, oils and greases, heat sealability.
Polyethylene terephtalate	Strength, durability, dimensional stability, low permeability to gases, odors and greases.

Edible coatings and films: The application of edible coatings appears to be one of the most innovative approaches to extend the commercial shelf life of fruits and vegetables by, among other mechanisms, acting as a barrier against gas transport and showing similar effects to storage under controlled atmospheres. Edible films and coatings are generally based on biological materials such as proteins, lipids and polysaccharides. The main polysaccharides that can be included in edible coating formulations are starch and starch derivatives, cellulose derivatives, chitosan, pectin, alginate and other gums. Carboxymethyl cellulose is a cellulose derivative that has received considerable attention with several examples of applications in many fruits and vegetables. Some polysaccharides have been used as edible coatings to improve the quality of different fresh cut fruit.

Active packaging: Active packaging is a promising and stirring area of food technology that is developing owing to advances in packaging technology, material science, biotechnology and new consumer demands. This technology can confer many preservation benefits on a wide range of ambient stable and chilled food products. The intention is to extend the shelf-life of fruits and vegetables, whilst at the same time maintaining nutritional quality and assuring microbial safety.

Active packaging employs a packaging material that interacts with the internal gas environment to extend the shelf-life of fruits and vegetables.

» **Ethylene scavenging:** A chemical reagent, incorporated into the packaging film, traps the ethylene produced by ripening fruit or vegetables. The reaction is irreversible and only small quantities of the scavenger are required to remove ethylene at the concentrations at which it is produced. Flexible packaging materials such as PE-LD and linear low density polyethylene (PE-LLD) when impregnated with potassium permanganate and cinnamic acid, respectively become ethylene scavengers. Fresh fruits and vegetables, such as mango, tomato, banana, and papaya exhibit more shelf-life of two to three weeks when packed in such ethylene scavenging films.

» **Oxygen scavenging:** The presence of oxygen in produce packages accelerates the spoilage of many fruits and vegetables. Oxygen can cause off flavour development, colour change, nutrient loss and microbial attack.

» **Humidity control:** Condensation or sweating is a problem in many kinds of packaged fruit and vegetables. When one part of the package becomes cooler than water is likely to condense in the cooler areas.

» **Carbon dioxide release:** High carbon dioxide levels are desirable in some fruits packages because they inhibit surface growth of microorganisms. Fresh strawberries are benefit from packaging in a high carbon dioxide atmosphere. However with the introduction of modified atmosphere packaging there is a need to generate varying concentrations of carbon dioxide to suit specific food requirements.

» **Sulphur dioxide:** Serious loss of table grapes can occur unless precautions are taken against mould growth. It is necessary to refrigerate grapes in combination with fumigation using low levels of sulphur dioxide. Fumigation can be conducted in the fruit cool stores as well as in the cartons. Carton fumigation consists of a combination of quick release and slow release systems which emit small amounts of sulphur dioxide. When the temperature of the cartooned grapes rises due to inadequate temperature control, the slow release system fails releasing its entire sulphur dioxide quickly. This can lead to illegal residues in the grapes and unsightly bleaching of the fruit.

Smart or intelligent packaging: Smart packaging uses features of high added value that enhance the functionality of the product, notably mechanical, electronic and responsive ink features, for example electronic and mechanical dispensers in which drugs are supplied and the prepared meal that automatically tells the microwave how it should be cooked. Smart packaging can be categorized into two types, those which incorporate integrated circuits (IC's) and do which do not incorporate IC's otherwise known as chipless smart packaging. Packaging that incorporate diagnostic indicators are also included in smart packaging. These can be used for such functions as monitoring vibration, acidity, tilt, shock, humidity, light, heat, time chemicals, virus or bacteria as they develop or as they are contacted. Fresh cut fruit and vegetables continues to be one of the fastest growing segments of food retailing and while conventional film packaging is suitable for lettuce and prepared salads, it cannot cope with the high respiration rates of precut vegetables and fruit, leading to early product deterioration. Novel breathable polymer films are already in commercial use for fresh cut vegetables and fruits. Packaging films that are acrylic side chain crystallizable polymers tailored to change phase reversibly at various temperatures from 0-68°C are available. As the side chain components melt, gas permeation increases dramatically, and by further tailoring the package and materials of construction, it is possible to fine tune the carbon dioxide to oxygen permeation ratios for particular products. The final package is smart because it automatically regulates oxygen ingress and carbon dioxide way out by transpiration according to the prevailing temperature. In this way, an optimum atmosphere is maintained around the product during storage and distribution, extending freshness and allowing shipping of higher quality products to the consumer. Intelligent packaging can change colour to let the customer know how fresh the product is and show if the product has been spoiled because of a change in temperature during storage or a leak in the packaging. Time temperature integrators (TTI's) are devices that show an irreversible change in a physical characteristic, usually colour or shape, in response to temperature history. The TTI's are expected to imitate the change of a certain quality parameter of the fruit and vegetable undergoing the same exposure to temperature. The TTI's presently on the market have working mechanisms based on different principles like biological, chemical and physical. Change in biological activity, such as microorganisms, spores or enzymes is the basic working principle, whereas the chemical or physical response towards time and temperature, such as an acid-base reaction, melting, polymerization, etc.

Fresh-Check-R-Life-Lines integrator is supplied as self-adhesive labels, which may be applied to packages of perishable fruits and vegetables to assure consumers at point-of-purchase and at home that the product is still fresh. It is a full history indicator whose working mechanism is based on the colour change of a polymer formulated from diacetylene monomers. It consists of a small circle of polymer surrounded by a printed ring for colour reference. The polymer, which starts lightly coloured, gradually deepens in colour to reflect the cumulative exposure to temperature.

Vitsab-R indicator is a full history integrator based on an enzymatic reaction. The device consists of a bubble like dot containing two compartments, one for the enzyme solution, lipase plus a pH indicating dye compound and the other for the substrate, consisting primarily of triglycerides. The dot is activated at the beginning of the monitoring period by application of pressure on the plastic bubble, which breaks the seal between compartments. The ingredients are mixed and as the reaction continue a pH change results in a colour change. The dot, initially green in colour, becomes progressively yellow as product approaches the end of shelf-life. The reaction is irreversible and will proceed faster as temperature is increased and slower as temperature is reduced.

Biodegradable packaging: Biodegradation is the process by which carbons containing chemical compounds are decomposed in the presence of enzymes secreted by living organisms. The use of bioplastic is to replicate the life cycle of biomass by conserving the fossil fuels, carbon dioxide and water production. Currently, there are several types of bio based polymers on the market, same coming from petrochemical monomer, like certain types of polyester, polyester amides and polyvinyl alcohol, produced by different manufacturer, used principally as films or moulding packing for fruits and vegetables. Four other bio based polymers are starch materials, cellulose materials, polylactic acid (Polyester, PLA), polyhydroxy acid (polyester, PHA). Until now, the polyhydroxy acid polymer is a very expensive polymer because it is commercially available in very limited quantities. Polylactic acid is becoming a growing alternative as a green food packaging materials because it was found that in many situations it performs better than synthetic ones, like oriented polystyrene (OPS) and PET materials. There are three requirements for the fast degradation process viz. temperature, humidity and type of microbes. In the short term, bio based materials will most likely be applied on fruits and vegetables, since biobased materials present opportunities for producing films with variable CO_2/O_2 selectivity and moisture permeability. However, to succeed, biobased packaging of fruits and vegetables must be in compliance with the quality and safety requirements of the products and meet legal standards. Additionally, the biobased materials should preferably preserve the quality of the product better and longer to justify any extra material cost.

Application of nanocomposites: Nanocomposite materials are composed of nanoscale structure that enhances the macroscopic properties of fruit and vegetable products. The common nano-composites used in the fruits and vegetables packaging industry are polymer clay nanoclay, silica nanocomposites of nanosilver. The effects of nanoclay in polymers are increased stiffness, strength, nucleating agent in foams, smaller cell size, higher cell density, and flame retardant. Nanosilver is composed of de-ionized water with silver in suspension which has excellent antibacterial properties. Silver nanoparticles interact well with other particles as they have large surface area relative to volume which increases their antibacterial efficiency as a result of which they are extensively used in the packaging industry.

One of the potential applications of nanotechnology in food packaging is polymer/clay nanocomposites, they have recently emerged due to their potential for improving properties of packaging materials such as increased mechanical, barrier and chemical properties with a small amount (less than 5% by weight) of nanoclays strengthening. The functional properties of biopolymer based edible films or coatings have been shown to act as barrier to solute and gas and enhance produce quality and shelf life. However, these films do not display good mechanical and water vapor barrier properties due to their hydrophilic characteristics. To overcome these issues, a new approach has been developed, which use hybrid materials consisting of polymers and layered silicates. Layered silicates, such as montmorillonite (MMT) clay mineral, result from the stacked arrangement of negatively charged silicate layers and contain a platelet thickness of about 1 nm with a high aspect ratio (ratio of length to thickness). The layered silicate filled polymer composites exhibit extraordinary enhancement of mechanical, thermal and physicochemical properties at a low level of filler concentration in comparison to pure polymer and conventional microcomposites. Most of the biopolymer-based nanocomposites have shown appreciable improvements in mechanical and barrier properties compared to the counterpart biopolymer films. Whey protein has received much attention for its potential use as an edible film and coating because it has been shown to make transparent films and coatings that can act as excellent oxygen barriers and provide certain mechanical properties.

Controlled Atmosphere Packaging: Controlled atmosphere packaging (CAP) is the enclosure of fruit and vegetables in a gas impermeable package inside which the gaseous environment with respect to CO_2, O_2, N_2, water vapor and trace gases has been changed, and is selectively controlled to increase shelf life. In this technology, the storage system consists of airtight storage chambers, O_2 regulatory unit, CO_2 absorbing unit equipment for monitoring as well as controlling the chambers and composition. Liquid N_2 generator is commonly installed to flush and chambers with liquid N_2 as and when required for maintaining optimum level of O_2.

For maintaining optimum level of CO_2 in the chambers use CO_2 absorbing materials such as hydrated lime, calcium or potassium hydroxides are generally used in scrubber. Refrigeration unit is employed for maintaining storage temperature. The storage life of various fruits and vegetables can be increased by 2 to 4 times the normal life by employing controlled atmosphere storage technology. However, CA stored produce deteriorate rapidly when exposed to normal atmospheres during marketing. It shortens the post storage life of the produce, which eventually affects marketing. The involvement of bulky and sophisticated equipment limits the use of CA technology during transport as well as retail storing of fruits and vegetables.

Modified Atmosphere Packaging (MAP): Modified Atmosphere Packaging (MAP) can be defined as the enclosure of fruits and vegetables in a package in which the atmosphere inside the package is modified or altered to provide an optimum atmosphere for increasing shelf life and maintaining quality of the fruits and vegetables. Polymeric films have been used to package fresh products, with a number of benefits, including control of water loss, protection from skin abrasion and reduced contamination of the produce during handling. They also provide a barrier to the spread of decay from one unit to another. These films will also affect the movement of respiratory gases depending on the relative permeability of the film. This can lead to the development of lowered O_2 and raised CO_2 levels within the package and, as with control atmosphere (CA) storage, this can reduce the respiration of the produce and potentially extend shelf-life. Many of the films used in MAP, singly do not offer all the properties required for a modified atmosphere pack. To provide packaging films with a wide range of physical properties, many of these individual films are combined through processes like lamination and co-extrusion. Polyethylene is most commonly used to provide a hermetic seal and also as a medium of control for characteristics like anti-fogging abilities, peel ability and ability to seal through a degree of contamination. High gas transmission films are obtained by modifying the film manufacturing process so that gases such as O_2, CO_2 and water vapour exit or enter the package in a controlled manner such that aerobic respiration needs are met and desirable CO_2 and moisture levels are maintained. For most plastic films, CO_2 permeability is 2-6 times greater than that of O_2, which provides a range of CO_2/O_2 permeability ratios. This allows the selection of films of suitable rations for some products. Typically, the key to successful MAP of fresh produce is to maintain an equilibrium MA (EMA) containing 2–10 per cent O_2/CO_2 within the package. For highly respiring produce such as mushrooms, beansprouts, leeks, peas and broccoli, traditional films like low-density polythene (PE-LD), poly vinyl chloride (PVC), ethylene vinyl acetate (EVAC), oriented polypropylene (OPP) and cellulose acetate are not sufficiently permeable. Such highly respiring produce is most suitably packed in highly permeable micro-perforated films. Films that have improved rates of gas transmission by virtue of their polymeric nature are usually blends of two or three different polymers, where each polymer of a blend performs a specific function such as strength, transparency and improved gas transmission to meet certain product descriptions.

The diameter of micro perforation generally ranges from 40 to 200 μm and by altering the size and thickness of micro perforations, gas permeability through a package can be altered to meet well defined product requirements. A very interesting development for the packing of fresh prepared produce involves the use of high O_2 (70–100 %) MAP, which has been recently shown to overcome the many disadvantages of current air packing and low O_2 MAP. High O_2 MAP has been demonstrated to inhibit enzymic discolorations, prevent anaerobic fermentation reactions and inhibit microbial growth with the result of extending prepared produce shelf-life. Different commodities have different amounts of internal air space (potatoes 1–2%, tomatoes 15–20%, apples 25–30%). A limited amount of air space leads to increase in resistance to gas diffusion. One of the primary effects of MAP is a lower rate of respiration, which reduces the rate of substrate depletion. Ethylene (C_2H_4) is a natural plant hormone and plays a central role in the initiation of ripening, and is physiologically active in trace amounts (0.1 ppm). C_2H_4 production is reduced by about half at O_2 levels of around 2.5 per cent. This low O_2 retards produce ripening by inhibiting both the production and action of C_2H_4. Mangoes stored in CA of 3–5 per cent O_2 and 5–8 per cent CO_2 at 10°C (ripe), shelf life extended up to 20 days, when compared to 5–7 days for control fruit. Broccoli stored in CA of $<$ 1 per cent O_2 and 10 per cent CO_2 at 1–2°C, shelf life extended up to 25 days when compared to 3–5 days after harvest. Produce retains fresh green florets and crispness. Basil stored in CA of 1–5 per cent O_2 and 5–15 per cent CO_2, at 0–5°C, shelf life extended up to 25 days when compared to 3–4 days.

Advantages of modified atmosphere packaging

- Shelf life will be increased by 50 to 400 per cent and reduced economic losses.
- Provides a high quality product or no need for chemical preservatives.
- Centralized, odourless, convenient packaging and portion control.
- Improved presentation clear view of product and all around visibility.
- Sealed packages are barriers against product recontamination and drip from package.

Disadvantages of modified atmosphere packaging

- Added costs for gases, packaging materials and machinery.

- Temperature control necessary.
- Different gas formulations for each product type.
- Special equipment and training required.
- Increased pack volume adversely affects transport costs and retail display space.
- CO_2 dissolving into the food could lead to pack collapse and increased drip.

Gases used in MAP: The three main gases used in MAP are CO_2, O_2, and N_2, either singly or in combination.

- **Carbon dioxide**: Carbon dioxide is the most important gas in the MAP of fruit and vegetables because of its bacteriostatic and fungistatic properties. It inhibits the growth of many spoilage bacteria, the degree of inhibition increasing with enhanceing concentration. It is particularly effective against aerobic spoilage bacteria such as Pseudomonas species. The solubility of CO_2 increases with decreasing temperature and therefore the antimicrobial activity of CO_2 is markedly greater at lower temperatures.
- **Oxygen:** Oxygen promotes several types of deteriorative reactions in fruits and vegetables including fat oxidation, browning reactions and pigment oxidation. Most of the common spoilage bacteria and fungi require O_2 for growth. For these reasons, O_2 is either excluded or the level set as low as possible. Exceptions occur where O_2 is needed for fruit and vegetable respiration.
- **Nitrogen**: Nitrogen is an inert gas with no odour or taste. It has a lower density than air and a low solubility in water and other food constituents, making it a useful filler gas in MAP to counteract package collapse caused by CO_2 dissolving in the food. Nitrogen indirectly influences the micro-organisms in perishable fruit and vegetable by retarding the growth of aerobic spoilage microbes but it does not prevent the growth of anaerobic bacteria.

Methods of creating modified atmosphere conditions

- Commodity generated or passive MA: Passive modification occurs as a consequence of the fruits and vegetables respiration or the metabolism of micro-organisms associated with the fruits and vegetables. First produce is packed in sealed plastic film the concentration of the O_2 reduced and CO_2 increased due to respiration thus generating MA. Film must allow O_2 to enter the package at a rate offset by the consumption of O_2 by the commodity and CO_2 must be vented from the package to offset the production of CO_2 by the commodity. This atmosphere must be established rapidly and without danger of the creation of injurious or high levels of CO_2.

» **Active packaging**: Active modification involves the pulling of a slight vacuum within the pack and then replacing the atmosphere with the desired gas mixture. This mixture can be adjusted by the use of adsorbers in the package to scavenge O_2, CO_2 or C_2H_4. Ethylene absorbers can help to ensure the delay of the climacteric rise in respiration. CO_2 absorbers can prevent the building up of CO_2 to injurious levels. Temperature control is extremely important with MAP, as this will influence the gas permeability properties of the film as well as the respiration rate of the product. One of the main drawbacks to MAP is the potential for O_2 levels to fall too low and cause the production of undesirable off-odours caused by fermentation of the tissues.

» Modified atmosphere conditions are created inside the packages by an active modification and also using O_2, CO_2 and ethylene scavengers within the package. These scavengers may be held in small sachets within the packages or impregnated in wrappers or into porous materials like vermiculate. The smart wrap, which contains a permeable membrane impregnated with an ethylene absorbent, an anti-fog material to avoid moisture condensation and a slow release fungicide to inhibit the mould growth.

» Inadequate (high) temperature and differences in the gas diffusion across plastic films as compared to those of physiological process such as respiration increases the anaerobic condition may occur in sealed plastic film packages. This risk is minimized using safe foil (eg. low melting point polymer) or variable aperture devices (eg. bimetallic strips) to regulate formation and/or size of the holes. Advances in microelectronics, biosensor and polymer sciences may develops the film that actively sense and respond in a controlled way of stimuli, such increase in temperature.

» **Oxygen absorbers**: Most commonly available O_2 absorbers include ferrous oxide (FeO). Iron is main active ingredient in powdered form becoming Fe_2O_3 and Fe_3O_4 and its hydroxides after absorption of O_2.

» **CO_2 absorbers**: They are hydrated lime, activated charcoal, magnesium oxide.

» **Ethylene absorbers/scrubber**: Compounds that can be used for ethylene absorption within polymeric film packages are potassium permanganate ($KMnO_4$) absorbed on celite, vermiculite, silica gel or alumina pellets. They oxidize ethylene to CO_2 and H_2O. Squalene and phenyl methyl silicon can also be held in small sachets within the packages or impregnated in the wrappers or into porous materials like vermiculite.

Films available for MA packaging: LD Polyethylene, HD Polyethylene, Polyethylene - Cast and Oriented, Rigid PVC, Ethylene Vinyl Acetate.

Moisture management: Water condensation in packages can be reduced by use of micro (pin-hole size) or macro-perforated films. In heat shrink wrap film (eg. citrus), condensation is not a problem because the film is in intimate contact with the fruit and assumes the same temperature as the fruit. In case of loose wrapped produce (eg. cut flowers with in carton liner and fruit in a consumer pack), condensation can be reduced by using simple newspaper or slats in spun bonded polythene sachets. Anti-fogging film and film with relatively high water permeability can be used (eg. cellophane and PVC). Water absorbent is incorporated into packages to absorb and hold the free water eg. Silica gel in dry fruits packing.

Insulation and dry ice technique: In absence of refrigeration, a certain level of temperature control during postharvest handling can be achieved with insulation (eg. polystyrene boxes) and heat sinks provided inside the packaging (eg. loose ice or ice packs). Use of external reflective and/or insulative covers (eg. thermal blankets) and heat sinks (eg. dry ice) can assist or provide an alternative to, temperature control management.

Table 75. As permeability and water vapour transmission rate (WVTR) of polymeric film available for packaging of MAP produce

Types of Film	Gas Transmission Rate (ml/m^2 24 hrs. atm) for 35 μ film at 25° C			WVTR (g/m^2/day/atm) at 38° C, 90% RH
	O_2	N_2	CO_2	
Cellulose acetate	325		13640	-
Ethylene vinyl alcohol (EVAL)	3-5	-	-	16-18
Ethyl cellulose	31000		77500	-
Methyl cellulose	1240	-	6200	-
Nylon 6	40	14	158-190	84-3100
Low-density polythene (PE- LD)	7800	2800	42000	18
High-density polyethylene (PEHD)	2600	650	7600	7-10
Plastised PVC	500-30000	300-10000	1500-40000	15-40
Polyacylonitrile	12	3	17	78

Types of Film	Gas Transmission Rate (ml/m² 24 hrs. atm) for 35 µ film at 25° C			WVTR (g/m²/day/atm) at 38° C, 90% RH
	O_2	N_2	CO_2	
Polypropylene cast (PPcast)	3700	680	10000	10-12
Polypropylene, oriented (OPP)	2000	400	8000	6-7
Polypropylene, oriented, PVdC Coated (OPP/PVdC)	10-20	8-13	35-50	4-5
Polybutylene	5000	-	-	8-10
Polyvinyl alcohol	<1		<1	-
Rigid poly vinyl chloride (PVC)	150-350	60-150	450-1000	30-40
Plasticized poly vinyl chloride (PVC-P)	500-30000	300-10000	1500-46000	15-40
Ethylene – vinyl acetate (EVAC)	12500	4900	50000	40-60
Polystyrene, oriented (OPS)	5000	800	18000	100-125
Polyurethane (PUR)	800-1500	600-1200	7000-25000	400-600
PVdC–PVC copolymer (Saran)	8-25	2-26	50-150	1.5-5.0
Polyamide (Nylon-6), (PA)	40	14	150-190	84-3100
Polyester	50-130	15-18	180-390	25-30
Onomer	4650-6975	-	-	25-35

Source: Chung and Yam, 1999; Day, 1993; Greengrass, 1993; Guilbert et al., 1996; Han, 2000; Park, 1999; Phillips, 1996, Packaging India, Feb-Mar 1998

Table 76. Recommended gas mixtures for MAP

Fruits	O_2 (%)	CO_2 (%)	N_2 (%)	Vegetables	O_2 (%)	CO_2 (%)	N_2 (%)
Apple	1-2	1-3	95-98	Artichoke	2-3	2-3	94-96
Apricot	2-3	2-3	94-96	Beans, snap	2-3	5-10	87-93
Avocado	2-5	3-10	85-95	Broccoli	1-2	5-10	88-94

Fruits	O_2 (%)	CO_2 (%)	N_2 (%)	Vegetables	O_2 (%)	CO_2 (%)	N_2 (%)
Banana	2-5	2-5	90-96	Brussels sprouts	1-2	5-7	91-94
Grape	2-5	1-3	92-97	Cabbage	2-3	3-6	81-95
Grapefruit	3-10	5-10	80-92	Carrot	5	3-4	91-95
Kiwifruit	1-2	3-5	93-96	Cauliflower	2-5	2-5	90-96
Lemon	5-10	0-10	80-95	Chilli peppers	3	5	92
Mango	3-7	5-8	85-92	Sweet corn	2-4	10-20	76-88
Orange	5-10	0-5	85-95	Cucumber	3-5	0	95-97
Papaya	2-5	5-8	87-93	Lettuce (leaf)	1-3	0	97-99
Peach	1-2	3-5	93-96	Mushrooms	3-21	5-15	65-92
Pear	2-3	0-1	96-98	Spinach	Air	10-20	-
Pineapple	2-5	5-10	85-93	Tomatoes	3-5	0	95-97
Strawberry	5-10	15-20	70-80	Onion	1-2	0	98-99

(Day, 1993; Exama *et al.*, 1993; Moleyar and Narasimham, 1994; Powrie and Skura, 1991; Smith and Ramaswamy, 1996).

Table 77. Effect of packaging materials and methods on the shelflife of fruits and vegetables

Food/Treatment	Packaging materials/methods	Shelf life
Peach, Cauliflower, Truffle	Tray: PP Cover: PE-LD/PET (40µm), 0- 14 micro-perforated package, all wrapped in PE	4 days at 4° C
Strawberry	Stretch PVC	8 days at 1° C
Minimally processed fruits (kiwi, banana and prickly pear)	1) PE/Al/PET 2) Coex. polyolefinic high permeable film	4-12 days at 5° C
Sweet cherry	5 % O_2 + 10 % CO_2	80 days at 1° C
	PE: 13-18 % O_2 + 2-4 % CO_2	40 days at 1° C
	70 % O_2 + 0 % CO_2	20 days at 1° C
	Air	30 days at 1° C
Cactus pear fruits	Cryovac MY 15, Plastic box	9 days at 4° C

Food/Treatment	Packaging materials/methods	Shelf life
Carrots, minimally processed	PP+cPP/OPP in:	
	5 % O_2 / 10% CO_2 / 85 % N_2	2 days at 4° C
	80 % O_2 / 10 % CO_2 / 10 % N_2	7 days at 4° C
Cabbage, shredded	OPP (30 μm)	9-10 days at 3° C
Cabbage, shredded	Glass jar; PE (30 μm); PP (30μm) in:	7 days at 0 and 10° C
	Air	
	100 % N_2,	
	MAP 1: 100 % N_2,	
	MAP 2: 5 % O_2 / 95 % N_2,	
	MAP 3:10 % O_2 / 90 % N_2,	
	MAP 4: 70% O_2 / 30 % N_2 and 100 % O_2	
Lettuce (shredded)	OPP (30 μm)	8 days at 4° C
Spinach Fresh-cut	PE-LD	8 days at 4° C & 90% HR
	OPP	6 days at 4° C & 90% HR
Lettuce	PE	10 days in dark at 7° C
Broccoli	PE-non-perforated	13 days at 4° C; 3 days at 20° C
	PE-2 holes (750 μm in diameter)	23 days at 4° C; 5 days at 20° C
	PE-4 holes (8.8 mm in diameter)	10 days at 4° C, 2.5 days at 20° C
	Unwrapped broccoli	10 days at 4° C, 2.5 days at 20° C
Fresh-cut vegetables, Fennels, cut lettuce salads and shredded carrots	PA/PE (102 μm)	-
	Air	less than 5 days at 4° C
	Vacuum	13.5 days at 4° C
	MAP: 70 % N_2 / 30 % CO_2	5.8 days at 4° C

Food/Treatment	Packaging materials/methods	Shelf life
Cherry tomatoes ozone treatment (5–30 mg/l ozone gas for 0–20 min)	CA 1: 5 % CO_2 / 17 % O_2 (7° C)	20 days at 7° C
	CA 2: 5 % CO_2 / 15 % O_2 (22° C)	10 days at 22° C
	PE-LD (50 μm); MAP	
	Glass jars	
Melon, minimally fresh processed	MAP passive in:	10 days at 5° C
	PP mikroperforated	
	BOPP	
	OPP	
	PP-makroperforated (control)	
Blueberry nectar Aseptically Filled (93–97° C for 30s)	PET 1: standard	9-12 months at 4-20° C
	PET 2: O_2-absorber	
	PET 3: 5 layers	
	PET 4: 3 layers	
Tomato	PE-LD (50 mm) MAP + O_2 absorber	8 days at 20° C

Source: M. SCETAR *et al.* (2010): Croatian Journal of Food Technology, Biotechnology and Nutrition 5 (3-4), 69-86

Table 78. Shelf-life of fruits and vegetables under MAP

Commodity	Shelf-life without MAP (days)	Room Temperature			Low Temperature			
		Shelf-life (days)	O_2 %	CO_2 %	Temperature (°C)	Shelf-life (days)	O_2 %	CO_2 %
Banana	6	11	12-15	2.7-5.0	13	42	9.7-16.0	4.7-7.0
Plantains	3	11	4.2	5.3	13	18	-	-
Carrots	4	10	12.0	4.0	-	-	-	-
Bell peppers	5	10-13	7.7	4.5	9	28	14.0	4.5
Chillies	3	5	8.4	8.0	12	23	15.0	2.5
Okra	2	6	7.0	8.0	-	-	-	-
Tomato	8	16-18	9.3	3.4	12	30-35	-	-

Table 79. Oxygen and carbon dioxide permeabilities of edible films

Film	Thickness (mm)	Permeability at 0% RH (10-15 l/m² s Pa)		Permeability Ratio (CO_2/O_2)
		O_2	CO_2	
Corn-zein	0.12- 0.31	0.36 (30° C)	2.67 (21° C)	7.5
Wheat gluten	0.23-0.42	0.20 (30° C)	2.13 (21° C)	9.5
Methyl cellulose Low level (MC (L))	0.04-0.07	2.17 (30° C)	69.00 (21° C)	31.6
Hydr-oxy-propyl cellulose Low level (HPC (L))	0.05	3.57 (30° C)	143.99 (21° C)	40.6
HPC/Lipids	0.15	3.44 (30° C)	81.75 (21° C)	23.7
Cozeen	0.09	0.89 (37.8° C)	5.25 (22.8° C)	5.9
Wheat gluten	0.14	0.09 (37.8° C)	0.03 (22.8° C)	0.3
Corn-zein	0.08	0.16 (25° C)	-	-
Wheat gluten	0.15	0.08 (25° C)	-	-

Source: M. SCETAR et al. (2010): Croatian Journal of Food Technology, Biotechnology and Nutrition 5 (3-4), 69-86

Table 80. Properties of major types of Ageless oxygen scavengers (Pocas, 2001)

Type	Function	aw	Absorption speed	Capacities
Z	Decreases O_2	< 0.65	1 – 4 days	20 – 2000 cc
S	Decreases O_2	< 0.65	0.5- 2 Days	500 – 2000 cc
FX	Decreases O_2	< 0.85	0.5 – 1 days	20 – 300 cc
E	Decreases O_2 Decreases CO_2	< 0.30	3 – 8 days	O_2: 25–200 cc CO_2: 250 – 2000 cc
G	Decreases O_2 Increases CO_2	0.3-0.5	1 – 4 days	NA

Table 81. Properties of some biodegradable plastics (Berkesch, 2005)

Material	Film preparation	Moisture barrier	Oxygen barrier	Mechanical properties
Starch/ Polyvinyl alcohol (PVAL)	Extrusion	-	+	+
Polyhydroxybutyrate/valerate (PHB/V)	Extrusion	+	+	+/-
Polylactic acid (PLA)	Extrusion	+/-	-	+

Source: M. SCETAR et al. (2010): Croatian Journal of Food Technology, Biotechnology and Nutrition 5 (3-4), 69-86

Table 82. Overview of potential applications for biobased packaging and coatings in MAP of fresh or minimally processed fruits and vegetables

Biopolymer	Product	MAP Improvement
Polysaccharides		
Xanthan gum	Baby carrots (coating)	Improve surface colour, texture, and flavour
Starch	Strawberries (coating)	Reduce weight loss, maintain texture and surface colour, delay maturation
Cellulose	Mandarin (coating)	Reduce weight loss, improve visual appearance, and delay ripening
	Apples (coating)	Reduce weight loss
	Mini-peeled carrots (coating)	Prevent white surface discolouration
Hydr-oxy-propyl cellulose	Guavas (coating)	Delay softening but surface blackening cannot be avoided
Methyl cellulose	Avocados (coating)	Decrease respiration rate, maintain higher firmness and greener colour, delay ripening
Carboxymethyl cellulose	Sweet cherries (coating)	Reduce weight loss, delay changes in firmness, skin colour, ascorbic acid content, etc.
	Tomatoes (coating)	Reduce weight loss, delay changes in firmness, ascorbic acid, lycopene, etc.
	Mangos (coating)	Reduce weight loss and delay changes in firmness and colour

Biopolymer	Product	MAP Improvement
Polysaccharides/Proteins		
Cellulose/soy protein	Apples (coating)	Reduce water loss
Cellulose/ wheat gluten	Mushrooms (packaging)	Delay cap opening
Starch/chitosan	Sliced carrots (coating)	Inhibit growth of total coliforms and lactic acid bacteria, slightly reduce counting of mesophilic aerobes, mold, and yeast
Proteins		
Chitosan	Cucumber (coating)	Reduce water loss, respiration rate, discolouration, and fungal infection
	Strawberries (coating)	Reduce decay caused by *Botrytis cinerea* and *Rhizopus stolonifer* delay change in colour and loss in firmness
	Tomatoes (packaging)	Reduce weight loss and formation of defective spots, and delay loss in firmness and colour
Corn zein	Broccoli (packaging)	Maintain original firmness and odour
	Tomatoes (coating)	Delay colour change, loss in firmness and weight
Wheat gluten	Mushrooms (packaging)	Delay cap opening, but browning cannot be avoided
	Sliced mushrooms (coating)	Delay cap opening
	Strawberries (coating)	Retention of firmness and maintenance of visual acceptance and taste

Source: Nathalie Gontard and Carole Guillaume (2010) Agro-polymers engineering and Emerging Technologies University of Montpellier II, Montpellier, France

Table 83. Active packaging principles applied to perishables

Principle	Application
Porosity control	Gas pressure release, gas composition balance
Polymer permeability control	Gas compensation balance, temperature compensation
Melting of waxes	Temperature compensation
Inorganic/organic oxidation	Oxygen scavenging, oxygen permeation barrier oxygen indicator, carbon dioxide generation, ethylene scavenging
Enzyme catalysis	Oxygen scavenging
Acid/ base reaction	CO_2 absorption, CO_2 generation, odour absorption
Adsorption	Taint removal, oxygen scavenging, ethylene scavenging, water removal
Absorption	Condensation control, drip collection
Hydrolysis	Sulfur dioxide release
Desorption	Ethanol release, hinokitiol release, water release
Organic reactions	Ethylene removal, oxygen barrier

Packaging of Important Horticultural Produce

Packaging is used to assemble the produce in convenient units for transportation, marketing and distribution of fruits, vegetables and flowers. Good packages must cope with long distance transportation, multiple handling, and changed conditions of storage, transport and marketing. Concept of packaging house or station is completely lacking in India and packing is mainly done on farm in unscientific and unhygienic manner. Use of traditional form of packaging like bamboo baskets and gunny bags are still prevalent and cause considerable post harvest losses during transportation. Use of modern methods of packaging such as ventilated corrugated fiber board boxes, plastic trays and high density films are extremely helpful to reduce post-harvest losses and enhance shelf life during storage. Plastic nets are most ideal packing materials for many fruits and vegetables. Cost effective waxing, shrink wrapping and modified atmosphere packing techniques should be standardized. Use of ethylene absorbents and sulphur dioxide releasing pad material should be extended to various fruit and vegetable crops. Pre-packaging can bring revolutionary progress in our trade practice and also benefit the consumer and the producer because of its low cost and ready availability.

Mango: Different types of packaging materials are used during transportation of mango, especially bamboo baskets and plastic crates. Nowadays, use the plastic crates through which the postharvest loss is minimized. This is also economic and reusable as compared to those of the presently practiced bamboo baskets.

- » **Local Market** – Packed in bamboo baskets with news papers as liners covered with paddy straw and then closed with bamboo lids. Ropes are used for laying the baskets.
- » **Distant Market** – Packed in wooden ventilated boxes with newspaper or paddy straw as liners (1.5 cm thick layers). Packing in 4-5 layers with alternate layer of liners and finally closed with wooden lid.
- » **Shipping** – Crates of light wood of capacity 10-30 kg are used. Paper cuttings are used as liners and moisture and ethylene absorption.
- » **Air Transport for export** – Corrugated fiber-board boxes with 2 per cent vents. These boxes have number of blocks. Tissue paper treated with biphenyl used as wrapper for individual fruits and kept in each block. Boxes are finally covered with lid.

Banana: Usually, necked banana bunches are transported through truck and rail by using dry leaves of banana as liners and covering. In this manner banana fingers or hands damaged and loose 20 to 25 per cent their quality. However, use of corrugated fiberboard cartons with foam made sleep sheets inside for banana hands. Banana bunches if packed in ventilated polythene or strong plastic bags or light jute bags and transport may cause less damage. For export 14 kg of banana fingers are packed in carton or card board container.

Litchi: Litchi fruits are packaged in bamboo baskets of different sizes (20-30 kg) and covered with gunny sacks. Large volumes of litchi leaves are placed inside the package as padding materials during long-distance transportation. Another important possible role of these leaves would be the elevation of CO_2 and reduction of O_2 concentrations inside the package which create a modified atmosphere suitable for storage of fruits. Wooden boxes and card board boxes with paper as liners are used for packing for distant places.

Jackfruit and pineapple: Packaging is not practiced in case of jackfruit and pineapple. However, layers of green leaves, vines and straw are used as padding materials to reduce vibration damage during transportation.

Papaya: Fruits are normally packed in baskets with paper cuttings. It is better to wrap the fruits in news paper and pack in strong basket or wooden boxes. Foam nets would be suitable packaging for papaya in India.

Guava, lime and mandarin: They should be packed in wooden boxes or lined or unlined corrugated fiberboard boxes. The boxes should be made in such a manner as to allow for proper aeration of the fruits. Sufficient quantity of straw should be put in the container to prevent fruits from rubbing against each other. Super grade fruit should be wrapped individually either in tissue paper or in any other suitable material before being packed into the container.

Citrus fruits: Packed in bamboo basket with using paddy straw as liners. Wooden boxes (48 cm x 35 cm x 35 cm) are also used for packing of loose skin oranges. Packing in wooden boxes causes less spoilage than jute bags and bamboo baskets. Kinnow mandarin is packed in gunny bags as well as in bamboo baskets with grasses as cushioning. Nagpur mandarin packed also in CFB boxes (50 cm x 30 cm x 30 cm). Mature green acid lime fruits and sweet orange are packed in gunny bags with rice straw as cushioning materials.

Grapes: Grape bunches are packed in bamboo basket using leaves or news paper as liners, losses or spoilage is more in such packing. Bunches are also packed in wooden and card board boxes coated with paraffin wax (10-12 kg capacity) using news paper as cushioning material. Sulphur dioxide releasing pads as an in packing material is used to control fungal diseases.

Pomegranate: Fruits are packed in bamboo baskets and wooden boxes with using paddy straw or paper or dry grasses as padding material.

Aonla: Fruits are packed in baskets particularly pigeon pea baskets of 40-45 kg capacity. Gunny bag of 40-50 kg capacity and wooden boxes of 20-25 kg capacity with news paper or polythene sheet as liners used for packing.

Apple: Generally packed in wooden card board boxes. Size varies with the grades like super grade (Box size: 45 x 30 x 28 cm), fancy/selected (Box size: 45 x 30 x 30 cm) and commercial/ janta Box size: 45 x 30 x 25 cm). Leaves, or paddy straw is used as padding material and news paper is used as liners, it is better to wrap the individual fruits in tissue paper and then pack in 3-4 layers according to the grades. Fruits are also packed in telescopic carton with tray packs, ventilated polythene may be used as pre-packing for better storage.

Pear: Fruits are packed in wooden boxes with using paddy straw as a cushioning material and then top is covered with news paper before covering the wooden lid.

Peach: Packed in small wooden boxes of 5-8 kg capacity. Paddy straw is used as cushioning material and news paper as liners.

Plum: Baskets and wooden boxes of 10 kg capacity are used. Proper size of boxes is 50 x 28 x 10 cm. Paddy straw is used as padding material and news paper as liners.

Cherry: Both baskets and wooden boxes are used. Size of wooden boxes is normally 5-10 kg capacity. Paddy straw is used as cushioning material and news paper as liners.

Brinjal and cucumber: Conventional packaging i.e. jute sacks are mainly used by merchant of brinjal and cucumber trades. Some traders use plastic sacks and crates as packaging materials for brinjal and cucumber marketing.

Okra: The usual means of packaging okra are plastic sacks and gunny bags. Recently introduced plastic crates for okra packaging for long distance transportation. Plastic crates are also helps reduce transport damage and thereby extend shelf life of produce. Packing in polythene bag for storage is better to maintain tenderness.

Onion and garlic: Perforated jute bags (9 x10 mesh per sq. inch) of 40 kg capacity and plastic woven bags having good ventilation are used for packing and transportation. A nylon net bag with good ventilation is also used for packing which causes less storage loss.

Tomatoes: Tomatoes should be packed in baskets or wooden boxes or lined or unlined corrugated fiberboard boxes. While packing, it should be ensured that the tomatoes are not unduly pressed when lid is closed. Packed in bamboo baskets or corrugated boxes of different shapes and sizes (15-20 kg capacity) and in wooden crates of 10-15 kg capacity. News paper is used as lining material in the container

Capsicum and chillies: Fresh capsicum and chillies should be packed in gunny bags or in bamboo baskets, or in corrugated boxes, lined or unlined. The containers should be so constructed as to allow for proper aeration of the packed material.

Potato: Potatoes are packed into gunny bags of 50 - 80 kg capacity.

Peas: Packed in gunny bag or in baskets of 30-50 kg capacity.

Cabbage and cauliflower: These vegetables are packed in baskets of different size and in gunny bags.

Ginger: Bulk packaging is done in jute bags and HDPE woven sacks are very useful for packing.

Mushroom: Plastic crates are used for packing of mushroom.

Definitions

Back haul: The return trip from delivering produce or any item to its destination. For efficiency, truckers try to limit empty unproductive (known as dead heads) back hauls.

Bulk container: A container designed to contain a relatively large quantity of produce upto 2000 kg. Bulk containers are used in conjunction with a shipping pallet but are normally separate from the pallet.

Bulk produce: Produce handled in generally truckload lots but not in small containers such as cartons, bags, etc. Bulk produce may be transported in gondolas, dump trucks, or refrigerated vans. Bulk produce is mostly used for processing.

Carton: A container of various construction, but usually made from corrugated fiberboard or possibly plastic, that generally contains 6.8 kg or less of fresh produce.

Cold chain: Most types of produce require continuous postharvest refrigeration for maximum quality maintenance. On farm refrigeration, refrigerated transport, the buyer's refrigerated receiving warehouse, and refrigerated retail displays all form the cold chain.

Container: Any type of box, carton, bag, or bin used to form a package of produce.

Controlled atmosphere package: An engineered package where the interaction between the produce and the packing material actively regulates a beneficial mix of environmental gases.

Count packing: Packing method in which a certain specified number of sized and graded items are placed in the carton.

Corrugated fiberboard: Common packaging material made from a layer of corrugated fiberboard sandwiched between two additional layers of fiber- board.

Face packing: A packing method where most of the container is loose or volume filled except for top layer. Items in the top layer are arranged in an orderly pattern for appearance.

Fiberboard: A paper material usually made by the kraft process having a thickness greater than .008 inches. Fiberboard may contain additional materials for strength and resistance to water.

Field packing: A packing method in which all harvesting, grading, and packing functions are per formed at the same time in the field or orchard.

Food service: An enterprise whose function is to supply food items, including fresh produce, to institutions, restaurants, and increasingly, grocery outlets.

Forced air cooling: Forced air cooling is a commonly used cooling method that utilizes specially constructed portable fans or rooms to draw chilled air horizontally through pallets or stacks of packaged produce. A properly designed forced air cooling system is fast, energy efficient and relatively inexpensive. It may be utilized on most types of produce.

Fresh cut (precut or ready cut): A value adding process where fresh produce is shredded, trimmed, sliced, or otherwise prepared for consumer use. Fresh cut produce is generally prepared by a processor or retailer and requires increased attention to sanitation, packaging, handling, the storage environment and labeling.

Hydro cooling: Cooling freshly harvested produce by flooding, immersing, or spraying with large quantities of cold water. Compared to other cooling methods, hydro cooling is fast and generally thorough. Hydro cooling is limited to those produce items that will tolerate liquid water.

Icing: Cooling fresh produce by the addition of crushed ice or slurry of crushed ice and water over the top of a load or to each individual package. This method is limited to those produce items that are not harmed by contact with ice.

Layer packing: Packing method where the entire package of produce is packed in orderly layers.

Lug: A sturdy container, often wholly or partially of wood, designed to have high stacking strength. Lugs are often used for soft fruit such as grapes, berries or tomatoes that are easily damaged by crushing.

Master (flat): A type of carton designed to contain smaller, usually consumer size, units of produce. Masters designed to contain 8 quarts or 12 pints or half pints are commonly used with strawberries and other small fruit.

Mixed load: A single truck load of fresh produce consisting of two or more products shipped together. The use of mixed loads reduces transportation costs but care must be exercised to prevent ethylene or odour contamination.

Modified atmosphere packaging: A method of packaging in which the produce is packed in a sealed container into which a specific mix of gases are introduced. The container only prevents the gases from escaping and does not regulate the mix.

Pack out rate: The portion, often expressed as a percentage, of harvested produce actually packed for shipment. That portion not packed is referred to as culls.

Package: Any type of filled box, carton, bag, or bin of product.

Packing house: A facility designed to wash, grade, or trim harvested produce but primarily to place the produce into containers suitable for sale.

Pallet bin: A bulk bin made integral to a pallet for the transport of bulk produce. Pallet bins may be made of wood, plastic or some other sturdy material and are primarily intended to be used between the field/orchard and packing house.

Plastic field boxes: These types of boxes are usually made of polyvinyl chloride, polypropylene, or polyethylene. They are durable and can last many years. Many are designed in such a way that they can nest inside each other when empty to facilitate transport, and can stack one on top of the other without crushing the fruit when full.

Polyethylene bags: Clear polyethylene bags are used to pack banana bunches in the field, which are then transported to the packinghouse by means of mechanical cableways running through the banana plantation. This technique of packaging and transporting bananas reduces damage to the fruit caused by improper handling.

Postharvest handling: Any operation performed after harvest constitutes postharvest handling. Among these operations are washing, grading, packing, storing, cooling, transporting, and marketing.

Precooling: Precooling is the practice of cooling bulk produce prior to grading and packaging although the term has been used to describe cooling at any time before transport. The practice of true precooling is gradually fading since precooled product has an opportunity to warm during subsequent operations.

Room cooling: Room cooling is the practice of storing bulk or packaged produce in a refrigerated room for an indefinite period. By the processes of conduction and convection, the heat is gradually removed from the produce. This system suitable for potatoes, cabbage and root crops only.

Shipper: An individual or company that transports produce from the grower to the buyer. Growers, packers, or buyers often assume their own shipping function.

Slip sheet: A sheet of material, roughly the size of a pallet, of corrugated fiberboard, plastic, or a combination of these materials designed as a replacement for a shipping pallet.

Stacking strength: The ability of a container to resist a specified vertical load without significant deformation.

Vacuum cooling: When warm produce is placed inside a closed container and reduced to a partial vacuum, a small portion of the water in the produce evaporates, causing a cooling effect. Vacuum cooling is fast and extra effective on those items with a large surface area to weight ratio such as lettuce and various greens.

Vibration fill packing: A packing method designed to reduce bruising and scuffing of the produce. After filling, the package is gently vibrated to maximize product density and stability.

Volume fills packing: Packing method where a specified volume of produce is packed in the container. Both the number of individual produce items and the weight may vary in volume fill packing.

Weight packing: Packing method where a specified minimum weight, but not necessarily numbers of produce items are packed in a container.

Wooden field boxes: These boxes are made of thin pieces of wood bound together with wire. They come in two sizes: the bushel box with a volume of 2200 in3 (36052 cm3) and the half-bushel box. They are advantageous because they can be packed flat and are inexpensive, and thus could be nonreturnable. They have the disadvantage of providing little protection from mechanical damage to the produce during transport.

11 Storage of Fruits and Vegetables

The ability to store harvested plant organs for extended periods of time has played a critical role in the development of agriculture. Simple baskets were used as early as 7000 BC to gather and store food until consumed. Advances in technology such as fired clay pottery would enable primitive societies to store crops in concealed environments, creating a simple modified atmosphere. Middle Eastern artisans specialized in making pottery of numerous shapes and sizes for varied usage (4500 BC), while pre-Neolithic, Middle Eastern societies held grain in underground pits 9,000–11,000 years ago. Ancient Egyptians and Samarians are thought to have stored some of their crops in sealed limestone crypts to prolong storage life about 2500 BC. Pits were and are still used by primitive societies for storing various types of fruits and vegetables. Silos for long-term grain storage were used by the Romans and continued to be popular until well into the nineteenth century. Many horticultural crops are seasonal in nature and have a relatively short harvesting season and they are also highly perishable. Hence, proper storage of these produce using appropriate methods would prolong their availability. Storage of fresh produce will also be helpful in checking market glut, providing wide selection of fruits, vegetables and flowers to the consumer through most part of the year i.e. especially during the off season. Storage of fresh produce is done to maintain freshness, quality, reduce the spoilage, extend their usefulness and increases profit to the producers/farmers. The basic principle of storage is to reduce the rate of physiological processes like respiration, transpiration, ripening and other biochemical changes. Proper storage also aims at controlling disease infection and preserving the commodity in its best quality for consumers. Storage time depends on the intrinsic characteristics and perishability of the product. Shelf life ranges from short periods for products such as raspberries to long periods for products such as onions, potatoes, garlic and pumpkins. Storage conditions also depend on specific product characteristics. For example, leafy vegetables tolerate temperatures close to 0° C, while most tropical fruits cannot tolerate exposure to temperatures below 10° C. In order to optimize storage conditions, only one crop should be stored in a room unless it is for a short period of time. Using the same storage area for different products can result in product damage because of incompatibility of temperature and relative humidity conditions, chilling and ethylene sensitivity, odour contamination and other problems affecting shelf life and quality.

DEVELOPMENT OF STORAGE SCIENCE

Some of the first experiments in which ripening of fruits were intentionally altered by changing the storage atmosphere were carried out in France by Jacques Berard around 1820. He demonstrated that fruit use O_2 and generate CO_2 while in storage, and if totally deprived of O_2, they did not ripen. Berard found that apples, apricots, peaches, pears, and prunes could be stored in an altered gas environment, and following their removal to ambient air and room temperature, were edible for extended periods. Refrigeration has a significant effect on the development of fruit and vegetable storage technology. The rationale for quickly lowering the temperature of harvested crops is essentially the same today as it was in antiquity, refrigeration greatly reduces food spoilage and waste. Although ice and snow have been used since the Roman era to preserve perishable food stuffs, it was not until the 1800s that natural ice from the colder climes of the northern hemisphere was commonly used to refrigerate foods worldwide.

Periods	Events
19th Century	Benjamin Nyce, an Ohio commercial storage operator, found that when he limited O_2 surrounding his fruit in an ice-refrigerated store, the produce was greatly improved. Nyce did not license the patent rights for his storage system despite strong commercial interest to implement the method, thus, it was not widely adopted. Ironically, Nyce's work resulted in a de facto prototype for modern controlled atmosphere (CA) storage, which, unfortunately, was not further developed for another 60 years.
	The San Jose Fruit Packing Company experimented with augmented CO_2 levels as a preserver of fruit. In a railroad car experiment, a load of grapes, peaches, pears, persimmons, and quinces stored in elevated CO_2 and without refrigeration were in good condition following 11 days of travel, and held up well in the retail chain. In other early experiments with storage gas modification, elevated CO_2 was found to be helpful, but could not completely compensate for lack of refrigeration. Temperature control was/is fundamental for preservation of fruit and vegetable quality, with modification of the O_2 and CO_2 atmosphere being an important but secondary quality preservation method.

20th Century	Kidd and West, working at the Low Temperature Research Station in Cambridge UK, conducted a number of experiments to determine the optimal atmospheric gas concentrations, i.e., lower O_2 and higher CO_2, for storing apples. It was known at that time respiration fueled the generation of metabolic heat, which was deleterious to fruit quality in long-term storage. Therefore, one of the goals of their research was to reduce the rate of storage respiration. Their studies led to assigning the term "climacteric" to indicate the burst of postharvest respiration in apples, which is now known to occur in many important fruit crops. In subsequent years, Kidd and West also became increasingly aware of the role of ethylene in the climacteric rise in apples.
	Following the seminal work of Kidd and West, CA storage became more common globally, although the UK was the first country to initially adopt commercial CA storage practices. This was soon followed by South Africa, the United States, Canada, Australia, New Zealand, Denmark, and the Netherlands. Today, CA storage is standard practice worldwide and is largely responsible for the high market place quality attained for many CA responsive fruits and vegetables. Interestingly, the challenge facing each major geographical region that adopted CA technology in the 1920s and 1930s is the same today: finding the optimal combination of temperature, O_2, and CO_2 for each species, cultivar, and even strain that leads to the highest quality retention for the longest duration. This goal is always tempered by the technology available and the economic viability of sustaining the storage environment for the desired length of time.

Goals of Storage

- Decrease product drying speed and moisture loss.
- Avoid physiological disorders and diminish physical damage.
- Slow down biological activity of the product by maintaining the lowest temperature that will not cause freezing or chilling injury by controlling atmospheric composition.
- Trim down the growth and spread of microorganisms by maintaining low temperatures and minimising surface moisture on the product.
- Reduce product susceptibility to damage from ethylene.

» Avoid glut and distress sale in the market, thus prolonging the market period.

» Making the fruits and vegetables availabilities in off-season.

Table 84. Classification of fresh horticultural crops according to their relative perishability and potential storage life in air at near optimum temperature and relative humidity

Relative perishability	Potential storage life (weeks)	Commodities
Very high	Less than 2	Apricot, blackberry, blueberry, cherry, fig, raspberry, strawberry, asparagus, bean sprouts, broccoli, cauliflower, green onion, leaf lettuce, mushroom, muskmelon, pea, spinach, sweet corn, tomato (ripe), most cut flowers and foliage, minimally processed fruits and vegetables.
High	2-4	Avocado, banana, grape (without SO_2 treatment), guava, loquat, mandarin, mango, melons (honeydew, crenshaw, persian), nectarine, papaya, peach, plum, artichoke, green beans, brussels sprouts, cabbage, celery, eggplant, head lettuce, okra, pepper, summer squash, tomato (partially ripe).
Moderate	4-8	Apple and pear (some cultivars), grape (SO_2 treated), orange, grapefruit, lime, kiwifruit, persimmon, pomegranate, table beet, carrot, radish, potato (immature).
Low	8-16	Apple and pear (some cultivars), lemon, potato (mature), dry onion, garlic, pumpkin, winter squash, sweet potato, taro, yam, bulbs and other propagules of ornamental plants.
Very low	More than 16	Tree nuts, dried fruits and vegetables

Tips for storage of high quality horticultural produce

» Store only high quality produce, free of damage, decay and of proper maturity (not over-ripe or under-mature) at proper temperature, relative humidity and ventilation, because many commodities are susceptible to damage from freezing or chilling.

» Do not over load storage rooms or stack containers closely and provide adequate ventilation in the storage room and keep storage rooms clean.

» Storage facilities should be protected from rodents by keeping the immediate outdoor area clean, and free from trash and weeds.

- Onion and garlic should not store in high humidity environments.
- Keep away from storing ethylene sensitive commodities with those that produce ethylene.
- Avoid storing produce known for emitting strong odours (apples, garlic, onions, turnips, cabbages and potatoes) with odour-absorbing commodities.
- Remove damaged or diseased produce to prevent the spread of problems.

Ideal storage temperature

There is no ideal storage for all fruits, vegetables and flowers, because their response to reduced temperatures varies widely. The importance of factors such as mould growth and chilling injuries must be taken into account, as well as the required duration of storage. Lowering the temperature to the lowest safe handling temperature is of paramount importance for enhancing the shelf life, reducing the losses and maintaining higher quality during marketing. Storage temperature for fruits, vegetables and flowers can range from –1 to 13° C, depending on their perishability. Extremely perishable fruits such as apricots, berries, cherries, figs, watermelons and flowers can be stored at –1 to 4° C for 1-5 weeks, less perishable fruits such as mandarin, nectarine, ripe or green pineapple can be stored at 5-9° C for 2–5 weeks, bananas at 10° C for 1-2 weeks and green bananas at 13° C for 1-2 weeks. Highly perishable vegetables can be stored up to 1-4 weeks such as asparagus, beans, broccoli, and brussels sprouts at –1- 4° C, cauliflower at 5-9° C for 2-4 weeks. Green tomato is less perishable and can be stored at 10° C for 3-6 weeks and non-perishable vegetables such as carrots, onions, potatoes and parsnips can be stored at 5-9° C for 12-28 weeks. Similarly, sweet potatoes can be stored at 10° C for 16-24 weeks. The storage life of produce is highly variable and related to the respiration rate there is an inverse relation between respiration rate and storage life in that produce with low respiration generally keeps longer. For example, the respiration rate of a very perishable fruit like ripe banana is 200 mL CO_2 $kg^{-1}h^{-1}$ at 15° C, compared to a non-perishable fruit such as apple, which has a respiration rate of 25 mL CO_2 $kg^{-1}h^{-1}$ at 15° C.

Principles underlying storability for fruits and vegetables

Fruits and vegetables are living organs: They are use oxygen and produce carbon dioxide during respiration, the process by which carbohydrate and other substrates, such as organic acids, proteins and fats, are metabolized. Respiration thereby provides the energy necessary for cells to maintain structure and for ripening processes such as colour and flavour development. Water loss or transpiration is a major factor affecting quality of fruits and vegetables. In addition to lower saleable weight, loss of water can affect quality in many ways, including wilting, shriveling, flaccidness, soft texture and loss of nutritional value. The maximum permissible losses can range from 3 per cent for lettuce to 10 per cent in onions. Water loss can be reduced by cooling products, maintaining a high relative humidity in the storage environment, controlling air circulation, and where permitted, the use of surface coatings or plastic film. Because active metabolism continues after harvest, a number of desirable and undesirable changes occur in storage and shelf life conditions. These changes include development of pigments for example, lycopene synthesis in tomato, anthocyanin synthesis in strawberry, and development of carotenoids (yellow and orange colours) in apricots and peaches. Other changes include softening to edible ripeness, loss of chlorophyll (green colour), and development of aroma and flavour characteristics. The same processes can be positive in some situations and negative in others, like loss of chlorophyll is desirable in tomatoes, but undesirable in cucumbers and broccoli. Conversion of starch to sugars is desirable for apples, but undesirable for potatoes, while conversely, conversion of sugars to starch is desirable for potatoes but undesirable for peas and sweet corn.

Table 85. Transpiration losses* for fruits and vegetables stored at various relative humidities

Crop	Storage temperature (° C)	Percent weight loss per day (*% Weight loss per 24 hours, Calculated from ASHRAE data)			
		95% RH	90% RH	85% RH	80% RH
Apples	0	0.011	0.022	0.033	0.044
Brussels sprouts	0	1.610	3.220	4.840	6.420
Cabbage	0	0.058	0.116	0.175	0.233
Carrots	0	0.315	0.630	0.945	1.260
Celery	0	0.460	0.920	1.38	1.840
Table grapes	0	0.036	0.064	0.096	0.128
Leeks	0	0.210	0.420	0.620	0.820
Lettuce	0	1.930	3.860	5.790	7.730
Parsnips	0	0.500	1.000	1.500	2.00
Peaches	0	0.150	0.300	0.450	0.600
Pears	0	0.018	0.036	0.054	0.072
Potatoes uncured	6	0.070	0.141	0.211	0.282
Potatoes cured	6	0.021	0.042	0.063	0.084
Tomatoes	6	0.060	0.119	0.180	0.240

Fruit and vegetable varieties: Growers usually select varieties on the basis of marketability and yield, because these factors directly affect the bottom line. Variety selection should also comprise resistance to postharvest diseases and physiological disorders. Differences in storage potential within specific products result from different physiologies and biochemistries of each variety. In some cases, genes which control processes such as less ethylene production, low respiration and slower softening are bred into the commodity. For example, the rin mutant has been bred widely into most commercial tomato varieties. Also, breeders have sometimes favoured slower ripening selections with better resistance to the handling abuses during harvest, handling and transport that result in bruising and skin damage.

Maturity at harvest: Decisions associated with harvest maturity and the way the commodity is handled and stored can greatly affect storage life of even the most long lived fruit or vegetable. The physiological stage that horticultural products are harvested for commercial utilization. On one hand, characteristics of the products such as sweetness and flavour that are desired by the consumer increase as they mature and ripen. At the same time, the storability of the product continues to decrease. The outcome of these opposing factors is that fruit intended for storage should be harvested earlier than one that is suitable for immediate consumption. Examples include an apple that is picked at full ripeness, with low starch content, high soluble solids and is highly aromatic but will have a short storage life, while one destined for long term storage must be harvested much earlier when the fruit has high starch content and less developed aroma volatiles. Strawberries that are fully red and flavoured will have a much shorter storage life than when harvested at the white tip stage of maturity. However, the fruit harvested at the white tip stage will have less intense flavour than the later picked one.

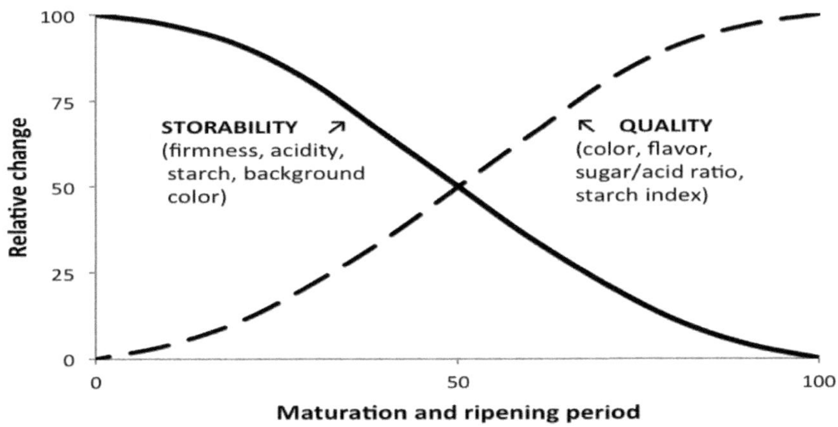

Figure 39. The relationship between the storability of a ripening horticultural product and the quality characteristics desired by the consumer.

Preharvest management affects storage quality: It is broadly accepted that the storability of fruits vegetables and flowers is affected by the mineral composition at harvest. Elevated calcium levels are usually associated with good storability, and on the contrary, low calcium with shorter storage life and susceptibility to storage disorders and pathogen infection. In tomato, for example, a disorder known as blossom end rot is associated with low calcium concentrations in the fruit. Nitrogen is often applied to increase yield, but usually negatively affects postharvest quality and increases storage rots in onion. High nitrogen causes the onions to develop thick necks that are prone to wounding when topped and therefore greater decay development. High levels of potassium, magnesium and boron, and low levels of phosphorus, can also lower storability.

Respiration of the product: Respiration also involves the production of water and the release of heat (energy). This heat results in higher temperatures around the commodity if it is not removed by refrigeration or ventilation. Because carbon dioxide is also a product of respiration, this gas can build up around the product if ventilation is inadequate, and conversely oxygen can be depleted, resulting in fermentation. Reduced oxygen and elevated carbon dioxide in the storage environment can be used purposefully however, to increase the storage life of certain fruits and vegetables. As a general rule, the rate of deterioration of the many different types of fruits and vegetables after harvest is associated with their respective respiration rates. The range of average respiration rates for each product type can be affected by seasonal growing conditions, variety and postharvest management, the storage life of products is broadly consistent – products with very high respiration rates such as asparagus, mushroom, parsley, peas, spinach and sweet corn deteriorate much more rapidly than those such as apple, beet, celery, garlic, grape, honeydew melon and onion, with low respiration rates. Different fruit and vegetable types have a wide range of respiration rates, from very low to very high.

Ethylene production: It is a gas, released from a liquid chemical formulation, ethephon (2-chloroethane phosphonic acid), that can exert its effects at very low concentrations, from parts per billion (ppb) to parts per million (ppm). Ethylene is naturally occurring in plant tissues, and is a vital part of normal ripening for many fruits. However, increased ethylene production can occur as a result of disease and decay, exposure to chilling temperatures, and wounding. In addition, ethylene is produced by internal combustion engines, smoke and other sources of pollution. Exposure of fruits and vegetables to ethylene can stimulate respiration. The two types of fruit have different respiration patterns during maturation and ripening is the non-climacteric type which does not produce appreciable ethylene, respiration rates gradually decrease over time, while in climacteric fruit a rush in respiration is associated with the onset of ethylene production. Climacteric fruit such as apples and pears have high ethylene production and high sensitivity, while other products (e.g. broccoli, cabbage, carrots and strawberries) can have low rates of ethylene production but are highly sensitive to ethylene. Most non-climacteric fruit, such as cherry, grape, berries and pepper, have low ethylene production and low sensitivity to ethylene. Exposure of ethylene to unripe climacteric fruit can cause earlier than desirable ripening, resulting in overly soft and mealy fruit. More commonly in the marketplace, ethylene exposure can result in undesirable yellowing of green vegetables such as cucumbers, parsley and broccoli, and many other negative effects. Exposure of vegetables and non-climacteric fruits also increases their respiration rates, meaning that the carbohydrate reserves are used more rapidly, as well as increasing water loss. Typically, exposure occurs from mixing ethylene producing and ethylene sensitive commodities in the same storage room. The exposure to ethylene is often exhibited as loss of flowers and leaves from the plant as ethylene increases abscission.

Table 86. Ethylene production and sensitivity of several commodities

Commodity	Ethylene production	Ethylene sensitivity
Climacteric fruit		
Apple, kiwifruit, pear, cherimoya	high	High (0.03 - 0.1 ppm)
Avocado, cantaloupe melon, passion fruit	high	Medium (> 0.4 ppm)
Apricot, banana, mango	medium	High (0.03 - 0.1 ppm)
Nectarine, papaya, peach, plum, tomato	medium	Medium (> 0.4 ppm)
Vegetables and non-climacteric fruit		
Broccoli, brussels sprouts, cabbage, carrot cauliflower, cucumber, lettuce, persimmon potato, spinach, strawberry	low	High (0.01- 0.02 ppm)
Asparagus, bean, celery, citrus, eggplant	low	Medium (0.04 -0.2 ppm)
Artichoke, berries, cherry, grape, pineapple, pepper	low	Low (> 0.2 ppm)

Table 87. Summary of the detrimental effects of ethylene related to quality

Ethylene effect	Symptom or affected organ	Commodity
Physiological disorders	Chilling injury	Persimmon, avocado
	Russet spotting	Lettuce
	Superficial scald	Pear, apple
	Internal browning	Pear, peach
Abscission	Bunch	Cherry tomato
	Stalk	Muskmelon
	Calyx	Persimmon
Bitterness	Isocoumarin	Carrot, lettuce
Toughness	Lignification	Asparagus
Off-flavours	Volatiles	Banana
Sprouting	Tubercle, Bulb	Potato, onion
Colour	Yellowing	Broccoli, parsley, cucumber
	Stem browning	Sweet cherry
Discolouration	Mesocarp	Avocado
Softening	Firmness	Avocado, mango, apple, strawberry, kiwifruit, melon

Storage temperature: The most fundamental postharvest tool available to the fruit and vegetable grower is temperature control. It is critical to decrease the temperature of fruits and vegetables as quickly as possible after harvest to slow down their metabolism. Benefits of lower temperatures include reduced respiration rates, water loss, sensitivity to ethylene, and decreased susceptibility of commodities to decay. In general, the lower the storage temperature, the longer the storage life of a given commodity. However, the lowest safe storage temperature is not the same for all commodities as many of them are sensitive to low temperatures and develop chilling injuries such as irregular ripening, failure to ripen, pits on the skin surface and increased susceptibility to decay. Therefore the optimum temperature for storage is higher for chilling sensitive than for non-chilling products. In non-chilling sensitive products, the longest storage life is associated with the lowest storage temperature. However, for chilling sensitive products, the storage life increases with decreasing storage temperature to reach a maximum from 7° C to 15° C depending on the product. At lower temperatures, the storage life decreases because of susceptibility to chilling injuries.

Table 88. Fruits and vegetables classified according to sensitivity to chilling injury

Sensitivity to chilling	Fruits and vegetables
Non-chilling sensitive commodities	Apples, apricots, artichokes, asparagus, beans (lima), beets, blackberries, blueberries, broccoli, brussels sprouts, cabbage, carrots, cauliflower, celery, sweet corn, cherries, currants, endive, garlic, grapes, lettuce, mushrooms, nectarines, onions, parsley, parsnips, peaches, pears, peas, plums, prunes, radishes, raspberries, spinach, strawberries, turnips
Chilling sensitive commodities	Bean (snap), cranberry, cucumbers, eggplant, muskmelons, peppers, potatoes, pumpkins, squash, sweet potatoes, tomatoes, watermelons, yams

Changes during storage: Several changes such as physical and biochemical occurs during storage.

Table 89. Physical and biochemical changes during storage

Parameters	Different changes in storage
A. Physical	
Weight loss	Increased. Loss of moisture due to transpiration caused weight loss.
Shriveling	Increased. Even 5 per cent loss of weight result shriveling.
Firmness	Firmness of fruit is due to calcium pectate. Break down of these substances reduced firmness and caused softness in fruits/vegetables.
Colour	Development in mango, banana, papaya, limes and lemon, etc. Discolouration in litchi, grape and strawberry.
Flavour	In the beginning of storage flavour developed in some fruits such as peach, pineapple, pear and guava and long storage results acidic flavour in many fruits.
B. Biochemical	
Hormones	Level of ethylene increased in climacteric fruits such as mango, pear, apple, banana, etc. Excess of ethylene stimulate softening and pathological disorders in many fruits and vegetables. Abscicic acid (ABA) increased in pear fruit and stimulates ripening of the fruit during storage which cut down the storage life.
Respiration	Increased level of ethylene during storage increased the rate of respiration which leads to rapid break down of glucose and other organic compounds.

Parameters	Different changes in storage
Pigments	Pigments such as chlorophyll and anthocyanin decreased and carotenoids increased in many fruits and vegetables.
Sugar	In the begging of storage increased (slightly break down) then decreased. In table variety of potato increased.
Pectic substances	An increase in pectinic acid and decrease in proto pectin.
Amino acid	Arginin and proline increased e.g. grape but total amino acid content decreased.
Enzyme	An increase in level of amylase, pectin esterase, malicenzyme, catalase, peroxidase and polyphenol oxidase.
Lipids	Palmitic and palmitolitic increased and linoleic decreased in avocado fruits.
Volatile compounds	Ethylene, esters; aldehyde and ketone increased and cause harmful effects.
Phenolic compound	Tanin, flavones and total phenols decreased.
Vitamins	Ascorbic acid (vit. C) decreased and β-carotene increased.

Factors affecting storage life

Storage life of fresh horticultural produce is affected by many factors like, pre harvest factors, maturity at harvest, harvesting and handling practices, pre-storage treatments, temperature and humidity in storage room and overall hygiene of the produce. Temperature and relative humidity are the most important among the above factors. Fresh horticultural produce continue to respire after harvest and temperature is able to regulate this physiological and biochemical activity. Higher the temperature, faster the respiration rate and leading to early senescence. Storage life of horticultural produce may be extended by temperature control, chemical treatments, atmosphere modification, mainly by regulating the physiological processes and controlling the post harvest diseases and pests. However, low temperature storage is the only known economical method for long term storage and quality maintenance of horticultural produce. All other methods will only useful in supplementing the low temperature storage.

Management of temperature and relative humidity in storage

Temperature management is the most effective tool for maintaining quality and safety and for extending the postharvest life of fresh horticultural commodities. It begins with the rapid removal of field heat by initial cooling and continues throughout the cold chain (refrigerated transportation, cold storage at wholesale distribution centers, refrigerated retail display, and cold storage at home). Management of relative humidity along with temperature is essential in reducing water loss during storage.

Temperature: Most perishable horticultural commodities last longest at temperatures near 0° C. At temperatures above the optimum, the rate of deterioration increases 2 to 3 fold for every 10° C rise in the temperature. Temperature influences how other internal and external factors influence the commodity, and has a dramatic effect on the spore germination and growth rate of pathogens. Temperatures outside the optimal range can cause rapid deterioration due to the following disorders:

» **Freezing:** In general, perishable commodities are high in water content (75 to 95%), and possess large, highly vacuolated cells. The freezing point of their tissues is relatively high (ranging from –3° C to –0.5° C), and the disruption caused by freezing usually results in immediate collapse of the tissues and total loss of cellular integrity during storage.

» **Chilling injury:** Some commodities (mainly those native to the tropics and subtropics) respond unfavorably to storage at low temperatures well above their freezing points, but below a critical temperature (between 5° C and 13° C depending on commodity and maturity stage) termed the chilling threshold temperature or lowest safe temperature. Chilling injury is manifested in a variety of symptoms including surface and internal discoloration, pitting, water soaking, failure to ripen, uneven ripening, development of off flavours and heightened susceptibility to pathogen attack.

» **Heat injury:** High temperatures are also very injurious to perishable products during storage. Organs removed from the plant, however, lack the protective effects of transpiration, and direct sources of heat, for example full sunlight, can rapidly heat tissues to above the thermal death point of their cells, leading to localized bleaching or necrosis (sunburn or sunscald) or general collapse.

Relative humidity: Water loss is directly proportional to the vapor pressure difference between the commodity and its environment. Vapor pressure difference is inversely related to RH of the air surrounding the stored commodity. RH can influence water loss, decay development, incidence and severity of some physiological disorders, and uniformity of fruit ripening. Condensation of moisture on the commodity (sweating) over long periods of time is probably more important in enhancing decay than is the RH of ambient air. An appropriate RH range for storage of fruits is 85 to 95 per cent while that for most vegetables varies from 90 to 98 per cent. The optimal RH range for dry onions and pumpkins is 70 to 75 per cent.

Some root vegetables, such as carrot, parsnip, and radish, can best be held at 95 to 100 per cent RH. Relative humidity can be controlled by one or more procedures like, adding moisture (water mist or spray, steam) to air by humidifiers, regulating air movement and ventilation in relation to the produce load in the cold storage room, maintaining temperature of the refrigeration coils within about 1° C of the air temperature, providing moisture barriers that insulate walls of storage rooms and transit vehicles, adding polyethylene liners in containers and using perforated polymeric films for packaging, wetting floors in storage rooms, adding crushed ice in shipping containers or in retail displays for commodities that are not injured by the practice, and sprinkling produce with sanitized, clean water during retail marketing and storage of commodities that benefit from misting, such as leafy vegetables, cool-season root vegetables, and immature fruit vegetables (snap beans, peas, sweet corn, and summer squash).

Principles of Storage of Horticultural Fresh Produce

Controlling of respiration: Respiration is a breakdown process hence storage method should provide a means to minimize this metabolic process. Cold storage, atmospheric modification, low pressure storage are the methods used based on this principle. The heat generated during respiration, usually know as respiratory heat, accumulates in the centre of the storage. The rate of respiration of stored produce increases if this heat is not removed from the storage room. So, proper ventilation will help in removing this heat thereby reducing the respiration rate. Reducing respiration rate will also help in delaying the ripening process in some fruits and vegetables thereby extending the storage life.

Controling of transpiration: Fresh produce carry on losing water even after harvest resulting in wilting or shriveling of produce. Five per cent loss of moisture is sufficient to make the produce shrivel making it unattractive for marketing. Relative humidity and temperature are the important factors that influence the loss of moisture from fresh produce. Water loss will also be high with increase in storage temperature. Fresh produce transpire more at high temperatures and low humidity. Hence, this process can be controlled by storing the produce at low temperatures and high relative humidity.

Controling of sprouting and rooting: Consumers do not prefer the sprouted or rooted vegetables for buying. Sprouting also makes the produce to lose moisture quickly, shrivel and become prone to microbial infection. Hence, prolonging the dormant period by creating unfavorable conditions is the principle for extending the storage life of this type of produce. Favourable conditions like warm temperature and high humid condition in the storage room enhance the growth of micro-organisms and increase the spoilage. Hence, storage methods should aim at retarding or control of the growth of these spoilage causing micro-organisms.

Methods of Fruits and Vegetable Storage

Traditional methods

The marketable life of fresh horticultural produce can be extended by prompt storage in an environment that maintains product quality. The desired environment can be obtained in facilities where temperature, air circulation, relative humidity, and sometimes atmosphere composition can be controlled. At earlier to achive desire environmental condition for long term storage of fruits and vegetables following methods were used.

In situ/ on site/ natural or field storage: In situ means holdup the harvesting until the crop is required and is employed for the root, tuber and rhizomes crops during winter season. Various roots (carrots, turnips, parsnips, sweet potato and cassava) tubers (potato) and rhizome crops (ginger) should be left in the soil until preparation for the market. After the ground begins to freeze in the late fall, cover the root crops with a foot or more of mulch-straw, hay, or dry leaves. Produce can be difficult to dig out of the frozen ground, but it will not be adversely affected until the temperature around the roots drops to -4.0° C or less. Carrots are damaged at about -4.0° C, but parsnips can stand somewhat lower temperatures. The land where crop is grown remains occupied and new crop cannot be planted there. This is similar to how citrus and some other fruits are left on the tree. In case of cassava, delayed harvest results in reduced acceptability and starch content and pre harvest losses. The crops should be protected from pest and disease attack, chilling and freezing injuries. If rodents are a problem, it may be wise to store produce in a buried container or an indoor storage area. Parsnips, horseradish, and turnips actually improve in flavour by light freezing. At temperatures between -2.5° C and 1.1° C, the starch changes to sugar. Other crops, such as beets, cabbage, Chinese cabbage, cauliflower, celery, endive, lettuce, kale, leeks, and onions can withstand the early light frosts and can be stored for several weeks under heavy mulch.

Sand and coir storage: In India, potatoes are traditionally stored longer periods of time, which involves covering the commodity underground with sand.

Bulk storage of dried bulb crops: Onions, garlic and dried produce are best suited to low humidity in storage. Onions and garlic will sprout if stored at intermediate temperatures. Pungent types of onions have high soluble solids and will store longer than mild or sweet onions, which are rarely stored for more than one month. For bulk storage of onions or garlic, ventilation systems should be designed to provide air into the store from the bottom of the room at the rate of 2 cubic feet /minute /cubic feet of produce.

Table 90. Storage condition for onion, garlic, dried fruit and vegetables

Commodity	Temperature (°C)	RH (%)	Potential storage duration
Onions	0-5	65-70	6-8 months
	28-30	65-70	1 month
Garlic	0	70	6-7 months
	28-30	70	1 month
Dried fruits and vegetables	<10	55-60	6-12 months

Clamp storage of root and tuber crops: This has been a traditional method for storing potatoes in Great Britain and India. Potatoes for processing are best kept at intermediate temperatures to limit the production of sugars which darken when heated during processing. Potatoes intended for consumption must also be stored in the dark, if kept in the light the tubers will produce chlorophyll and develop the toxic alkaloid solanine. Potatoes stored for use as seed are best stored in diffused light. The chlorophyll and solanine that accumulate will aid to protect the seed potatoes from insect pests and decay organisms.Tropical root and tuber crops must be stored at temperatures that will protect the crops from chilling injury. The width of the clamp is about 1.0–2.5 meter, at the side of the field. The dimensions are marked out and the potatoes piled on the ground in an elongated conical heap. Sometimes straw is laid on the soil before the potatoes. The central height of the heap depends on its angle of repose, which is about one third the width of the clump. At the top, straw is bent over the ridge so that rain will tend to run off the structure. Straw thickness should be from 15 to 25 cm when compressed. After 2 weeks, the clamp is covered with soil to a depth of 15–20 cm, but this may vary depending on the climate.

Table 91. Storage condition for root and tuber crops

Commodity	Temperature (°C)	RH (%)	Potential storage duration
Potatoes (Fresh market)	4-7	95-98	10 months
Seed potatoes	0-2	95-98	10 months
Cassava	5-8	80-90	2-4 weeks
	0-5	85-95	6 months
Sweet potato	12-14	85-90	6 months
Ginger	12-14	65-75	6 months

Pits or trenches: Pits are dug at the edges of the field where the crop has been grown. Usually pits are placed at the highest point in the field, especially in regions of high rainfall. The pit or trench is lined with straw or other organic material and filled with the crop being stored, then covered with a layer of organic material followed by a layer of soil. Holes are created with straw at the top to allow for air ventilation, as lack of ventilation may cause problems with rotting of the crop.

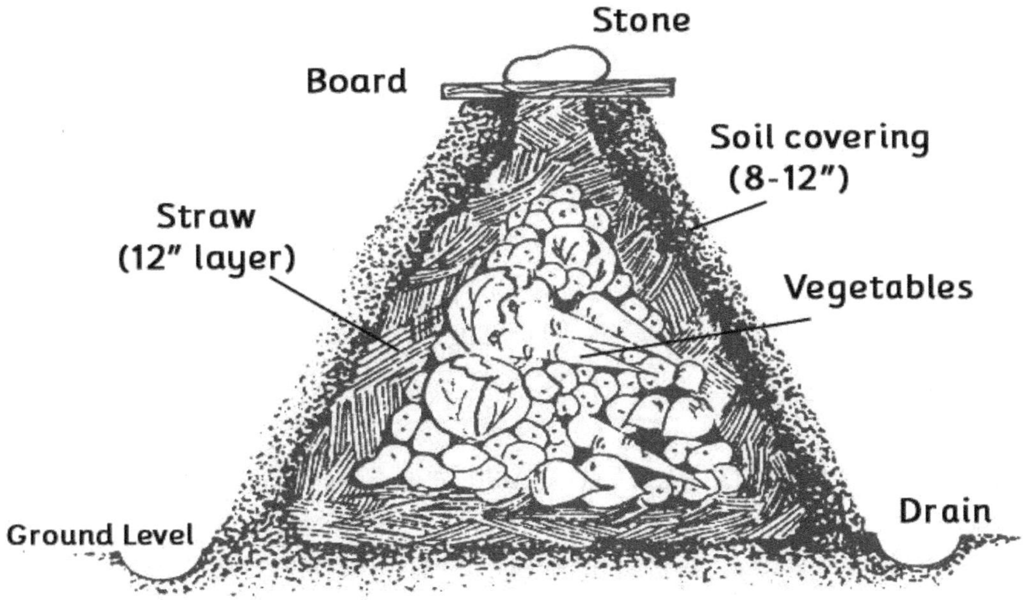

Zero energy cool chambers (ZECC): Zero energy cool chambers are based on principal of evaporative cooling and developed for short term storage of fruits and vegetables. The degree of cooling depends on the original humidity of the air and the efficiency of evaporating surface. The raw materials used for construction of cool chambers are brick (400 brick) and river bed sands (10 bags) to hold about a quintal of fruits or vegetables. The floor of the storage space is made with a single layer of bricks over which a doubled wall rectangular structure is erected with approximately 7.5 cm space between the bricks which are filled with sand. The outer dimensions of the chamber should be about 165 x 115 x 67.5 cm. The top of storage space is covered with gunny cloth in a bamboo frame structure. The chamber should be constructed under a shed with a lot of aeration and should be closer to water source. After construction, the whole structure is made wet by sprinkling water once once in the morning and once at evening maintain the temperature (with ±2° C to ±5° C internal temperature then ambient) and humidity (90%) through out the year. Direct contact of water with fruits and vegetables should be avoided. Fruits and vegetables should be placed in crates or in suitable baskets and then in the chamber. Storage life of different commodities can be increased by 2 to 3 times as compared to ambient conditions especially during summer.

Table 92. Storage life of different commodities in zero energy cool chambers

Vegetables	Shelf life (days)		Fruits	Shelf life (days)	
	Ambient condition	ZECC		Ambient condition	ZECC
Leafy veg.	1	2	Ber	5	9
Okra	3	5	Karonda	7	12
Tomato	2	3	Ripe banana	3	5
Brinjal	4	6	Mango	4	6
Bottlegaurd	4	7	Citrus	6	10
Chilli	3	5	Guava	3	5
Watermealon	7	15	Jamun	1	3
Roundgourd	4	6	Lasoda	6	11

Windbreaks: Windbreaks are constructed by driving wooden stakes into the ground in two parallel rows about 1 meter apart. A wooden platform is built between the stakes about 30 cm from the ground, often made from wooden boxes. Chicken wire is affixed between the stakes and across both ends of the windbreak. This method is used in Britain to store onions.

Cellars: These underground or partly underground rooms are often beneath a house. This location has good insulation, providing cooling in warm ambient conditions and protection from excessively low temperatures in cold climates. Cellars have traditionally been used at domestic scale in India to store apples, cabbages, onions, and potatoes during winter.

Night ventilation: In hot climates, the variation between day and night temperatures can be used to keep stores cool. The storage room should be well insulated when the crop is placed inside. A fan is built into the store room, which is switched on when the outside temperature at night becomes lower than the temperature within. The fan switches off when the temperatures equalize. The fan is controlled by a differential thermostat, which constantly compares the outside air temperature with the internal storage temperature. This method is used to store bulk onions and garlic.

Natural ventilation: Amongst the wide range of storage systems, natural ventilation is the simplest. It takes advantage of the natural airflow around the product to remove the heat and humidity generated by respiration. Structures that provide some form of protection from the external environment and gaps for ventilation can be used. Produce is placed in bulk, bags, boxes, bins, pallets, etc. Although natural storage is widely practiced, it leaves products exposed to pests and diseases as well as to adverse weather conditions that can have a detrimental effect on quality. This system can only be used with crops that store well under natural conditions such as potatoes, onions, sweet potatoes, garlic and pumpkins.

Improved storage methods

At present time to achive desire environmental condition for long term storage use different more scientific methods like cold storage, controlled atmosphere storage, modified atmosphere storage, solar drive cold storage, hypobaric storage and jacketed storages etc.

Cold storage (Refrigeration, Chilling and Freezing): Low temperature storage is the best known, effective and most widely used method for maintain quality, freshness, reduce losses and extend the storage life of perishable commodities like fruits, vegetables and flowers. Cold storage also helps in preserve the seasonal produce and selling during off season to fetch higher returns and reduce losses due to wastage. The cold storage facilities now available are mostly for a particular commodity like potato, orange, apple, grapes, pomegranates, flowers, etc. which results in poor capacity utilization.

Table 93. Number of cold storages and its capacity in states (2018)

S. No	Name of the states	Number	Capacity (In 000MT)	S. No	Name of the states	Number	Capacity (In 000MT)
1	Andaman & Nicobar	3	0.810	18	Madhya Pradesh	302	1281.41
2	Andhra Pradesh & Telangana	452	1836.36	19	Maharashtra	603	979.60
3	Arunachal Pradesh	2	6.00	20	Manipur	3	7.10
4	Assam	37	163.26	21	Meghalaya	4	8.20
5	Bihar	306	1415.60	22	Mizoram	3	3.97
6	Chandigarh	7	12.46	23	Nagaland	4	7.35
7	Chhattisgarh	98	484.33	24	Orissa	177	566.32

8	Delhi	97	129.86	25	Pondicherry	3	0.085
9	Goa	29	7.70	26	Punjab	672	2201.39
10	Gujarat	890	3516.00	27	Rajasthan	167	561.30
11	Haryana	352	791.78	28	Sikkim	2	2.100
12	Himachal Pradesh	65	125.97	29	Tamil Nadu	173	347.58
13	Jammu & Kashmir	55	182.53	30	Tripura	14	45.48
14	Jharkhand	58	236.70	31	Uttar Pradesh	2368	14500.77
15	Karnataka	209	602.46	32	Uttrakhand	47	162.82
16	Kerala	199	81.70	33	West Bengal	511	5940.51
17	Lakshadweep	1	0.015	34	NHB/State	3	20.20
					Total	**7916**	**36229.68**

India has around 7916 cold storage facilities, with a capacity of 36.22 million tones. However, some 75-80 per cent of these refrigerated warehouses are suitable only to store potatoes, a commodity that produces only 20 per cent of agricultural revenue. It is reported that only 10-11 per cent of the fruits and vegetables produced in India use cold storage. Storage capacity needs to be increased by 40 per cent to avoid wastage. There is more wastage of fruits and vegetables in the southern and western regions of India due to the tropical and humid climate. India needs expansion of cold storage infrastructure in an affordable, reliable and sustainable way to increase the contribution of agriculture to the economy.

Factors for Effective Cold Storage of the Produce

Product quality: Fresh horticultural produce intended for storage should be free from physical damage, of optimum maturity and free from infections.

Temperature: Low temperature storage is recommended for perishables as it retards respiration and metabolic activity, aging, softening and colour changes, moisture loss, spoilage due to diseases and undesirable growth (sprouting). Maintenance of uniform temperature constantly, continuously and also adoption of optimum low temperature for each specific produce are very essential.

Relative humidity: The relative humidity of the air in storage rooms directly affects the keeping quality of the produce held in them. If it is too low, wilting or shrivelling is likely to occur, if it is too high, it may favour the development of decay. For most of the fresh fruits and vegetables (except onion, garlic) the relative humidity in cold storage should be kept in the range of 85 to 95 per cent.

Air circulation and package spacing: This is required to remove respiratory heat and additional CO_2 and ethylene from storage. Air must be circulated to keep a cold storage room at an even temperature throughout the storage. Entry of outside air and proper spacing of containers on pellets are also important.

Respiration rates, heat evolution and refrigeration: When the storage of fresh produce is alive and carry on all activities of living tissues, the most important being respiration. During this process, energy is released in the form of heat which varies with the commodity and the temperature. This vital heat expressed in BTU (British thermal units) is of paramount importance in calculating the refrigeration load of the commodity.

Weight loss in storage: Loss of water from harvested horticultural crops is a major cause of deterioration in storage. Some loss can be tolerated but losses great enough to cause wilting or shrivelling must be avoided. Under good handling conditions with recommended humidity and temperature, moisture loss can be held under control.

Sanitation and air purification: Good air circulation alone is of considerable value in minimizing surface moulds. Accumulation of odours and volatiles may contribute to off flavours and hasten deterioration.

Refrigeration Cycle

Principle of refrigeration: The refrigeration cycle begins with the refrigerant in the evaporator. At this stage the refrigerant in the evaporator is in liquid form and is used to absorb heat from the product. When leaving the evaporator, the refrigerant has absorbed a quantity of heat from the product and is a low pressure and low temperature vapour. This vapour is then drawn from the evaporator by the compressor. When vapour is compressed it rises in temperature. Therefore, the compressor transforms the vapour from a low temperature vapour to a high temperature vapour, in turn increasing the pressure. This high temperature, high pressure vapour is pumped from the compressor to the condenser; where it is cooled by the surrounding air or in some cases by fan assistance. The vapour within the condenser is cooled only to the point where it becomes a liquid once more. The heat, which has been absorbed, is then conducted to the outside air. At this stage the liquid refrigerant is passed through the expansion valve. The expansion valve reduces the pressure of the liquid refrigerant and therefore reduces the temperature. The cycle is complete when the refrigerant flows into the evaporator, from the expansion valve, as a low-pressure, low-temperature liquid. For determination of refrigeration load, the following factors should be quantified

- » Field heat (Temperature of produce at harvest)
- » Heat of respiration of the produce
- » Conductive heat gain – building floor, wall, roof ceiling etc.
- » Convective heat gain – air mixing during opening of door
- » Equipment load – fans, lights, forklifts and personnel etc.
- » Service and defrost factors of the facility – hot weathers

Components of refrigeration system: There are five basic components of a refrigeration system, these are: Evaporator, Compressor, Condenser, Expansion valve, Refrigerant to conduct the heat from the product.

Evaporator: The purpose of the evaporator is to remove unwanted heat from the product, via the liquid refrigerant. The liquid refrigerant contained within the evaporator is boiling at a low-pressure. To enable the transfer of heat, the temperature of the liquid refrigerant must be lower than the temperature of the product being cooled. Once transferred, the liquid refrigerant is drawn from the evaporator by the compressor via the suction line. When leaving the evaporator coil the liquid refrigerant is in vapour form.

Compressor: The purpose of the compressor is to draw the low-temperature, low-pressure vapour from the evaporator via the suction line. Once drawn, the vapour is compressed. When vapour is compressed it rises in temperature. Therefore, the compressor transforms the vapour from a low-temperature vapour to a high-temperature vapour, in turn increasing the pressure. The vapour is then released from the compressor in to the discharge line.

Condenser: The purpose of the condenser is to extract heat from the refrigerant to the outside air. The condenser is usually installed on the reinforced roof of the building, which enables the transfer of heat. Fans mounted above the condenser unit are used to draw air through the condenser coils. The temperature of the high-pressure vapour determines the temperature at which the condensation begins. As heat has to flow from the condenser to the air, the condensation temperature must be higher than that of the air; usually between

-12° C and -1° C. The high-pressure vapour within the condenser is then cooled to the point where it becomes a liquid refrigerant once more, even as retaining some heat. The liquid refrigerant then flows from the condenser in to the refrigerant line.

Expansion valve: Within the refrigeration system, the expansion valve is located at the end of the refrigerant line, before the evaporator. The high-pressure refrigerant reaches the expansion valve, having come from the condenser. The valve then reduces the pressure of the refrigerant as it passes through the orifice, which is located inside the valve. On reducing the pressure, the temperature of the refrigerant also decreases to a level below the surrounding air. This low-pressure, low-temperature refrigerant is then pumped in to the evaporator.

Low temperature (chilling) injuries

Majority of tropical and sub tropical horticultural produce are injured when stored at very low temperatures due to chilling injury. Chilling injury occurs when commodities of tropical and subtropical origin, such as mango, banana and tomato are held at temperatures above their freezing point and below 5 to 15° C depending on the commodities. However, the critical temperature at which chilling injury occurs varies among commodities. Chilling injury is completely different to freezing injury (which results when ice crystals form in plant tissues at temperatures below their freezing point). Both susceptibility and symptoms of chilling injury are product and even cultivar specific. Moreover, the same commodity grown in different areas may behave differently in response to similar temperature conditions.

Symptoms of chilling injury

Skin pitting: It is a common chilling injury symptom that is due to collapse of cells beneath the surface. The pits are often discoloured and high rates of water loss from damaged areas may occur, which accentuates the extent of pitting.

Browning or blackening of flesh tissues: Chilling-induced browning in fruit typically appears first around the vascular (transport) strands e.g. avocado. Browning can result from the action of the polyphenoloxidase enzyme on phenolic compounds released from the vacuole during chilling, but this mechanism has not been proven in all cases.

Rotting: Chilling injury causes the release of metabolities (e.g. amino acids, sugars) and mineral salts from cells. Leakage of metabolites and ions, together with degradation of cell membranes, provides substrates for growth of pathogenic organisms, especially fungi. Such pathogens are often present as latent infections or may contaminate produce during harvesting and postharvest operation.

Other symptoms: Some other symptom like water soaking in leafy vegetables and papaya, slow de-greening of citrus fruit, development of off-flavour or odour (low O_2 levels), immature picked fruits may fail to ripen or ripen unevenly or slowly after chilling injury (e.g. tomato). Some physiological disorders like black heart in pineapple, woolliness in peach and flesh browning in plum are also develop. Symptoms of chilling injury normally occur while the produce is at low temperature. However, they sometimes chilling injury appear when the produce is removed to a higher temperature and deterioration may then be quite rapid, often within a matter of hours.

Table 94. Lowest safe storage temperature and symptoms of chilling injury in produce

Produce	Lowest safe storage temperature (°C)	Symptoms
Apple	2-3	Soft scald, brown core
Avocado	5-12	Pitting, browning of pulp and vascular strands
Banana	12-13	Brown streaking on skin
Ber	7.5-8.5	Surface pitting, appearance of brown streaks on the peel
Guavas	4-5	Pulp injury, decay
Lemon	10-12	Pitting of flavedo, membrane staining, red blotches
Lime	7-8	Pitting
Mango	12-13	Dull skin, brown areas
Melon	7-10	Pitting, surface rots, browning, water loss, decay
Papaya	7-15	Pitting, water-soaked areas
Pineapple	6-15	Brown or black flesh
Asparagus	4-5	Gray tips and loss of sheen
Beans	5-10	Opaque discolouration and rusty spots
Casaba	7-10	Pitting, surface decay, failure to ripen
Cucumber	7-8	Dark-coloured, water-soaked areas, water loss, decay
Eggplant	7-8	Surface scald, pitting and brown spotting
Okra	7-8	Discoloration, water-soaked areas, pitting, decay
Onion	15-18	Water soaked scales
Pepper	7-8	Sheet pitting, alternaria rot on pods and calyxes, darkening of seed

Potatoes	3-4	Sweetening
Squashes	12-13	Pitting, water loss and decay
Sweet Potatoes	12-13	Decay, pitting, internal discoloration; hard core when cooked
Tomato ripe	10-12	Water soaking and softening decay
Tomato mature green	12-13	Poor colour when ripe, Alternaria rot
Pumpkins	15-18	Pitting, water loss and decay, especially alternaria rot
Watermelons	4-5	Pitting, objectionable flavour

Management of chilling injury

Maintaining critical temperature: The safest way to manage chilling injury is to determine the critical temperature for its development in a particular produce and then not expose the commodity to temperatures below that critical temperature.

Modified atmosphere storage: In modified atmospheres or controlled atmospheres, gases are removed or added to create an atmospheric composition around the commodity that is different from that of air (78.08% N_2, 20.95% O_2, and 0.03% CO_2). Usually this involves reduction of O_2, and/or elevations of CO_2 concentrations.

Maintaining high RH: High relative humidity both in storage at low temperature and after storage can minimize expression of chilling injury symptoms, particularly pitting (e.g. film-wrapped cucumbers).

Mechanism of chilling injury

The critical temperature, below which chilling injury occurs is an integrated genotypic but expressed in phenotypic characteristic of the particular organ. The cellular events of chilling injury can be separated into primary and secondary events. Primary events are transiently reversible, but become irreversible, particularly with the onset of cell death and tissue necrosis.

Primary events in chilling injury: Low temperature-induce changes in the properties of cell membranes due to changes in the physical state of membrane lipids (membrane phase change). Production of reactive oxygen species (eg. hydrozen peroxide) that oxidize leading to altered enzymatic activities and structural proteins (e.g. tubulin) are disrupted.

Secondary events in chilling injury: The physical changes in membrane lipids alter the properties of their parent membranes.

» Ion and metabolites moves across affected membranes and activities of membrane bound enzymes are disrupted.

» The overall consequence of membrane disturbance is breakdown of sub-cellular compartmentation, which is readily measured as increased ion leakage from tissues.

» Changes in the relative activities of enzymes lead to imbalanced metabolism and can ultimately result in cell death.

» Accumulation of toxic compounds (e.g. acetaldehyde).

» Structural proteins of the cell cytoskeleton (eg. tubulin) dissociate in chilling-sensitive tissues at low temperatures.

Effect of cold storage on subsequent behaviour of horticultural produce

At refrigerated temperatures, aging and decay are retarded, resulting in longer life. As the potential life is used up in storage, the stored produce cannot stay for longer period after removal as freshly harvested produce. In some cases in post storage period, the produce has to be ripened properly. Removal of refrigerated stored produce to higher temperature should be done by a gradual warming to avoid sweating resulting in loss of quality. Not all the produce can be stored together because of difference in their temperature requirements. But at times, it may be necessary to store different produce together provided the optimum low temperatures do not differ much. Cross transfers of odours, ethylene, and strongly scented produce should be avoided in mixed storage. Based on the compatibility of produce, without any deleterious effect it can be stored.

Table 95. Compatibility groups of fruits and vegetables

Group	Temperature	Crops	Status of commodities
Group 1	0-2° C and 90-95% RH	Apple, apricot, asian pear, grapes, litchis, plum, prunes, pomegranate, mushroom, turnip, peach	Produce ethylene
Group 2	0-2° C and 95-100% RH	Asparagus, leafy greens, broccoli, peas, spinach, cabbage, carrot, cauliflower, cherries	Sensitive to ethylene
Group 3	0-2° C and 65-70% RH	Garlic, onions dry	Moisture will damage these crops

Group 4	4-6° C and 90-95% RH	Cantaloupes, guava, mandarin, tangerines	-
Group 5	8-10° C and 85-90% RH	Beans, potatoes (with cipc treatment), cucumber, brinjal, okra, pepper. potato, brinjal, beans, okra, peppers, olive	Sensitive to chilling and ethylene
Group 6	13-15° C and 85-90% RH	Mangoes, banana, tomato ripe, grapefruit, coconuts, guava, limes, papaya, pineapple, tomato, pumpkin, ginger	Sensitive to chilling injury
Group 7	18-21° C and 85-90% RH	Pears, tomato, watermelon, yams and sweet potato	-

The following factors determine compatibility in mixing various fruits and vegetables together during transport and storage:

Temperature: Fruits and vegetables are generally divided into two groups:

» Non-chilling sensitive commodities (such as apple, broccoli, grape, lettuce, and pear) that are best kept at temperatures above their freezing points (-2.0 to -0.5° C) and up to 2° C.

» Chilling sensitive commodities (such as avocado, banana, citrus fruits, mango, melons, and tomato) that are best kept at 5° C to 15° C, depending on the commodity, cultivar, maturity ripeness stage, and storage duration.

Relative humidity (RH): With the exception of a few commodities (such as dry garlic and onion bulbs, pumpkins, winter squash, nuts, dried fruits and vegetables), fresh produce should be kept at 90-95 per cent relative humidity to minimize water loss. If the dried fruits, nuts, and vegetables are packaged in moisture proof containers, they can be mixed with other produce kept at 90- 95 per cent RH. Conditions that result in water condensation on the surface of produce should be avoided to reduce decay incidence.

Ethylene: Many ripening fruits (such as apple, pear, peach, banana, and tomato) produce ethylene gas, which can be detrimental to ethylene sensitive commodities (such as avocado, broccoli, cabbage, carrot, kiwifruit, lettuce, persimmon and watermelon). Symptoms of ethylene induced disorders include yellowing of broccoli and cucumber, russet spotting on lettuce, softening of fruits, bitterness of carrots, tissue maceration in watermelons and calyx abscission of eggplant. Exposures of ethylene sensitive commodities to ethylene are cumulative and must be avoided throughout the postharvest handling system. Continuous scrubbing of ethylene to below 1 ppm from the circulating air in the storage facility can facilitate mixing ethylene producing and ethylene sensitive commodities at the distribution centers and retail levels.

Odour volatiles: Commodities with strong odours (such as garlic, leek, onion, pepper, and potato) should not be mixed with commodities that can absorb these odours (such as apple, avocado, citrus fruits, grape and pear).

Sulfur dioxide: Some table grapes are shipped with SO_2 generating pads to control decay caused by *Botrytis cinerea*. These grapes should be stored alone because SO_2 can damage most other fruits and vegetables.

Smart fresh (1- methylcyclopropene): It is suitable to mix commodities that had been treated with the ethylene action inhibitor, SmartFresh (1-MCP) with untreated commodities since 1-MCP does not migrate from treated to untreated produce.

Organic produce: Organically grown produce should be handled and stored separately from conventionally grown produce to avoid any potential contamination by pesticide residues due to direct contact. Handlers of organic produce are required to keep a record of cleaning dates and products used for cleaning the storage room in which the organic produce is kept.

Cold chain

The sequence of farm activities which starts from the field and ends on the consumer's table, involving precooling, transport, storage, distribution and storage in home under low temperature condition (refrigeration) until consumed. The harvested produce has to be graded either for export or local trade in nearby packing houses, packed in containers to be precooled to the storage temperature and then transported in refrigerated trucks to the cold storage for long term storage or to the wholesale market in reefer containers. The wholesale market should have the facility for the cold storage. Once the produce is precooled after harvest, it should not be exposed to undesirable temperatures at any stage of storage and handling to maintain its harvest fresh quality till consumption.

Table 96. Optimum cold storage conditions and approximate storage life of fruit and vegetables

Fruits/ Vegetables	Temperature (°C)	RH (%)	Approximate storage life (weeks)
Fruits			
Apple	0-2	85-90	20-30
Apricot	-0.5 to 0	90-95	1-3

Artichoke	0-1	95-100	2-3
Chilling tolerant varieties avocado Chilling sensitive varieties avocado	4.4-5.0 12.5-13.0	85-90 85-90	4-5 2-3
Cavendish green banana Cavnedish ripe banana Ney Poovan green banana Ney poovan ripe banana	12-13 12-13 12-13 8-10	85-90 85-90 85-90 85-90	3-4 1-5 2-3 1-2
Barbados cherry	0-1	85-90	7-8
Ber	5-6	85-90	4-5
Black berry	-0.5 to 0	90-95	2-3 days
Blue berries	-0.5 to 0	90-95	2-3
Cherries Sweet	-1 to 0.5	90-95	2-3
Cherries Sour	0-1	90-95	1-2
Coorg mandarin (main crop) Coorg mandarin (rainy season) Sathgudi orange (Moosambi)	8-9 8-9 8-9	85-90 85-90 85-90	8-9 6-7 15-16
Custard apple	15-16	85-90	1.5-2
Coconut	0-1.5	80-85	4-8
Date	6-7	85-90	2-3
Fig	1-2	85-90	6-7
Grapes	-0.5 to 0	90-95	2-8
Grapefruit	10-15	85-90	6-8
Guava	10-12	85-90	2-5
Jackfruit	11-12	85-90	6-7
Jerusalem artichoke	-0.5 to 0	90-95	17-21
Kiwi fruit	-0.5 to 0	90-95	13-21
Lemon	10-13	85-90	4-26
Lime	9-10	85-90	6-8
Litchi	2-3	85-90	8-10
Mandarin	4-7	90-95	2-4
Mango	12-13	90-95	2-3

Alphonso, (Mango mature green) Banganapalli,(Mango mature green)	12-13 12-13	85-90 85-90	4-5 5-6
Melon (Others)	7-10	90-95	2-3
Nectarine	-0.5 to 0	90-95	2-4
Olives, fresh	5-10	85-90	4-6
Orange	0-9	85-90	8-12
Papaya green	9-10	85-90	3-4
Papaya turning	9-10	85-90	2-3
Passion fruit	6-7	85-90	3-4
Peach	-0.5 to 0	90-95	2-4
Pear	-1.5 to 0.5	90-95	8-30
Persimmon	-1 to -0.5	90-92	13-17
Pineapple (green)	9-10	85-90	4-6
Pineapple (25% yellow)	6-7	85-90	1-2
Plum	-0.5 to 0	90-95	2-5
Pomegranate	7-8	85-90	10-12
Quince	-0.5 to 0	90-95	8-13
Raspberries	-0.5 to 0	90-95	2 days
Sapota mature	20-22	85-90	1-2
Strawberry	0-1	85-90	1-2
Water melon	10-15	90-95	2-3
Vegetables			
Asparagus	0-2	95-98	3-4
Snap beans Winged beans	8-10 10-12	85-90 85-90	3-4 8-10
Bean (dry)	4-10	40-50	26-43
Beetroot	0-1	90-95	8-10
Brinjal	10-11	90-95	1-2
Broad beans	0-2	90-98	1-2
Broccoli	0-1	95-100	3-4
Cabbage(wet season)	0-2	90-95	4-6

Cabbage(dry season)	0-2	90-95	12-13
Cactus leaves	2-4	90-95	2-3
Cantaloupe (half slip)	2-5	95-98	2-3
Cantaloupe (full slip)	0-2	95-98	1-2
Capsicum(green)	7-8	85-90	3-5
Carrot topped	0-2	90-95	20-24
Carrot (bunched)	0-1	95-100	1-2
Cassava	0.5-1.0	85-86	4-8
Cauliflower	0-2	90-95	7-8
Celery	0-2	90-95	7-8
Chicory	0-1	95-100	2-3
Chinese cabbage	0-1	95-100	8-13
Coriander leaves	0-2	90-95	4-5
Chow chow	12-13	90-95	2-3
Cucumber	10-11	90-95	1-2
Garlic(bulbs) dry	0-1	65-70	28-36
Ginger	8-10	75-80	16-20
Gourd, bottle	8-9	85-90	4-6
Gourd, snake	18-20	85-90	1-2
Green onions	0-1	95-100	3-4
Horse radish	-1 to 0	98-100	43-51
Kohlrabi	0-1	98-100	8-13
Leek	0-1	95-100	8-13
Lettuce, leaf	0-1	95-98	0.5-1
Lima bean	3-5	95-98	0.5-1
Mushroom	0-1	95-98	1.0-1.5
Muskmelon, Honey dew	7-8	85-90	4-5
Okra	10-11	90-95	1.0-1.5
Onion, Red	0-1	65-70	20-24
Onion, white	0-1	65-70	16-20

Pea, green	0-1	90-95	2-3
Parsley	0-1	95-100	4-8
Parsnip	0-1	95-100	17-26
Pepper (bell)	7-13	90-95	2-3
Potato (early)	7-16	90-95	1-2
Potato (late)	4.5-13	90-95	21-42
Pumpkin	12-15	70-75	24-36
Radish, topped	0-1	90-95	3-5
Squash	12-15	70-75	8-24
Sweet Potato	10-12	80-90	13-20
Sweet corn	0-1.5	95-98	0.5-1
Spinach	0-1	90-95	10-14
Summer squash	5-10	95-98	1-2
Taro	7-10	85-90	17-21
Mature green tomato Red ripe tomato	12-13 5-6	85-90 85-90	4-5 1-2
Turnip	0-1	90-95	8-16
Watermelon	12-15	80-90	1-2
Yam	16-20	60-70	8-30

Cooling Concept

As a form of energy, heat always seeks equilibrium. The rapid removal of at least 7/8 of the field heat from the crop by a compatible cooling method. The time required to remove 7/8 of the field heat is known as the 7/8 cooling time. Removal of 7/8 of the field heat during cooling is strongly recommended to provide adequate shipping life for shipment to distant market, also 7/8 of the heat can be removed in a relatively short amount of time. Removal of the remaining 1/8 of the field heat will occur during subsequent refrigerated storage and handling with little detriment to the product. The rate of heat transfer, or the cooling rate, is critical for efficient removal of field heat in order to achieve cooling. In the case of cooling, the field heat from the product is transferred to the cooling medium. The efficiency of cooling is dependent on time, temperature and contact. In order to achieve maximum cooling, the product must remain in the precooler for sufficient time to remove heat. The cooling medium (air, water, crushed ice) must be maintained at constant temperature throughout the cooling period. The cooling also must have continuous, intimate contact with the surface of the individual vegetables.

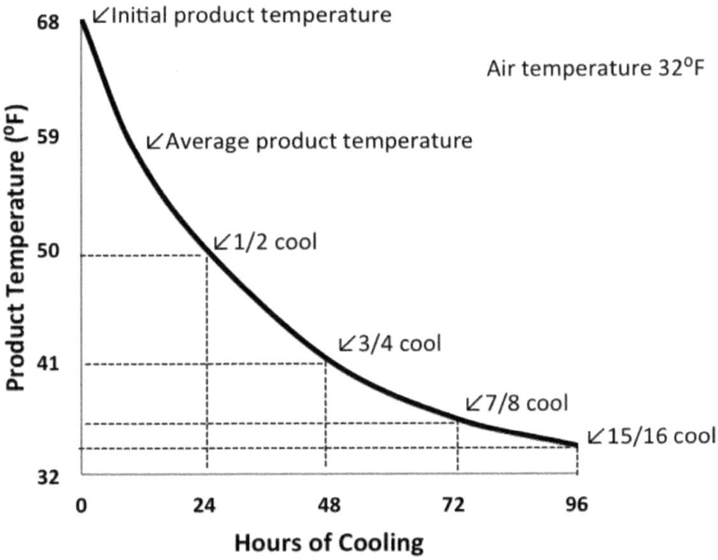

Figure 40. A cooling curve for a perishable product in room cooling

Different methods of cooling

Cooling methods will differ according to product type and scale of operation. The the majority of growers and storage operators is used room cooling method. Forced-air cooling, icing, hydro-cooling and vacuum cooling are used extensively in major fruit and vegetable growing regions. The cooling rate is not only dependent upon time, temperature and contact with the commodity, it is also dependent upon the cooling method being employed. The various cooling media used to cool produce have different capacities to remove heat. The typical cooling curve for a fruit or vegetable in a cold storage room illustrates the concept of "half-cooling" or "seven-eighths-cooling" times. Half-cooling refers to the time taken to reduce the initial product temperature from that when first placed in the cold room halfway to the set temperature in the room. For example, if the product temperature was 20° C when placed in the room and air temperature in the room was set at 0 or 10° C, half cooling would refer to the time that it took for the product temperature to reach 10 and 15° C, respectively. It will then take the same time period to reduce the product temperature by half again, and so on. So, seven-eighths cooling is three times as long as half-cooling.

Cold storage management

Temperature management is one of the most difficult aspects of small farm operations. In contrast to large apple storages, refrigeration units are often in short supply for many growers, and it is unusual to find the ideal situation of more than one cold storage unit. Therefore, it is common for fruits and vegetables to be stored at incorrect temperatures storage at a lower than recommended temperature for each product can result in chilling injury, while storage at a higher than optimum storage temperature will compromise its storage life. Ethylene sensitive and high ethylene producers are included at the lowest temperature of 0-2° C. While less than ideal, the reduced ethylene production and action at these low temperatures make it less likely that injury will occur within a short storage period. Ethylene should be kept below 1ppm, but depending on the volume of each product can be unrealistic to achieve, e.g. in the presence of apples, and therefore the 7 day safe period may be reduced. Regardless, longer storage periods are potentially damaging and fruits should be stored separately from vegetables.

The selection of the cooling unit is very important when a cooler is designed. If the temperature difference between the air and the cooling unit is too large, then the condensers will accumulate ice from the moisture in the air. This drying of air can dehydrate many fruits and vegetables. The atmosphere typically should be humid enough to prevent water loss. Ideally then a cooler should be able to maintain a RH of 90 to 95 per cent at 0° C, but these types of coolers are more expensive and less common than for dry goods. Mechanical refrigeration is the preferred method for obtaining precise temperature control for fruits and vegetables, a new technology known as CoolBot™ has been developed that may be useful for smaller farmer operations. It is a technology that allows the lower temperature limit of an air conditioner to be modified so that rooms without mechanical refrigeration units can be converted into cold rooms. The controller operates so as to force the air conditioner unit to produce low air temperatures without building up ice on the evaporator fins. Two temperature sensors and a small heater are used. The heater is fastened to the air conditioning unit's temperature sensor, and thus 'tricks' the air conditioner into continuing to cool below its normal set point. The CoolBot's first sensor measures the temperature of the cool-room, and once the desired room temperature is achieved, turns the heater off, so that the air conditioner sensor cools and its compressor turns off. The controller's secondary sensor is attached to the evaporator fins, and is adjusted to sense the presence of ice. When ice starts to build up, the controller turns the heater off, so that the air conditioner compressor turns off until the ice is melted. Since the controller prevents the evaporator from falling below the selected frost point, moisture condensed on the evaporator fins is returned to the cold room air, resulting in high RH which will reduce water loss in stored perishables. The air conditioning unit must be sized appropriately for the dimensions of the cooler, and the room must be well insulated and without air gaps.

Minimizing water loss during storage

Most fresh fruits, vegetables and flowers are composed of 90-95 per cent water, which is vital for normal biochemical processes, is responsible for many textural qualities (i.e., firmness and crispness), and represents most of the saleable weight of these commodities. After harvesting cannot replace lost water. Thus, reducing water loss after harvest is important for maintaining quality and the saleable weight of the product. Water escapes from fresh vegetables through evaporation into the surrounding air. Commodities with large surface area (e.g. leaves) are much more prone to water loss than compact, round commodities (e.g. potato). In addition, commodity natural barriers to water loss, such as a waxy cuticle, abscission layer, or even pubescent hair, help prevent water loss. Also, cut or abrasions that break these natural barriers will increase water loss. The lower RH of the air, the faster water is lost from fresh produce. Therefore, in most instances, maintaining RH between 85 to 95 per cent is recommended. Another important fact to consider is that warmer air holds much more water vapor than colder air. Thus, air at 0° C with 100 per cent RH will be at only about 50 per cent RH after warming to 10° C. Water loss can be minimizing fluctuations in commodity and air temperatures during handling and shipping operations.

» Handle fresh produce at early morning for storage and quickly cool commodities and maintain.

» Cure/dry certain root, bulb, citrus fruits and tuber vegetables to allow natural tissues that restrict water loss to develop at the side of wounds.

» Use waxes and other surface coating to further restrict water loss during storage.

» Add moisture to the air through the use of humidifiers and maintain RH at the highest recommended level within causing commodity or container deterioration.

» Design cooling system so that the evaporator coils run within 0° C of the air temperature.

Importance of cold storages

The importance of cold storage for fruits and vegetables has to achieve following objectives.

Seasonal production: The demand for fresh fruits and vegetables continuous almost throughout the year and their supplies are only seasonal. For e.g. apples grown in Jammu and Kashmir and Himachal Pradesh find markets all over the county. Therefore, cold storages have more importance in storing and supplying the seasonal fruits and vegetables.

Spoilage: A crucial problem faced in the case of fruits and vegetables is that of the huge enormous losses that occur on account of spoilage. The loss of spoilage on different stages on marketing estimated between 20 per cent to 30 per cent of the total production. There is an urgent need to minimize the present spoilage loss through cold storages.

Losses in transit: Most of the fruits and vegetables are of extremely perishable nature and cannot be kept long under ordinary conditions of storage. Refrigerated trucks and air-cooled wagons which will help in minimizing losses is transport.

Better distribution and fair prices: The prices paid by the consumer to the producer is very low ranging from 40 to 60 per cent one of the principal reasons for this is prevalence of various malpractices in the fruits and vegetables in terminal markets like Mumbai, Chennai, Delhi, Calcutta etc. These markets receive large quantities of fresh fruits from outside the state. The producer never knows at what price his product was in the nearest cold stores and gets fair prices with better distribution.

Stabilizing market prices: Besides the role of stabilizing market prices and also distributing both on demand and time basis. The farmers get opportunity of producing cash crops to get fair prices at the same time, the consumers get the supply of perishable fruits and vegetables with minimum prices i.e. cold storages stabilizing market prices of fruits and vegetables.

Maintain quality: Cold storages will help to maintain the quality of product for some period of time. It gives a chance to protect the quality of perishable goods for long time whenever it needed to store.

Wastage reduction: The cold storage system will greatly help in reducing the wastage of perishable goods. The country lacks an efficient supply chain for the distribution of the fruits and vegetables. Cold storage helps in increasing the utilization life and processing helps in preservation of fruits and vegetables which can be consumed during off season.

Controlled Atmosphere Storage (CA Storage)

The storage of fruits and vegetables in CA storage is one of the most advanced methods of storage. It was first suggested by W. R. Philips in Canada. Controlled atmosphere storage involves maintaining an atmospheric composition that is different from air composition (about 78% N_2, 21% O_2, and 0.03% CO_2) generally, O_2 below 8 per cent and CO_2 above 1 per cent are used. Atmosphere modification should be considered as a supplement to maintenance of optimum ranges of temperature and RH for each commodity in preserving quality and safety of fresh fruits, ornamentals, vegetables, and their products throughout postharvest handling.

Physiological basis of CA storage: Air contains about 20.9 per cent O_2, 78.1 per cent N_2, 0.003 per cent CO_2 and trace amount of other gases including Ne, He, CH_4 and water vapour. In CA storage, oxygen is reduced and CO_2 is increased, which slow down the ripening process and reduce respiration rates.

Biological basis of CA effects: Exposure of fresh horticultural crops to low O_2 and/or elevated CO_2 atmospheres within the range tolerated by each commodity reduces their respiration and ethylene production rates however, outside this range respiration and ethylene production rates can be stimulated, indicating a stress response. This stress can contribute to incidence of physiological disorders and increased susceptibility to decay. Elevated CO_2 induced stresses are additive to and sometimes synergistic with stresses caused by low O_2, physical or chemical injuries, and exposure to temperatures, RH, and/or C_2H_4 concentrations outside the optimum range for the commodity.

The shift from aerobic to anaerobic respiration depends on fruit maturity and ripeness stage (gas diffusion characteristics), temperature, and duration of exposure to stress inducing concentrations of O_2 and/or CO_2. Up to a point, fruits and vegetables are able to recover from the detrimental effects of low O_2 and high CO_2 stresses (fermentative metabolism) and resume normal respiratory metabolism upon transfer to air. Plant tissues have the capacity for recovery from the stresses caused by brief exposure to fungistatic atmospheres (>10% CO_2) or insecticidal atmospheres (<1% O_2 and/or 40 to 80% CO_2). Post climacteric fruits are less tolerant and have lower capacity for recovery following exposure to reduced O_2 or elevated CO_2 levels than pre climacteric fruits. The speed and extent of recovery depend on duration and levels of stresses and underlying, metabolically driven cellular repair.

Elevated CO_2 atmospheres inhibit activity of ACC synthase (key regulatory site of ethylene biosynthesis), while ACC oxidase activity is stimulated at low CO_2 and inhibited at high CO_2 concentrations and/or low O_2 levels. Ethylene action is inhibited by elevated CO_2 atmospheres. Optimum atmospheric compositions retard chlorophyll loss (green colour), biosynthesis of carotenoids (yellow and orange colours) and anthocyanins (red and blue colours), and biosynthesis and oxidation of phenolic compounds (brown colour). Controlled atmospheres slow down the activity of cell wall degrading enzymes involved in softening and enzymes involved in lignification, leading to toughening of vegetables. Low O_2 and/or high CO_2 atmospheres influence flavour by dropping loss of acidity, starch to sugar conversion, sugar inter conversions, and biosynthesis of flavour volatiles. When produce is kept in an optimum atmosphere, retention of ascorbic acid and other vitamins are retained, resulting in better nutritional quality. Severe stress CA conditions decrease cytoplasmic pH and ATP levels and reduce pyruvate dehydrogenase activity, while pyruvate decarboxylase, alcohol dehydrogenase, and lactate dehydrogenase are induced or activated. This causes accumulation of acetaldehyde, ethanol, ethyl acetate, and/or lactate, which may be detrimental to the commodities if they are exposed to stress CA conditions beyond their tolerance. Specific responses to CA depend on cultivar, maturity and ripeness stage, storage temperature and duration, and in some cases ethylene concentrations. N_2 is an inert component of CA storage, it replacing with argon or helium may increase diffusivity of O_2, CO_2, and C_2H_4, but they have no direct effect on plant tissues and are more expensive than N_2 as a CA component.

Super-atmospheric levels of O_2 up to about 80 per cent may accelerate ethylene induced degreening of nonclimacteric commodities, ripening of climacteric fruits, increasing respiration and ethylene production rates, and incidence of some physiological disorders (such as scald on apples and russet spotting on lettuce). At levels above 80 per cent O_2, some commodities and postharvest pathogens suffer from O_2 toxicity. Use of super-atmospheric O_2 levels in CA will likely be limited to situations in which they reduce the negative effects of fungistatic, elevated CO_2 atmospheres on commodities that are sensitive to CO_2 induced injury.

Essential features of CA storage

» Mechanical refrigeration is used to maintain temperature of -1 to 3° C.

» The CA storage room is constructed gas tight.

» Nitrogen gas is introduced into the storage by cylinder to reduce the oxygen level after room is filled and sealed.

» Excess CO_2 is removed by dry hydrated lime, ethanolamine, aluminium calcium silicate, activated carbon, magnesium oxide, activated carbon are other CO_2 scrubbers.

» Atmospheric composition is crop specific. However, the most common gas combinations are 2-5 per cent oxygen and 3-10 per cent carbon dioxide. The storage room atmosphere samples are taken daily for CO_2 and O_2 monitoring.

Beneficial effects of CA (Optimum composition for the commodity):

» Reduction of produce sensitivity to ethylene action at O_2 levels below 8 per cent and/ or CO_2 levels above 1 per cent.

» Alleviate certain physiological disorders such as chilling injury of various commodities, russet spotting in lettuce, and some storage disorders including, scald of apples.

» Retardation of senescence (including ripening) and associated biochemical and physiological changes, particularly slowing down rates of respiration, ethylene production, softening, and compositional changes.

» CA can have a direct or indirect effect on postharvest pathogens (bacteria and fungi) and consequently decay incidence and severity. For example, CO_2 at 10 to 15 per cent significantly inhibits development of botrytis rot on strawberries, cherries, and other perishables.

» Low O_2 (<1%) and/or elevated CO_2 (40 to 60%) can be a useful tool for insect control in some fresh and dried fruits, flowers, vegetables and in dried nuts.

Limitations of CA storage

» Causes certain physiological disorders such as black heart in potatoes, brown stain of lettuce, brown heart in apples and pears and chilling injury of tropical commodities.

» Irregular ripening of fruits, such as banana, mango, pear and tomato, can result from exposure to O_2 levels below 2 per cent or CO_2 levels above 5 per cent for more than 2 to 4 weeks.

» Development of off flavours and off odours at very low O_2 concentrations (as a result of anaerobic respiration) and very high CO_2 levels (as a result of fermentative metabolism).

» Timely non availability of gas, costly and technical knowhow is required.

» Increased susceptibility to decay when the fruit is physiologically injured by too low O_2 or too high CO_2 concentrations.

Commercial application of CA storage

Several refinements in CA storage have been made in recent years to improve quality maintenance. These include creating nitrogen by separation from compressed air using molecular sieve beds or membrane systems, low O_2 (1.0 to 1.5%) storage, low-ethylene (<1 μL L^{-1}) CA storage, rapid CA (rapid establishment of optimal levels of O_2 and CO_2), and programmed (or sequential) CA storage for example, storage in 1 per cent O_2 for 2 to 6 weeks followed by storage in 2 to 3 per cent O_2 for the remainder of the storage period. Other developments, which may expand use of atmospheric modification during transport and distribution, include improved technologies for establishing, monitoring, and maintaining CA using edible coatings or polymeric films with appropriate gas permeability to create a desired atmospheric composition around and within the commodity. Modified atmosphere packaging (MAP) is widely used in marketing fresh-cut produce.

Applications of CA to cut flowers are very limited because decay caused by *Botrytis cinerea* is often a limiting factor to postharvest life, and fungistatic CO_2 levels damage flower petals and/or associated stem and leaves. Also, it is less expensive to treat flowers with anti-ethylene chemicals than to use CA to minimize ethylene action. Commercial use of CA storage is greatest on apples and pears worldwide, less on cabbages, sweet onions, kiwifruits, avocados, persimmons, pomegranates, nuts, dried fruits and vegetables. Atmospheric modification during long-distance transport is used with apples, asparagus, avocados, bananas, broccoli, cane berries, cherries, figs, kiwifruits, mangos, melons, nectarines, peaches, pears, plums, and strawberries. Continued technological developments in the future to provide CA during transport and storage at a reasonable cost (positive benefit/cost ratio) are essential to greater applications on fresh horticultural commodities and their products

Table 97. Classification of horticultural crops according to their CA storage potential at optimum temperatures and RH.

Storage duration (months)	Commodities
>12	Almond, Brazil nut, cashew, filbert, macadamia, pecan, pistachio, walnut, dried fruits and vegetables
6 to 12	Some cultivars of apples and European pears
3 to 6	Cabbage, Chinese cabbage, kiwifruit, persimmon, pomegranate, some cultivars of Asian pears
1 to 3	Avocado, banana, cherry, grapes, mango, olive, onion (sweet cultivars), some cultivars of nectarine, peach, plum and mature-green tomato
<1	Asparagus, broccoli, cane berries, fig, lettuce, muskmelons, papaya, pineapple, strawberry, sweet corn, fresh-cut fruits and vegetables, some cut flowers

Table 98. Recommended conditions for storage of some fruits and vegetables under CA storage

Commodity	Temperature Range (°C)	CO_2 Range (%)	O_2 Range (%)
Apple	0-5	1-2	2-3
Banana	12-15	2-5	2-5
Strawberry	0-5	15-20	10-11
Kiwi fruit	0-5	4-5	2-3
Nuts and dry fruit	0-25	0-100	0-1
Tomato (Mature green)	12-20	0	3-5
Tomato (Partially ripe)	8-12	0	3-5
Lettuce	0-5	0	2-5

Modified Atmosphere Storage (MAS)

Modified Atmosphere (MA) storage refers to a change in the atmosphere around the product, typically a reduction of oxygen levels from 21 per cent in ambient air, and an increase in carbon dioxide from 0.03 per cent in ambient air. MA can be developed passively by product respiration or by active means where desired gas composition is injected into a bag, often in a package (MA packaging, MAP). Subsequently, the atmosphere in the bags is then a function of factors such as product type and temperature that affect respiration rates, permeability of the plastic film to oxygen and carbon dioxide diffusion, and the ratio of product to the bag. MA storage implies a lower degree of control of gas concentration in atmosphere surrounding the commodity. The MA and CA differ only in degree of control, CA is more exact. Advances in the manufacture of polymeric films with wide range of gas permeability have stimulated interest in creating and maintaining modified atmospheres within flexible film packages.

Biochemical and physiological basis of MA: The rate of respiration and metabolism doubles for every 10° C rise in temperature. Respiration can be therefore reduced by decreasing the temperature, O_2 level and/or increasing the CO_2 level in the storage atmosphere. Both O_2 and CO_2 levels exert independent effects on respiration. The net effect may be additive or synergistic. When O_2 concentration is reduced below 10 per cent, respiration rate is decreased. However, when O_2 concentration falls below 2 per cent, anaerobic respiration may set in, thereby leading to the accumulation of ethanol and acetaldehyde. The desirable effect of MA on plant tissues is also attributed to lower pH, due to dissolution of CO_2 in tissues. Ethylene action and biosynthesis are also effected besides water loss and chilling injury.

Table 99. Modified storage atmosphere condition and storage temperature

Commodity	Temperature range °C	Modified Atmosphere	
		Per cent O_2	Per cent CO_2
Apple	0-5	1-2	0-3
Apricot	0-5	2-3	2-3
Avocado	5-13	2-5	3-10
Banana	12-16	2-5	2-5
Blackberry	0-5	5-10	15-20
Blueberry	0-5	2-5	12-20
Cherimoya and atemoya	8-15	3-5	5-10
Cherry (sweet)	0-5	3-10	10-15
Chilli	5-12	3-5	0-5
Cranberry	2-5	1-2	0-5
Custard apple	12-20	3-5	5-10
Durian	12-20	3-5	5-15
Fig	0-5	5-10	15-20
Grape	0-5	2-5 (5-10)	1-3 (10-15)
Grapefruit	10-15	3-10	5-10
Kiwifruit	0-5	1-2	3-5
Lemon and lime	10-15	5-10	0-10
Litchi	5-12	3-5	3-5
Mango	10-15	3-7	5-8
Nectarine	0-5	1-2	3-5
Nuts and dried fruits	32-50	0-1	0-100
Olive	5-10	2-3	0-1
Orange	5-10	5-10	0-5
Papaya	5-15	2-5	5-8
Peach (Clingstone)	0-5	1-2	3-5
Peach, freestone	0-5	1-2 (4-6)	3-5 (15-17)
Pear, Asian	0-5	2-4	0-3
Pear (European)	0-5	1-3	0-3

Persimmon	0-5	3-5	5-8
Pineapple	8-13	2-5	5-10
Plum	0-5	1-2	0-5
Pomegranate	5-10	3-5	5-10
Rambutan	8-15	3-5	7-12
Raspberry	0-5	5-10	15-20
Strawberry	0-5	5-10	15-20
Asparagus	0-5	Air	5-10
Bean green	5-10	2-3	4-7
Broccoli	0-5	1-2	5-10
Brussels sprouts	0-5	1-2	5-7
Cabbage	0-5	3-5	5-7
Cantaloupes	2-7	3-5	10-20
Cauliflower	0-5	2-5	2-5
Cucumber	8-12	3-5	0
Leek	0-5	1-2	3-5
Lettuce	0-5	2-5	0
Mushrooms	0-5	3-21	5-15
Okra	8-12	3-5	0
Onion (green)	0-5	1-2	10-20
Onion (bulb)	0-5	1-2	0-10
Peas	0-5	2-3	2-3
Pepper (bell)	8-12	3-5	0-5
Potato	4-12	None	None
Radish (topped)	0-5	1-2	2-3
Spinach	0-5	Air	10-20
Sweet corn	0-5	2-4	10-20
Tomato (green)	12-20	3-5	3-5
Tomato (partially ripe)	8-12	3-5	0
Tomato (ripe)	10-15	3-5	3-5

Environmental factors affecting MA storage

Temperature and relative humidity: Ambient temperatures of the surrounding atmosphere affect the commodity temperature. Temperature changes also affect the permeability of the film, which increases with increase in temperature. CO_2 permeability responds more than O_2 permeability. Relative humidity has little effect on permeability of most film packages. Most common films are good barriers to moisture and vapour because they maintain high internal humidity even in dry, ambient conditions.

Light: Green vegetables consume large amount of CO_2 and reduce O_2 through photosynthesis and would antagonize the process of respiration which aids in maintenance of specified MA within the package. Greening of potatoes can cause loss in quality unless light is excluded. Hence, opaque packages should be used for such commodities.

Sanitation factors: The high humidity maintained within MA packages may enhance the growth of plant pathogens. So care must be taken to ensure proper sanitation and to avoid conditions favourable to growth and reproduction of such micro organisms. Fungicidal treatment of packaged vegetables is thus very important.

Table 100. Differences between CA and MA storage

CA Storage	MA Storage
High degree of control over gas concentration	Low degree of control over gas concentration
Longer storage life	Less storage life
More expensive technology	Less expensive technology
Atmosphere is modified by adding gas	It is created by either actively (addition or removal of gas) or passively (produce generated)
Specific temperature should maintain	May or may not be maintained

Table 101. Maximum storage life (days) in normal atmosphere storage (NA), controlled atmosphere (CA) and low-pressure storage (LP)

Commodity	Maximum storage time (days)		
	Normal atmosphere storage	Controlled atmosphere	Low-pressure storage
Apple	200	300	300
Asparagus	14-21	Slight benefit-off odours	28-42
Avocado (Lula)	30	42-60	102
Banana	14-21	42-56	150
Carnation (flower)	21-42	No benefit	140
Cherry (sweet)	14-21	28-35	56-70
Cucumber	9-14	14+ (slight benefit)	49
Green pepper	14-21	No benefit	50
Lime (Persian)	14-28	Juice loss, peel thickens	90
Mango (Haden)	14-21	No benefit	42
Mushroom	5	6	21
Papaya (Solo)	12	12+ slight benefit	28
Pear (Bartlett)	60	100	200
Protea (flower)	<7	No benefit	30
Rose (flower)	7-14	No benefit	42
Spinach	10-14	Slight benefit	50
Strawberry	7	7+ off flavour	21
Tomato (mature-green)	7-21	42	84

Limitation of Modified Atmosphere (MA) and Controlled Atmosphere (CA) Storage

Reducing oxygen concentrations around fruits and vegetables slows down respiration rates until a point known as the anaerobic compensation point is reached, when respiration rates dramatically increase. This respiration is associated with fermentation and injurious accumulations of acetaldehyde and ethanol that result in damage. Similar damage due to fermentation can occur if the carbon dioxide concentration is too high. Apples are the major commodity stored in CA storage worldwide. The reason for limited use of CA to relatively few products, despite its benefits on quality of many others, is that it requires significant capital investment. Structures must be air tight and refrigerated, with precise temperature control and equipment to modify the atmospheres. The volume of these storage rooms is also necessarily large to maximize the value of the equipment. Therefore, the return on investment requires long lived horticultural products that are stored for months and not days or even weeks. In addition, extension of storage life is often too small to warrant investment. If the storage potential of a particular product is a week, for example, then adding on a few days extra, is often not commercially significant. However, newer tent-based CA storage systems, with lower cost oxygen and carbon dioxide control monitors, are now available, and may be more useful for shorter lived products. The tents and controllers are placed in cold storage rooms and have been used successfully for local market strawberries for example.

Other atmosphere based technologies such as hypobaric storage, in which commodities are stored at low atmospheric pressures provide potential for non-chemical storage, but practical cost-affordable systems are not available. Dynamic CA (DCA), where the oxygen around commodities is lowered to levels close to the anaerobic compensation point where fermentation occurs is commercially available. The metabolism of the fruit is monitored by measuring fluorescence signals or ethanol concentrations in the fruit, and the oxygen concentrations adjusted to avoid damage hence the term dynamic.

Table 102. Products which are incompatible in long-term storage

Products			Effects
Apple or Pears	with	Celery Cabbage Carrots Potato Onion	Ethylene from apples and pears damages or causes off flavours in vegetables
			Potatoes cause earthy flavour in fruit.
			Potatoes are injured by cold temperatures
			High humidity causes root growth in onions.
			Ethylene causes bitterness in carrots
Celery	with	Onions or carrots	Odour transfer occurs between products
Meat, egg and dairy product	with	Apple and citrus	Fruit flavours are taken up by the meat, eggs, and dairy products.
Leafy greens and flowers	with	Apple, pear, peaches, tomatoes and cantaloup	Ethylene produced by the fruit crops damages leafy greens and flowers.

Cucumbers peppers and green squash	with	Tomatoes Apples Pears	Ethylene from tomatoes, apples, and pears causes loss of green colour. This is aggravated by storage temperatures of 7⊠10° C which are too warm for apples and pears.

Each commodity has a safe range of oxygen and carbon dioxide based on the tolerance of each to low oxygen (Table 106) and high carbon dioxide (Table 107), and these levels can be affected by growing region and by variety.

Table 103. Fruits and vegetables classified according to their tolerance to low O_2 concentrations

Minimum O_2 concentration tolerated (%)	Commodities
0.5	Tree nuts, dried fruit and vegetables
1	Some cultivars of apples and pears, broccoli, mushrooms, garlic, onion, minimally processed fruits and vegetables
2	Most cultivars of apples and pears, kiwifruit, apricot, cherry, nectarine, peach, plum, strawberry, papaya, pineapple, olive, cantaloupe, sweet corn, green bean, celery, lettuce, cabbage, cauliflower and brussels sprouts
3	Avocado, persimmon, tomato, peppers, cucumber, artichoke
5	Citrus fruits, green pea, asparagus, potato, sweet potato

Table 104. Fruits and vegetables classified according to their tolerance to elevated CO_2 concentrations

Minimum CO_2 concentration tolerated (%)	Commodities
2	Apple (Golden Delicious), Asian pear, European pear, apricot, grape, olive, tomato, sweet pepper, lettuce, endive, chinese cabbage, celery, artichoke and sweet potato
5	Apple (most cultivars), peach, nectarine, plum, orange, avocado, banana, mango, papaya, kiwifruit, cranberry, pea, chilli, eggplant, cauliflower, cabbage, brussels sprouts, radish, carrot

10	Grapefruit, lemon, lime, persimmon, pineapple, cucumber, summer squash, snap bean, okra, asparagus, broccoli, parsley, leek, green onion, dry onion, garlic, potato
15	Strawberry, raspberry, blackberry, blueberry, cherry, fig, cantaloupe, sweet corn, mushroom, spinach, kale, Swiss chard

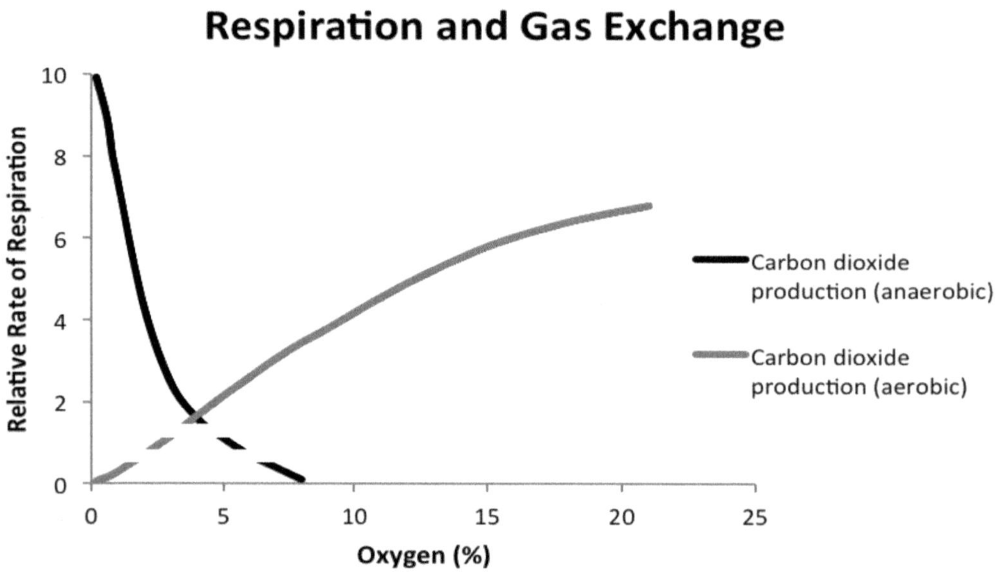

Figure 41. A schematic representation of the effects of oxygen concentration on aerobic and anaerobic respiration rates of fresh fruits and vegetables. Modified from Kader and Saltveit (2003).

Solar driven cold stores

In tropical countries, solar energy is utilized in refrigeration cycle. In Sudan, such stores have been developed having single stage ammonia/water absorption refrigerator with 13 kw peak cooling power and were designed to keep 10 tonnes of horticultural products (volume 50 m^2) at a minimum temperature of 5° C, as tested on bananas. This system is however costly when compared to conventional cold stores operated by electricity.

Jacketed storages

These are double walled storages where heat conducted through the floor, walls and ceiling is intercepted and removed by the refrigeration system before it reaches the storage space. The walls, ceiling and floor act as cooling surfaces. Humidity close to 100 per cent is maintained.

These jacketed storages built in Canada are 10 per cent more costly than conventional storages.

Low pressure storage / hypobaric storage

Fruits can be stored under low pressure of 0.2 – 0.5 atmospheric pressure and temperature of 15 - 24° C under airtight chamber. Pressure is reduced by sucking air and creating vacuum. When presser reduced from the 1 atm to 0.1atm the effective O_2 concentration reduced from 21 to 2.1 per cent. In apples, low pressure reduces level of ethylene to 0.01ppm which does not stimulate ripening. Released ethylene is removed out of storage and volatiles such as CO_2, acetaldehyde, acetic acid, ester etc. are removed /reduced. Hypobaric stores are expensive to construct because of the low internal pressures required, and this high cost of application appears to limit hypobaric storage to high value produce such as cut flowers. Secondly control of gases during the storage cannot be manipulated.

Table 105. Comparative storage life (in days) of produce stored in refrigeration and under hypobaric conditions.

Fruits	Cold storage	Hypobaric storage	Vegetables	Cold storage	Hypobaric storage
Pine apple (ripe)	9-12	40	Green pepper	16-18	50
Grapefruit (ripe)	30-40	90-120	Cucumber	10-14	41
Strawberry (ripe)	5-7	21-28	Beans	10-13	30
Sweet cherry	14	60-90	Onion (green)	2-3	15
Banana (unripe)	10-14	90-150	Lettuce	14	40-50
Avocado (unripe)	23-30	90-100	Tomato (mature green)	14-21	60-100
Apple (unripe)	60-90	300	Tomato (breaker stage)	10-12	28-42
Pear (unripe)	45-60	300			

General recommendations for storage: The University of California (Thompson *et. al.*, 1999) recommends three combinations of temperature and relative humidity

Temperature (°C)	RH (%)	Crops
0 – 2	90 – 98	Leafy vegetables, crucifers, temperate fruits and berries
7 – 10	85 - 95	Citrus, subtropical fruits and fruit vegetables
13 - 18	85 – 95	Tropical fruits, melons, pumpkins and root vegetables

Note: ethylene level should kept below 1 ppm during storage

Five different storage conditions for fruit and vegetables storage are as 0° C temperature and 90-100 per cent RH, 7-10° C temperature and 90-100 per cent RH, 13° C temperature and 85-90 per cent RH, 20° C temperature and 80-85 per cent RH and ambient conditions.

Influences of preharvest factors on storage quality

Preharvest factors greatly influence both the condition of the crop at harvest and the crops storage and nutritive potential. The postharvest quality of fresh horticultural commodities markedly depends upon the quality attained at the time of harvest. Many preharvest factors are known to affect storage quality, including, genotype and cultivar selection, stage of maturity when harvested, temperature, light intensity, rainfall amounts, soil texture and fertility, fertilizer type and application rates, disease and insect pressure, application of growth regulator and pesticide. Evaluating which preharvest conditions exert the most influence on postharvest fruit quality is difficult as they interact during the growing season and can change in degree of influence from year to year. Nonetheless, the goal in managing preharvest factors is ultimately to harvest the crop at the highest degree of quality and to sustain that quality throughout the storage period. The harvest of any crop, whether fruit, vegetable, or flower, is associated with mechanical stresses and because the plant part is separated from the parent, removal from a supply of water, nutrients, hormones and energy. When harvested, the plant part has an altered ability to respond to stresses in the environment. Stress interrupts, restricts, or accelerates normal metabolic processes in an adverse or negative manner, and therefore is usually considered as potentially injurious to any plant system. However, consideration of postharvest systems is complicated as many storage regimens beneficially utilize stress conditions such as temperature and atmosphere modification to maximize storage potential of fresh crops. Therefore, it could be concluded that, preharvest factors greatly influence both the condition of the crop at harvest and the crop's storage and nutritive potential.

Table 106. Potential risks of microbial contamination and recommended preventive measures

Production step	Risks	Prevention
Production field	Animal fecal contamination	Avoid animal access, wild, production or even pets.
Fertilizing	Pathogens in organic manure	Use inorganic fertilizers. Proper composting
Irrigation	Pathogens in water	Underground drip irrigation Check microorganisms in water

Harvest	Fecal contamination	Personal hygiene. Portable bathrooms. Risk awareness
Field transportation	Pathogens in containers and tools	Use plastic bins. Cleaning and disinfecting tools and containers
Pack house	Fecal contamination	Personal hygiene. Sanitary facilities. Avoid animal entrance. Eliminate places may harbor rodents.
	Contaminated water	Alternative methods for precooling. Use potable water. Filtration and chlorination of recirculated water.
Storage and transportation	Development of microorganisms on produce	Adequate temperature and relative humidity. Cleaning and disinfection of facilities. Avoid repackaging. Personal hygiene. Do not store or transport with other fresh products. Use new packing materials
Sale	Product contamination	Personal hygiene. Avoid animal access. Sell whole units. Cleaning and disinfection of facilities. Discard garbage daily.

Key words

Refrigeration: It is the process of removing heat from an enclosed space or commodity. Main function is to lowering the temperature and maintaining the lower temperature.

Cooling: It refers to any natural or artificial process by which heat is dissipated.

Cryogenics: Cryogenics is a process of artificially producing extremely cold temperature by using cryogenic refrigerants such as liquid nitrogen.

Cold: It is a condition in which absence of heat. To decrease the temperature, heat must be removed rather than adding cold.

Refrigeration ton/tone: It is the unit used to quantify the refrigeration load.

One tonne of refrigeration: It is defined as the energy removed from the one metric tonne (1000kg) of water to freeze within 24hr at 0°C. One tonne of refrigeration =13898kj/hr = 3.861kw. One tonne of refrigeration is about 10 per cent larger than 1 ton of refrigeration (3.517 kw).

Capacity requirement: One ton (3.5 kw) of refrigeration required to cool 18 T produce. Variation in whole cold storage should not exceed ± 1°C, whereas it should not exceed ± 0.5°C in one position 1kg of melting ice absorbs 325kj of heat.

12

Postharvest Disease and Insect Management

Losses due to postharvest disease may occur at any time during postharvest handling, from harvest to consumption. When estimating postharvest disease losses, it is important to consider reductions in fruit quantity and quality, as some diseases may not render produce unsalable yet still reduce product value. For example, blemished fruit may not be sold as fresh fruit but may still be suitable for processing, in which case, it brings a lower price. It is also important to take into account costs such as harvesting, packaging and transport when determining the value of produce lost as a result of postharvest wastage. Aside from direct economic considerations, diseased produce poses a potential health risk. A number of fungal genera such as Penicillium, Alternaria and Fusarium are known to produce mycotoxins under certain conditions. Generally speaking, the greatest risk of mycotoxin contamination occurs when diseased produce is used in the production of processed food or animal feed. In most cases, fresh produce which is obviously diseased would not be consumed. Losses due to postharvest disease are affected by a great number of factors including: commodity type, cultivar susceptibility to postharvest disease, the postharvest environment (temperature, relative humidity, atmosphere composition, etc.), produce maturity and ripeness stage, treatments used for disease control, produce handling methods, postharvest hygiene. Virtually all postharvest diseases of fruit and vegetables are caused by fungi and bacteria. In some root crops and brassicas, viral infections present before harvest can sometimes develop more rapidly after harvest. Postharvest diseases are often classified according to how infection is initiated. The so-called 'quiescent' or 'latent' infections are those where the pathogen initiates infection of the host at some point in time (usually before harvest), but then enters a period of inactivity or dormancy until the physiological status of the host tissue changes in such a way that infection can proceed. The dramatic physiological changes which occur during fruit ripening are often the trigger for reactivation of quiescent infections. Examples of postharvest diseases arising from quiescent infections include anthracnose of various tropical fruit caused by *Colletotrichum* spp. and grey mould of strawberry caused by *Botrytis cinerea*.

The other major group of postharvest diseases is those which arise from infections initiated during and after harvest. Often these infections occur through surface wounds created by mechanical or insect injury. Wounds need not be large for infection to take place and in many cases may be microscopic in size. Common postharvest diseases resulting from wound infections include blue and green mould (caused by *Penicillium* spp.) and transit rot (caused by *Rhizopus stolonifer*). Bacteria such as *Erwinia carotovora* (soft rot) are also common wound invaders. Many pathogens, such as the banana crown rot fungi, also gain entry through the injury created by severing the crop from the plant.

CAUSES OF POSTHARVEST DISEASE

Many of the fungi which cause postharvest disease belong to the phylum Ascomycota and the associated Fungi Anamorphici (Fungi Imperfecti). In the case of the Ascomycota, the asexual stage of fungus (the anamorph) is usually encountered more frequently in postharvest diseases than the sexual stage of the fungus (the teleomorph). Important genera of anamorphic postharvest pathogens include Penicillium, Aspergillus, Geotrichum, Botrytis, Fusarium, Alternaria, Colletotrichum, Dothiorella, Lasiodiplodia and Phomopsis. Some of these fungi also form ascomycete sexual stages. In the phylum Oomycota, the genera Phytophthora and Pythium are important postharvest pathogens, causing a number of diseases such as brown rot in citrus (*Phytophthora citrophthora* and *Phytophthora parasitica*) and cottony leak of cucurbits (*Pythium* spp.). Rhizopus and Mucor are important genera of postharvest pathogens in the phylum Zygomycota. *Rhizopus stolonifer* is a common wound pathogen of a very wide range of fruit and vegetables, causing a rapidly spreading watery soft rot. Genera within the phylum Basidiomycota are generally not important causal agents of postharvest disease, although fungi such as *Sclerotium rolfsii* and *Rhizoctonia solani*, which have basidiomycete sexual stages, can cause significant postharvest losses of vegetable crops such as tomato and potato. While diseases caused by these pathogens are primarily field diseases, the development of symptoms often accelerates after harvest. The chief causal agents of bacterial soft rots are various species of Erwinia, Pseudomonas, Bacillus, Lactobacillus and Xanthomonas. Bacterial soft rots are very important postharvest diseases of many vegetables, although they are generally of less importance in most fruit. This is because most fruit have a low pH which is inhibitory to the majority of bacterial plant pathogens.

Table. 107 Example of common postharvest disease and pathogens of fruits and vegetables

Disease	Pathogen	Disease	Pathogen
Pome Fruits		**Stone Fruit**	
Blue mould	*Penicillium expansum*	Rhizopus rot	*Rhizopus stolonifer*
Grey mould	*Botrytis cinerea, Botryotinia fuckeliana*	Brown rot	*Monilinia fructicola Sclerotinia fructicola*
Bitter rot	*Colletotrichum gloeosporioides Glomerella cingulata*	Grey mould	*Botrytis cinerea, Botryotinia fuckeliana*
Alternaria rot	*Alternaria spp.*	Blue mould	*Penicillium expansum*
Mucor rot	*Mucor piriformis*	Alternaria rot	*Alternaria alternata*
Black rot	*Physalospora obtusa*	**Berries**	
Grapes and small fruit		Rhizopus rot	*Rhizopus stolonifer*
Grey mould	*Botrytis cinerea, Botryotinia fuckeliana*	Grey mould	*Botrytis cinerea, Botryotinia fuckeliana*
Blue mold	*Penicillium spp.*	Blue mold	*Penicillium spp.*
Rhizopus rot	*Rhizopus stolonifer*	Cladosporium rot	*Cladosportum spp.*
Citrus Fruit			
Blue mould	*Penicillium italicum*	**Avocado**	
Green mould	*Penicillium digitatum*	Bacterial soft rot	*Erwinia carotovora*
Black centre rot	*Alternaria citri*	Anthracnose	*Colletotrichum gloeosporioides Colletotrichum acutatum Glomerella cingulata*
Stem end rot	*Phomopsis citri Diaporthe citri*		
Brown rot	*Phytophthora citrophthora Phytophthora parasitica*	Stem end rot	*Dothiorella spp. Lasiodiplodia theobromae Phomopsis perseae*
Banana			
Anthracnose	*Colletotrichum musae*	**Mango**	

Crown rot	*Fusarium spp* *Verticilliu spp.* *Colletotrichum musae*	Anthracnose	*Colletotrichum gloeosporioides* *Colletotrichum acutatum* *Glomerella cingulata*
Black end	*Colletotrichum musae* *Nigrospora sphaerica*	Stem end rot	*Dothiorella spp.* *Lasiodiplodia theobromae* *Phomopsis mangiferae* *Botryosphaeria spp.*
Ceratocystis fruit rot	*Thielaviopsis paradoxa* *Ceratocystis paradoxa*		
Papaya		Rhizopus rot	*Rhizopus stolonifer*
Anthracnose	*Colletotrichum spp.*	Black mould	*Aspergillus niger*
Phomopsis rot	*Phomopsis caricae-papayae*	Alternaria rot	*Alternaria alternata*
Black rot	*Mycosphaerella caricae* *Phoma caricae-papayae*	Grey mould	*Botrytis cinerea,* *Botryotinia fuckeliana*
Rhizopus rot	*Rhizopus stolonifer*	Blue mould	*Penicillium expansum*
Phytophthora fruit rot	*Phytophthora palmivora*	Mucor rot	*Mucor circinelloides*
Peaches and plums		**Pineapple**	
Brown rot	*Monilinia fructicola*	Water blister	*Thielaviopsis paradoxa* *Ceratocystis paradoxa*
Rhizopus rot	*Rhizopus stolonifer*	Fruitlet core rot	*Penicillium funiculosum* *Fusarium moniliforme* var. subglutinans *Gibberella fujukurot* var subglutinans
Gray mold	*Botrytis cinerea*		
Blue mold	*Penicillium spp.*		
Alternaria rot	*Alternaria spp.*		
Gilbertella rot	*Gilbertella persicaria*		
Cucurbits		Yeasty rot	*Saccharomyces spp.*
Bacterial soft rots	*Erwinia spp.* *Bacillus polymyxa* *Pseudomonas syringae* *Xanthomonas campestris*	Bacterial brown rot Tomato, Eggplant and Capsicum	*Erwinia ananas*

		Bacterial soft rots	*Erwinia spp.* *Bacillus polymyxa* *Pseudomonas syringae* *Xanthomonas campestris*
Grey mould	*Botrytis cinerea, Botryotinia fuckeliana*		
Fusarium rot	*Fusarium spp.*		
Alternaria rot	*Alternaria spp.*	Grey mould	*Botrytis cinerea, Botryotinia fuckeliana*
Charcoal rot	*Macrophomina phaseolina*		
Cottony leak	*Pythium spp.*	Fusarium rot	*Fusarium spp.*
Rhizopus rot	*Rhizopus spp*	Alternaria rot	*Alternaria spp.*
Brassicas		Cladosporium rot	*Cladosporium spp.*
Bacterial soft rot	*Erwinia spp.,* *Bacillus spp.* *Pseudomonas spp.* *Xanthomonas campestris*	Cottony leak	*Pythium spp.*
	Rhizopus rot		*Rhizopus spp*
	Watery soft rot		*Sclerotinia spp.*
Grey mould	*Botrytis cinerea, Botryotinia fuckeliana*	Sclerotium rot	*Sclerotium rolfsii* *Athelia rolfsii*
Alternaria rot	*Alternaria spp.*	**Leafy Vegetables**	
Watery soft rot Phytophthora rot	*Sclerotinia spp.* *Phytophthora porri*	Bacterial soft rot	*Erwrinia spp.,* Pseudomonas spp. Xanthomonas campestris
Onions			
Bacterial soft rot	*Erwinia spp.,* *Pseudomonas spp.* *Lactobacillus spp.*	Grey mould Watery soft rot	*Botrytis cinerea, Botryotinia fuckeliana* *Sclerotinia spp.*
Black mould rot	*Aspergillus niger*	**Carrots**	
Fusarium basal rot	*Fusarium oxysporum* f. sp. cepae	Bacterial soft rot	Erwrinia spp., Pseudomonas spp.
Smudge	*Colletotrichum circinans*	Rhizopus rot	Rhizopus spp

Potatoes		Grey mould	
Bacterial soft rot	*Erwinia spp.*		Botrytis cinerea, Botryotinia fuckeliana
Fusarium tuber rot	*Fusarium spp.*	Watery soft rot	Sclerotinia spp.
Dry rot	*Fusarium spp.* *Gibberella spp.*	Sclerotium rot	Sclerotium rolfsii Athelia rolfsii
Wet rot	*Pythium spp.*	Chalara	Chalara thielavioides
Slimy soft rot	*Clostridium spp.*	Thielaviopsis rots	Thielaviopsis basicola
Gangrene	*Phoma exigua* var exigua and var foveata	**Sweet potatoes**	
		Bacterial soft rot	*Erwinia chrysanthemi*
Black scurf	*Rhizoctonia solani* *Thanatephorus cucumeris*	Fusarium surface rot Black rot	*Fusarium oxysporum* *Ceratocystis fimbriata*
Silver scurf	*Helminthosporium solani*	Ring rot	*Pythium spp.*
Skin spot	*Polyscytalum pustulans*	Java black rot	*Diplodia gossypina*
Fruit and vegetables		Fusarium root rot	*Fusarium solani*
Alternaria rot	*Alternaria alternata*	Rhizopus soft rot	*Rhizopus nigricans*
Buckeye rot	*Phytophthora sp.*	Charcoal rot	*Marcrophomina spp.*
Gray mold	*Botrytis cinerea*	**Fruit and vegetables**	
Soft rot	*Rhizopus stolonifer*	Watery soft rot	*Sclerotinia spp.*
Sour rot	*Geotrichum candidum*	Cottony leak	*Pythium butleri*
Bacterial soft rot	*Erwinia spp.* *Pseudomonas spp.*	Fusarium rot Ripe rot	*Fusarium spp.* *Colletotrichum spp.*

Host physiological status

The development of postharvest disease is intimately associated with the physiological status of the host tissue. To create the right environment for minimizing postharvest losses due to disease it is important to understand the physiological changes that occur after produce is harvested. All plant organs undergo the physiological processes of growth, development and senescence. Growth and development generally only occur while the organ is attached to the plant (with the exception of seed germination and sprouting of storage organs), but senescence will occur regardless of whether the organ is attached or not (albeit at different rates). When an organ such as a fruit is harvested from a plant, it continues to respire and transpire depleting both reserve food, water and lead to senescence.

Mode of infection

Infection of fruit and vegetables by postharvest pathogens can occur before, during or after harvest. Infections which occur before harvest and then remain dormant until some point during ripening are particularly common amongst tropical fruit crops. Anthracnose, which is the most serious postharvest disease of a wide range of tropical and sub-tropical fruit such as mango, banana, papaya and avocado, is an example of a disease arising from dormant infections established prior to harvest. Anthracnose is also an important postharvest disease of a number of vegetables (e.g. bean) and temperate fruit (e.g. strawberry) caused by various species of Colletotrichum. Some species (e.g. *Colletotrichum musae*) are host-specific, whereas others can attack a wide range of fruit and vegetables (e.g. *Colletotrichum gloeosporioides*). The infection process begins when conidia germinate on the surface of host tissue to produce a germ tube and an appressorium. Although it is known that there is a dormant phase in the life cycle of the fungus, it is not entirely clear whether the ungerminated or the germinated appressorium represents the quiescent stage. It may be that the fungus behaves differently on different hosts, or perhaps some researchers have been unable to detect appressorial germination due to the limitations of the techniques used. Appressoria germinated to produce infection hyphae prior to the onset of quiescence. In any case, the fungus ceases growth soon after appressorium formation and remains in a quiescent state until fruit ripening occurs. During ripening, the fungus resumes activity and colonises the fruit tissue, leading to the development of typical anthracnose symptoms. Natural antifungal compounds present in fruit tissue may be involved in regulating the quiescence of *Colletotrichum infectrons*. In avocados, antifungal dienes are present in the peel of unripe fruit at concentrations inhibitory to *Colletotrichum gloeosporioides*, the avocado anthracnose pathogen. During ripening, levels of these dienes decline to sub-fungitoxic levels coincidentally with the development of anthracnose symptoms on fruit. Treatments which stimulate the production of diene compounds also delay symptom development, suggesting that these compounds have a role in regulating quiescence. Grey mould of strawberry caused by *Botrytis cinerea* is another important postharvest disease sometimes arising from quiescent infections established before harvest. Conidia of *Botrytis cinerea* on the surface of necrotic flower parts germinate in the presence of moisture. The fungus colonises the necrotic tissue and then remains quiescent in the base of the floral receptacle. Several months later when fruit are harvested, infections develop as a stem end rot in ripe fruit. Many postharvest diseases develop from the stem end of fruit. The mode of infection involved in this group of diseases can however vary considerably. In the example of *Botrytis cinerea*, lesions occurring at the stem end of fruit arise from quiescent floral infections. Stem end rots of citrus caused by *Lasiodiplodta theobromae* and *Phomopsis citri* result from quiescent infections in the stem button of fruit. These infections can be initiated at any stage of fruit development, and remain quiescent until the button begins to separate from the fruit during abscission. In other stem end rot diseases, infection occurs during and after harvest through the wound created by severing the fruit from the plant (e.g. banana crown rot). Endophytic infection, whereby the fungus symptomlessly and systemically colonises the stem, inflorescence and fruit pedicel tissue, is important in a number of stem end rot diseases of tropical fruit. Mango stem end rot caused by *Dothiorella dominicana* is one example of a postharvest disease arising from endophytic colonisation of fruit pedicel tissue. In this case, the fungus colonises the pedicel and stem end tissue of unripe fruit, where it remains quiescent until fruit ripening commences. A considerable number of postharvest diseases develop from infections which occur shortly before harvest. Such infections may not be visible at the time of harvest, yet at the same time are not necessarily inactive. Symptoms may develop more rapidly after harvest, particularly if storage conditions favour pathogen development. Examples of postharvest diseases which can arise from late season infections include brown rot of peach (caused by *Monilinia fructicola*), grey mould of grape (caused by *Botrytis cinerea*), yeasty rot of tomato (caused by *Geotrichum candidum*) and sclerotium rot of various vegetables (caused by *Sclerotium rolfsii*). Many common postharvest pathogens are unable to directly penetrate the host cuticle. Such pathogens therefore infect through surface injuries or natural openings such as stomata and lenticels. Injuries can vary in size from microscopic to clearly visible and may arise in a number of ways. Mechanical injuries such as cuts, abrasions, pressure damage and impact damage commonly occur during harvesting and handling. Insect injuries may occur before harvest yet remain undetected at the time of grading, providing ideal infection sites for many postharvest pathogens. Infection of lychee fruit by yeasts commonly occurs through insect injuries which are difficult to see at harvest. Some chemical treatments used after harvest, such as fumigants used in insect disinfestations and disinfectants such as chlorine, may also injure produce if applied incorrectly. Various types of physiological injury such as chilling and heat injury can predispose produce to infection by postharvest pathogens. Tropical fruit in particular are very sensitive to low temperatures, many developing symptoms of chilling injury below 13° C (depending on storage duration). Symptoms of chilling injury include abnormal ripening, peel and flesh discolouration, water-soaking and pitting. Chilling injury can increase the susceptibility of produce to postharvest disease considerably. For example, the incidence of alternaria rot in papaya, apple and various vegetable crops is increased by exposure to excessive cold. High temperatures can also increase susceptibility of harvested produce to disease. For example, hot water dipping of mangoes for excessive times and/ or temperatures can result in increased levels of stem end rot (caused by *Dothiorella* spp.).

The natural resistance of fruit and vegetables to disease declines with storage duration and ripeness. Weak pathogens which normally require a wound in order to infect can become a problem in produce that has been stored for long periods of time. Treatments which help to maintain the natural vitality of fruit and vegetables aid in delaying the onset of disease in stored produce.

Postharvest Treatments of Fruits and Vegetables

The post-harvest treatments play an important role in extending the storage and marketable life of horticultural perishables. The most important postharvest treatments include:

Washing with chlorine solution: Chlorine treatment (100–150 ppm available chlorine) can be used in wash water to help control inoculums build up during packing operations. Maintain pH of wash water between 6.5 and 7.5 for best results.

Ethylene inhibitors/Growth regulator/fungicide treatments: 1-MCP (1-methyl cyclopropene), Amenoethoxyvinyl glycine (AVG), silver nitrate, silver thiosulfate, cycloheximide, benzothiadiazole etc. are some of the chemicals which inhibit ethylene production and/or action during ripening and storage of fruits. The growth regulators or fungicidal application such as GA_3 can be effectively used to extend/enhance the shelf life of fruits.

Calcium application: The post-harvest application of $CaCl_2$ or $Ca(NO_3)_2$ play an important role in enhancing the storage and marketable life of fruits by maintaining their firmness and quality. Calcium application delays aging or ripening, reduces postharvest decay, controls the development of many physiological disorders and increases the calcium content, thus improving their nutritional value. The post-harvest application of $CaCl_2$ (2–4%) or $Ca(NO_3)_2$ for 5–10 min dip extend the storage life of pear up to 2 months, plum up to 4 weeks and apple up to 6 months at 0–2° C with excellent colour and quality. Calcium infiltration reduces chilling injury and increase disease resistance in stored fruit.

Hot water treatment: Fruits may be dipped in hot water before marketing or storage to control various post-harvest diseases and improving peel colour of the fruit.

Table 108. Hot water treatment temperature and holding time for different fruits

Commodity	Pathogens	Temperature (°C)	Time (min)	Possible injuries
Apple	*Gloeosporium spp.* *Penicillium expansum*	45	10	Reduced storage life
Grapefruit	*Phytophthora citrophthora*	48	3	Slight spotting
	Penicillium digitatum	52	5-10	
	Phytophthora spp.			
Green Beans	*Pythium butleri* *Sclerotinia sclerotiorum*	52	0.5	Reduced storage life
Lemon	*Penicillium digitatum* *Phytophthora spp.*	52	5-10	Poor de-greening
Mango	*Anthracnose* *Collectotrichum gleosporioides*	52	5	No stem rot control
Melon	Fungi	57-63	0.5	Slight spotting
Orange	*Diplodia spp.* *Phomopsis spp.* *Phytophthora spp.*	53	5	Poor de-greening
Papaya	Fungi	48	20	Softening of fruits
Papaya*	*Anthracnose* *Colletotrichum gleosporioides*	42 49	30 20	
Peach	*Monolinia fruticola* *Rhizopus stolonifer*	52	2.5	Motile skin
Bell Pepper	*Erwinia spp.*	53	1.5	Slight spotting

*Papaya anthracnose control requires both treatments, 30 minutes at 42°C followed by 20 minutes at 49°C.

Table 109. Hot forced-air treatment for different fruits

Commodity	Pathogens	Temperature (°C)	Time (minute)	RH (%)	Possible injuries
Apple	*Gloeosporium spp.* *Penicillium expansum*	45	15	100	Deterioration
Melon	Fungi	30-60	35	low	Marked breakdown
Peach	*Monolinia fruticola* *Rhizopus stolonifer*	54	15	80	
Strawberry	*Alternaria spp.* *Botrytis spp.,* *Rhizopus spp.* *Cladosporium spp.*	43	30	98	

Source: Barkai-Golan and Phillips, 1991

Vapor heat treatment: This treatment proved very effective in controlling infection of fruit flies in fruits after harvest. The boxes are stacked in a room, which are heated and humidified by injection of steam. The temperature and exposure time are adjusted to kill all stages of insects (egg, larva, pupa and adult), but fruit should not be damaged. A recommended treatment for citrus, mangoes, papaya and pineapple is 43° C in saturated air for 8 hours and then holding the temperature for further 6 hours. Vapor heat treatment is mandatory for export of mangoes.

Fumigation: The fumigation of SO_2 is successfully used for controlling post- harvest diseases of grapes. This is achieved by placing the boxes of fruit in a gas tight room and introducing the gas from a cylinder to the appropriate concentration. However, special sodium metabisulphite pads are also available which can be packed into individual boxes of a fruit to give a slow release of SO_2. The primary function of treatment is to control the *Botrytis cinerea*. The SO_2 fumigation is also used to prevent discoloration of skin of litchis.

Irradiation: Ionizing radiation can be applied to fresh fruits and vegetables to control micro-organisms and inhibit or prevent cell reproduction and some chemical changes. It can be applied by exposing the crop to radiations from radioisotopes (normally in the form of gamma-rays measured in Grays (Gy), where 1 Gy = 100 rads.

Waxing: Waxing of fruits or vegetables is a common post-harvest practice. Food grade waxes are used to replace some of the natural waxes removed during harvesting and sorting operations and can help reduce water loss during handling and marketing. It also helps in sealing tiny injuries and scratches on surface of fruits and vegetables. It improves cosmetic appearance and prolongs the storage life of fruits and vegetables. The most important goals of postharvest handling are keeping the product cool, to avoid moisture loss and slow down undesirable chemical changes, and avoiding physical damage such as bruising, to delay spoilage.

Control Methods of Post Harvest Disease

Traditional strategies for postharvest disease control: Fungicides are used broadly for postharvest diseases manage in fruit and vegetables. Timing of application and type of fungicide used depend primarily on the target pathogen and when infection occurs. For postharvest pathogens which infect produce before harvest, field application of fungicides is often necessary. This may involve the repeated application of protectant fungicides during the growing season, and/or strategic application of systemic fungicides. In the control of mango anthracnose (caused by *Colletotrichum gloeosporioides*) in India, for example, trees are sprayed regularly with a protectant fungicide such as mancozeb during flowering and fruit development. If rain occurs during flowering, a systemic fungicide is applied to inactivate infections already established and to guard against new infections. In the postharvest situation, fungicides are often applied to control infections already established in the surface tissues of produce or to protect against infections which may occur during storage and handling. In the case of quiescent field infections present at the time of harvest, fungicides must be able to penetrate to the site of infection to be effective. In some tropical fruit crops such as mango, the penetration of systemic fungicides is enhanced by applying them as heated dips. In the case of infections which occur during and after harvest, fungicides can be used to interrupt pathogen development. In general, fungicides for the control of wound-invading pathogens should be applied as soon as possible after harvest. If infection is well advanced at the time of postharvest treatment, control will be difficult to achieve. The usual approach with controlling wound pathogens is to maintain a certain concentration of the fungicide at the injury site which will suppress (though not necessarily kill) pathogen development until the wound has healed. In this sense, most of the fungicides which are used postharvest are actually fungistatic rather than fungicidal in their action under normal usage. Disinfectants such as sodium hypochlorite can be used to kill pathogen propagules on the surface of fruit, but are unable to control pathogens once they have gained entry to host tissue. Postharvest fungicides can be applied as dips, sprays, fumigants, treated wraps and box liners or in waxes and coatings. Dips and sprays are very commonly used and depending on the compound, can take the form of aqueous solutions, suspensions or emulsions. Fungicides commonly applied as dips or sprays include the benzimidazoles (e.g. benomyl and thiabendazole) and the triazoles (e.g. prochloraz and imazalil). The benzimidazole group of fungicides is very useful for the control of many important postharvest pathogens such as Penicillium and Colletotrichum. Fumigants, such as sulphur dioxide for the control of grey mould (caused by *Botrytis cinerea*) of grape and various postharvest diseases of lychee, are sometimes used for disease control. Other fumigants used in certain situations include carbon dioxide, ozone and ammonia. Fruit wraps or box liners impregnated with the fungicide biphenyl are used in some countries for the control of Penicillum in citrus. Fungicides to control postharvest diseases of citrus and some other fruit are often applied to the fruit in wax on the packing line.

Maintenance of host resistance to infection through manipulation of the postharvest environment

The ability to control the postharvest environment provides a tremendous opportunity to delay senescence. Temperature is perhaps the single most important factor influencing disease development after harvest. Temperature not only directly influences the rate of pathogen growth, but also the rate of fruit ripening. The development of many postharvest diseases is closely associated with fruit ripeness, so treatments which delay ripening tend also to delay disease development. Low temperature storage of fruit and vegetables is used extensively to delay ripening and the development of disease, although the temperatures commonly used for storage are not lethal to the pathogen. For this reason, cool stored produce which is transferred to ambient temperatures for ripening and/or sale may rapidly breakdown with postharvest disease. Temperatures used to store produce depend largely on the chilling sensitivity of the produce. For example, many temperate fruit and vegetables (e.g. apples, peaches and broccoli) can be stored at 0° C, whereas many tropical fruits cannot be stored below 10° C without developing symptoms of chilling injury. The rate of fruit respiration can be reduced by increasing CO_2 and decreasing O_2 levels in the storage environment. Storage atmosphere can also have a direct effect on pathogen growth, although levels of CO_2 or O_2 required to achieve this are often damaging to the produce if applied for extended periods. A notable exception to this is strawberry, which for extended storage periods can tolerate the high levels of CO_2 (20-30%) required inhibiting the development of grey mould (caused by *Botrytis cinerea*). Short-term exposure to very high levels of CO_2 has shown some potential for delaying the onset of anthracnose (caused by *Colletotrichum gloeosporioides*) symptoms in avocado, although the treatment is not currently in commercial use. The relative humidity of the storage environment can have a major influence on the development of postharvest disease. High humidities are often used to minimise water loss of produce, however it can increase disease levels, particularly if free moisture accumulates in storage containers. The humidity chosen for storing produce is frequently a trade-off between minimising water loss and decreasesing disease.

Hygiene practices: Maintenance of hygiene at all stages during production and postharvest handling is critical in minimising sources of inoculum for postharvest diseases. To most effectively reduce inoculum, a good knowledge of the life cycle of the pathogen is essential. Sources of inoculum for postharvest diseases depend largely on the pathogen and when infection occurs. In the case of postharvest diseases which arise from preharvest infections, practices which make the crop environment less favourable to pathogens will help reduce the amount of infection which occurs during the growing season. For example in tree crops, pruning and skirting can increase ventilation within the tree canopy, making conditions less favourable for fungi and bacteria. Removal of dead branches and leaves entangled in the tree canopy is also an important way to minimise inoculum build-up. In many diseases, overhead irrigation can encourage pathogen spread and infection, trickle or micro-sprinkler irrigation systems may be more appropriate. As many pathogens are soil-borne, minimising contact of leaves and fruit with the soil is desirable.

Inoculum for infections occurring after harvest commonly originates from the packing shed and storage environments.

» Water used for washing or cooling produce can become contaminated with pathogen propagules if not changed on a regular basis and if a disinfectant such as chlorine is not incorporated. Water temperature can also be an important factor in the transfer of inoculum in some situations. For example, tomatoes harvested during hot weather may have a higher temperature than the water used to wash them. In this scenario, inoculum present in the washing water can be taken in by the fruit tissue, causing higher levels of diseases such as bacterial soft rot.

» Reject produce which has not been discarded from the packing shed or storage environment provides an ideal substrate for postharvest pathogens.

» Packing and grading equipment, particularly brushes and rollers, which is not cleaned and disinfected on a regular basis can also be a major source of inoculum.

» Containers used for storing and transporting fruit can harbour pathogen propagules, particularly if recycled a number of times without proper cleaning.

Preharvest factors: A wide range of preharvest factors influence the development of postharvest disease. These include the weather (rainfall, temperature, etc.), production locality, choice of cultivar, cultural practices (pesticide application, fertilisation, irrigation, planting density, pruning, mulching, fruit bagging, etc.) and planting material. These factors may have a direct influence on the development of disease by reducing inoculum sources or by discouraging infection. Alternatively, they may affect the physiology of the produce in a way that impacts on disease development after harvest. For example, the application of certain nutrients may improve the 'strength' of the fruit skin so that it is less susceptible to injury after harvest and therefore less prone to invasion by wound pathogens.

Prevention of injury: As many postharvest pathogens achieve entry through wounds or infect physiologically damaged tissue, prevention of injury at all stages during production, harvest and postharvest handling is critical. Injuries can be mechanical (e.g. cuts, bruises and abrasions), chemical (e.g. burns), biological (e.g. insect, bird and rodent damage) or physiological (e.g. chilling injury, heat injury). Injuries can be minimized by careful harvesting and handling of produce, appropriate packaging of produce and controlling insect pests in the field, storing produce at the recommended temperature and applying postharvest treatments correctly. Where injuries are present, the process of wound healing can be accelerated in some instances through manipulation of the postharvest environment (e.g. temperature and humidity) or by application of certain chemical treatments. Wound healing has been shown to be associated with resistance to certain postharvest diseases such as bacterial soft rot of potatoes caused by *Erwinia spp.*

Cold treatments

Certain fungi and bacteria in their germination phase are prone to cold and infections can be decreased by treating produce with a few days of storage at the coldest temperature the commodity can withstand without incurring damage (0° C for apple, pear, grape, kiwifruit, persimmon and stone fruits). Rhizopus stolonifer and *Aspergillus niger* (black mould) can be killed when germinating by 2 or more days at 0° C and pathogen growth can be nearly stopped by storage at temperatures below 5° C. Cold treatments can also serve to control some insect pests and are currently allowed for the control of fruit flies, the false codling moth, melon fly, pecan weevil and lychee fruit borer. Treatment for control of fruit flies (*Ceratitis capitata*) requires 10 days at 0° C or below, or 14 days at 1.7° C or below, so treatment is only suited to commodities capable of withstanding long-term low-temperature storage such as apples, pears, grapes, kiwifruit and persimmons.

Controlled/modified atmosphere treatments

The CO_2 enriched air (15 to 20 per cent) can be used as a fungistat to control decay causing pathogens, such as *Botrytis cinerea* on strawberry, blueberry, blackberry, fresh fig and table grapes during transport. Insecticidal atmospheres (0.5% or lower O_2 and/or 40% or higher CO_2) have been shown to be an effective substitute for methyl bromide fumigation to disinfest dried fruits, nuts and vegetables. The effectiveness of insecticidal atmospheres depends upon the temperature, relative humidity, duration of exposure and life stage of the insect.

» The first and third instars of the green headed leaf roller (*Planotortrix excessana*) and the first and fifth instars of the brown headed leaf roller (*Ctenopseustis obliquana*) and the light brown apple moth (*Epiphyas postvittana*) are completely killed in 2 months when apples are stored at 0.5° C in 3 per cent O_2 and 3 per cent CO_2.

» The eggs of the apple rust mite (*Aculus schlechtendali*) and the European red mite (*Panonychus ulmi*) are killed in 5 months when apples are stored at 2.8° C in an atmosphere of 1 per cent O_2 and 1 per cent CO_2.

» Codling moth larvae (*Cydia pomonella*) are killed in 3 months when apples are stored at 0° C, 1.5-2.0 per cent O_2 and less than 1 per cent CO_2.

» In kiwifruit, the adult two-spotted spider mite (*Tetranychus urticae*) is killed by 40° C, 0.4 per cent O_2 and 20 per cent CO_2 in only 7 hours.

» When persimmons are stored at 20° C, 0.5 per cent O_2 and 5 per cent CO_2, the third instar of leaf rollers (*Planotortrix excessana*) is killed in 4 days and the larvae and adult mealy bug (*Pseudococcus longispinus*) is killed in 7 days.

» Sweetpotato weevil (*Cylas formicarius elegantulus*) has been controlled at ambient temperature in stored tropical sweetpotatoes by treatment with low oxygen and high carbon dioxide atmospheres. At 25° C, storage in 2 to 4 per cent oxygen and 40 to 60 per cent carbon dioxide results in mortality of adult weevils in 2 to 7 days.

» Codling moth (*Cydia pomonella*) in stone fruits can be controlled at 25° C by using atmospheres of 0.5 per cent oxygen and 10 per cent carbon dioxide for 2 to 3 days (adult or egg) or 6 to 12 days (pupa).

» Treatments with 45 per cent CO_2 at 0° C are being developed for several surface pests, including omnivorous leaf roller (*Platynota sultana*), western flower thrips (*Frankliniella occidentalis*) and Pacific spider mite (*Tetranychus pacificus*) on harvested table grapes. This treatment requires 13 days at 0° C to 2°C and could be conducted in a marine container during transport.

Heat treatments: Heat treatments applied after harvest can be used to control certain postharvest diseases. Heat works by either killing the pathogen (and/or its propagules) or by suppressing its rate of development following treatment. However, commodities vary greatly in their physiological tolerance of heat treatments. For example, most temperate fruit types are quite susceptible to heat injury, particularly at the temperatures required to achieve disease control. Hot water is a more efficient medium for heat transfer than hot air, but is also more likely to cause injury to the commodity. Hot water is often used in combination with fungicides to control postharvest diseases in commodities such as mango, papaya and rockmelon. Hot air treatments commonly used for fruit fly disinfestation in various harvested commodities can also give some control of postharvest diseases. In mangoes, an effective treatment is 46.4° C for 65 to 90 minutes, depending on fruit size, variety and country of origin. Fruit should not be handled immediately after heat treatment. Whenever heat is used with fresh produce, cool water showers or forced cold air should be provided to help return the fruits to their optimum temperature as soon as possible after completion of the treatment. Brief hot water dips or forced-air heating can be effective for disease control, especially for reducing the microbial load for crops such as plums, peaches, papaya, cantaloupe and stone fruits, sweetpotatoes and tomatoes.

Ionising radiation: Ionising radiation is another physical treatment that can be used after harvest to reduce disease in fruit and vegetable. The commodities must be able to tolerate the doses of ionising radiation required to achieve disease control. Strawberries can tolerate the doses of radiation required to effectively control grey mould caused by *Botrytis cinerea.* In other commodities, however, abnormal ripening, tissue softening and off-flavours can result from applying ionising radiation at doses lethal to the target pathogens. Poor consumer acceptability of fresh produce irradiation coupled with high treatment costs pose additional limitations to the widespread use of this technology at the present time.

Chemical controls

Fruits and vegetables can be washed with hypochlorite solution (25 ppm available chlorine for two minutes) then rinsed to control bacterial decay. Alternatively, these commodities can be dipped in hypochlorite solution (50 to 70 ppm available chlorine) then rinsed with tap water for control of bacteria, yeasts and moulds. The effectiveness of the treatment will be reduced if organic matter is allowed to build up in the wash water. The effectiveness of chlorine increases as pH is reduced from 11 to 8, but at lower pH chlorine becomes unstable. Sulfur is used on bananas as a paste (0.1% active ingredient) to control crown rot fungi. Sulfur dioxyed (SO_2) is used as a fumigant (with a 10 ppm residue tolerance) on grapes to control Botrytis, Rhizopus and Aspergillus fungi. Bisulfites are usually contained within a paper or plastic pad that can be placed inside a carton to release SO_2 for control of moulds on grapes during transportation and storage. The use of bicarbonate salts for prevention of postharvest decay has been used on fresh peppers, melons, potatoes, carrots and citrus fruits. These salts are very inexpensive, safe to use, readily available and accepted as certified organic and chemical free for marketing purposes. Bicarbonate of sodium or baking soda ($NaHCO_3$), and Potassium bicarbonate ($KHCO_3$) are also used for post harvest treatment of fruits and vegetables. Bacterial soft rot (*Erwinia spp.*) of cabbage can be controlled by using lime powder or a 15 per cent solution of alum (15 g aluminum potassium sulphate in 100 ml of water). After treatment of the butt-end of the cabbage heads, the produce should be allowed to dry for 20 to 30 minutes before packing. On instances when fungicides must be applied to produce, a simple tray with holes punched in the bottom can be used hold the commodity while it is sprayed.

A back-pack sprayer is a useful tool for applying fungicides and other postharvest chemical treatments. When fruit is packed for export, fungicides are often applied to meet the requirements of international quality standards and to reduce deterioration during transport. The cascade applicator illustrated below (Figure 42) was developed to apply fungicide uniformly and effectively by using a liquid curtain to drench the fruit.

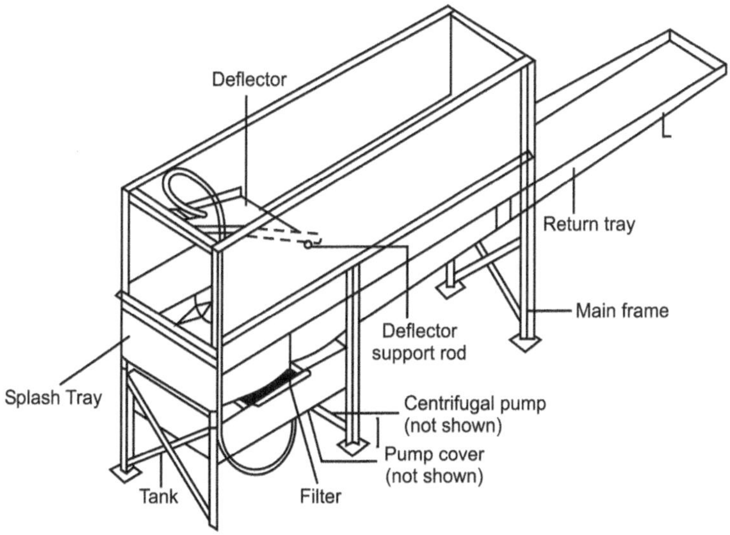

Figure 42. Fungicide applicator (Source: Overseas Div., AFRC, 1974)

Fruit in a perforated plastic tray is introduced on a roller conveyor belt (not shown) into the applicator. Inside a simple fan shaped deflector creates a curtain of liquid fungicide. The fruit passes under the curtain where it is drenched, then out of the applicator to drain on a tilted return tray. The tank holds up to 50 litres of fungicide solution and a pump is mounted at the level of the tank outlet. A filter is fitted on the top of the tank to remove foreign matter from the return flow of fungicide from the applicator box and the return tray.

Biological control and plant growth regulators

Biological control products (antagonistic organisms) are currently used as complementary tools (to chemical and/or heat treatments) for the management of postharvest decays together with other strategies as part of an integrated pest management programme for a few fruits and vegetables. Plant growth regulators can be used to delay senescence of citrus fruits and consequently delay their susceptibility to decay.

Table 110. Commercially available biological control materials and plant growth regulators

Category	Organism/ Product	Year	Crop	Decay organisms or function	Methods of application	Residue tolerance (ppm)
Biocontrol	Pseudomonas syringae (Bio-Save)	1995	Citrus	*Penicillium digitatus, Penicillium italicum, Geotrichum citri- aurantii*	Dip or spray	Exempt
			Cherries	*Penicillium expansum, Botrytis cinerea*	Drench	Exempt
			Apples, Pears	*Penicillium expansum, Botrytis cinerea, Mucor piriformis*	Dip or drench	Exempt
			Potatoes	*Fusarium sambucinum, Helminthosporium solani*	Dip or spray	Exempt
Biocontrol	Candida oleophila (Aspire)	1995	Pome fruits	*Decay pathogens*	Any type of application	Exempt
			Citrus	*Decay pathogens*	Any type of application	Exempt
PGR	Gibberellic acid (Pro Gibb)	1955	Citrus	Delays senescence (delays onset of decay)	Storage wax	Exempt
PGR	2,4-D (Citrus Fix)	1942	Citrus	Delays senescence of buttons (delays onset of decay)	Storage wax	5

Emerging technologies for postharvest disease control

Increasing consumer concerns over the presence of pesticide residues in food have prompted the search for non-chemical disease control measures. Fungicides used after harvest is of particular concern because they are applied close to the time of consumption. A number of new approaches to control postharvest diseases are currently under investigation, including biological control, constitutive or induced host resistance and natural fungicides.

Biological control: In recent years, there has been considerable interest in the use of antagonistic microorganisms for the control of postharvest diseases. Such organisms can be isolated from a variety of sources including fermented food products and the surfaces of leaves, fruit and vegetables. Once isolated, organisms (whether they be bacteria, yeasts or filamentous fungi) can be screened in various ways for inhibition of selected pathogens. In most reported cases, pathogen inhibition is superior when the antagonist is applied prior to infection taking place. For this reason, control of quiescent field infections (e.g. *Colletotrichum spp.*) using postharvest applications of antagonists is often more difficult to achieve than control of infections occurring after harvest (e.g. *Penicillium spp.*). Unless an antagonist has eradicant activity or has some effect on host defense responses, field applications are often necessary to achieve control of quiescent infections. Postharvest applications of *Bacillus subtilis* and *Bacillus licheniformis*, suppress anthracnose (caused by *Colletotrichum gloeosporioides*) and stem end rot (caused by *Dothiorella spp.* and other fungi) development in avocado. A number of modes of action are involved in these pathogen-antagonist interactions, including site exclusion, nutrient and space competition and antibiotic production. Field applications of a non-antibiotic producing strain of *Bacillus spp.* have shown potential for the control of anthracnose in avocado. Antagonists which act against postharvest pathogens by competitive inhibition at wound sites include the yeasts *Pichia guilliermondi*, *Cryptococcus laurentii* and *Candida spp.* While the potential for biological control of postharvest diseases clearly exists, future success relies on the ability to achieve consistent results in the field and after harvest. It will be necessary to enhance the efficacy of biological control agents postharvest disease and commercialise the technology involved. Ideally, an antagonist should be effective against a broad spectrum of pathogens on a wide range of fruit and vegetables, it should be unique and be able to be produced on inexpensive growth media. Formulations incorporating such antagonists should have a long shelf-life and be able to be manufactured at low cost. Most antagonists do not satisfy all of these criteria. Many are quite specific in their activity against pathogens and for this reason may not be particularly appealing to prospective investors.

Constitutive and induced host resistance: Plants possess various biochemical and structural defense mechanisms which protect them against infection. Some of these mechanisms are in place before arrival of the pathogen (i.e. constitutive resistance), while others are only activated in response to infection (i.e. induced resistance). Compared to intact plants, relatively little is known about host defense responses in harvested commodities, although there is growing interest in this area. Preformed antifungal diene compounds present in fungitoxic concentrations in unripe fruit decline to sub-lethal concentrations in ripe fruit allowing the pathogen to develop once fruit ripening commences. Levels of the diene compounds can be increased by applying various treatments such as challenge inoculation with either pathogenic or non-pathogenic strains of Colletotrichum or by treatment with certain antioxidants or high concentrations of CO_2. In contrast to preformed antimicrobial compounds, phytoalexins are only produced in response to pathogen invasion, although in some cases they can be elicited by certain chemical and physical treatments. For example, non-ionising ultraviolet-C radiation is known to induce production of phytoalexins in various crops. UV treatment of carrot slices induces production of 6-methoxymellen which is inhibitory to *Botrytis cinerea* and *Sclerotinia sclerotiorum*.

Natural fungicides: Many compounds produced naturally by microorganisms and plants have fungicidal properties. Chitosan is not only an elicitor of host defense responses but also has direct fungicidal action against a range of postharvest pathogens. Antibiotics produced by various species of Trichoderma have potent antifungal activity against *Botrytis cinerea*, *Sclerotinia sclerotiorum*, *Corticium rolfsii* and other important plant pathogens. There are many other natural compounds which have been isolated and shown to possess considerable antifungal activity. Although these compounds may be more desirable than synthetic chemicals from a consumer viewpoint, their potential toxicity to humans needs to be evaluated before useable products are developed. On a positive note, the high specificity of many of these compounds also means that they biodegrade readily.

Use of ozone: In recent years, increasing attention has been focused on the safety of fruits and vegetables, and in particular on the intervention methods to reduce and abolish human pathogens from fresh produce. Traditional technology utilizes water with or without a sanitizing agent to wash fresh fruits and vegetables. Chlorine is the most widely used sanitizing agent available for fresh produce, but it has a limited effect in killing bacteria on fruit and vegetable surfaces. The most that can be expected at permitted concentrations is a 1^{-log} to 2^{-log} population reduction. Research and commercial applications have verified that ozone (O_3) is a high-energy molecule can replace traditional sanitizing agents and provide other benefits. The potential utility of ozone in the produce industry depends on the fact that as an oxidizing agent, it is 1.5 times stronger than chlorine and is effective over a much wider spectrum of microorganisms than chlorine and other disinfectants. Ozone kills bacteria such as Escherichiacoli, Listeria, and other food pathogens much faster than traditionally used disinfectants, such as chlorine, and is free of chemical residues. Its half life in water at room temperature is only 20 min, and it decomposes into simple oxygen with no safety concerns about consumption of residual ozone in the treated food product. Fresh fruits and vegetables are washed first by ozonated water, and the wash water can be recaptured and treated by a combination of ozonation and filtration. The treated wash water is free of bacteria, colour, and suspended solids and can be recycled to reduce water usage. Ozone can also destroy pesticides and chemical residues, such as chlorinated by-products. Gaseous ozone is a strong sanitation and fumigation agent and can be used to sanitize fruits and vegetables in the storage room and during shipping to prevent bacteria, mold, and yeast on the fruit and vegetable surface and to control insects. It can eliminate undesirable flavour produced by bacteria and chemically remove ethylene gas to slow down the ripening process, thus allowing extended distribution. Thus, ozone can successfully replace traditional sanitizing agents to control fruits and vegetables pathogens.

Applications of ozone: Many applications in the produce industry appear suitable for the use.

Process water sterilization: Normally, processing water is disinfected and sterilized using chlorine. However, chlorine cannot reduce the level of organic compounds and will produce chlorinated compounds. Ozone has been proven to be an ideal replacement for chlorine for disinfection and sterilization of process water. In fact, it is more effective than chlorine against microorganisms, including chlorine resistant Cryptosporidium and Giardia. Ozone can also destroy chlorine byproducts, pesticides, and toxic organic compounds in the process water without any toxic residues. Practical applications of ozone to process water range from 0.5 to 5 ppm (depending on the water source), with less than 5 minute contact time. Ozone is also used to remove iron, manganese, and sulfur and to control taste and odour of fresh water. This application will continuously maintain highquality water free of microorganisms and toxic chemicals for the fruit and vegetable industry.

Fruit and vegetable washing: To maintain or even improve the safety of fresh produce is to wash vegetables and fruits using ozonated water. Two types of washing systems spray and flume can be used to reduce microbial counts on the surface of produce. Ozone is particularly effective against *E. coli*, the food pathogen of most concern to the produce industry.

Fruit and vegetable storage: Ozone can be employed in cold storage of produce to guard against mold and bacteria at a very low concentration. It can not only destroy mold and bacteria in the air and on the surface of produce but also deodorize. Gaseous ozone to prevent microbial activity on produce surfaces and extend the shelf life of fruits and vegetables including apples, potatoes, tomatoes, strawberry, broccoli, pears, cranberries, oranges, peaches, grapes, corn, and soybeans. The shelf life of strawberries, raspberries, currents, and grapes could be doubled if 2–3 ppm of gaseous ozone is applied continuously for a few hours per day. Shelf life of apple could be increased by several weeks by applying 2–3 cm^3 of ozone m^{-3} of air a few hours a day. The shelf life of potatoes could be extended to as long as 6 months at 6–14 °C and 93–97 per cent relative humidity with 3ppm of ozone, without affecting the potato quality. One of the important effects of ozone in cold storage is to slow down the fruit and vegetable ripening process. During ripening, many fruits, such as bananas and apples, release ethylene gas, which speeds up the ripening process. Ozone is very effective in removing ethylene through chemical reaction to extend the storage life of many fruits and vegetables.

$$H_2C{=}CH_2 + 2O_3 \rightleftarrows 2CO_2 + H_2O$$

13 Postharvest Care and Handling of Cut Flowers

Each plant material has its own best harvest stage and this can vary depending on the use. Materials for preserving usually are harvested more mature than those for fresh and wholesale markets. Ingeneral spike type flowers harvest when one-fourth to one-half of the individual florets are open, wheres daisy type flowers harvest when flowers are fully open. For local markets, the material can be more mature than national wholesale market. The best time is the coolest part of the day and when there is no surface water from dew or rain on the plants. Also, harvesters need enough light to see what they are harvesting. This usually is in the cool of the morning after the dew has dried. Late afternoon or evening also has possibilities because the plants have stored carbohydrates from the day which will provide reserve food for the plant material.

This is most important on plants that produce multiple flowers/crops per season harvest the longest stem possible without sacrificing future production. Very vigorous plants can be cut back to fewer nodes, while less vigorous plants should have more nodes left. Most stems should be at least 35 to 45 cm long. The base cut is slanted or squared, for cutting use sharp, clean utensils. Sharp cutting utensils will not crush the xylem and block the flow of water up the stem. Clean utensils will not introduce harmful microbes to the cut stems. Cutting tools should be cleaned daily with disinfectant a 1:10 solution of chlorine bleach in water. Flowers with sticky sap require special treatment instantly after harvest. To prevent the flow of the sticky sap, which can block the xylem, dip the cut ends in boiling water for 10 seconds or sear with a flame, immediately after harvest. Poppies, mignonette and poinsettia are examples of flowers with sticky sap.

Table 111. Optimal stage of development for harvest of fresh cut flowers

Common Name	Species	Stage of Development
African marigold	*Tagetes erecta*	Fully open flowers
Allium, Ornamental onion	*Allium spp.*	One-fourth to one-third florets open
Annual Gaillardia, Blanket flower, Indian blanket flower	*Gaillardia pulchella*	Fully open flowers
Astilbe	*Astilbe hybrids*	One-half florets open
Bachelor's button	*Centaurea spp.*	Flowers beginning to open
Bearded iris	*Iris Bearded cvs.*	Coloured buds
Bee-balm, Fragrant balm, Oswego tea	*Monarda didyma*	Almost open flowers
Black-eyed susan, Yellow oxeye Daisy, English bulls-eye	*Rudbeckia spp.*	Fully open flowers
Calendula, Pot marigold	*Calendula officinalis*	Fully open flowers
Canterbury bells	*Campanula spp.*	One-half florets open
China aster, Annual aster	*Callistephus chinensis*	Fully open flowers
Clarkia, Farwell to spring	*Clarkia unquiculata*	One-half florets open
Climbing lily, Glory lily	*Gloriosa superba*	Almost fully open flowers
Cockscomb	*Celosia argentea var. cristata*	One-half florets open
Columbine	*Aquilegia hybrids*	One-half florets open
Columbine meadow rue	*Thalictrum aguilegiifolium*	One-half florets open
Common grape hyacinth	*Muscari botryoides*	One-half florets open
Common mignonette	*Reseda odorata*	One-half florets open
Common stock	*Matthiola incana*	One-half florets open
Common garden, Late tulips	*Tulipa cvs.*	Half-coloured buds
Common foxglove, Finger flower, Purple foxglove	*Digitalis purpurea*	One-half florets open
Common sunflower	*Helianthus annuus*	Fully open flowers
Coreopsis, Tickseed, Lance coreopsis	*Coreopsis grandiflora*	Fully open flowers
Daffodil, Narcissus, Jonquil	*Narcissus* cvs.	"Goose neck" stage

Common Name	**Species**	**Stage of Development**
Dahlia	*Dahlia* cvs.	Fully open flowers
Daylily	*Hemerocallis* cvs.	Half-open flowers
Delphinium	*Delphinium spp.*	One-half florets open
Dutch iris	*Iris x hollandica*	Coloured buds
English daisy, True daisy	*Bellis perennis*	Fully open flowers
Fleabane	*Erigeron hybrids*	Fully open flowers
Freesia	*Freesia hybrids*	First bud beginning to open
Garden forget-me-not, Woodland forget-me-not	*Myosotis sylvatica*	One-half florets open
Glads	*Gladiolus cultivars*	1 to 5 buds showing colour
Globe thistle	*Echinops ritro*	Half-open flowers
Goldenrod	*Solidago spp.*	One-half florets open
Joseph's coat, Amaranth, Fountain plant, Tampala	*Amaranthus tricolor*	One-half florets open
Kaffir, Lily, Clivia	*Clivia miniata*	One-fourth florets open
Larkspur, Annual delphinium	*Consolida ambigua*	2 to 5 florets open
Lily-of-the-valley	*Convallaria majalis*	One-half florets open
Lisianthus	*Eustoma grandiflorum*	5 to 6 open flowers
Love-in-a-mist	*Nigella damascena*	Open flowers
Lupine	*Lupinus* cvs. *Russell*	One-half florets open
Montebretia	*Crocosmia x crocosmiiflora*	One-half florets open
Nasturtium	*Tropaeolum majus*	Fully open flowers
Nerine	*Nerine bowdenii*	Oldest buds almost open
Pansy	*Viola x wittrockiana*	Almost open flowers
Peony	*Paeonia cvs.*	Coloured buds
Perennial gaillardia, Blanket	*Gaillardia x grandiflora*	Fully open flowers
Perennial & Annual baby's breath	*Gypsophila spp.*	Flowers open but not overly mature
Pincushion flower	*Scabiosa spp.*	Half-open flowers
Poppy anemone	*Anemone coronaria*	Buds beginning to open

Common Name	Species	Stage of Development
Ranunculus	*Ranunculus asiaticus*	Buds beginning to open
Sea holly	*Eryngium spp.*	Fully open flowers
Showy stonecrop sedum, Live-forever	*Sedum spp.*	Fully open flowers
Siberian squill, Blue squill	*Scilla siberica*	Half-open flowers
Snapdragon	*Antirrhinum majus*	One-third florets open
Speedwell	*Veronica spp.*	One-half florets open
Statice, Sea-lavendar	*Limonium spp.*	Most fully open flowers
Summer phlox, Garden phlox, Fall phlox	*Phlox paniculata*	One-half florets open
Sunflower heliopsis, Hardy zinnia, Orange sunflower, False sunflower	*Heliopsis helianthoides*	Fully open flowers
Sweet pea	*Lathyrus odoratus*	One-half florets open
Sweet william	*Dianthus barbatus*	One-half florets open
Sweet violet, English violet, Garden violet, Florists violet	*Viola odorata*	Almost open flowers
Tall gayfeather, Blazing star, Button snakeroot	*Liatris spicata*	One-half florets open
Tiger, Asiatic, Oriental lilies	*Lilium spp.*	Coloured buds
Torch-lily, Common poker plant, Flame flower	*Kniphofia uvaria*	Almost all florets are showing colour
True monkshood	*Aconitum napellus*	One-half florets open
Tuberose	*Polianthes tuberosa*	Majority of florets open
Yarrow	*Achillea filipendulina*	Fully open flowers
Zinnia	*Zinnia elegans*	Fully open flowers

Quality Loss in Ornamentals

Cut or intact, ornamentals are complex plant organs, in which loss of quality of stems, leaves, or flower parts may result in rejection in the marketplace. In some ornamentals loss of quality may result from one of several causes, including wilting or abscission of leaves and/or petals, yellowing of leaves, and geotropic or phototropic bending of scapes and stems.

Causes of quality loss

Growth, development and aging: In plants, death of individual organs, and of the whole plant itself, is an integral part of the life cycle. Even in the absence of senescence of floral organs or leaves, the continuing growth process can result in quality loss, for example in spike-type flowers that bend in response to gravity.

Flower senescence: The early death of flowers is a common cause of quality loss and reduced vase life for cut flowers. Flowers can be divided into several categories in terms of their senescence. Some flowers are extremely long-lived, especially in the daisy and orchid families. Others are short-lived, including many of the bulb crops, like Tulip, Iris, and Narcissus.

Wilting of ornamentals: Extended life for cut and potted ornamentals depends absolutely on a continuing and adequate supply of water. Failure of water supply, whether through obstruction of the cut stems, or through inadequate watering of pots, results in rapid wilting of shoot tips, leaves, and petals.

Leaf yellowing and senescence: Yellowing of leaves and even of other organs (buds, stems) commonly is associated with the end of display life in some cut flowers like alstroemeria. Leaf yellowing is a complex process that may be caused by a range of different environmental factors and stability of pigment in leaves.

Shattering: Loss of leaves, buds, petals, flowers, or even branchlets, a process called 'shattering', or 'abscission', is also a common problem in cut flowers. Very often, this problem is associated with presence of ethylene in the air, but other environmental factors may also be involved.

Factors Affecting Postharvest Quality of Cut Flowers

There are many environmental, physiological and bio-chemical factors affecting post harvest quality of cut flowers. Understanding these factors allows the grower and shipper to develop and implement optimum postharvest handling technologies for maintain quality for long duration.

Flower maturity: Minimum harvest maturity for a cut-flower crop is the stage at which harvested buds can be opened fully and have satisfactory display life after distribution. Many flowers are best cut in the bud stage and opened after storage, transport or distribution. This technique has many advantages, including reduced growing time for single-harvest crops, increased product packing density, simplified temperature management, reduced susceptibility to mechanical damage and reduced desiccation. Many flowers are presently harvested when the buds are starting to open (rose, gladiolus), although others are normally fully open or nearly so (chrysanthemum, carnation). Flowers for local markets are generally harvested much more open than those intended for storage and/or long-distance transport.

Temperature: Respiration of cut flowers, an integral part of growth and ageing, generates heat as a by-product. Furthermore, as the ambient temperature rises, the respiration rate increases. For example, a flower held at 29° C is likely to respire (and therefore age) up to 45 times as fast as a flower held at 2° C. The rate of ageing can be dramatically reduced by cooling the flowers. Rapid cooling and proper refrigeration are thus essential for maintaining quality and acceptable vase life of cut flowers. The optimum temperature for storage of most of the common cut flowers is near the freezing point (0° C). Some tropical crops such as anthurium, bird-of-paradise, some orchids and ginger are injured, however, at temperatures below 10° C. Low temperature cause chilling injury and develop symptoms of darkening of the petals, water soaking of the petals (which look transparent) and, in severe cases, collapse and drying of leaves and petals.

Food supply: Starch and sugar stored in the stem, leaves and petals provide much of the food needed for cut-flower opening and maintenance. These carbohydrate levels are the highest when plants are grown in high light and with proper cultural management. Carbohydrate levels are, in fact, generally the highest in the late afternoon after a full day of sunlight. However, flowers are preferably harvested in the early morning, because temperatures are low, plant water content is high and a whole day is available for processing the cut flowers. The quality and vase life of many cut flowers can be improved by pulsing them after harvest with a solution containing sugar. Pulsing is done by standing the cut flowers in solution for a short period, usually less than 24 hours and often at low temperature. Typical examples are rose, where bloom size and colour are enhanced and in gladiolus flowers open further up the spike are bigger and have a longer vase life by the addition of sugar. Sugar is also an important part of the bud-opening solution used to open bud-cut flowers before distribution and as part of the vase solution used at the retail and domestic level.

Light: The presence or absence of light during storage is generally not a concern, except in cases where yellowing of foliage is a problem. The leaves of certain cultivars of chrysanthemum, protea, marguerite daisy and other crops turn yellow if stored in darkness at warm temperatures. The blackening of leaves of cut flowers of *Protea nerifolia* can be prevented by maintaining the flowers in high light. The problem is induced by low carbohydrate status in the harvested inflorescence.

Water supply: Cut flowers especially those with leafy stems, have a large surface area, so they lose water and wilt very rapidly. They should be stored at relative humidity above 95 per cent to minimize water loss, particularly during long-term storage. Water loss is dramatically reduced at low temperatures, another reason for prompt and efficient cooling of cut flowers. Even after flowers have lost considerable water (for example during transportation or storage) they can be fully rehydrated using proper techniques. Cut flowers will absorb solutions without difficulty provided there is no obstruction to water flow in the stems. Air embolism, bacterial plugging and physiological plugging are factors that can reduce solution uptake.

Air embolism: Air embolism occurs when small bubbles of air (emboli) are drawn into the stem at the time of cutting. These bubbles cannot move far up the stem, so the upward movement of solution to the flower is restricted. Emboli can be removed by either recutting the stems under water (removing about 1 inch), ensuring that the solution, is acid (pH 3 or 4), or by placing the stems in a vase solution heated to 41° C (warm, but not hot).

Bacterial plugging: As stated before the quality and vase life of flowers can be improved by supplying more food (as sugar) after harvest. Sugar can, however, also act as food for the growth of detrimental fungi and bacteria in the water and this growth can be further enhanced by organic materials that leak out of the cut stem. Substances produced by the bacteria and the bacteria themselves, can plug the water-conducting system. For this reason, it is important that buckets should be cleaned and disinfected regularly and that flower-holding solutions should contain germicides to prevent the growth of microorganisms. An acidic solution also inhibits bacterial growth.

Physiological plugging: When a plant is cut or injured the cells at the cut surface may respond to the damage by closing off the wound. In nature, this response helps prevent infection of the plant through injured surfaces. Physiological plugging in cut stems seems to result from deposition of pectic materials in the xylem elements. Obviously, such deposits drastically reduce water flow and lead to early wilting.

Quality of water: Hard water frequently contains minerals that make the water alkaline. Alkaline water, which does not move readily through cut-flower stems, can substantially reduce vase life. This problem can be overcome either by removing minerals from the water (by using a deionizer) or by making the water acid. Commercial flower preservatives may not contain enough acid to acidify some very alkaline waters. In that case, more acid should be added to the water. Sodium (Na), present in high concentrations in soft water, is, for example, toxic to carnations and roses. Fluoride (F) is very toxic to gerbera, gladiolus and freesia;

fluoridated drinking water contains enough F (about 1 ppm) to damage these cut flowers.

Ethylene: Certain flowers, especially carnations and some rose cultivars, die rapidly if exposed to minute concentrations of ethylene gas. A number of cut flowers produce ethylene as they age. In carnations and sweet peas, this ethylene is involved in the death of the flower. In other flowers, such as calceolaria, snapdragon and delphinium, ethylene causes flower abscission or shattering. Levels of ethylene above one hundred parts in one billion parts of air (100 ppb) in the vicinity of most cut flowers can cause damage and therefore should be avoided. Storage and handling areas should be designed not only to minimize contamination of the atmosphere with ethylene but with adequate ventilation to remove any ethylene that does occur. Treatment with the anionic thiosulfate complex of silver (STS) reduces the effect of ethylene (exogenous or endogenous) on some flowers. Finally, refrigerated storage is beneficial in that ethylene production and ethylene sensitivity of the product are reduced greatly when temperatures are low.

Growth tropisms: Certain responses of cut flowers to environmental stimuli (tropisms) can result in quality loss. Most important are geotropism (bending away from gravity) and phototropism (bending towards light). Geotropism often reduces quality in spike-flower crops like gladiolus and snapdragon, because the flowers and spike bend upward when stored horizontally. These flowers should be handled upright whenever possible.

Mechanical damage: Bruising and breakage of cut flowers should be avoided. Flowers with torn petals, broken stems or other obvious injuries are undesirable for aesthetic reasons. Disease organisms can more easily infect plants through injured areas. In fact, many disease organisms can only enter a plant at an injured area. Additionally, respiration and ethylene evolution is generally higher in injured plants, further reducing storage and vase life.

Diseases: Flowers are very susceptible to disease, not only because their petals are fragile, but also because the secretions of their nectaries often provide an excellent nutrient supply for even mild pathogens. To make matters worse, transfer from cold storage to warmer handling areas can result in condensation of water on the harvested flowers. The most commonly encountered disease organism, grey mould (*Botrytis cinerea*), can germinate wherever free moisture is present. In the humid environment of the flower head, it can even grow (albeit more slowly) at temperatures near freezing. Proper management of greenhouse hygiene, temperature control and minimizing condensation on the harvested flowers all reduce losses caused by this disease. Some new fungicides, such as Ronalin and Rovral have been approved for use on cut flowers and are very effective against grey mould.

POSTHARVEST MANAGEMENT TECHNIQUES

Systems for harvesting and marketing cut flowers vary according to individual crops, growers, production areas and marketing systems. All involve a series of steps like harvesting, grading, leaf removal, bunching, recutting, sleeving, hydration, special treatments, packing, pre-cooling, cold storage, transportation and delivery to market are done to prepare the flowers for market are collectively called handling. Not all of these are done to all flowers, and whether they are used or not depends on the market the flowers are going to be sold to. Management systems should be selected so as to maximize postharvest life of the flowers, which usually requires prompt pre-cooling and proper temperature management throughout the harvesting chain. Increasingly, producers are trying to reduce the number of separate steps in the marketing chain. For example, some field flower growers cut, grade, bunch and pack their product in the field. The packed boxes are then taken directly to the precooler. Such systems, where appropriate, reduce damage to the flowers and may decrease labour costs. Flowers for local retail markets often are packed this way since they are marketed immediately after harvest. Flowers also can have these steps performed in the field and then be transported to a packing shed where recutting, special treatments, precooling and dry packing can be performed. All the handling steps can be done in a packing shed. It often makes for a better flow of activities if they are all done in the same place.

Harvesting: Harvesting is normally done by hand using shears or a sharp knife. Simple mechanical aids are used to harvest some crops, for example the hook-shaped comma which permits chrysanthemum harvest without stooping and rose shears which grip the flower stem after it has been cut, allowing it to be withdrawn single-handedly from the bench. The harvested flowers should not be placed on the ground because of the danger of contaminating the flowers with disease organisms. All the activities harvesting, grading and packing should be done in dry condition that is to say without the use of chemical solutions or water. If this is not possible, however, clean buckets containing, clean water containing a biocide and sufficient citric acid should be used instead.

Grading: Grading starts with deciding which flowers to harvest. Marketable flowers are free of blemishes, including both leaves and petals. The flowers can be grouped or graded by stem length if there are differences and also by developmental stage. More mature ones should be sold as soon as possible, while others can be held in cold storage for later sales. The grade standards for cut flowers are depending on nature of flower and stem length. Vase life or usefulness, straightness of stems, flower size, freedom from defects, maturity, uniformity and foliage quality are among the factors which should, also be used in cut flower grading.

Mechanical grading systems should be carefully designed to ensure efficiency and to avoid damaging the flowers.

Bunching: Damage through multiple handlings can be reduced if grading, sizing and even bunching are carried out in the field or greenhouse. Flowers should be graded and bunched before being treated with chemicals or being placed in storage. The bunched and packaged flower depends on the market and their use. At present time mixed bunches and single type bunches both are popular. Larger flowers such as lilies, gladiolus and sunflowers often are sold as single stems. Flowers are normally bunched, except for anthuriums, orchids and some other specialty flowers. The number of flowers in the bunch varies according to growing area, market and flower species. Groups of 10, 12 and 25 are common for single-stemmed flowers. Spray-type flowers are bunched by the number of open flowers, by weight or by bunch size. Most of flowers are bunched by 10 or 5 some, like roses, carnations, lilies-of-the- valley are bunched in 25 and sweet violets are bunched in 100 with a collar of leaves underneath the flowers. Large, expensive to grow flowers can be sold by single stems. Bunches are held together by string or paper-covered wire or elastic bands and are frequently sleeved soon after harvest to separate them, protect the flower heads, prevent tangling and identify the grower or shipper. Materials used for sleeving include paper (waxed or unwaxed) and polyethylene (perforated, unperforated and blister). Sleeves can be preformed (although variable bunch size can be a problem), or they can be formed around each bunch using tape, heat sealing (polyethylene) or staples.

Pre shipping handling: Proper pre-shipping handling is important in order to get flowers to the market in good shape. The flowers should be well hydrated but not wet when packed. Most spike flowers like snapdragons and gladiolus need to be packed upright to prevent the tips from curving. Special boxes or hampers are made for these types of flowers. Once bunched, flowers should be hydrated, placed in water for awhile before they are packed dry. The hydrating step should include a step where, after the flowers are bunched, the stems are recut under water to eliminate any air bubbles in the xylem that can block the uptake of water. These air bubbles can occur when the flowers were harvested. Once recut, the flower can be placed in a general holding solution used to hydrate the flowers or receive a special treatment such as silver thiosulfate.

Chemical solutions: The various chemical solutions used after harvest to improve the quality of cut flowers usually have specific purposes. Fresh flower preservatives are chemicals added to water to make flowers last longer. They contain a germicide, a food source, a pH adjuster, water, and sometimes surfactants and hormones. Germicides are used to control bacteria, yeasts and molds. These microorganisms harm flowers by producing ethylene, blocking the xylem, producing toxins and increasing sensitivity to low temperatures. Bacterial counts of 10 to 100 million per 1 milliliter impairs uptake, while counts of 3 billion per 1 milliliter causes wilting. The 8-HQC is the most common preservatives and sucrose is ordinary source of energy (food). It provides energy to sustain flowers longer and to open in the bud stage. One to 2 per cent sucrose is the standard amount in preservatives but never use sucrose without a germicide, because it is the primary food source for microorganisms. Surfactants (eg.Tween 20 and Triton) and wetting agents reduce water tension so water is taken up more easily. Water is the most important component of floral preservatives. There are some properties or characteristics like pH, temperature, soluble salts, alkalinity and hardness effect on flower shelf-life. Water with a low pH (3.5 to 5.0) is taken up by the flowers quicker and more easily and fewer microbes can grow. The lower pH inhibits the growth of xylem blocking microbes. Citric acid, an organic acid, usually is used to acidify water to adjust the pH of the water to 3.5 to 5.0. The best temperature is warm 38° C. Warm water has less dissolved gases in it, dissolved gas bubbles can cause blockages in the xylem just like microbes. The total soluble salts are made up of alkalinity or hardness and salinity it expressed as ppm or mg liter^{-1} or microsiemans (1 ppm = 1 mg/liter = 700μ siemans). Low soluble salt content is best. They can interfere with water uptake because they change the osmotic potential of the water. They can burn leaves and petals because they can accumulate in the tips. Alkalinity is measured as mg liter^{-1} Calcium Carbonate, $CaCO_3$ equivalents as negative ion salts in water. The active ions are bicarbonate (HCO_3^-), carbonate, (CO_3^{--}) and hydroxide (OH^-). Water with less than 100 mg liter^{-1} $CaCO_3$ equivalents is best. If it is higher, the pH of the water may not be able to be adjusted with a floral preservative and will make the floral preservatives ineffective since the acidifiers in them may be completely "buffered out" by the alkalinity of the water. To make the floral preservative more effective, an acid such as citric acid will need to be added instead of adding more floral preservative.

Table 112. Toxic soluble salt concentrations for selected species

Species	Sensitivity concentration	Species	Sensitivity concentration
Glads	700 ppm	Mums	200 ppm
Roses	200 ppm	Carnations	200 ppm

Hardness often is used interchangeably with alkalinity but it measures the positive ion salts in the water like magnesium, Mg++ and calcium, Ca++. The standard practice to decrease hardness in water is to soften it by exchanging the magnesium and calcium ions with sodium ions. Sodium ions, though, can be toxic to the flowers at high levels. This method of decreasing water hardness should not be used for cut flowers. Flowers in the lily family and other monocots are more sensitive to fluoride than others. Fluoride toxicity is more of a problem at a lower pH, which is best for holding flowers. Species that are more sensitive to fluoride are listed below-

Table 113. Fluoride sensitivity for select species

Species	Sensitivity Concentration	Species	Sensitivity Concentration
Freesias	1 ppm	Mums	5 ppm
Gladiolus	1 ppm	Snapdragon	5 ppm
Gerberas	1 ppm	Roses	5 ppm

The soluble salt content, fluoride levels and alkalinity are so bad a reverse osmosis, deionizing or distilltion water system will be needed to purify the water. Pure water has no contaminants, no soluble salts, no fluoride, and no buffering effect so preservative pH adjustors work well. Besides the standard floral preservative solutions for holding flowers there are some specific solutions and treatments that serve different needs.

» A harvesting solution often will simply be water acidified with citric acid to a pH of 3.5 to 5.0.

» A conditioning, hardening, or hydrating solution is used to restore the turgor of wilted flowers and dry packed flowers. It usually is warm water with a germicide, acidified to pH 3.5 to 5.0 with citric acid and a wetting agent, i.e. Tween- 20 at 0.01 to 0.1 percent.

» Impregnation is a treatment that protects stems against the blockage of water vessels by microbes. Stems are dipped in 1000 ppm silver nitrate solution for 10 minutes and should not be recut after treatment.

» Pulsing or loading is a type of treatment used to extend the vaselife of flowers held in water, stored wet or dry for long periods or shipped long distances. It is called a pulse because it is only done for a short period of time, or called loading because the flowers are loaded up with food for a long storage period.

» The higher concentration of sugar can act like a soluble salt causing petal and leaf injury, the treatment is only a few hours or a day. The temperature should be 18-24° C and light intensity should be 2000 lux.

» Bud opening solutions are used to open flowers harvested in a tight bud stage. Stems are placed in solutions containing higher concentrations of sugar, plus a germicide and hormonal compounds that facilitate bud growth and development.

Rehydration: Wilted flowers, placed in water to restore turgidity, should be rehydrated with deionized water containing a germicide. Wetting agents (0.01 to 0.1%) can be added and the water should be acidified with citric acid, HQC or aluminum sulfate to a pH near 3.5. No sugar should be added to the solution and rehydration should be carried out in a cooler.

Pulsing: The term pulsing means placing freshly harvested flowers for a relatively short time (a few seconds to several hours) in a solution specially formulated to extend their storage and vase life. Sucrose is the main ingredient of pulsing solutions and the proper concentration ranges from 2 to 20 per cent, depending on the crop. Some cut flowers are also pulsed with silver thiosulfate (STS) to reduce the adverse effects of ethylene. Flowers can be pulsed for short periods at warm temperatures (eg. 10 minutes at 21° C) or for long periods at cool temperatures (eg. 20 hours at 2° C). Short pulses (10 seconds) in solutions of silver nitrate have proved valuable for some crops. Chinese asters and maidenhair fern respond well to solutions containing 1000 ppm silver nitrate. Other flowers are damaged by these high concentrations, but respond well to 100-200 ppm (eg. gerberas). The function of the silver nitrate is as a germicide (eg. chrysanthemums). In all cases, residual silver nitrate solution should be rinsed from the stems before packing.

Bud opening: Bud-cut flowers must be opened in bud-opening solutions before they are sold to the consumer. These solutions contain a germicide and sugar. Foliage of some flowers (especially roses) can be damaged if the sugar concentration is too high. Buds should be opened at relatively warm temperatures (21-27° C), moderate humidities (60-80% R.H.) and reasonably high light intensities (100-200 foot candles).

Tinting: Artificial colouration of flowers (tinting) is carried out in two ways: application via the stem (for carnations) and by dipping the flower heads (for other flowers, principally daisies). In tinting carnations, proprietary dye solutions (combinations of food type dyes with adjuvants designed to increase the uptake of the solution) are mixed in a bucket and warmed to about 41° C. The carnations to be tinted (usually White sim) are allowed to dry somewhat (overnight in the packing shed at 18° C) to increase their rate of solution uptake. Dyeing is stopped before the flowers reach the desired colour, because dye still in the stem is flushed into the flower by the vase solution. Flowers that cannot be treated in this way are tinted with different proprietary tinting solutions containing aniline dyes dissolved in organic solvents such as rubbing alcohol (isopropanol). The heads of the flowers are dipped in a vat of the dye and shaken to remove surplus solution and then placed on a rack to dry before storage or packing.

Packing: There are many shapes of packing containers for cut flowers, but most are long and flat. This design restricts the depth of the flowers in the box, which may in turn reduce physical damage of the flowers. In addition, flower heads can be placed at both ends of the container for better use of space. With this kind of flower placement, whole layers of newspaper have traditionally been used to prevent the layers of flowers from injuring each other. The use of small pieces of newspaper to protect only the flower heads, however, is probably the better practice, since it allows for more efficient cooling of flowers after packing. It is critically important that containers be packed in such a way that transport damage is minimized. Most packers secure the product by using enough flowers and foliage in the box so that it holds itself firmly. To avoid longitudinal slip, packers use one or more cleats. These are normally foam or newspaper covered wood pieces that are placed over the product, pushed down and stapled into each side of the box. Padded metal straps, high density polyethylene blocks and cardboard tubes can also be used as cleats. The heads of the flowers are normally placed 3 to 5 inch (7 to 12 cm) from the end of the box to eliminate the danger of petal bruising at the time of transport. When flowers are packed into boxes, the bunches are sleeved or wrapped and then packed tightly so the bunches do not move or vibrate in transit (causes bruising). The standard flower box is 12 x12 x 48 inches. There are smaller sizes, too, called half or quarter boxes that are 6 x 12 x 48 inches and 6 x 6 x 48 inches, respectively.

Gladioli, snapdragon and some other species are often packed in vertical hampers to prevent geotropic curvature which reduces their acceptability. Cubic hampers are used for upright storage of daisies and other flowers. Specialty flowers such as anthurium, orchid, ginger and bird-of-paradise are packed in various ways to minimize friction damage during transport. Frequently the flower heads are individually protected by paper or polyethylene sleeves. Cushioning materials such as shredded paper, paper wool and wood wool may be distributed between the packed flowers to further reduce damage. Sleeving or wrapping the bunches helps prevent the different bunches and flowers from becoming tangled. Columbine, larkspur, delphinium, baby primrose, forget-me-nots and buddleia are flowers that should be wrapped or sleeved prior to marketing to prevent tangling. The packing shed may be an ultra modern air conditioned building or an open air covered porch.

The handling space should

» Be shaded or covered to keep temperatures lower and prevent direct sunlight on the flowers.

» Have a clean water source for preparing harvest, treatment and holding solutions, and for use in cleaning the area.

» Have ample space so all handling activities can be performed smoothly, such that workers are not crossing over each other.

» Have a cold storage or at least a cooler, shaded place to store the flowers until they are ready for market.

Precooling and Storage

Precooling is a step that rapidly brings the temperature of the flowers down from the field temperature to a proper storage temperature. A low temperature slows the respiration rate of the flowers which in turn helps them last longer. By far the most important part of maintaining the quality of harvested flowers is ensuring that they are cooled as soon as possible after harvest and that optimum temperatures are maintained during distribution. Most flowers should be held at 0-2° C. Chilling-sensitive flowers (anthurium, bird-of-paradise, ginger, tropical orchids) should be held at temperatures above 10° C. Individually, flowers cool and warm rather rapidly (half-cooling times of a few minutes) and once packed, cut flowers are difficult to cool. Their high rate of respiration and the high temperatures of most greenhouses and packing areas result in heat build-up in packed flower containers unless measures are taken to ensure temperature reduction. It is therefore necessary to cool the flowers as soon as possible after packing. The individual flowers can be cooled quickly, it is also true that individual flowers brought out of cool storage into a warmer packing area will warm quickly and develop condensation prior to packing. The simplest method of ensuring that packed flowers are adequately cooled and dry is, therefore, to pack them in the cool room. Although this method is not always popular with packers and may slow down packing somewhat, it ensures a cooled, dry product. Forced-air cooling of boxes with end holes or closeable flaps is the most common and effective method for pre-cooling cut flowers. Cool air is sucked or blown through the boxes. Care must be taken to pack in such a way that air can flow through the box and not be blocked by the packing material. In general, packers use less paper when packing flowers for pre-cooling. The half-cooling time for forced-air cooling ranges from 10 to 40 minutes, depending on product and packaging.

The precooling of flowers is a very important step for individuals selling to a large wholesale market, distant markets and if their crop is to be stored for a long time such as peonies. Individuals who sell at a local retail market usually do not need to worry about this step since their flowers will be in the customers home the day they are picked. If the packages are to remain in a cool environment after pre-cooling, vents may be left open to assist removal of the heat of respiration. Flowers that are to be transported at ambient temperatures can be packed in polyethylene caskets, foam-sprayed boxes or boxes with the vents resealed. Ice that is used after pre-cooling is only effective if it is placed so that it intercepts heat entering the carton (i.e. it must surround the product) and care must be taken to ensure that the ice does not melt onto the flowers or cause freezing damage. Special care needs to be taken with tropical flowers shipped in a mixed load. The flowers should be packed in plenty of insulating material (an insulated box packed with shredded newsprint, for example). These flowers should not be pre-cooled. If they are to be shipped by truck, they should be placed in the middle of the load, away from direct exposure to cooling air. Cold storage is recommended for all flowers that will not be in the market immediately and any flowers sold wholesale. As stated before, low temperatures slow the respiration rate of the flowers and prolong the vaselife of the flowers. In general, temperatures should be 0 to 5° C and have a relative humidity of 85 to 90 percent, for most flowers. Flowers should never be stored with fruits and vegetables. Some fruits and vegetables produce ethylene that can dramatically shorten the life of the flowers. Once flowers are bunched into marketable units they should be placed in cold storage. As a new grower using local retail markets, a refrigerated cold storage may not be available or affordable. Since most of their flowers will be sold within hours a cool place such as an air conditioned room, cellar or basement could be used. Flower storage life and vaselife are considered to be two different things. The customer wants to know the vaselife, how long will the flowers last in home while the grower needs to know both to determine how long flowers can be kept in cold storage and to be able to tell customers how long the flowers will last. Table 117 lists several species of common cut flowers and their vaselives. If flowers have to be stored before marketing, a cool place (preferably a refrigerated cold storage, especially for flowers) should be used. There are many flowers that are not commonly found in the wholesale market because they do not store well, ship well or last long. These should only be used for local markets. These include foxglove, garden phlox, lupine, clarkia, stevia, common stocks, candytuft, cornflower, feverfew, blue laceflower, english daisy, calendula, pot marigold, sweet violets and gaillardia.

Table 114. Recommended storage temperature and approximate storage life and vaselife for fresh cut flowers and florist greens

Cut flowers	Storage Temperature		Approximate Period	
	°C	°F	Storage Life	Vaselife
Allium	0 to 2	32 to 35	2 weeks, dry 4 week, wet	10 to 14 days -
Anemone	4 to 7	40 to 45	2 days	4 to 8 days
Aster, China	0 to 4	32 to 40	1 to 3 week	5 to 10 days
Buddleja	4	40	1 to 2 days	1 to 2 days
Calendula	4	40	3 to 6 days	-
Candytuft	4	40	3 days	-
Clarkia	4	40	3 days	5 to 10 days
Columbine	4	40	2 days	-
Coreopsis	4	40	3 to 4 days	-
Cornflower	4	40	3 days	6 to 10 days
Cosmos	4	40	3 to 4 days	4 to 6 days
Dahlia	4	40	3 to 5 days	7 to 14 days
Daisy, English	4	40	3 days	-
Daisy, Marguerite	2	36	1 to 2 week	4 to 7 days
Daisy, Shasta	4	40	7 to 8 days	-
Delphinium	4	40	1 to 2 days	4 to 12 days
Feverfew	4	40	3 days	-
Forget-me-not	4	40	1 to 2 days	-
Foxglove	4	40	1 to 2 days	-
Freesia	0 to 5	32 to 33	10 to 14 days	4 to 12 days
Gaillardia	4	40	3 days	-
Gladiolus	2 to 5	35 to 42	5 to 8 days	6 to 10 days
Godetia	10	50	1 week	-
Gypsophila	4	40	1 to 3 weeks	5 to 10 days
Iris, bulbous	-0.5 to 0	31 to 32	1 to 2 weeks	3 to 6 days
Lilac, forced	4	40	4 to 6 days	7 to 10 days

Lily	0 to 1	32 to 34	2 to 3 weeks	7 to 10 days
Lily-of-the-valley	-0.5 to 0	31 to 32	2 to 3 weeks	4 to 11 days
Lupine	4	40	3 days	-
Marigolds	4	40	1 to 2 weeks	-
Narcissus	0 to 0.5	32 to 33	1 to 3 weeks	4 to 8 days
Poppy	4	40	3 to 5 days	-
Peony, tight buds	0 to 1	32 to 34	2 to 6 weeks	2 to 7 days
Phlox	4	40	1 to 3 days	2 to 7 days
Ranunculus	0 to 5	32 to 41	7 to 10 days	5 to 7 days
Snapdragon	4	40	1 to 2 weeks	5 to 7 days
Snowdrop	4	40	2 to 4 days	-
Squill	0 to 0.5	32 to 33	2 weeks	-
Statice	2 to 4	35 to 40	3 to 4 weeks	4 to 8 days
Stock	4	40	3 to 5 days	5 to 8 days
Strawflower, fresh	2 to 4	35 to 40	3 to 4 weeks	7 to 10 days
Sweet pea	-0.5-0	31 to 32	2 weeks	3 to 7 days
Sweet william	7	45	3 to 4 days	5 to 9 days
Tulip	-0.5 to 0	31 to 32	2 to 3 weeks	3 to 6 days
Violet	1 to 5	34 to 41	3 to 7 days	-
Zinnia	4	40	5 to 7 days	6 to 10 days
Florist greens				
Asparagus(plumosa)	2 to 4	35 to 40	2 to 3 weeks	6 to 14 days
Asparagus(sprenger)	2 to 4	35 to 40	2 to 3 weeks	6 to 14 days
Dagger and wood ferns	0	32	2 to 3 weeks	-
Hedera	2 to 4	35 to 40	2 to 3 weeks	-
llex (holly)	0 to 4	35 to 40	3 to 5 weeks	5 to 14 days
Juniper	0	32	1 to 2 months	-
Leatherleaf (baker fern)	1 to 4	34 to 40	1 to 2 months	7 to 15 days
Magnolia	2 to 4	35 to 40	2 to 4 weeks	-
Mistletoe	0	32	3 to 4 weeks	-

Cooling time calculation

The time necessary to reach a desired temperature is expressed in terms of a typical cooling curve. The seven-eighths cooling time, or the time required to reduce the temperature of the flowers seven-eighths of the way from the initial temperature to the temperature of the cooling air, is considered the target for a commercial cooler. Note that the rate of cooling becomes very slow as the temperature of the flowers nears the temperature of the refrigerated air. Consequently, flowers are seldom completely cooled to the temperature of the cooling air. Half an hour of cooling is required to reach 3.0°C (seven-eighths cooling). More than 2 hours cooling would be required for the flowers to approach the temperature of the cooling air (0°C). Care must be taken to pack in such a way that air can flow through the box and not be blocked by the packing material. In general, packers use less paper when packing flowers for pre-cooling. The half-cooling time for forced-air cooling ranges from 10 to 40 minutes, depending on product and packaging. Flowers should be cooled for three half-cooling times (by which time they are 7/8 cool).

Retail handling

In order to ensure maximum quality and satisfaction to the final customer, it is vital that retail florists handle flowers properly on receipt. On arrival, packed boxes should be unpacked or placed immediately in a cool-room. After unpacking, flower bunches should be re-cut (preferably under water) and placed in a cool-room (in a warmed re-hydration solution) for at least a few hours. If flowers are to remain in the store for several days, they should be displayed in a clean bucket or vase containing a vase preservative. Chilling-sensitive flowers (anthurium, bird of paradise, ginger, tropical orchids) should be held at temperatures above 10° C. Other flowers should be held at 0-2° C. Cut flowers that are held during storage with their bases in water (wet storage) are usually kept only for 1 to 3 days, while those stored dry are often kept for longer periods. The airflows and pressures needed to cool a full-sized box of different types of flowers are depicted in the following table. In designing a precooler, the airflow (cfm) required can be estimated by multiplying the required cfm per box by the number of boxes that will be cooled, and adding 25 per cent to allow for air leaks.

Table 115. Static pressures, airflows, and cooling times required to cool standard boxes of specific flowers

Flower type*	Static Pressure			
	0.5 inch		1.0 inch	
	Air flow (Cfm box^{-1})	7/8 cooling time (Minutes)	Air flow (Cfm box^{-1})	7/8 cooling time (Minutes)
Carnations	70	48	90	40
Chrysanthemums	80	62	130	58
Gypsophila	170	10	260	8
Roses	140	34	200	25
Statice	150	40	210	18

*Carnation box size 48x 21 x12 inches, with two 2-inch diameter vent holes in each end, 51 lb gross weight
*Chrysanthemum box size 57x 21 x12 inches with two 2-inch diameter vent holes in each end, and 33 lb gross weight; 45-lb box will allow 50 per cent less air through the box.
*Gypsophila box size 42x 21 x12 inches with two 3-inch diameter vent holes in each end
*Rose box size 48 x 21 x12 inches, with two 2-inch diameter vent holes in each end, 20 bunches per box.
*Statice box size 42x 21 x12 inches with two 3-inch diameter vent holes in each end and grow weight of 75 lb.

Tropical flowers

Special care needs to be taken with tropical flowers shipped in a mixed load because these flowers should not be pre-cooled. The flowers should be packed in plenty of insulating material like insulated box with shredded newsprint. If they are to be shipped by refrigerated truck, they should be placed in the middle of the load, away from direct exposure to cooling air. Tropical flowers are well suited to air transport, since they are usually of high value, and are damaged by low temperatures.

Cooling for aquapacks

A marked trend in recent years has been to pack flowers vertically in boxes (Proconas™, aquapacks) that have water (hopefully containing a least a biocide) in the bottom. Considerations of weight and the potential for spills mean that when such boxes are used for air transport, they are packed without addition of solution (which may be added after transport). Pre-cooling of these boxes represents a particular challenge. Palletized Proconas™ can be placed against a modified plenum designed to draw air down through the flowers and out of the ventilation grids and handholds in the hamper bottoms. Padding above the air gap in the plenum seals air from the bottom of the boxes and the pallet. A strip of canvas or sheet of fibreboard around the bottom of the boxes and the pallet ensures that air is drawn down through the boxes.

Vacuum cooling

Vacuum coolers have been installed in flower marketing terminals (mostly near airports but also by a few growers) as a means of removing heat from packed flowers. Vacuum cooling is based on the fact that water boils at lower temperatures at lower pressures. At about $1/10^{th}$ atmosphere pressure, water boils at 0° C. When flowers or plants are placed at these low pressures, the water in their cells boils, removing heat. Although these systems cool leaves, stems, and growing media promptly and efficiently, flower petals often cool somewhat more slowly, because water cannot escape readily from them. Essentially the same seven eighths precooling time (or possibly 10-15 minutes less) is required to vacuum cool flowers as for the forced-air cooling systems. When used at or near airports, these systems serve as a useful 'backstop' for flowers that have not been properly cooled by growers and/or shippers. However, in some cases it can be many hours before packed flowers arrive at the airport or trucking dock, which suggests that they may have already deteriorated significantly before being placed in a vacuum cooler. Some advantages of vacuum cooling methods include that paper and plastic packaging materials in the boxes do not affect the efficiency of cooling, free water is removed, boxes can be stacked in any manner in the precooler, and it can work equally well with dry packs and wet packs for cut flowers and greens and for potted plants. The main disadvantage is the high capital cost of the equipment. In a centralized facility this high cost can be offset by the high throughput of boxes that these systems can achieve.

Irradiation

Flowers that are shipped across international or even state boundaries are often subject to quarantine inspection for the presence of harmful diseases and pests. For many years it has been suggested that ionizing irradiation might be used to sterilize or kill insect pests without damaging the flowers. Some proponents of irradiation have even suggested that the irradiation treatment may improve the life of the flowers. The nature of ionizing radiation is such that it leaves no residue after treatment and if insects could be controlled without any effect on flowers or foliage this would provide an excellent quarantine treatment. Insects are relatively more susceptible to ionizing radiation than plants. Doses as low as 0.2 Gy are sufficient to sterilize most harmful insects. Although flowers do not suffer immediate and obvious damage at radiation doses below 0.7 - 1 Gy, the limited research that has so far been carried out that effects on flower quality and vase life can be detected at radiation doses close to those required to control insects. For example, at 0.2 Gy, vase life of irradiated wax flowers was more than halved. These data indicate the need for careful evaluation of effects on a range of flower species before the adoption of radiation as a measure for insect disinfestation in ornamentals.

Vase solutions

Vase solutions normally contain a low concentration of sugar (0.5 to 2.0%), a chemical to keep the pH of the solution low and a germicide. This treatment is usually provided by the retail seller or the customer for increase self-life. Preservatives are supplied to the trade in bulk or may be purchased in individual sachets intended for 1 quart of solution. A typical preservative solution contains 1 to 2 per cent sugar, a biocide (HQC or a mixture of quaternary ammonium compounds) and 300 ppm citric acid. Various other chemicals are sometimes incorporated, including aluminum sulfate (which acts by forming the flocculant aluminum hydroxide to trap particles in the water and also lowers the pH), slow release chlorine compounds (good biocides, but sometimes toxic) and 6-benzyl adenine (a plant growth regulator which has been shown to increase the vase life of some flowers).

Ethylene damages of cut flowers

Ethylene gas cause damages in some cut flowers species by causing flowers to drop prematurely, flower buds to not open, and flower petals to close. For prevent ethylene damage the flowers keep away from ethylene by preventing ethylene pollution, remove ethylene from the atmosphere and inhibit the effect of ethylene on flowers. Silver thiosulfate acts on both ethylene receptors and production sites in the flower and it can inhibit the effect of ethylene. This protects the flowers from ethylene in the environment and it stops the flower from producing ethylene itself. Other chemicals are MCP (1-methylcyclopropane) a gas which acts only on receptors, but is not available commercially. Some specific measures to prevent ethylene damage on flowers are:

Protect plants against pest and diseases infestation.

- Harvest flowers at optimum stage and prevent form pollination.
- Avoid physical injury to flowers during handling and cool flowers immediately after harvest.

- Keep storage and handling facilities clean, and remove diseased and dying plant material.
- Do not use internal combustion engines in any handling work or production areas.
- Have good air circulation and ventilation in handling and storage areas.
- Do not store flowers with ethylene producing fruits and vegetables.
- Do not store newly harvested flowers in bud stages with fully open flowers.
- Use ethylene scrubbers in cold storage area and use silver thiosulfate treatment on sensitive species.
- The handling area and cold storage should be cleaned and sanitized after each use.
- Equipment, cutting utensils, containers and handling surfaces should be cleaned and disinfected with a 1:10 bleach solution.

Ethylene sensitive species: Snapdragons, Dutch iris, Sweet peas, Freesia, Larkspur, Baby's breath, Delphinium, Alstroemeria.

Preparation of silver thiosulfate (STS) solution

1. Dissolve 0.079 g $AgNO_3$ in 500 ml of deionized water.
2. Dissolve 0.462 g $Na_2S_2O_3X\ 5H_2$ in 500 ml deionized water.
3. Pour $AgNO_3$ solution into $Na_2S_2O_3X\ 5H_2$ solution while stirring. The concentration of silver is 0.463 mM (463 ppm).
4. The silver thiosulfate solution is now ready to use. If not used immediately, the solution may be kept in a dark glass or plastic container at 18 to 24° C , in total darkness for up to four days.

Flowers in the supermarket

Flowers intended for supermarket sales require special care throughout the distribution chain. First, only flowers which have adequate vase life to remain saleable for the intended period of time should be used. Some short-lived flowers such as iris, tulip and narcissus should therefore be avoided. Flowers with common postharvest problems (roses, gypsophila) are a danger too. It is particularly important that ethylene-sensitive flowers (notably carnations) be treated with STS if they are to be marketed through produce stores.

Drying of flowers

A number of cut flower crops are sold in both fresh and dried forms. For some, such as statice and strawflower, the drying process can be as simple as hanging bunches of harvested flowers upside down (to keep the stems straight) in a warm, dry location. Others, such as gypsophila and silver-dollar eucalyptus, which become too brittle if dried in this fashion, are placed in a freshly prepared solution of 30 per cent glycerol. The glycerol moves through the plant in the course of a day or so. If the dried foliage or flowers are to be coloured, dye can be added to the glycerol. After the glycerol treatment, the flowers are hung, upside down, in a warm dry environment to dry.

Quality control of cut flowers

Because there are no official grade standards for cut flowers, it is vital that the industry itself police the quality of flowers that are marketed. Each of the individual product sections below suggests quality features that should be considered by producers, wholesalers and retailers. The most important quality parameter, however, is freshness or vase life. This parameter is difficult to assess visually, but because of its importance, the producers and shippers alike should consider setting up a quality control program that would involve evaluation of the vase life of representative flowers on a continuing basis. Supply of small vases (use plastic baby bottles with disposable liners) and a reasonably little space where the temperature is fairly constant.

Logistics for air transport of cut flowers

Because of the dramatic effects of transport temperatures on subsequent vase life and the propensity for overheating of packed flowers, they should be transported at close to the optimal temperature (0° C for most species). Passive and active refrigeration systems are usefully be applied to the transportation of cut flowers. Envirotainers™, as an example, provide a dry-ice refrigeration system that could provide controlled temperatures suitable for use with cut flowers and would allow transport for a considerable distance at costs that would be more than recompensed by the improved quality of the flowers on arrival. Passively refrigerated and insulated containers provide alternative means for providing some control of temperature during transport. If product is properly cooled before being palletized or packed in an LD-3 container, insulation alone certainly will improve temperature management during the transportation chain. Given the lack of temperature control in the majority of aircraft carrying cut flowers, and the extreme response of cut flowers to temperature abuse, the logistics for air transport of cut flowers must focus on doing everything possible to maintain the cold chain. Flowers should be properly cooled at the grower's operation, and transported to the airport in refrigerated (or at least well insulated) trucks. In some airports vacuum coolers that been installed to reduce temperature of flowers before they are air-freighted. This expensive equipment certainly provides a means of rescuing product that is warm on arrival at the airport, but this is not an optimal procedure. Prior heat exposure and the additional water loss from the flowers during the vacuum cooling process undoubtedly compromise flower quality and vase life. Perhaps the weakest link in the flower postharvest chain is at the airport. Freight forwarders may be inundated with late arrivals just prior to aircraft leaving. Pallets are assembled hastily, there is no time for pre-cooling of warm flowers, and boxes are handled roughly. Pallets may be at ambient temperature for up to 4 hours while the aircraft is being loaded. The producing and transportation industries need to work together to set standards for temperature management, pallet construction, and temperature maintenance during loading.

Temperature management before loading

Flowers would be packed and cooled at the production location and loaded into LD-3 containers or made into pallets on LD-9 sheets before being transported in a refrigerated vehicle to the freight forwarder or air cargo facility. Whenever active refrigeration is not available, flowers should be kept cold by the following measures: · Consolidation into pallets to reduce the surface/volume ratio · Reduction in radiative heat gain by holding in a shaded location · Reduction of air infiltration by closing precooling vents (essential), and using pallet covers. Pallet covers should preferably be white or foil backed, and has the additional benefit of reducing water loss during air transport. Flowers cooled to 0° C and packed in a large pallet will gain about 1° C per hour due to the heat generated by the process of respiration. As the temperature rises, increase respiration rate, and the heat gain increases too, so that by the time the flowers are at 10° C they may gain 3° C per hour, and once they are at 20° C, the heat gain can be as much as 10° C per hour. Flower boxes unloaded from aircraft pallets have been probed at temperatures as high as 55° C, indicating the runaway danger of inadequate temperature management and delays in transport.

Pallet construction

Flowers are delicate, and box strength is compromised if corners don't line up. It's important that the pallet be constructed so that the corners are square and line up above each other. One of the most vexing problems of the flower industry is the wide array of box sizes, which may make construction of a square and well aligned papallet difficult. Walking or kneeling on flower boxes should not be permitted. If it's necessary to provide trestles and planks to allow construction of pallets without workers kneeling or walking on the boxes, that would be money well spent. Preferably, the pallets should be constructed from the side (base to top for each successive layer). Standardization of boxes so that they easily fit the standard LD-9 and LD-3 footprint would help a good deal. Immediately after pallet construction, the palletized flowers should be covered with an insulated cover (or at least with shrink- or cling-wrap polyethylene film). This will reduce heat gain as a result of air movement through the pallet.

Recommendation for Cut Flowers

The recommendations for different cut flowers are based on climatic condition as well as post harvest life of flowers.

Anthurium (*Anthurium andraeanum*)

The elegant blooms of this tropical aroid are produced and sold throughout the world. The true flowers are found on the "spadix", the upright organ in the centre of the "spathe", which is the decorative petal-like organ surrounding the spadix. Although anthuriums are sensitive to low temperatures, they have a long vase life when properly handled. The end of their vase life is usually the result of inability to draw water from the vase solution and is associated with loss of glossiness and then blueing of the spathe. Most of the water lost by the flower evaporates from the spadix. Application of wax to prevent this water loss, or pulsing with silver nitrate, to improve water relations of the flower, can extend their vase life considerably.

Maturity: Maturity of anthurium flowers is determined by the proportion of open flowers on the spadix. In immature anthuriums, the spadix is smooth. Flower opening starts at the base of the spadix and proceeds upwards; spadices with open flowers are noticeably rough. Although producers in some countries harvest anthuriums when as little as 20 per cent (1/5) of the spadix is rough. Hawaiian growers harvest flowers when only 1/4th of the spadix is still smooth (3/4th of the flowers are therefore open). Harvesting anthuriums when more mature increases overall vase life.

Grading: Although there are no formal grade standards for anthuriums, top quality implies long stems, uniformity of colour and size, proper maturity, glossiness of the spathe and freedom from any damage or disease.

Table 116. US standards of grading of Anthuriums and masterbox capacities

Grade	Average length and width of spathe	Stems box-1
Miniature	Under 7.5	-
Small	7.5 to 10.0	20
Medium	10.0 to 12.5	16
Large	12.5 to 15.0	6
Extra large	Over 15.00	12

Pre-treatments: The short vase life problem of flower is associated with bacterial contamination of the cut stem bases. If anthuriums are placed in water after harvest, a biocide (eg. 50 ppm hypochlorite) should be added. Either pulse the recut stems for 10 to 20 minutes in 1000 ppm silver nitrate (Rinse the stems with fresh water after the treatment) or dip the whole flower in an emulsion of carnuba wax (3% FMC wax). After dipping, place the flower stems in water while the wax dries.

Chemical solutions: Pre-treated flowers are not harmed by vase preservatives but derive no additional benefit from them. Anthuriums can have a vase life of up to 3 weeks if properly treated. Even after a week of storage, vase life can be adequate, if proper techniques are used to handle the flowers.

Packing: Anthuriums are normally packaged individually. They are commonly packed in moist shredded newsprint or other shredded paper. Major damage during transportation is the result of spadices puncturing the spathe of neighbouring flowers in the pack. Many producers now sheathe the flowers in small plastic bags and pack the anthuriums more densely in the box.

Cooling and storage: Anthuriums are very sensitive to chilling injury. Holding the flowers for any length of time at temperatures below 10° C will induce purpling, browning and then necrosis. Anthuriums should therefore never be pre-cooled with other flowers, nor held in low temperature cool-rooms. Anthuriums shipped in mixed loads at low temperatures should be protected from chilling exposure by appropriate insulation (for example wrapping the flowers in newsprint and packing them in an insulated box). Anthurium flowers can be stored for at least one week if packed in moist shredded newsprint and held at 16° C. Anthuriums respond favorably to storage in a controlled atmosphere. Vase life was increased by 50 per cent when flowers were stored in air at 13° C or in 2 per cent oxygen at room temperature for up to 2 weeks.

Chrysanthemum

It is the second most important of the three major cut flowers. Chrysanthemums (both standard or foot ball and pompons) have a long postharvest life when properly handled. The chief postharvest problems in these flowers are failure to draw water (which results in premature leaf wilting) and leaf yellowing.

Maturity: Chrysanthemums are normally harvested fully open, or nearly so. It has been shown, however, that these flowers can be harvested as quite tight buds and opened satisfactorily with simple bud-opening solutions. Bud-cut standards can be harvested at stage 2 (inflorescence about 2" across) or at stage 3 (inflorescence about $3^{1/2}$" across) when ray florets are just beginning to unfurl, or at stage 4 (inflorescence about 5" across) when fresh weight is about half that of fully developed inflorescences. Buds cut tighter than stage 2 develops smaller flowers at full bloom and are difficult to open. Pompon varieties can be harvested when most of the petals on the most mature flower are still upright. The flowers can be opened after storage or transportation.

Harvesting: Stems should be cut (with a knife, shears, or the specially-designed comma) at least four inches above the soil line to avoid taking woody plant tissue. Pinched pompons can be pulled out of the soil and then cut to correct length. Leaves are removed from the lower third of the stems.

Grading and bunching: The Society of American Florists has suggested the following grades for fully open standard chrysanthemums:

Label colour/parameter	Blue	Red	Green	Yellow
Minimum flower diameter (mm)	140	140	121	102
Mininimum length overall (cm)	76	76	61	60

Pompons are graded into 8 to 12 oz. bunches containing several stems. Standards or disbuds of equal sizes are graded into groups of 10 or 12. Each bunch of 5-8 pompons should be wrapped with plastic to prevent flowers from becoming entangled. Standards and spider mums can be wrapped individually with thin wax paper to avoid bruising and entangling florets. Some growers place nets over spider mums in the greenhouse before the buds open.

Chemical solutions

Rehydration: Stems should be placed in water containing a germicide soon after harvest.

Use 25 ppm silver nitrate in solution or a 10 second to 10 minute dip in 1000 ppm silver nitrate followed by good quality water that is low in salts.

Bud-opening: Bud-cut, standard chrysanthemums should be placed in a bud-opening solution containing 2 to 3 per cent sucrose and a germicide. Higher sugar concentrations injure the leaves. Physan is a common, effective germicide, but it discolours the stem portion in the solution; therefore only 1½ to 3 inches of solution should be used. After the buds are open, the injured portion of the stem can be removed. Silver nitrate at 25 ppm + citric acid at 75 ppm are very effective but more expensive to use than physan. Silver nitrate is, however, absorbed into the stem and becomes a lasting germicide throughout the life of the flower. HQC at 200 ppm may also be used as a germicide. The temperature during bud opening should be 20-22° C. Lower temperatures delay opening and higher temperatures reduce quality by producing shaggy florets. Light is necessary for about 16 hours a day to maintain the chlorophyll (green pigment) in the leaves. The light intensity at mid plant-height should be about 100 foot candles, about the light intensity of a well lighted office. Cool-white fluorescent lamps provide good light quality. Stage 2 buds require about 7 days for opening, while stage 3 buds are ready in about 6 days. The flowers should be treated as normal flowers for shipment after the buds are opened.

Prevention of leaf yellowing: Immersion in solutions of the cytokinin 6-benzyl adenine has been shown to be effective in preventing premature leaf yellowing in some pompon cultivars that are prone to this problem. This treatment is not yet used commercially.

Storage

Bud-cut stems: Stage 2 buds should not be stored, because they are too immature. Stage 3 buds can be stored up to 2 weeks and stage 4 buds up to 3 weeks at 0-1° C. Bud-cut stems that are held in cold storage beyond recommended durations can develop flat-topped inflorescences.

Fully open inflorescences: Mature, wrapped blooms can be stored dry for 3 to 4 weeks at -0.5° C. Storage at 2-3° C should not exceed 2 weeks. Yellowing of leaves can occur at 5° C in the dark but is less likely to occur at 1° C. Yellowing of leaves, fading of dark-coloured flowers and pink colouration forming in white or yellow flowers are indications of ageing in chrysanthemum. Extreme ageing is indicated by the wilting of florets. Pink colouration sometimes forms in white or yellow petals on plants grown at low night temperatures but this is an accumulation of anthocyanins, not a sign of ageing.

Handling after storage or transportation: Proper rehydration is vital for good vase life of chrysanthemums that have been stored or shipped long distances. Remove chrysanthemum bunches from the boxes, recut stems (remove about 1 inch) and place in tepid water 38°C containing 0.1 per cent Tween 20 and 75 ppm citric acid. This solution brings back turgidity within 2 hours if the room is cool and the light is subdued. After the stems are rehydrated, they can be moved to water containing 100 ppm Physan and placed in a cool room. It is possible to use a solution containing 5 to 10 ppm sodium hypochlorite (bleach) instead of Physan. Solutions containing these chlorine compounds should be replenished every 2-3 days. Sugar is not necessary in the vase solution, as it is not beneficial to open chrysanthemum flowers.

Carnation (*Dianthus caryophyllus*)

It is one of the most important of the commercial cut-flowers, the carnation has benefitted enormously from the use of STS, which can increase its vase life two to threefold. Carnations can be stored longer than any other flower and can be opened to high quality flowers from very tight buds. These features should make production of carnations competitive in this country, despite the intense competition from imported flowers.

Maturity: The maturity at which carnations are harvested depends on the proposed marketing procedure. Star-stage buds (stage 1) are too immature for most purposes except long-term storage. Buds with petals straight up (stage 2) will open quickly. Flowers for immediate use are normally harvested between stages 3 and 4.

Harvesting: Many pickers place cut flowers on the top of wires for later collection into bunches. Flowers collected into canvas slings can be taken to the shed by mechanical devices ranging from overhead cables to tractor-hauled trailers designed to hold the slings.

Grading and bunching: Both standard and miniature carnations are graded by stem strength, stem length, bloom diameter and freedom from defects. Stem strength is determined by holding the stem horizontally at a point one inch above the minimum length for the grade. If the deviation of the flower head is more than 30 degrees from the horizontal. (with the natural curvature down), the flower is considered defective. Other defects include slabsides, bullheads, blown heads, singles, sleepy appearance, splits, discolouration and damage from insects and diseases.

Table 117. The Society of American Florists has suggested the grades for standard carnations

Grade name		Fancy	Standard	Short
Label colour		**Blue**	**Red**	**Green**
Minimum flower diameter (mm)	Tight	50	44	None
	Fairly tight	62	56	None
	Open	75	69	None
Minimum stem + flower length (cm)		55	43	30

There are 25 standard carnations stems per bunch; each bunch is tied at the base and atleast one other place below the flower heads. Instead of different coloured labels, some growers indicate different grades by colour and/or number of rubber bands on bunches. A miniature carnation bunch contains 30 buds total, at least 7 of which are open. With standard carnations, flower heads are alternated (5 high, 5 low) at the top of the bunch to produce a neat and compact bunch and reduce the risk of neck breakage. The size of the cartons is 122cm long, 50cm wide, 30cm hight should have polythene lining to help in maintaining high relative humidity inside the package.

Storage: Flowers for storage should be of the highest quality, absolutely free from pests and diseases. Place them at 33-35°F (1° C) in a box lined with polyethylene and newspaper. Open flowers (stage 3 or 4) can be stored 2 to 4 weeks, while bud-cut flowers, harvested at stage 2, can be safely stored up to 4 or 5 weeks. There are methods available for storing stage 1 buds (or younger) for up to 4 months.

Chemical solutions: After bunching, all carnations should be pulsed overnight in a cooler with a solution containing 6 g l^{-1} STS and 10 per cent sucrose. A one to two hour pulse with this solution at room temperature is also quite effective.

Bud opening: Carnation buds can be opened in a solution containing 7 per cent sucrose and 200 ppm Physan. The buds should have been treated, first, with 6 g l^{-1} STS overnight at 0° C.

Gerbera (*Gerbera jamesonii*)

Cut gerbera flowers, known for their remarkable variety in colouration and form, are increasing in importance in international commerce. One of the major problems in postharvest handling of these flowers is their tendency to conk (bend 4 to 6 inches below the flower head), resulting in an unmarketable flower. This bending has been variously attributed to harvesting of the flower before the stem has hardened sufficiently, microbial plugging of the stem and subsequent water stress, incorrect cutting of the stem, or poor genetic stock. The tendency to conk varies with variety and also varies throughout the year for any given variety.

Maturity and harvesting: Most gerbera varieties should be harvested when the 2 outer rows of disk florets have begun to open, but some cultivars can be harvested later, particularly those types that close at night. Flowers are harvested by twisting the stems off near the point of attachment to the rhizome; this is thought to encourage subsequent flower production. Flower stems should be cut to an appropriate size for packing soon after harvest and can be recut several times thereafter to improve vase life.

Grading and bunching: There are no industry-wide standards in India for grading and bunching gerberas. Some producers merely pull the flowers into protective transparent plastic caps and bind the stems into bunches of 10 with rubber bands. This may, however, damage the stems and is not recommended. Other producers pack individual flowers horizontally in shallow cardboard containers especially designed to support gerberas. Flower stems are passed through slits in the bottom of a cardboard tray so that the flower heads face up showing their colours while the stems pass under the box. Several rows of flowers can be arranged in each box. The boxes are then hung so that the stems dangle downward and can be placed in a storage solution. Afterward, 2 trays of flowers are packed horizontally into a fibre board box in such a way that the flowers are facing upward and easily seen, while the stems are underneath the flower boxes.

Chemical solutions: Present industry practice is to place cut gerbera stems in a 40 ppm sodium hypochlorite solution immediately after harvest to improve vase life. Recent research has indicated that a rapid pulse treatment with silver nitrate is sufficient to greatly alleviate postharvest problems with many of the gerbera cultivars that are relatively short-lived (eg. Roeland, 2609). After the dip, rinse the flowers in good quality water. This treatment causes only minimum phytotoxicity (brown damage to the stem). The use of 6 per cent sugar + 200 ppm 8-HQC as a preservative has shown to be of some benefit but can cause stem elongation during storage and may reduce overall flower quality.

Storage: To improve their post-storage quality, gerberas can be pretreated with AgNO3 before storage. Generally, gerberas should not be stored longer than 1 week at 2°C, even this short storage period can reduce subsequent vase life. Storage in chemical solutions is considered preferable to dry storage or storage in water.

Gladiolus (*Gladiolus hybrids*)

Gladiolus an important commercial cut flower despite a substantial decline in production in recent years, gladiolus responds well to proper postharvest management.

Maturity and harvesting: Traditional harvest in India is done at the stage when the bottom two or three florets on the spike are showing colour. For long-distance transportation, an even earlier harvest stage can be recommended if it is combined with pulsing to ensure proper opening of the flowers at their destination. Local market flowers are cut when the first floret is open.

Harvesting is carried out so as to leave as many leaves on the plant as possible. A knife is run down between the leaves with the back of the knife down. When the knife blade is as low as the cutter believes it should go, it is pulled upward and out, severing the stem, which can then be pulled out of the leaves.

Grading and bunching: Gladioli, like most spike-type flowers, are very sensitive to high temperature, so flowers are held at low temperature during storage and transportation. Quality factors for gladiolus include stem straightness and strength, freedom from damage and disease and maturity. Spikes are graded into three lots, based on overall quality, variety, colour of spike, length and number of florets per spike. After grading, they should be bunched in units of 12 and held together by rubber bands, packed in ribbed cardboard boxes of 120 x 60 x 30 cm. The bundles of 12 spikes each are first wrapped in corrugated paper for protection from sudden temperature fluctuations, bruisisng and moisture loss.

Table 118. The Society of American Florists has suggested the grades for standard gladiolus

Parameters	Fancy	Special	Standard	Utility
Spike length (cm)	>107	>96	>81	<81
No. of florets/spike	16	15	12	10

Chemical solutions: Gladioli respond well to vase solutions containing sugar, even if pretreated prior to shipping. Flowers hold overnight at 21° C in a preservative solution containing 200 ppm 8-HQC with 20 per cent sugar in dark room. Gladioli are very sensitive to fluoride (as little as 1 ppm) in the water, so tap water should not be used to prepare vase solutions. Symptoms include deterioration of the petal margin (bleaching, water soaking, then necrosis), failure of florets to open and develop normally, burning of the floret sheath and marginal leaf scorch.

Packing: Because of their response to gravity, gladioli are traditionally packed in tall glad hampers clearly marked for upright stacking. Since the advent of pre-cooling, some shippers have packed gladiolus in normal flower boxes. This practice is fairly safe if the flowers will remain refrigerated throughout the marketing chain and will be removed from the box on arrival. Excessive moisture on the foliage should be avoided so as to minimize the risk of botrytis.

Storage: Traditionally gladioli store at 7° C because of the danger of chilling injury with their flowers. In India producers commonly handle the flowers at 1° C. One beneficial aspect of low temperature handling and transportation is that the onset of their normal negatively geotropic response is delayed. This allows them to be packaged in the standard Californian box. For longer storage, California gladioli are best stored upright at 1° C.

Orchids: (Many species in the plant family Orchidaceae, principally of the genera Cymbidium, Cattleya, Vanda, Dendrobium, Paphiopedilum and Phalaenopsis).

Apart from their exotic forms and colours, one of the principal attractions of cut orchid flowers is their outstanding longevity. Even out of water, an orchid flower will last for an evening in a corsage. Spikes of cymbidiums will often last a month in a vase. The principal orchid genera grown for cut flowers are Cattleya, Cymbidium, Paphiopedilum and Phalaenopsis. With few exceptions, these flowers are handled in the same manner.

Harvesting: Orchid flowers are usually harvested 3 to 4 days after opening, because flowers cut prematurely will fail to develop normally off the plant. Early and late in the season, individual flowers are cut from the spike as they develop, because prices are high at these times. In mid-season, the whole spike is cut. Virus diseases can be spread from plant to plant during harvest, so cutting tools should be sterilized before being used on the next plant or disposable razor blades should be used.

Grading and bunching: Orchid flowers are graded as star 1 or star 2. Grade star 1 flowers are essentially perfect in form and condition. Grade star 2 flowers have slight flaws such as a spot on the flower or poor shape. Cut orchids should be moved quickly from the greenhouse to packing benches. There the stems are recut in air and the flowers inserted immediately into water tubes (orchid tubes). Each water tube holds one orchid and 12 to 24 flowers are packed in each carton. Each flower is wrapped in shredded wax paper to prevent mechanical injury to the blooms. Box inserts hold the individual water tubes stationary. Shredded wax paper is tucked around and between the flowers for additional protection. Whole spikes are normally placed in a water tube and laid in a box with shredded paper as is done with individual blooms.

Table 119. Orchids graded based, number of opened flowers and buds

Grade	Number of flowers	Number of buds
A	3-5	2-3
B	6-8	2-3
C	9-11	2-3
D	>11	2-3

Chemical solutions: Although commercial preservatives are used in the water tubes by some growers. Without vase solutions, Cymbidium and Paphiopedilum have vase lives of about 1 month and cut Cattleya and Phalaenopsis blooms last 7 to 10 days. Ethylene-sensitive orchid species can be protected from this gas by pulse treatment with STS. Treatment depends on individual species, but a 1 hour treatment with 4 oz. of concentrate per gallon of water (25 g l^{-1}) is adequate.

Storage: Orchid flowers are not commonly stored, because the flowers generally last longer on the plants. However, cut flowers generally be stored for up to 2 weeks at 5-7°C. The optimal storage conditions depend on the species and clone. Cymbidium and Paphiopedilum flowers can be held up to 3 weeks at 1°C but longer storage results in chilling injury. Stored flowers are saleable 3 to 4 days after removal from storage. Cymbidium and Paphiopedilum flowers can last 6 to 7 weeks on the plant. Some orchid species are very sensitive to ethylene, so harvesting, packing and storage areas must be maintained free of ethylene gas. This means keeping vehicles with internal combustion engines and open-flame heaters away. Fruit and other produce should never be stored in cool-rooms intended for cooling or holding orchids.

Rose (*Rosa spp.*)

Most of the commercial cut roses will easily last in the vase for 10 days. This is partly because poor water uptake by certain cultivars of purchased roses (especially Cara Mia) all too often results in the symptom called "bent neck" in which the flower neck wilts and the bud fails to open. Recently, too, scientists found that many commercial cultivars are quite sensitive to ethylene gas. The cut flower industry has an important stake in overcoming the poor postharvest reputation of the cut rose. All that is required is proper postharvest care for those cultivars susceptible to bent neck and STS pre-treatment of those that are sensitive to ethylene, especially if they are to be sold in supermarkets or other ethylene polluted areas.

Maturity and harvesting: Roses are harvested at different levels of maturity, depending on marketing and cultivar. For long-distance transport or storage, roses should be harvested with some of the sepals reflexed (stage 3). Yellow roses should be harvested earlier (at about stage 2). The marketing life of roses harvested at stage 4 will be reduced, unless extra care is taken with their postharvest handling.

Shears provided with auxiliary jaws to hold the bloom after harvest should be used. The cut is normally made so as to leave 2 five- foliate leaves (below the tip (A). When stem length is an important consideration, the cut should be made below the hook and the stem trimmed (B). Cuts which leave only trifoliate leaves may result in weak breaks that may even be blind (C). In times of high light, a knuckle cut may provide extra shoots in sparse beds (D).

Grading and bunching: Objective grading is based on stem length; subjective grading is based on flower maturity, stem straightness, stem caliper and quality of flower and foliage. Dislocation and defects on the outer guard petals are not normally a cause for down-grading, because these petals are usually removed by retail florists. Leaves and thorns may be removed manually or mechanically if desired. This operation has little effect on vase life. There are 12 or 25 stems per bunch.

Table 120. The Society of American Florists has suggested the stem-length grades for rose

Label colour	Blue	Red	Green
Hybrid teas	22”	14”	10”
Sweethearts	14”	10”	6”

Table 121. The flowers are graded on based on length of stem

Grade	Extra Class	Class I	Class II
Standard	70-80 cm	60-70 cm	50-60 cm
Spray	60 cm	50 cm	40 cm

Chemical solutions

Bacteria growing in the stems the single most important factor adversely affecting the vase life of roses. Consequently, effective bacteriocides are of most benefit. The following solutions are useful for increase vase-life of rose.

Rehydration: Most, but not all rose cultivars should be rehydrated after cutting with a solution containing 500 ppm citric acid (Sonia rehydrates well with water alone) and a bacteriocide.

STS pretreatment: STS applied so as to put 0.5 to 1 μmol of Ag^+ per stem will protect roses from the concentrations of ethylene normally encountered in the environment. It is the best to calculate the dose being given; a ½ hour dip in a solution containing 1 oz of concentrate per gallon of solution (1 mM) is quite effective.

Vase solution: Roses do not react well to vase solutions containing silver nitrate or Physan. Salts of 8-hydroxyquinoline (8-HQC, 8-HQS) make excellent germicides for rose preservatives, but some cultivars (eg. Ilona) are damaged by 8-HQS. Sugar (2%) is required for best opening and to prevent blueing of red roses, above 2 per cent may cause leaf damage. For roses, a good vase preservative is one that contains 200 ppm 8-HQC and 2 per cent sugar.

Storage: Before packing, roses are routinely wrapped. Pre-cooled roses which are intended for air transport are frequently packed in polystyrene caskets to insulate them from warm outside temperatures. Ice is frequently added either as flake ice or as frozen gel-ice packages. Roses intended for long-term transport are pre-cooled, packed in ventilated cartons and placed in refrigeration. Roses can be kept up to 2 weeks in dry storage if they are sleeved in polyethylene to reduce water loss and the temperature is maintained close to 0° C.

Bird-of-paradise (*Sterilitzia reginae*)

Bird-of-paradise inflorescence consists of a boat-shaped bract containing a series of 4 or 5 flowers, so when an exposed flower withers, another one can be pulled out. Slender, elongated waxed paper bags are placed over the expanding bracts a week or more before the orange flowers are ready to emerge. The bags protect the brittle flowers by holding them inside or next to the bracts. The bag also helps prevent Botrytis mould, rain and hail damage, aphid attacks and sunburn of the flowers.

Maturity: Inflorescences bend as they reach maturity, assuming a 90° angle to the stem. The first emerging orange flowers can barely be seen through the white paper bags. The stems are pulled at this point. Stems can be harvested in the tight bud stage before the first flowers emerge. At this stage; the inflorescence is swollen and there is a slight orange crack on the upper surface. The inflorescence is mature at the tight bud stage and is at the optimum stage for harvest, from the point of view of ease of handling and flower longevity (Fig. 90).

Harvesting: Flower stems are generally pulled off rather than cut. A side to side pulling motion is often necessary to loosen the stem at the base of the plant, although the inflorescences can break off if jerked too vigourously.

Grading and bunching: Flower stems are sorted into atleast 3 grades according to stem length and inflorescence size. For premium grade flowers, the field bags are replaced with new bags. Five stems are firmly tied together at two points with the inflorescences facing in same direction. The stem ends are evenly trimmed. A paper wrapper is placed around the bunched inflorescences to protect the flowers.

Chemical solutions: Flower stems should be treated after harvest with the fungicide to prevent grey mould (Botrytis). Flower longevity can be substantially increased by pulsing buds or flowers for 24 hours (48 hours is even better) with a solution containing 10 per cent sucrose, 250 ppm 8-hydroxyquinoline citrate (8-HQC) and 150 ppm citric acid.

Packing: The waxed paper on each inflorescence and the paper wrapper on each bunch give ample protection from handling injury and dessication. Because they are very heavy, bird of paradise stems should be packed in shallow cartons.

Storage: Store stems at 8° C and 85-90 per cent RH for extended. Chilling induced brown lesions form on the flower parts and bracts of such stored flowers. Strelitzia flowers can have a vase life of upto 14 days and stems harvested in the tight bud stage will last 4 weeks in storage if treated with a fungicide, wrapped to prevent desiccation and stored at 8° C and 85-90 per cent RH.

Chemical Treatment in Cut Flowers

A diverse array of treatment chemicals have been recommended for use with cut flowers. Before adopting any particular chemical for use in a commercial operation, it is strongly recommended that evaluate several different materials. High price is not necessarily equated with effectiveness. The preparation of the solutions listed below requires a little more effort than purchasing a pre-mixed formulation, but they will certainly be less expensive.

Components

In preparing a treatment solution, the flower handler is normally concerned with several different important functions; water (specifically, its quality), pH (acidity), food source and biocide.

Water: Hard water is generally very damaging to cut flowers. It does not flower easily through the stem and the flower therefore has difficulty in obtaining sufficient water. Researchers have shown that some level of minerals in the water is often beneficial, but in many cases, deionized water is much better than tap water for cut flowers. Evaluate the effect of water by comparing the life of the flowers that handle in preservative prepared with tap water with that of flowers in preservative prepared with deionized water.

Acidity: Flowers prefer acidity (pH) of 3.2-3.5. If the solution is not in this range, add more acid to the preservative formulation. Typical hard waters can be brought to the desirable pH by the addition of 300 - 500 ppm (parts per million) of citric acid. Aluminum sulfate and 8-HQC (common constituents of preservative solutions) also increase the acidity of the preservative solution.

Food: The best concentration of sugar, in a preservative solution depends on purpose of the solution. Rehydration solutions should not contain sugar and other chemicals. Pulse solutions and bud opening solutions must contain the correct concentration for the flower being treated. General preservative solutions usually contain 1.5 – 2 per cent sugar. This is at the upper limit for roses which may develop severe leaf scorch at higher concentrations but is less than ideal for many other flowers.

Biocides: Some biocides used for cut flowers are truly effective in killing the microorganisms present in vase water. 8-hydrdxyquinoline citrate (8-HQC) and 8-hydroxyquinoline sulfate (8-HQS), which are the most commonly, used biocides, prevent rapid growth of bacteria, but do not kill them. Physan-20, a mixture of quaternary ammonium compounds (200 ppm.), is an effective biocide, but it is toxic to some flowers (causes bleaching of the stems and may reduce vase life). This material is recommended for handling carnations, gypsophila and even chrysanthemums. The most effective biocide is silver nitrate, it can be used as a quick dip (100 to 1000 ppm, depending on the flower being treated), or as a component of the preservative solution (25 to 50 ppm, normally). It is more expensive and causes staining of clothes and blackening of the skin. Preservatives containing silver nitrate must be prepared with demineralized water. The chloride present in most tap water will reduce the effectiveness of the silver nitrate. The best biocide for commercial application is probably hypochlorite, the main ingredient in household bleach or in liquid swimming pool chlorinizers. Most flowers can tolerate low concentrations (20 - 50 ppm) of chlorine in a preservative. The problem is that it quickly disappears from the solution and must be replenished regularly.

Table 122. Germicides used in floral preservatives

Germicide types	Common name	Recommended concentration
8-hydroxyquinoline sulphate	8-HQS	200–600 ppm
8-hydroxyquinoline citrate	8-HQC	200–600 ppm
Silver Nitrate	$AgNO_3$	10–200 ppm
Silver Thiosulfate	STS	0.2–4 ppm
Thiobendazole	TBZ	5–300 ppm
Quarternary ammonium salts	QAS	5–300 ppm
Slow-release chlorine compounds	Cl	50–400 ppm
Aluminum sulphate	$Al_2(SO_4)_3$	200–300 ppm

Preparation of biocide solution

The concentrations are often given in parts per million (ppm). Because of the variations in density of the different materials obtained from different sources, it is safer to weigh the chemicals when making up the solution. Liquid chemicals (Physan-20 and bleach) can be measured with a medicine measure. The solutions are listed in the order in which they might normally be used in a flower handling operation.

Rehydration: Make fresh daily

- 1.75g citric acid and 3.5g bleach (5%) in 1 litre of water or
- 1g citric acid and 0.7g 8-HQC in 1 litre of water or
  (For hard to re-hydrate flowers, like stored mums)
- 1g citric acid, 0.7g 8-HQC and 0.7g Triton- X 100 or Tween 20 in 1 litre of water. Do not leave the flowers in this solution more than three hours.
- The pH should be 3.2-3.5, if it's too low or too high, adjusts the amount of citric acid to suit.

Pulsing with sugar: Sugar pulses will improve vase life and opening of many flowers. It is more suitable with spike-type flowers such as gladiolus and tuberose which need food for opening tight buds. In addition to the sugar, pulsing solutions should contain an acidulant and a biocide. Required amount of sugar + 1 g citric acid + 0.7 g 8-HQC per litre of water suits most flowers, whereas, required amount of sugar + 3.5g Physan-20 per litre of water and add citric acid to bring pH to 3.5 is better for carnations and gypsophila than others.

Pre-treatment with STS

Pretreatment with STS is essential for flowers sensitive to ethylene, including sweet pea, carnation, alstroemeria, delphinium, lily and snapdragon. The flowers can be treated in a short pulse, or long pulse.

Preparing the concentrate: The STS concentrate is prepared by first dissolving 20g of anhydrous sodium thiosulphate or 30g of prismatic sodium thiosulphate into a clean plastic bucket containing one quart of water. Four grams of silver nitrate is then dissolved into a separate bucket containing a quart of water. The stock solution is completed by slowly pouring the silver nitrate solution into the sodium thiosulphate solution, stirring rapidly as the solutions are mixed. Add 1g Physan-20 to the solution. This stock solution should be stored in a dark bottle. The working solution of STS is prepared differently depending on the treatment technique.

Short term pulsing: Add 15g of STS concentrate to 125ml of water. Pulse the flowers for 10 to 20 minutes at room temperature. Flowers treated in this way must be immediately rinsed and placed in a preservative solution for at least 1 hour.

Long term pulsing: Add 3.5g of STS concentrate to 125ml of water. Pulse the flowers for 1 hour at room temperature or overnight in the cooler. With carnations, a 10 per cent sugar, 3.5g/125ml STS treatment in the cooler overnight has proved quite effective.

Pre-treatment with silver nitrate: Used as a pulse pretreatment (a 10 second dip is usually enough), silver nitrate is a very effective treatment for some flowers. Use 0.35g per liter to prepare a 100 ppm solution (3.5g per 125ml for a 1000 ppm solution). This material must be prepared with demineralized, distilled or deionized water. Use rubber gloves when handling the solution and make only as much is needed at one time. It should be held out of direct sunlight and should keep organic trash (leaves, petals etc.) out of the solution to keep its activity high.

Ethylene inhibitors: Ammonia ethoxy vinyl glycine (AVG) and Methoxy vinyl glycine (MVG) are costly. Whereas another ethylene inhibitor like silver thiosulphate is commonly used and amino ethoxy acetic acid (AOA) is cheap but less active. These ethylene inhibitors prevent premature or undesirable abscission and sleepiness. They also reduce the action of ambient ethylene as well as catalytic production of ethylene by flowers.

Growth regulators: Play a major role in senescence of cut flowers, by affecting maintenance of flowers and production of ethylene usually coincides with the onset of respiratory climacteric in flower crops like carnation, hibiscus etc.

Dimension of different flowers

Alstroemeria: Due to varietal differences and climatic conditions in the growing areas the number of peduncles per stem may vary.

Grade	1	2	3
Colour designation	Blue	Yellow	Red
Minimum length (cm)	75.0	65.0	60.0
Minimum flower diameter (cm)	3.75	3.75	3.75
Stem strength	20°	20°	20°
Stem deviation curvature (cm)	7.5	7.5	7.5
Minimum flower per stem	3.0	3.0	3.0

Anthurium: Overall length equals tip of spathe to base plus stem

Grade	1	2	3	4
Colour designation	Blue	Yellow	Red	Green
Minimum length (cm)	60	50	40	30
Minimum spathe diameter (cm)	15	12	10	6
Stem strength	NA	NA	NA	NA
Stem deviation curvature (cm)	2.5	2.5	2.5	2.5
Minimum flower per stem	1	1	1	1

Spray Asters: Due to climatic condition in the various groeing regions and plant maturity at harvest, the weight per bunch may vary.

Grade	1	2	3
Colour designation	Blue	Yellow	Red
Minimum length (cm)	70	60	50
Minimum stems per bunch	10	10	10
Minimum weight per bunch (gm)	300	230	180
Stem strength	5°	10°	10°
Stem deviation/curvature	1/2”	1/2”	1/2”
Minimum flower per stem	NA	NA	NA

Birds of paradise: Due to varietal difference and climatic conditions in the various growing areas, bract size may vary.

Grade	1	2	3	4
Colour designation	Blue	Yellow	Red	Green
Minimum length (cm)	90	75	60	50
Minimum bract length (cm)	21	18	15	12
Stem strength	0	0	0	0
Stem deviation/ curvature (cm)	2.5	2.5	2.5	2.5
Minimum bracts per stem	1	1	1	1
Minimum flower per stem	4	4	3	3

Carnations: Due to varietal differences and climatic conditions in the various growing areas, flower diameter may vary.

Grade	1	2	3	4
Colour designation	Blue	Yellow	Red	Green
Minimum length (cm)	60	50	45	38
Minimum flower diameter (cm)	6.9	5.6	5	5
Stem strength	20°	20°	20°	20°
Stem deviation/ curvature (cm)	1.25	2.50	2.50	2.50
Minimum flower per stem	1	1	1	1

Chrysanthemums and gerberas: Due to varietal differences and climatic conditions in the various growing area, flower diameter may vary.

Grade	Chrysanthemums		Gerberas		
	1	2	1	2	3
Colour designation	Blue	Yellow	Blue	Yellow	Red
Minimum length (cm)	70	60	50	40	38
Minimum flower diameter (cm)	10	10	9.0	7.50	7.50
Stem strength	20°	20°	10°	10°	10°
Stem deviation/ curvature (cm)	2.50	2.50	2.50	2.50	2.50
Minimum flower per stem	1	1	1	1	1

Gladiolus: Due to varietal differences and climatic conditions in the various growing areas, flower diameter and flower count per stem may vary.

Grade	1	2	3	4
Colour designation	Blue	Yellow	Red	Green
Minimum length (cm)	115	105	90	80
Minimum flower diameter (cm)	8.75	8.25	7.50	6.25
Stem strength	15°	15°	15°	15°
Stem deviation/ curvature (cm)	5	5	7.5	10
Minimum flower per stem	12	12	10	6

Gypsophola: Stem strength should be measured after hydration.

Grade	1	2	3	4
Colour designation	Blue	Yellow	Red	Green
Minimum length (cm)	65	55	45	35
Minimum weight per bunch (gm)	300	210	150	90
Minimum stems per bunch	10	10	10	10
Minimum laterals per stem	3	2	2	1
Stem strength	5°	5°	5°	5°
Stem deviation/ curvature (cm)	2.5	2.5	2.5	2.5
Minimum flower per stem	NA	NA	NA	NA

Hybrid lilies: Due to varietal differences and climatic conditions in the various growing areas, flower diameter and flower count may vary.

Grade	1	2	3	4	5
Colour designation	Blue	Yellow	Red	Green	Orange
Minimum length (cm)	60	60	55	50	40
Minimum flower diameter (cm)	10	10	10	7.50	6.25
Stem strength	12°	12°	12°	12°	12°
Stem deviation/ curvature (cm)	0”	0”	1/2”	1”	1”
Minimum flower per stem	6	4	3	3	1

Hybrid tea roses:

Grade	1	2	3	4	5	6	7	8	9	10
Colour designation	Blue	Yellow	Red	Green	Orange	Violet	White	Light blue	Grey	Brown
Minimum length (cm)	70	65	60	55	50	45	40	35	30	25
Minimum flower diameter (cm)	NA	NA	NA	NA	NA	NA	NA	NA	NA	NA
Stem strength	20°	20°	20°	20°	20°	20°	20°	20°	20°	20°
Stem deviation/ curvature (cm)	2.5	2.5	2.5	2.0	2.0	2.0	1.5	1.5	1.5	1.5
Minimum flower per stem	1	1	1	1	1	1	1	1	1	1

Iris and limonium: Due to varietal differences and climatic conditions in the various growing areas, flower length may vary.

Grade	Iris			Limonium		
	1	2	3	1	2	3
Colour designation	Blue	Yellow	Red	Blue	Yellow	Red
Minimum length (cm)	60	45	35	70	60	50
Minimum flower diameter (cm)	NA	NA	NA	-	-	-
Stem strength	10°	10°	10°	25°	25°	25°
Stem deviation/ curvature (cm)	2	2	2	5	5	5
Minimum flower per stem	1	1	1	-	-	-
Minimum stem per bunch	-	-	-	5	7	10
Minimum weight per bunch	-	-	-	300	280	250

Liatris: Due to varietal differences and climatic conditions in the various growing areas, flower length may vary.

Grade	1	2	3	4
Colour designation	Blue	Yellow	Red	Green
Minimum length (cm)	90	80	70	50
Minimum spike diameter (cm)	18	15	12	10

Stem strength	5°	5°	5°	5°
Stem deviation/ curvature (cm)	1.25	1.25	1.25	1.25
Minimum flower per stem	NA	NA	NA	NA

Miniature carnations: Due to varietal differences, the number of flowers per stem may vary.

Grade	1	2	3	4
Colour designation	Blue	Yellow	Red	Green
Minimum length (cm)	60	55	50	45
Stem strength	35°	35°	35°	35°
Stem deviation/ curvature (cm)	4	2.5	2.5	2.5
Minimum flower per stem	5	4	3	4

Solidaster and spray roses: Due to climatic conditions in the various growing regions and plant maturity at harvest, the weight per bunch may vary.

Grade	Iris			Limonium		
	1	2	3	1	2	3
Colour designation	Blue	Yellow	Red	Blue	Yellow	Red
Minimum length (cm)	70	60	50	60	50	40
Stem strength	10°	10°	10°	20°	20°	20°
Stem deviation/ curvature (cm)	2.5	1.25	1.25	2.5	2.5	2.5
Minimum flower per stem	1	1	1	4	3	3
Minimum stem per bunch	10	10	10	-	-	-
Minimum weight per bunch (gm)	300	250	225	-	-	-

Tulips and statice: In tulips stem strength and stem deviation should be measured after hydration and in statice, due to varietal differences, minimum weight for novelty colours will also vary

Grade	Tulips		Statice		
	1	2	1	2	3
Colour designation	Blue	Yellow	Blue	Yellow	Red
Minimum length (cm)	46	30.5	60	50	50
Minimum flower diameter (cm)	NA	NA			

Stem strength	NA	NA	20°	15°	15°
Stem deviation/ curvature (cm)	2	2	2.50	2.50	2.50
Minimum flower per stem	1	1	NA	NA	NA
Minimum stems per bunch	-	-	10	10	10
Minimum weight per bunch (gm)	-	-	400	230	150

Postharvest Problems of Cut Flowers

Bent neck: It is a common postharvest problem particularly in roses due to water deficiency in the neck tissue, which is controlled by the transpiration rate, an increase in the stem flow resistance of cut blooms due to vascular occlusions of metabolic origin, appearance of plugging materials like pectin, cellulose and microbes. Extreme temperature during shipping or storage also causes bent neck. Use of de-aerated and micro-organism-free water and wetting agents reduce the extent of bent neck.

Limp neck: It is a type of disorder in roses caused by water stress in the area just below the flower head. Insufficient stored energy of flower heads may be one of another reasons. The affected flower buds bend down due to the weight of the top.

Bull head: Thrips infestation, insufficient carbohydrate supply to developing buds and hard pruning cause bull head in roses. Affected roses have shorter shelf life.

Blind shoot: Blind shoot in roses is due to failure of the shoot to initiate flower buds or due to flower bud atrophy or bud abortion. Inadequate carbohydrate supply, unfavourable temperatures and light conditions during bud initiation at the early stages of development are associated with blind shoot production.

Calyx splitting: Calyx splitting in carnation is a factor of quality. The sepals beneath the flowers are unable to form a cylindrical calyx tube, which supports the base of the petals. During bud opening, the calyx may split either half or completely. Reasons are genetic, high doses of nitrogenous fertilizers, high day temperature, lower night temperature and high density planting.

Sleepiness: A disorder in carnation, it is caused by exposure of the flowers to ethylene gas or water stress. The extent of this disorder is much higher when flowers are stored for long periods or when they are kept at high temperatures.

Topple: Topple in gladiolus is characterized by breaking of spikes after opening of the florets in the vase. The spikes harvested from calcium-deficient soil have higher incidence of topple.

Negative geotropism: A postharvest disorder of gladiolus and antirrhinum, it occurs during storage and shipping. The spikes curve upward if they are placed horizontally. Differential distribution of auxins is responsible for this phenomenon. Transport of cut flowers in upright position is the remedy.

Flower bud blasting: It is a disorder found in tulip and is caused by dry storage or forcing of bulbs. It has been reported that a change in the plants hormonal status, which regulates the distribution in carbohydrates and other organic materials within the plant, is the main reason for blasting.

Petal discolouration: Low night and day temperatures are correlated with accumulation of pigments and is accompanied by blackening of petals and often stimulated by penetration of ultraviolet radiation through the materials covering the greenhouse. Bluing during storage is attributed to breakdown of proteins, release of free ammonia and a concomitant increase in pH.

Flower bud and petal abscission: This disorder is caused by ethylene evolution, formation of an abscission layer, a rising activity of cell wall hydrolytic enzymes, shaking, wounding and high temperatures, pollination and fertilization causing ethylene production. Remedial measures are use of ethylene inhibitors, spray with NAA (30 to 50 ppm), application of MH (200 to 400 ppm) and citric acid (500 ppm) through cut stem. Treatment with Amino ethoxy acitic acid or AIB reduces abscission.

Foliage discolouration: Placing cut flowers in the dark at high temperatures causes yellowing, whereas oxidation of native phenols, mainly leuco-anthocyanins, results in darkening of foliage. Insufficient carbohydrate supply and water stress cause leaf blackening in *Protea nerifolia*. Ethylene has been implicated in the yellowing of some species, since endogenous ethylene enhances it and AVG reduces it. Exogenous ethylene has been found to increase leaf yellowing in potted plants like *Dieffenbachia, Dracaena, Euphorbia, Hibiscus, Pelargonium, Rhaplidophora and Yucca.* Spray with cytokinin and STS pulsing and treatment with GA_3, AOA or AIB had been found to reduce leaf yellowing and darkening.

Petal wilting: The wilting of petals following ethylene exposure is a characteristic of only a limited number of plant families and the Caryophyllaceae is the family which displays petal wilting in response to ethylene. This family includes the commercial important flowers like carnations and gypsophila. STS pulsing, inclusion of AOA, cytokinin and AIB (α-amino isobutyric acid) in the vase solution reduces petal wilting.

Bud drying: In Asiatic hybrid lilies, the apparent desiccation and shrivelling of flower buds followed by abscission is a common problem. This problem is solved by STS pulsing.

Premature leaf in-rolling and bract discolouration: This occurs in *Heliconia psittacorum* and vascular blockage is the reason behind it. Application of GA_3 (0.1 mM) and ABA (0.0001 to 0.001 mM) delays leaf rolling by 2-3 days.

Water stress: Water stress causes wilting of Cymbidium and Phalaenopsis. Low rate of transpiration associated with low uptake of water are major factors involved in such problems.
Leaf wilting: A lower rate of water uptake caused by bacterial plugging is prevalent in chrysanthemum. Placement of the stems in cold water overcomes the problem.

Scape bending: This phenomenon of cut gerbera is caused by excessive bacterial growth in vase solution which reduces longevity.

Lack of flower opening: An interaction between ethylene and gibberellins is involved in controlling flower opening of Pharbitis nil. In Petunia hybrida, gibberellins from the anthers were found to regulate flower development. Flower opening in Iris is controlled by elongated growth of the pedicel, which is inhibited by low concentrations of ethylene and stimulated by STS or gibberellins. In lily flowers, the opening of flowers in the vase is often inhibited by endogenous ethylene, which is counteracted by STS treatment. In Euphorbia fulgens, ethylene treatment reduces flower opening, while STS treatment promotes growth and opening of flower buds. In freesia, STS treatment promotes growth and opening of flower buds. Lack of flower opening in gladiolus hybrids, Gypsophila and roses are overcome by the addition of sucrose in the vase solution. Gibberellins are involved in opening of cut roses.

Low water uptake: The rate of water uptake will depend on the transpiration pull and on the temperature and composition of the solution. The ionic composition of the vase solution is also a determinant of the rate of water uptake. The rate of water uptake of freshly cut flowers may initially be high when the plant has a low water potential at cutting. The rate of uptake will reach a steady state corresponding to the rate of transpiration but, depending on the species, during vase life the rate of transpiration also declines but tends to be higher than the water uptake rate. This results in a negative water balance (= rate of uptake - rate of transpiration), a decrease in water potential and in stomatal closure. When the rate of water uptake remains lower the rate of transpiration of the flowers or leaves, or both, may show turgor loss.

$$P = \Psi r - (-\Psi w) = \frac{SRT}{V} \Psi w$$

$(\Psi w = \Psi r + \Psi p + \Psi m)$

Ψr = osmotic potential

Ψp = pressure potential

Ψm = matrix potential

Ψw= water potential of a cell

P = turgor pressure

S = solute content in osmoles

R = gas constant

T = absolute temperature

V = volume

The rate of change of turgor (P) with a change in water potential depends upon the elasticity of the cell wall and the osmotic potential. The reduction of water uptake in cut roses is a result of occlusion in a large variety of xylem conduits.

14

Postharvest Handling of Spices

India is known as land of spices since time immemorial and it's a fact to share that spice were considered more valuable than gold in old times as it was believeed as unattainable luxury. Its aromatic and culinary properties along with medicinal value make them crops of every house hold in India and few part of the world. Spices are used for flavour, colour, aroma and preservation of food or beverages. Spices may be derived from many parts of the plant like bark, buds, flowers, fruits, leaves, rhizomes, roots, seeds, stigmas and styles or the entire plant tops. The term herb is used as a subset of spice and refers to plants with aromatic leaves. Spices are often dried and used in a processed but complete state. Another option is or used to prepare extracts such as essential oils by distilling the raw spice material (wet or dry), or to use solvents to extract oleoresins or other standardized products. The most important spices traditionally traded throughout the world are products of tropical environments. The major exception to this group are the capsicums (chilli peppers, paprika) and coriander which are grown over a much wider range of tropical and sub-tropical environments. In terms of world trade value, the most important spice crops from the tropical regions are pepper, capsicums, nutmeg/ mace, cardamom, allspice/pimenta, vanilla, cloves, ginger, cinnamon and cassia and turmeric. Coriander, cumin, mustard and sesame seeds and the herbs sage, oregano, thyme, bay and the mints are the most important spice crops from sub tropical environments. Seed spices are mainly cultivated in the states of Rajasthan and Gujarat and it also has sizeable area in the states of Madhya Pradesh, Haryana, Punjab, Uttar Pradesh, Andhra Pradesh and Karnataka. Gujarat, Rajasthan, parts of Madhya Pradesh and Andra Pradesh alone covers nearly 83.0 percent of the country's area, which estimated around 14.0 lakh hectare with a production of 12.0 lakh tonnes representing 69.0 per cent of national production. State of Rajasthan boundaries have diverse climatic conditions, the western hot arid regions are known for quality cumin cultivation, south eastern humid regions have country's highest area under coriander cultivation for seed purpose, north western and central regions have high area under fenugreek, kasuri methi and dill production, whereas south western region well-known for fennel seed production. This large acreage of seed spices in the state is around 8.86 lakh hectare which is 50 per cent of the country. Rajasthan state as a synonym for seed spices very often the western arid/semi arid region is cited as bowl of seed spices. Though 17 seed spices are grown in the country, coriander, cumin, fennel and fenugreek are pillars of economic importance. Ajwain seed, dill seed, celery, nigella and poppy seed too contribute a minor share. Chilli is the major spice crop contributing 23 per cent, seed spices accounts for 21 per cent, turmeric 17 per cent and pepper 2 per cent of total spices production in the country.

Table 123. Status of seed spices area in the country and leading states

Commodity	Total area in India (lakh ha)	Leading states in seed spices cultivation				
		Leading States	Area (Lakh ha)	Production (Lakh tonnes)	% Share in area	% Share in production
Cumin	7.60	Rajasthan Gujarat	5.00 2.27	3.37 1.98	99.0	99.0
Coriander	6.62	Rajasthan Madhya Pradesh	1.81 2.00	2.20 2.00	62.0	67.0
Fennel	0.74	Gujarat Rajasthan	0.44 0.45	0.79 0.56	95.0	98.0
Fenugreek	2.18	Rajasthan Madhya Pradesh	1.29 0.40	1.78 0.26	86.0	84.0
Ajwain	0.24	Rajasthan Telanagana	0.15 0.018	0.10 0.05	75.0	75.0

Table 124. Status of seed spices acreage (Lakh ha) in India and Rajasthan

Main seed spice crops	Area (India)	Area (Rajasthan)	National Share in area by Rajasthan	National ranking of the Rajasthan
Cumin	7.60	5.00	68.77	1st
Coriander	6.62	1.81	27.34	1st
Fennel	0.74	0.45	50.56	2nd
Fenugreek	2.18	1.29	59.17	1st
Ajwain	0.24	0.15	62.5	1st
Total	17.38	8.86	50.58	

Table 125. Status of seed spices production (lakhs tonnes) of India and Rajasthan

Main seed spice crops	Production (India)	Production (Rajasthan)	State production share (%)	National productivity (t ha^{-1})	State productivity (t ha^{-1})
Cumin	5.35	3.37	62.99	0.74	0.67
Coriander	6.09	2.20	36.12	0.92	1.22
Fennel	1.25	0.56	44.80	1.40	1.24
Fenugreek	2.20	1.78	80.90	1.01	1.38
Ajwain	0.14	0.10	71.42	0.58	0.67
Total	15.03	7.02	53.29	0.87	0.90

ECONOMIC IMPACT OF SPICES

There are around 40 to 50 spices of global economic and culinary importance, many other species that are used in traditional cooking in the region of their natural occurrence but have yet to reach any significant trade. The major spices of international trade have well known stable long-term markets but most species tend to be commodities and as such there are competition and price fluctuations. There is a very considerable domestic consumption of spices in many spice producing countries. The supply side of the industry has always been vibrant and has been punctuated by periodic relocations of the major production areas. To remain competitive, countries such as India are moving into the value added sector producing spice essential oils, oleoresins, powders, speciality extracts and blends. In addition, India has established spice Agri-Export Zones and they are actively developing capabilities in quality management, improved packaging and technology innovation in production and processing. The USA is the world's biggest importing country of essential oils followed by Japan and European Union. Most of the oils go into ordinary industries and the number of mainstream users continues to decline through mergers and acquisitions. Aromatherapy is seen as a potential high value growth area, on the other hand, it represents less than 1-2 per cent of the total essential oil trade. Use of essential oil for perfume and fragrance applications is predicted to reduce. Derivatives of dual purpose and industrial oils will be imported from processing factories located in oil producing countries rather than from imported oils. In the EU countries, 55-60 per cent of the total spice and herb use is for industrial consumption, 35-40 per cent by the retail sector and 10-15 per cent by the catering sector. The high industrial sector use reflects the growing popularity of ready-to-use spice mixtures. Another reason is the increasing consumption of processed foods and ready to eat dishes, which often rely on spices and herbs to retain and enhance food flavour.

The ever more stringent regulations governing the international trade in spices and their derivatives for culinary use in food processing will mean that spice producers who ignore such specifications will eventually lose their markets to quality producers. The European Union has aflatoxin regulations to detect contaminated consignments and restrict their import while adulteration of spices with colourants has led to product ban. Some two hundred essential oils are produced and traded internationally in volumes that range from 20,000 - 30,000 tonnes for orange oil to less than 100 kg for some specialty flower extracts. Prices vary widely but for the majority of oils traded in volume fall within the range of US$40- $60 kg^{-1} and for specialist minor oils the price can be many hundreds of US$ kg^{-1}. The citrus oils (orange, lemon, tangerine, lime, manadarin, grapefruit etc.) plus the mint oils have by far the largest number of important applications in tonnages as well as variety of flavours. The seasoning oils of spices and herbs (clove bud, coriander, cinnamon, garlic, nutmeg, onion etc) are traded in lesser quantities and have more selected applications. The majority of other oils is used in even lower quantities and has more targeted applications. The growth markets for uses of essential oils are predicted to be in the rapidly developing countries (dominated by China and India), Eastern Europe and Russia and less developed countries. Long-term the amounts consumed in these growing markets is predicted to exceed the current consumption of the industrialized countries. Demand is predicted to be greatest for low value fragrance oils used in soaps, detergents and related products with smaller growth in the flavour oils, mainly for non alcoholic beverages (citrus and spice oils) and oral care products (mint oils). A growth in use of flavour oils in the processed food industries will be confined mainly to developing countries with growing middle class populations.

Export Potential of Spices

Spices are quoted valuable as gold and silver, 1498 Vasco De Gama arrival at Calicut started Indian spice trade in world market. India commands a formidable position in the World Spice Trade with a share of 48 per cent in terms of volume and 44 per cent in terms of value. Spices export from India has crossed 2 billion US$ in the year 2011-12. The export earnings showed a spectacular growth during the period from 2005-06 to 2014-15 as the earnings increased from 592.9 million US$ to 2432.85 million US$. The volume of spices exported increased from 350,363 metric tons in 2005-06 to 893,950 metric tons in 2014-15. In the export basket of spices, seed spices (raw) account for about 30 per cent of total volume and 19 per cent of total foreign exchange earnings. During 2014-15, India exported 270,150 metric tons of raw seed spices valued at Rs 2815.60 crores. For the period of 2017-18, India is exporting only 15 percent of its production annually and full fills the 50-60 percent of world demand. The total export of seed spice crops is Rs 3738 crore, out of which cumin alone contributes Rs 2418 crore annually. Among seed spices, cumin is the highest export earner with an export of 155,500 MT valued at Rs 1838.20 crores whereas coriander is next. Cumin contributes 48 per cent of volume and 65 per cent of value of the total seed spices export. Indian seed spices especially cumin is having more demand because of its high volatile oil content. In the world market, India faces stiff competition from Syria which exports 80 per cent of its production. Other countries giving stiff competition are Turkey and Iran, as the bulk of their production is exported and their prices are much lower than India's prices.

Table 126. Export status of major seed spices and its national contribution (source year: 2016-17)

Commodity	Qty exported (Lakh tones)	Value earned (Billion rupees)	% quantity share in total raw spices exported	% value share in total value earned by exporting raw produce	Highest quantity exported in last five years (Lakh tones)	Highest value (Rs) earned in last five
Cumin	1.19	1963.2	13.45	16.05	1.55 (2015)	1963.2 (2017)
Coriander	0.303	292.07	3.42	2.39	0.46 (2015)	498.12 (2015)
Fennel	0.351	308.75	3.97	2.52	0.35 (2017)	308.75 (2017)
Fenugreek	0.346	182.76	3.92	1.49	0.35 (2014)	233.8 (2016)
Total	2.191	2746.79	24.76	22.46		

Table 127. Major seed spice importing countries from India (2016-17)

Countries	Quantity (Metric tonnes)	Value (Rs. Lakhs)
United States of America	9700.0	58200.0
United Kingdom	1875.63	11495.14
Germany	965.47	7905.81
Sweden	1034.26	7804.79
Japan	815.64	6517.07

Existing market chain of seed spices

The market is pre-dominantly of raw seed and powdered masala, limited demand do exists for seed spice use in value added products and herbal medicine. The main component of value addition in basic farm product is mainly seed cleaning, grading and packaging. Present market chain in seed spices are demonstrated in figure 43.

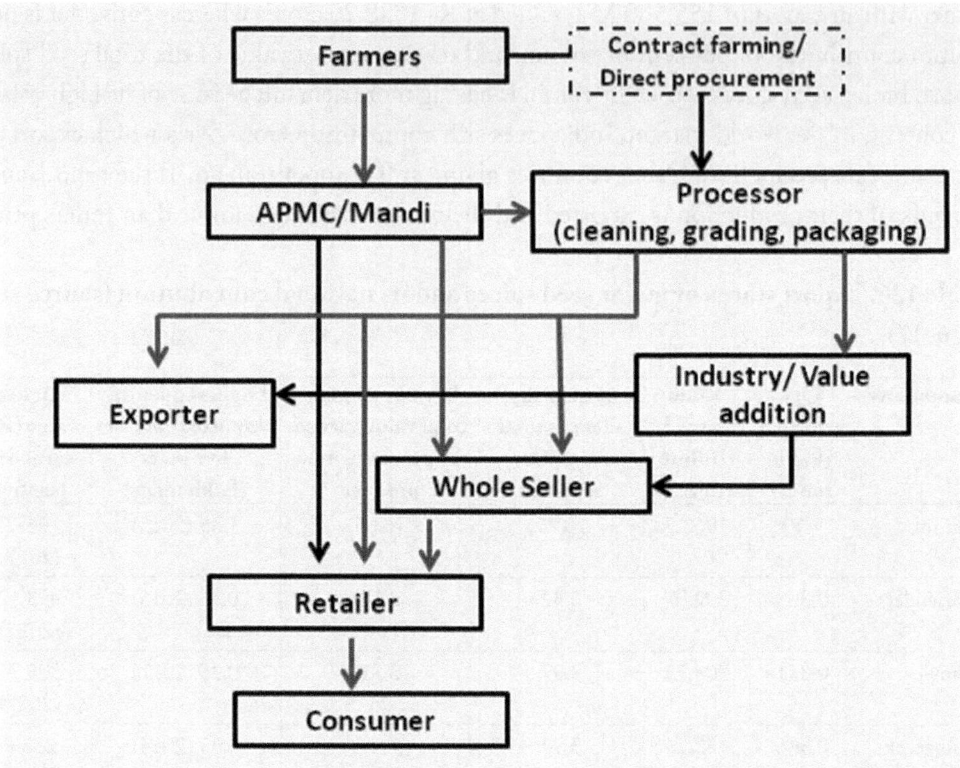

Figure 43. Present market chain in seed spices (red lines shows the dominant network)

Direct business model for seed spices

In Rajasthan, seed spices are cultivated in more than 8.86 lakh hectare area contributing nearly 65 per cent of national seed spice production. Seed spices are the wealth of the state, being a low volume and high value crop it has immense potential for high economic gains from seed spice cultivation. The farmers profit can be increased by introducing innovative models of market value chain linking farmers and other stakeholders in an enterprise mode. The below model (Figure 44) is a direct business model for minimizing market chain length and maximizing farmers profit in seed spice trading to national and international markets.

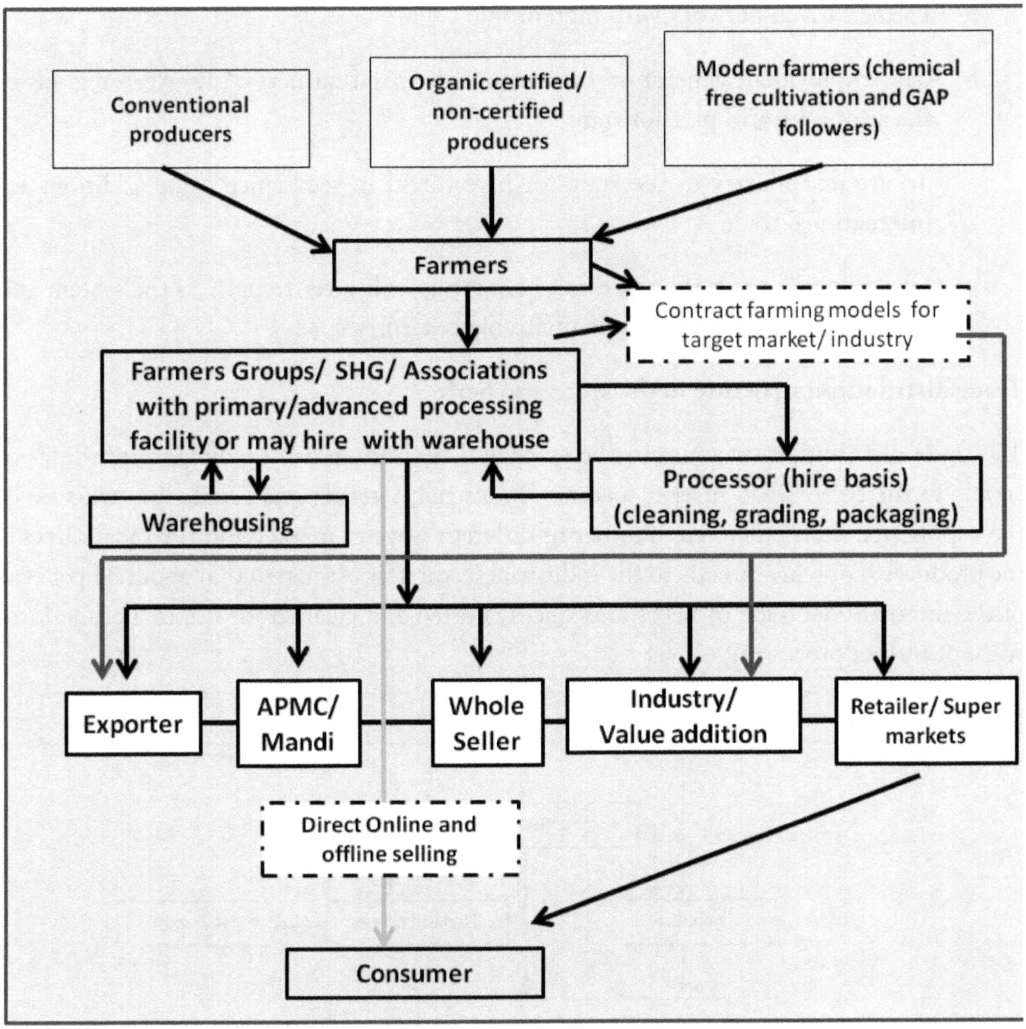

Figure 44. Direct business model for seed spices

Key feature of direct business model

Looking to the pre-dominance of the state share in national seed spice production and battery of middlemen involved in seed spice market/value chain. An advanced market/value chain model i.e. Direct Business Model is envisaged for increasing the benefit to the real producer and ultimate consumer by taking farmers in centre. This model envisage-

» Direct linking of farmers and other stakeholders

» Farmers are operative as a group in form of associations, self help groups (SHG's) etc.

- Farmer driven network with high profit
- Self employment generation for resource deprived farmers of western regions of the arid/semi-arid parts of state
- To attract industry in the state for investment in seed spice value additions for international trade
- Clustering of association and self help group will give strength to the system and will allow to compete in market with big enterprises

Trade distribution structure in the spice and herb

The trade distribution structure in the spice and herb trade can be divided into lines of supply to the three broad market sectors – industrial, catering and retail. The structure of the supply tree shows there are number of different ways to market and the most direct is the producer supplying directly to the industrial sector. It is estimated that about 85 per cent of the international trade of herbs and spices is dried and cleaned for use in a crude form without further processing.

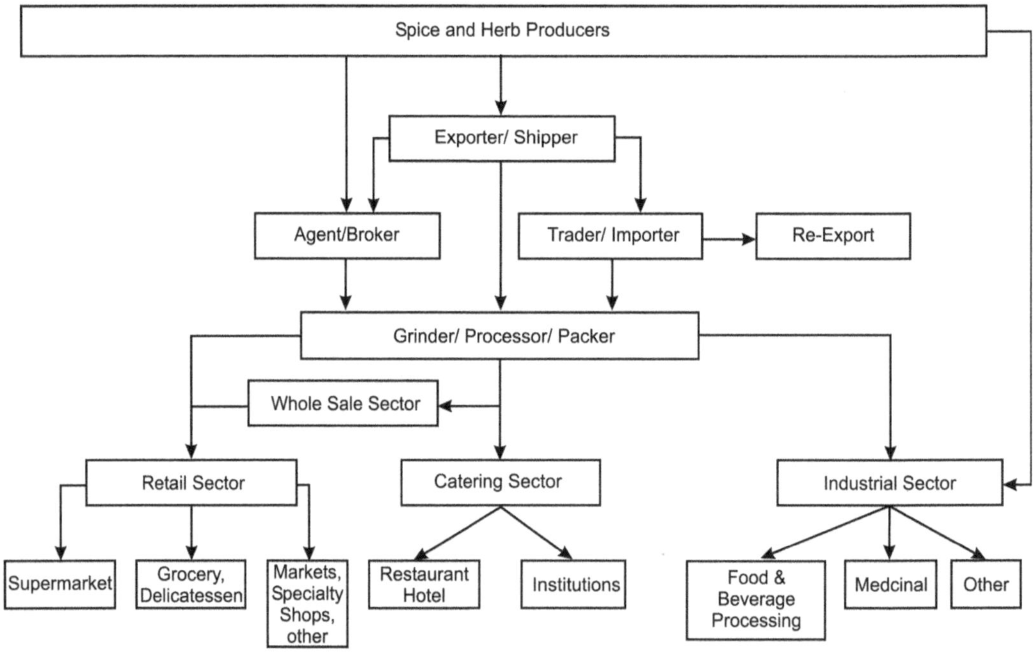

Figure 45. Trade distribution structure for spices and spice products

The major spice trading countries are China, Madagascar, Indonesia and India while Guatemala, Brazil, Vietnam and Sri Lanka are significant traders. The value of that trade varies annually and fluctuates about $US 2.5 billion. Closely linked to the spice and herb trade is the extraction of flavour and aroma compounds in essential oils. The trade distribution for essential oils and oleoresins can also have a number of routes to market. The most simple is the small oil producer selling to the local market or to tourists while the large-scale producers would normally deal through fragrance and flavour house formulators. There are many manufacturing industries where the extracts of spices, herbs and aromatic resins are used for flavour, aroma or product formulation.

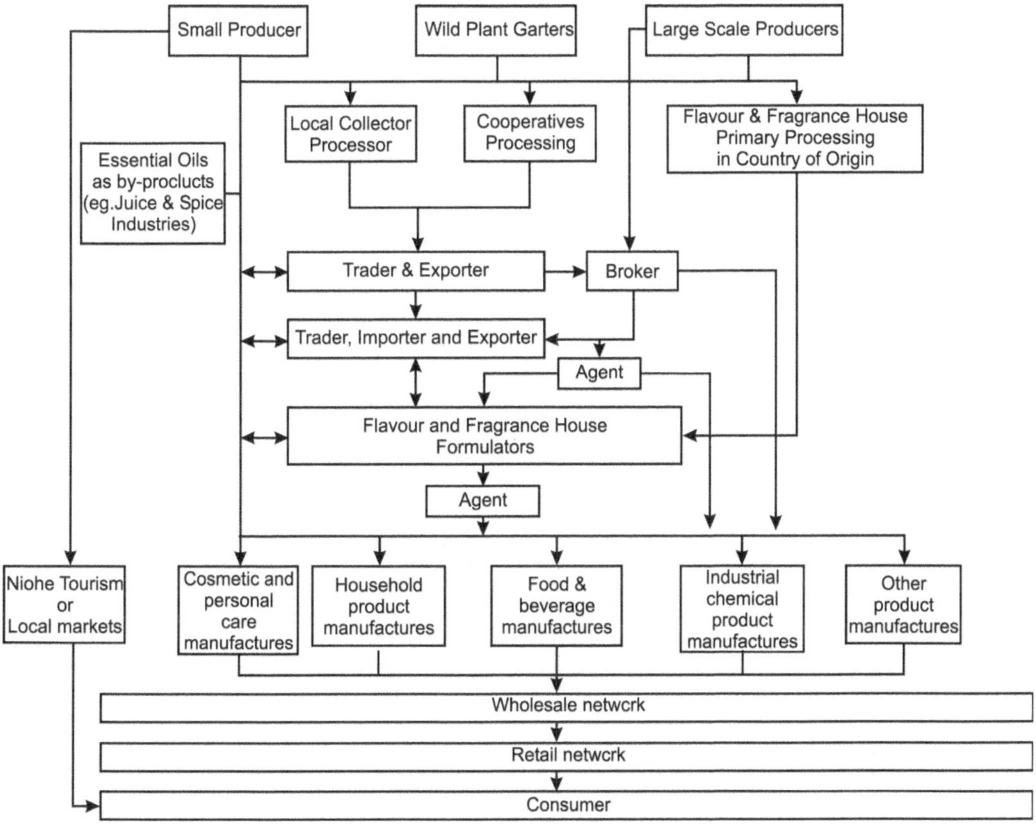

Figure 46. Trade distribution structure for essential oils and oleoresins

The steam distilled essential oils or solvent-soluble oleoresins are primarily extracted from the herbs and spices in their raw form, but other plant parts such as leaves or rhizomes can also be extracted. The production tree for essential oils shows there are four broad sectors – flavour industries, personal care, pharmaceutical and industrial.

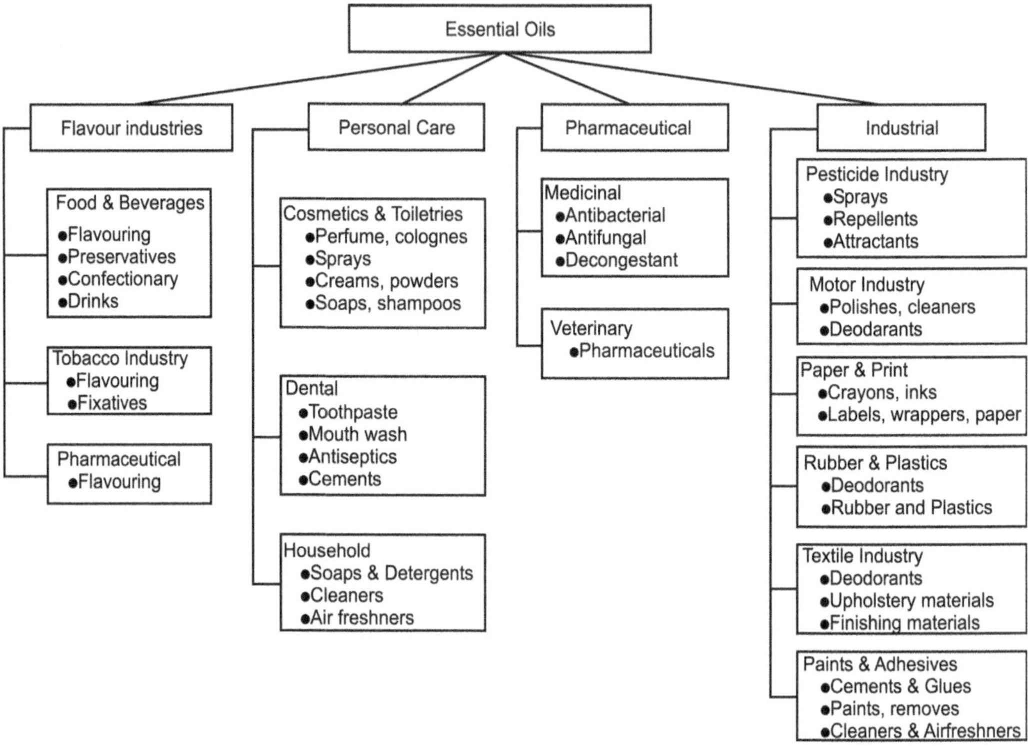

Figure 47. Industries and product categories that use essential oils

The spice flavours for food, beverage, or industrial use may come from different extraction processes and these pathways are outlined in Fig. 48.

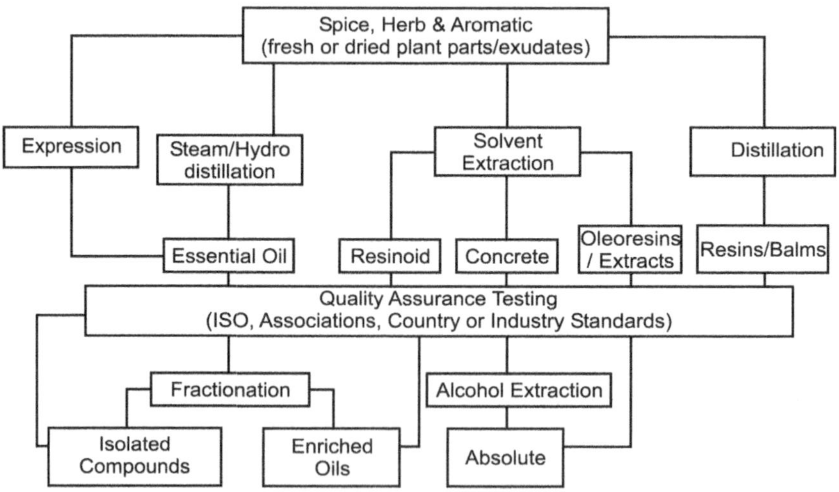

Figure 48. Extraction processes used and products from spice, herb and aromatic plants

Requirements for Export and Quality Assurance

Spices and herbs seasonings are valued for their distinctive flavours, colours and aromas but they can be heavily contaminated with micro-organisms because of the environmental and processing conditions under which they are produced. The microbial load has to should be reduced before they can be safely incorporated into other food products. High temperature treatment can cause significant loss of flavour, aroma and volatile oils. Steam also results an increase in moisture, loss of volatile flavour, aroma components and colour changes. Until recently, most spices and herbs were fumigated with sterilizing gases such as ethylene oxide to destroy contaminating micro-organisms. However, the use of ethylene oxide was prohibited by an EU directive in 1991 and has been banned in a number of other countries because it is a carcinogen. Irradiation has since emerged as a viable alternative and its use results in cleaner, better quality herbs and spices compared to those fumigated with ethylene oxide. Irradiation, a cold dry process, such as irradiation is ideal to kill the micro-organisms. Irradiation of herbs and spices is now widely practiced on a commercial scale. The use of irradiation alone as a preservation technique will not solve all the problems of postharvest food losses, but it can play an important role in reducing the dependence on chemical pesticides. A code of good irradiation practice for the control of pathogens and other micro-flora in spices, herbs and other vegetable seasonings has been developed by the International Consultative Group on Food Irradiation (ICGFI) under the aegis of FAO, IAEA and WHO. The purpose of the irradiation is to disinfect the spices and herbs seasonings of micro-organisms and/or insect pests. It is important to note irradiation is not used for the preservation of these ingredients and the process of preservation is obtained through proper drying, packaging and storage. There is still unwillingness by some countries and market segments to adopt irradiation and there have been recent developments to sterilize botanical powders such as spices by contact with an oxidant such as nascent atomic oxygen (United States Patent 6,682,697). The International Organization for Standardization (ISO), a worldwide network of national standards institutes working in partnership, develops voluntary technical standards for a wide range of products that are traded internationally. The ISO standards help raise the levels of quality and provide assurance of minimum standards as well as detailing standardized analytical methods. For example, the traditional method for evaluating pungency (heat level) of capsicum products, the organoleptic scoville test, has now been replaced by HPLC (High Performance Liquid Cehromatography) where specific compounds are identified and levels quantified.

Quality specifications for export

The export of seed spices constitutes only a small share of the total production. Steps to increase production and productivity of seed spices and to develop new value added products would augment the export. Since quality is an important factor for export, care should be taken at all stages of cultivation to prevent contamination and deterioration of quality.

Commercial requirements: These requirements vary from country to country and crop to crop. It depends on many factors like customer needs, acceptability and country specifications. The commercial specifications of seed spices crops are colour, appearance, taste, pungency, texture, shape, volatile oil and packaging. The general commercial requirements for seed spices crops are given below.

Table 128. General commercial requirements of seed spices for export

Coriander	Yellow-brown or green colour seed, small seeded, large seeded, free flowing, high volatile oil (minimum, 0.25%)
Fennel	Light green and light brownish green colour seed, bitter or sweet, small or large seeds, aromatic, free flowing, volatile oil (minimum 1% for Germany, 3% for Netherlands, 2% for UK)
Cumin	Light brown colour seed, large seed, aromatic, volatile oil (minimum, 3%)
Fenugreek	Yellow seeded, green seeded, small or large seeds.
Ajwain	Brownish grey seed, free flowing, aromatic, volatile oil (minimum, 2%)
Celery, Aniseed	Brownish grey colour seed, free flowing, aromatic volatile oil (minimum 1.5%)
Dill	Brown colour seed, free flowing, aromatic volatile oil (minimum, 2%)

Table 129. ASTA, cleanliness specifications for seed spices

Crop	Whole insect dead (by count)	Excreta mammalian (mg kg^{-1})	Excreta other (mg kg^{-1})	Mould (%wt)	Insect deliled/ infected (%wt)	Extraneous foreign matter (%wt)
Coriander	4	6	20	1	1	0.5
Cumin	4	6	10	1	1	0.5
Fennel	2	4	4	2	1	0.5
Anise	4	6	6	1	1	0.5
Celery	4	6	6	1	1	0.5
Dill	4	6	6	1	1	0.5
Fenugreek	4	6	6	2	1	0.5

Table 130. Other quality standards for seed spices as per ESA and ISO

Crop	Ash level % w/w (minimum)	Acid insoluble ash % w/w (maximum)	Moisture content % (maximum)	Volatile oil % (minimum)
Anise	9	2.5	12	1.0
Celery	12	2.0	11	1.5
Coriander	7	1.5	12	0.3
Cumin	14	3.0	13	1.5
Dill	10	2.0	12	1.0
Fennel	9	2.0	12	1.5
Fenugreek	7	2.0	12	-

Mesh / particle size: Particle size as per standardized sieve for each seed spice crop 95 per cent of seed should pass through 25 mesh 30 mesh sieve.

Health specifications: The micro contaminants such as pesticide residues, microbial counts, aflatoxins, heavy metals etc. are the health specifications. The importing countries have limits the maximum level of these contaminants. Maximum Residue Level = 0.05 to 0.2 per cent, microbial counts = 1 X 102 g^{-1} to 1 X 106 g^{-1} (Salmonella, *E. coli*, yeast, moulds), aflatoxin = 5 ppb to 10 ppb (Maximum), heavy metals like arsenic (5 mg kg^{-1}), copper (20 mg kg^{-1}), lead (10 mg kg^{-1}), zinc (50 mg kg^{-1}).

Quality Management in Seed Spices

Developed countries are the major markets for seed spices export and they have their own stringent food laws and regulations. The main objectives of the law are to protect the health and safety of their citizens. The importers prescribe grade specifications for various spices depending upon the end use. Spices exported to any country should conform to the cleanliness specification specified by the country. These countries set limits for cleanliness specification such as number of dead insects, amount of mammalian excreta in the sample. If these requirements are not fulfilled by the exporting country, the consignment may be detained for reconditioning or rejected. Definition of quality as per International Organization for Standardization (ISO), the totality of characteristics of an entity that bear on its ability to satisfy stated and implied needs. In many instances needs can change with time, this implies a periodic review of requirements for quality. Needs are usually translated into characteristics with specified criteria and may include, for example, aspects of performance, usability, dependability (availability, reliability, maintainability), safety, environment, economics and aesthetics. Other definitions of quality as published by the International Trade Centre are Fitness for purpose or use, conformance to requirements. From the many definitions being used in different contexts, it is clear that quality is in fact a complex and multifaceted concept, embracing the sum of all the characteristics of a product or service that contribute to its superiority and excellence. Requirement for quality is defined as the expression of the needs or their translation into a set of quantitatively or qualitatively stated requirements for the characteristics of an entity to enable its realization and examination. It is crucial that the requirements for quality should reflect the stated and implied needs of the customer. The term requirements cover market-based and contractual as well as an organization's internal requirements. They may be developed, detailed and updated at different planning stages. Quantitatively stated requirements for the characteristics include, for instance, normal values, rated values, limiting deviations and tolerances. A product in this context spice and spice products results from a series of processes, and transforming inputs into outputs. Many factors influence product quality are the raw material, production machines and equipment, production processes and the workers. Among the contributors to product quality are the following-

» Continuous or upgrading of the product to meet the evolving requirements of the market.

» Building quality into the design of the product to meet the requirements and opportunities of the market.

» Providing support throughout the life cycle of the product so as to maintain its design characteristics and value for the customer.

In developing countries like India, it is necessary to promote standardization and quality control by giving wide publicity to standards, quality control techniques and the implementation of quality management systems. This means actually spelling out the benefits to be gained from their application and emphasizing the overall economics resulting from simplification, variety reduction and quality management. Publicity through the mass media, workshops, seminars and training programmes is also is a prerequisite. Many factors peculiar to a country, such as its level of technological maturity, economic development, climatic and environmental conditions, inevitably lead to explicable and justifiable differences between the national standards of different countries. These will have to be carefully considered and the differences weighed in comparison to international standards if the latter are to serve for reference purposes in preference to national standards. While many governments with a view to promoting trade have thus directed their policies towards enforcing the minimum or acceptable quality requirements needed for quality improvement, significant national differences have become apparent in their ability to reach acceptable quality levels. In many developing countries, governments have also found that unassisted market forces cannot be fully relied upon to ensure the best-practice quality management or the quality requirements for a particular product, whether it is a safety, social needs, protection of the environment or protection simply of the export market. It will therefore be clear from the above that quality requirements for varying needs are the prime objective of a national or international standard. This applies whether it remains a voluntary standard for adoption by interested parties or it is of a mandatory, compulsory or regulatory nature, in the interest of society as a whole.

Standardizing Bodies

The most remarkable historical development in the reality of standardization has been the evolution of the authority, which makes voluntary standards effective instruments for guiding commerce and industry, thus constituting an economic force in national life and also in international life. This is brought about mainly by following the consensus principle in preparing standards. Standards exist at various levels such as Company, Trade, National and International.

International standards bodies

The main function of most international standardizing organizations is to produce international standards. In addition, they promote the voluntary adoption of these standards worldwide, their use in conformity assessment and accreditation; the health, safety and environmental benefits resulting from them and the use of these standards in regulatory and procurement activities by government authorities and in the market place. International Standardizing Organizations stay ahead of the market demand for the services they provide. The objectives of International Organization for Standardization (ISO) is to formulate and promote the development of international standards and related activities at the international level, with a view to facilitating the international exchange of goods and services by the reduction of technical barriers to trade. The image of ISO was tremendously enhanced when it developed and issued the ISO 9000 series of standards on quality management systems in 1987. The ISO 14000 series of standards the world's first series of international standards on environmental management systems looks all set to become the next revolutionary series of international standards. Codex Alimentarius Commission (CAC), the Food and Agriculture Organization (FAO) of the United Nations and the World Health Organization (WHO) recognized the need for a mechanism for formulating international standards for food, and established the CAC in 1962. The Commission through these standards has helped to upgrade standards in food manufacturing, processing, safety and quality all over the world, thereby ensuring quality and safety in the growing, harvesting and distribution of food.

National standards bodies

National standards bodies are established to create national standards and codes of practice, lay down procedures for quality control and testing, and establish certification and marking schemes. National standards are adopted and made available to the public by a national standards organization which may be governmental, non-governmental, or a government supported body. Examples of standards bodies are the Bureau of Indian Standards (BIS), American National Standards Institute (ANSI) and the Association Francaise de Normalization (AFNOR). Government of India has prescribed standards for almost all the important spices viz., pepper, ginger, cardamom, chillies, turmeric, garlic, onion, coriander, fennel, curry powder, fenugreek, cumin and celery are graded compulsorily under law before export. The remaining spices are being brought under this scheme in a phased programme. The grades adopted for various spices are those prescribed under the Agricultural Produce Grading and Marketing Act 1937 (APGM Act) and these grades are popularly known as Agmark grades. Importers of seed spices from India would be well advised to place orders on the basis of Agmark grades or ascertain from exporters, the specific Agmark Grades of the seed spices, proposed to be exported. Spice and spice products free from bacterial and fungal contamination, zero pesticide residues are preferred. The Indian supply of seed spices not only faces above problems but also lacks in qualities like size, weight, colour, brokenness, insect, immaturity/ over maturity, extraneous matter, mycotoxins, heavy metals, non-permitted colours and animal filth.

Table 132. Grade specification of seed spices powder

Crop	Moisture % by wt.	Total Ash % by wt.	Acid insoluble in HCL (%)
Coriander	10.0	7.0	1.5
Cumin (standard)	12.0	8.0	1.5
Cumin (general)	12.0	9.5	1.5
Fennel	12.0	9.0	2.0
Fenugreek	10.0	7.0	2.0

Table 133. AGMARK grade specification of coriander

Grade	Extraneous matter (%) by wt.	Split seeds (%) by wt.	Damaged blackened & insect bored seeds (%)	Shriveled & immature seeds (%) by wt.	Moisture % by seed wt.
Special	1.5	10.0	1.0	1.0	10.0
Good	3.0	15.0	2.0	2.0	10.0
Fair	7.0	30.0	4.0	3.0	10.0

Table 134. AGMARK grade specification of cumin

Grade	Extraneous matter (%) by wt.	Other seeds % by wt.	Damaged Discoloured weeviled seeds by wt.	Shrivelled & Immature seed % by wt.
Special	1.5	0.5	1.5	1.5
Good	2.5	1.0	2.5	3.0
Fair	3.5	1.5	3.5	4.0
Average	5.0	-	5.0	5.0

Table 135. AGMARK grade specification of fennel

Grade	Inorganic extraneous matter (%) by wt.	Organic extraneous matter	Damaged discoloured weeviled seeds by wt.	Shrivelled and Immature seed % by wt.	Moisture (%)
Special	0.25	1.5	1.5	2.0	10.0
Good	0.25	2.5	4.0	4.0	10.0
Fair	0.25	3.5	6.0	6.0	10.0

Table 136. AGMARK grade specification of fenugreek

Grade	Extraneous matter	Damaged discoloured weeviled seeds by wt.	Shrivelled and Immature seed % by wt.	Moisture (%)
Special	1.0	0.5	1.5	10.0
Good	2.0	1.0	3.0	10.0
Fair	4.0	2.5	5.0	10.0

Macro-cleanliness: The standards of most importing countries carry specifications of macro-cleanliness. These define the maximum percentages of extraneous matter and foreign matter permissible in a product. The term extraneous matter is normally taken to mean parts of the same plant (stems, stick, dry leaves etc.) which differ from the product. Foreign matter consists stones, dirt, wire, string, excreta, animal manure etc. Macro-cleanliness can be ensured largely by suitable post-harvest interventions particularly at the time of harvesting, drying and storage. Most quality conscious exporters and food companies have invested in sophisticated machinery a combination of sophisticated and indigenous machinery for cleaning the products. The European Spice Association (ESA) has fixed the limit on extraneous matter in spices at 1 per cent and foreign matter at 2 per cent. In addition, it has specified limits for ash, acid insoluble ash (AIA), volatile oil (VO) and water.

Microbial load: Both ASTA and ESA prescribe limits for *Salmonella* and *E. coli* contamination. There are strict limits for mould in import requirements for most products. The ASTA standards reproduced here, state the permissible limit in terms of maximum percentage of mould by weight. The ESA's minimum quality and sanitary requirements give specifications on yeast and mould and other microbial contaminants.

Aflatoxins (mycotoxins): Aflatoxins are highly toxic metabolites including B_1, B_2, G_1, G_2, M_1 and M_2 produced by the fungi *Aspergillus flavus* and *Aspergillus parasiticus*. Aflatoxins contamination has become a major health hazard in food products and importing countries have imposed stringent requirements in regard to it. Maximum limit for total aflatoxin has been set in almost all developed countries at between 1 ppb and 2ppb.

Trace metal contamination: Trace metal contamination is a comparatively new health issue in developed countries. The metals currently considered hazardous are mercury, cadmium, arsenic, chromium and lead. Some developing countries which had been using lead as a solder for tin containers have been able to replace it with another soldering material to comply with the regulations of importing countries. It should be noted that copper is an ingredient in fungicides widely used in the agricultural sector (for example, the Bordeaux mixture) and zinc is a component of preparations containing micro-nutrients. In view of the restrictions on the presence of these metals in food products, agricultural scientists will be obliged to re-examine current recommended practices and suggest proper options in the light of the new health regulations in importing countries.

Pesticide residues: Regulations on pesticide residues are likely to become the single most important non-tariff barrier to trade in spices. Since these regulations are non-discriminatory in nature, it will not be possible to negotiate differential standards. Pesticide residues are acquired in the pre-harvest stage and it is not possible to eliminate them completely during processing. It is therefore necessary to evolve practical and cost-effective pre-harvesting strategies to address this concern.

Pre-Harvest Operations

To produce a high quality product that will attract a premium market price, which benefitted to growers and the industry. Pre-harvest operations involve the preparation of the facilities for the harvest material, which will ensure the crop is stored and dried quickly under hygienic conditions. The main reasons for low quality product are harvesting the crop when it is not mature, poor drying systems where there is a high risk of moisture retention and microbial contamination (dirt floors) and frequent rain during the drying process. Many growers of spices use traditional methods and high moisture retention, microbial contamination and contamination with extraneous matter are common processing problems. Poor storage facilities and unhygienic and improper storage methods also contribute to contamination with mammalian and other excreta, as well as moulds or other microbes. In order to overcome these problems completely, capital investment is necessary, particularly for mechanised handling after harvest. International sanitary and phyto-sanitary agreements define measures to be taken to protect against risks arising from additives, contaminants, toxins or disease causing organisms in food or foodstuffs. In particular, there are problems with mould, high moisture and aflatoxin contents. All personnel (including field workers) involved in the propagation, cultivation, harvest and postharvest processing stages of plant production should maintain appropriate personal hygiene and should have received training regarding their hygiene responsibilities. Only properly trained personnel, wearing appropriate protective clothing (such as overalls, gloves, helmet, goggles, face mask), should apply agrochemicals.

Maturity of spices crops

The selection of seed or fruit to harvest which is at the correct maturity or ripeness, based on colour, is critical to obtain a product of even high quality. Allspice (Pimenta) fruit is harvested 3-4 months after flowering when mature but not ripe (green) fruit is most strongly flavoured. Small or over-ripe berries should be removed as they detract from the appearance of the finished product. Colour is important for vanilla pod harvest and daily picking of mature vanilla pods is essential to obtain the correct maturity for processing, as immature pods produce an inferior product and over-mature pods split at curing. The right picking stage is when the distal end of the pod turns yellow and fine yellow streaks appear on the pods. Vanilla pods, when green, are almost odourless but they develop a faint phenolic odour as the beans reach a stage of harvest maturity. The beans are ready to harvest 6-9 months after flowering. Immature fruits of coriander contain higher volatile oil than ripe fruits but have a disagreeable aroma. The crop should be harvested when the majority of the umbels are in a ripe chestnut colour. Black pepper spikes from the vines are hand-picked when one or two berries begin to turn bright orange or red on the flower spike. These are taken to the threshing floor where the berries are separated from the spikes. White pepper is harvested fully ripe; black pepper is harvested fully mature and near ripe, canned pepper at 4-5 months growth, dehydrated green pepper at 10-15 days before full maturity, oleoresin at 15-20 days before maturity, pepper oil at 15-20 days before maturity and pepper for powder is harvested when the berries are fully mature with maximum starch. In quality terms, the volatile oil percentage decreases with maturity but piperine and starch per cent increase.

Blanching and primary processing of black paper

Blanching with hot water for one minute is required to maintain better colour of the produce. This process cleans the fruit and removes other contaminants and gives a shining black colour to the fruit by activating the phenolase enzymes responsible for producing the black colour. Blanching allows the fruit to dry more rapidly. Mechanization of these processes has been undertaken by spraying boiling water onto the fruits as they move on a conveyor belt fitted with mechanical brushes. The blanching time can be adjusted by the speed of the conveyor. The washed and blanched berries pass through sieves where the fruits are graded into selected sizes. The graded peppers are air-dried to remove the surface moisture and then placed in a drying yard or in a drier.

Continuous harvesting of spice crop

Nutmeg and Mace trees bear fruits throughout the year and fleshy drupe turns yellow when ripe and process run continuously. The pulpy outer husk (pericarp) splits into two halves exposing a purplish-brown seed surrounded by a red aril. The fruits are often allowed to split and fall to the ground before harvesting when they should be collected as soon as possible to prevent discolouration and the risk of mould or insect damage. A long pole may be used to take partially opened pods directly from the tree and this ensures a better quality aril but can result in damage to flowers and younger fruit. The frequency with which nutmegs are harvested is dependent on the availability of labour, level of production and the market price, but a daily harvest to every 2-3 days in off-peak times is typical.

Maturity indices in seed spices

Seed spices crops are valued for their peculiar flavour and harvesting stage plays a major role in detaining the flavour which adds to the quality. The essential oil content present in fruits of seed spices chiefly contribute towards the flavour and fragrances. Thus it becomes necessary to harvest the crop at right stage.

Parameters for deciding maturity indices

Closing of flower initiation: At the time of maturity flower initiation bring to a close.

Type of the umbel: It depends on the type of umbel to be harvested.

Colour of the leaves: At time of maturity the colour of leaf changes from green to yellow or brown.

Colour of the pods: At the time of maturity the colour of pods changes from green to yellow or brown.

Colour of seeds: Colour of seed changes from green to brown or yellow or black.

Colour of fruits: At time of maturity the colour of fruits change from green to yellow or brown.

Stage of harvest: The stage of harvest depends on the targeted market/ consumer's choice.

Ajowain: The crop of Ajowain matures in 130-170 days and harvesting is usually done from February to April. Flower production ceases when the seeds start maturing and become greyish brown in colour. Mature crop should be harvested with sharp sickle in the morning hours to avoid shattering.

Anise seed: The crop matures in 120-140 days. At the time of maturity, tips of fruit turn greyish green and then fruiting umbels are cut off, tied in bundle and stacked for 4-5 days dried and then threshed out.

Caraway: The annual crop of caraway matures in 130-150 days whereas biennial crop in 450 days. Plants should be harvested when the umbels turn brown, over mature crop is prone to shattering and therefore, it should be harvested well before this stage. The harvested plants should be properly stacked and cured or dried till the harvest is completed.

Celery: Celery crop matures in 150 days after transplanting, if not harvest on time may start seed shattering. Spraying of polyvinyl acetate (PVA) glue on umbel prevented shattering and increased seed yield upto 33 per cent.

Coriander: Coriander crop matures in about 100-150 days of sowing depending upon varieties and climate. The coriander crop grown exclusively for green leaves is harvested after 35-40 days of sowing. For green coriander seeds, the crop is harvested when seeds are green and have attained full size. Delay in harvesting should be avoided, otherwise shattering and splitting of seeds may occur. Under ripe coriander fruits (seeds) have an unpleasant flavour, therefore harvesting should take place when between half or two thirds of the fruits are ripe.

Cumin: Cumin is harvested in about 100-110 days after seed sowing. The quality and quantity of volatile oil improved with ripeness. The mature plants are uprooted and stocked for drying. The seeds are seperated manually by beating with sticks. The clean dried seeds are filled in gunny bags. Barrier liners are used to protect loss of aroma and flavour.

Dill: When dill is grown for vegetable purposes it can be harvested at 4-5 leaves stage. Dill crop takes about 70-80 days for flowering and 130-150 days for harvesting and crop becomes ready for harvest by end of March. The plants become pale yellow at maturity and the seeds are light to dark brown in colour. The percentage of essential oil on a fresh weight basis also increased gradually till the seed ripened after which it started declining. The seeds are better dried in shade rather than in sun to avoid loss of dill oil, the yield of oil was 2.88 per cent when seeds were dried in shade, but only 1.03 per cent when dried in the sun.

Fennel: Fennel takes 170-175 days to mature. The umbels attains a slight greenish yellow colour are appropriate for harvesting. At this stage volatile oil content was found to be the highest and fruit fetches better prices in the market. A good quality fennel for chewing purposes commonly known as Lucknow saunf or Aabu saunf is produced by harvesting the umbel 30 to 40 days after pollination. At this stage, the size of the grains is half of the mature fruit and fibre content is 11 per cent as compared to 15 per cent crude fibre in normal fruit.

Fenugreek: Fenugreek comes to harvest within 90-100 days after sowing. The harvesting time is judged when the colour of leaves and pods turn yellow, delay in harvesting may cause seed shattering. Harvesting in the early hours of the day is preferable. After harvesting plants are allowed to dry under the sun.

Nigella: The nigella crop takes 140-160 days to reach maturity. The crop should be harvested when the seed attained full maturity in capsule and have turned to full black colour. Delay in harvesting may cause shattering of seeds.

Harvesting of Spices

Harvesting is the primary process of collecting the target crop product from the field, where it is open to the vagaries of the climate and the growing environment and placing that product in controlled processing and stable storage conditions environment. The harvesting requirements will differ for the final product sought, and there are specific needs such as maturity and evenness, that will dictate the harvesting management and timing. Plants should be harvested during the optimal season or time period to ensure the production of plant materials and finished spice products of the best possible quality. The best time for harvest should be determined according to the quality and quantity of biological target constituents. During harvest, care should be taken to ensure that no foreign matter is mixed with the harvested plant materials. When possible, plant parts should be harvested under the best possible conditions, avoiding dew, rain or exceptionally high humidity. If harvesting occurs in wet conditions, the harvested material should be transported immediately to a drying shed and drying started to expedite drying so as in order to prevent any possible deleterious effects due to increased moisture levels, which promote microbial fermentation and mould. Cutting devices, harvesters and other machines should be kept clean and adjusted to reduce contamination from soil and other materials. They should be stored in an uncontaminated dry place, free from insects, rodents, birds and other pests and inaccessible to livestock and domestic animals. Soil can have a high microbial content and contact between the harvested crop and the soil should be avoided so as to minimize the microbial load on the harvested plant materials. Where necessary, large drop cloths, preferably made of clean muslin, should may be placed on the soil surface before the plants are used as an interface between the harvested plants and the soil. If the underground parts (such as rhizomes, roots) are harvested used, any adhering soil should be removed from the plant material as soon as possible it is harvested. Mechanical damage or compacting of raw plant materials as a consequence of overfilling or stacking of bags may result in fermentation, composting or rot and other damage should be avoided. Rotting plant materials should be discarded during harvest, postharvest inspections and processing, in order to avoid contamination and loss of product quality.

Table 137. Crop duration and stage of maturity of seed spices

S. No.	Seed spices crop	Crop duration (days)	Stage of Maturity	Level of shattering
1	Ajowain	130-170	Flowering closes and seeds in umbel are about to attain brown colour	Medium
2	Anise	100-110	Tips of fruits greyish green in colour.	Low
3	Caraway	Annual- 130-150 Biennial- 450-460	Oldest seeds start turning brown	High
4	Celery	150 (after transplanting)	Pick the mature umbels periodically when they turn brown	High
5	Coriander	90-135	Central umbels are about to attain yellow colour	Medium
6	Cumin	100-110	Maturity of the umbel is indicated by browning of seeds	Medium
7	Dill	130-150	Umbels turn brown	High

8	Fennel	210-240	Harvest before fruits are fully ripe i.e., fruits turn to yellow colour cut the stems along with umbel.	Low
9	Fenugreek	140-150	The pods become yellowish and lower leaves start shedding	Very high
10	Nigella	140-160	Harvest when the seed have attained full maturity in capsule and have turned to full black colour	Medium

Harvesting of leaves and stems

The aromatic herbs such as basil, bay leaves, mints, oregano, parsley, rosemary, sage, tarragon and thyme are often used fresh to garnish food, as dry seasonings or for extracted essential oils. The leaf and stem should be cut when the growth has matured to an elongated flower stem but without full flower or significant senescence of the lower leaves. Harvesting the plant material should be avoided when it is wet and if it is to be dried the plant material should be evenly spread onto the drying racks or drier to ensure there is no sweating, fermentation and microbial invasion.

Harvesting flowers and buds

When flowers and buds are the source of a spice or essential oil, the harvest time can be throughout the year in the tropics or during a limited flowering season in temperate regions. Clove clusters are picked when the unopened flower buds are full-sized, the calyx base has developed the characteristic pink flush, but no buds have opened or petals fallen to expose the stamens. The harvested clove clusters are taken to the storage centre and the stems, which are about 25 per cent of the total dry weight, are removed and separated. At this time the flowers are drier, contain more oil and the oil is of higher quality. Flowers are harvested early in the day. Only fully developed yellow flowers are picked as green flowers produce poor quality oil. In order to avoid fermentation, the flowers should not be held in a mass and distillation should be undertaken as soon as possible. Saffron (*Crocus sativus*) is an example of a flower-derived spice which has a specific harvesting time. There can be up to 12 flowers per corm which should be hand-picked daily, after the flower has opened but has not withered. The stigma is hand-separated or airflow-separated from the flower parts at the drying and processing facility.

Harvesting roots and rhizomes

The common spices which are derived from roots, bulbs and rhizomes are turmeric, ginger, onion, garlic and horseradish. Turmeric and ginger, a perennial herb grown as an annual crop, is ready for digging when the lower leaves turning yellow. The length of time to harvest maturity is dependent on cultivar. Hand digging is the most common form of harvesting the rhizomes although diggers and lifters can be used. The leaves must be cut prior to mechanical lifting or after hand digging. Care is necessary to avoid damage to the rhizome (splitting or bruising) as injuries can result in fungal infection and rejection. Rhizomes are lifted whole, washed, sun dried and the rhizome fingers (which are kept as seed material) are separated from the mother rhizome. The maturity of onions for harvest is based on top senescence when the bulb is converted from active growth to dormancy by under cutting and the beginning of the drying process. Garlic can be harvested at the bulbs are well mature and the tops have dried completely. Horseradish, a perennial that is grown for its enlarged root, is ready for harvest after the leaves have been killed by frost. Horseradish that is harvested while it is still actively growing does not store as well.

Harvesting bark, wood and resins

Cinnamon and Cassia are generally harvested for wood or bark products and managed under coppice plantation systems. Harvesting of the shoots is undertaken two years after coppicing during the rainy season. The rough outer bark of the selected shoots is first scraped off and the young tender inner bark is peeled off carefully from the uniform brown colour of bark. The inner bark curls naturally into the well-known quills. The best pieces of the peeled bark are packed with small pieces and tightly rolled to preserve the flavour and dried. The coppiced shoots are left for fermentation for 24 hours, dried in the shade for one day and in the sun for four days. The smaller quills are inserted into larger ones to form compound quills. The products are graded as quills, quillings, featherings, scraped chips and powders. The finest quality bark is obtained from shoots (<1.25 m by 1.25 cm diameter) with uniform brown thin bark harvested at six-month intervals.

Table 138. Suggested harvesting equipment for seed spices

S.No.	Name of equipments	Capacity	Expected cost (Rs.)	Remarks	Source
1	Self propelled reaper	0.2 ha/hrs	70000.0	Operated by 6 HP diesel engines. Can be used after optimization of operating parameters.	CIAE Bhopal
2	Power tiller operated reaper	0.12 ha/hrs	25000.0	Front mounted, perated by 8HP power tiller.	CIAE Bhopal
3	Self propelled reaper	1.15 ha/hrs	125000.0	Operated by 6 HP diesel engines. Cutter bar 76.2 mm	CIAE Bhopal

Post production operation

Spice and herb production can be sub-divided into a number of activities and although there is a route to market through by-product extraction, most spices and herbs have a series of postharvest operations, which follows a logical sequence. The postharvest processing tree shows each stage in the process.

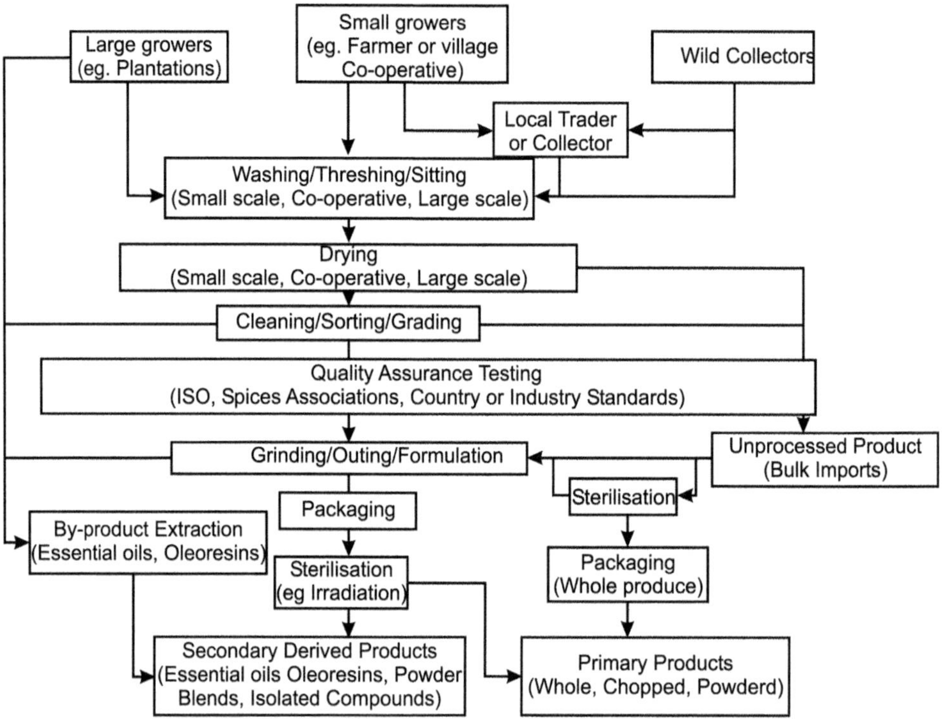

Figure 49. Postharvest processing and products produced from spices and herbs

Threshing

Threshing is the process of removing and separating the fruit or seed from the unwanted flower stems or plant stalks as well as removing damaged or immature material. This process can be undertaken by hand, assisted by sieves and screens, by use of winnowing or by mechanical shakers and sorters. Pepper is an example where spikes need to be washed prior to threshing to remove dust and dirt. Harvested spikes are spread onto a clean floor and threshed (decorned) manually by trampling or by using a mechanical thresher. The berries are also graded at this stage by removing undersized berries and put into a size category.

After harvesting, it is common practice to put the harvested crop in open yard under sun to attain the optimum moisture level for threshing. The manual threshing is still being followed in seed spices, which produces lots of broken and contaminates the spices with straw, broken and leaves etc. Adoption of mechanized threshing can improve the quality and reduce the post harvest losses (Table 3). The method of spreading in open yard for drying can be replaced by whole crop dryers as used in hay drying so that better efficiency of threshing is achieved without loss in volatile oil due to atmosphere exposure.

Table 139. Suggested threshing equipment for seed spices

S. No.	Name of equipments	Capacity	Expected cost (Rs.)	Remarks	Source
1	Paddle thresher	Depending upon the crop		Operated by one man power	Commercially available
2	Axial flow thresher	200-500 kg hrs-1	50000.0	Can be operated with tractor or 7.5 HP electric motor	Commercially available
3	Groundnut Stripper	Up to 20 kg hrs-1	15000.0	Manual operated can be used for coriander after modification	CIAE Bhopal
4	Power operated stripper	Up to 70 kg hrs-1	20000.0	Operated by 3 HP motor	CIAE Bhopal

Drying of Produce

The moisture content of seed spices is normally more than 20 per cent at the harvest, which is reduced to 14 per cent for threshing. However it needs to be further reduced to 10 per cent or less depending upon type of seed spices for safe storage. In practice the plants after harvest are piled into small stacks in the field to achieve a moisture level suitable for threshing. The aim of drying is to reduce the moisture content of the product from actively growing in the field to a level that prevents deterioration of the product and allows storage in a stable condition. Drying is a two stage process: firstly the transfer of heat to the moist product to vapourize the water in the product and secondly mass transfer of moisture from the interior to the product surface where it evaporates. The most important and immediate management concern is to ensure the harvested crop will not rot or become grossly invaded with yeasts, bacteria and mould (producing aflatoxins) or become contaminated by pests. This is the start of the preservation process, which for most spice crops requires drying that will enable the long-term crop storage and the opportunity for further processing. The drying phase of postharvest management can include four preliminary stages the selection of high quality produce from the field, cleaning the crop by washing and disinfection, preparing the crop for drying by peeling or slicing, pre-treating with anti-oxidants, blanching or sulfurizing. In some cases, washing prior to processing is desirable to remove field contaminants (dust, soil) using anti-microbial solutions to reduce the microbial populations to a low level prior to the drying process. The most basic method of drying is to spread the crop on a surface exposed to the sun. In this case, the process is aided by a cover system that prevents wetting with rainfall. An improved method, speeding up the drying, is to use a fuel source (wood, oil/ diesel, gas or electricity) to heat the drying room. Solar drying systems together with solar powered fans are also available. The drying process should dry the crop as quickly as possible, at temperature levels which do not drive off the volatile flavour compounds. The drying temperature regime will be specific to each crop as will be the final moisture percentage for storage. High crop loss and low product quality result from inadequate drying, long drying times, fungal spoilage, insect infestations, bird and rodent damage and contamination plus the effects of sunlight and the weather. Even in the most favourable climate it is often not possible to get the moisture content of the product low enough for safe storage. In the tropics the high relative humidity of the air prevents drying of harvested crop products during the wet season.

The objective of a drier is to supply the product with more heat than is available under ambient conditions. A relatively small amount of heating can greatly enhance the moisture carrying capability of the air. For example, heating of air from a temperature of 20° C at 59 per cent relative humidity (RH) to a temperature of 35° C at 25 per cent RH increases the moisture holding capability three times. In a drier the major requirement is the transfer of heat to the moist product by convection and conduction. The absorption of the heat by the product supplies the energy necessary for the vapourization of water from the product. Spices and essential oil crops derived from leaves or flowers are relatively thin and therefore relatively easy to dry due to their small diffusion thickness. Conversely in thick and fleshy materials such as roots, the drying process requires much more careful control of temperature, temperature ramp rate and airflow rate. If the temperature is too high, elevated too quickly or the airflow rate is too high when drying thick fleshy products, case hardening may result and only the outer surface will dry. Solar-energy drying systems are put forward as technology for small rural farmers and enterprises in developing countries. Solar driers can be classified into two generic groups, passive or natural air circulation solar dyers and active or forced convection solar-energy driers. Forced convective driers employ motorised fans for circulation of the drying air. The electricity for the fan can come from a solar photoelectric panel and battery. Each group can also be sub-divided into three subgroups:

» Integral types (direct solar driers where the crop is placed in a drying chamber with transparent walls and the solar radiation falls directly on the crop, coupled with convection air flow from the heated surrounding air).

» Distributed type (indirect heating, where solar radiation heats a solar collector external to the drying chamber).

» A mixed type where there is both direct and indirect heating.

The solar drier has two major disadvantages: a limited ability to process crops when the weather is poor; and drying can only occur during the day time. This not only limits production and extends the drying time but also may have an adverse effect on production and product quality particularly fleshy crop products such as roots and stems that typically have drying times of several days. This has led to the development of hybrid systems with auxiliary heating systems such as burners using biomass, biogas or fossil fuels. Coriander is artificially dried in some countries including the Russia at temperatures of 80-90° C. Temperatures of over 100° C and direct sunlight results in the loss of volatile oils. Mechanized drying could enhance the quality and reduce the post harvest losses. Green house drying may also be a good option for drying the seed spices as most of the seed spices are harvested during the period of February to April. The threshed seed spices can be dried in tray type using electricity or agricultural waster fired dryer as source of energy (Table 143).

Table 140. Suggested drying system for different seed spices

S. No.	Name of equipments	Capacity	Expected cost (Rs.)	Source
1	IGSI Grain dryer	6.5 q batch^{-1}	25000.0	IGSI, Hapur
2	Portable batch dryer	-	30000.0	PAU Ludhiana
3	Bin dryer	3-5 q batch^{-1}	60000.0	TNAU, Coimbatore
4	Tunnel type green house dryer	100kg batch^{-1}	-	CIAE Bhopal
5	Cardamom dryer can be used for all seed spices	50 kg 10 hour^{-1}	25000.0	UAS GKVK Bangalore
6	Tray dryer	50-100 kg batch^{-1}	30000.0	CIAE Bhopal
7	Solar cabinet dryer	3-5 kg batch^{-1}	10000.0	
8	Low cost poly solar dryer	-	15000.0	CPCRI, Kasargod
9	Tray dryer	50-100 kg hrs^{-1}	50000.0	

Drying Seeds and Fruits

Immediate drying: After threshing and blanching, black pepper fruits are dried in the sun for 7-10 days till the moisture is about 10 per cent and the seeds have the characteristic wrinkled appearance of black pepper. Pepper drying in the sun has to be turned over periodically to aid uniform drying and discourage mould infestation. The main disadvantage of sun drying is the lack of uniformity and potential contamination by microorganisms. Improvements in black pepper need to focus on the drying process. In areas where there is rainfall all year round and a high humidity, the quality of black pepper produced is dependent on drying. Mechanical driers have only had limited success at farm level and the cost of drying using fossil fuels is relatively high but other forms of drying such as solar-driers or solid fuel driers have not been found as effective. Nutmeg and Mace fruits are dried in their shells in the sun and turned each day to prevent fermentation. The nuts are sufficiently dry when they rattle and drying takes about one week. Once dried, the nuts can be stored for a considerable time. The separated mace is flattened by hand and dried on mats in the sun. This takes between 2-4 hours. In Grenada mace is cured in dark storage for 4 months. This produces a brittle pale yellow mace that has a premium price.

Chillies (Capsicums, Chilli peppers and Paprika) have a thick waxy skin, which prevents rapid drying. Sun drying of fruit is the most common method and is undertaken by spreading the crop in a dry area exposed to the sun or under cover. Weather exposure can bring high losses and contamination is also a problem. Fruits are spread in thin layers on a hard dry surface. The use of black plastic under the fruit speeds the drying time. The fruit is regularly stirred to ensure uniform drying and to reduce discolouration and fungal growth. At night or in wet conditions the fruits are heaped and covered. Drying will take up to 15 days dependent on the sunshine hours and weather conditions. A 100 kg of fruit produces about 25-35 kg of dried spice. Large traders use racks that can be moved in or out of the sun until the fruits are dried below 10 per cent whilst retaining colour and pungency. This system reduces the drying time by half and produces a more uniform product. A chemical emulsion 'Dipsol' can be sprayed on fresh fruits to accelerate sun drying. Artificial driers such as energy-efficient heat pump driers are well suited for drying chillies as they operate at low temperatures. This can be essential for chilli drying under changeable weather conditions as the spice quality becomes less desirable when it becomes brown instead of bright red. The drying temperatures should be below 60° C.

Programmed drying: The enzymatic development of flavour compounds

Vanilla pods, during the curing process, develop flavour as a result of naturally-induced enzymatic action. The final vanilla quality depends mostly on the glucovanillin content in the green bean. This means that the quality of cured beans will be higher if the curing starts with mature beans that have the highest glucovanillin concentration. A number of procedures have been developed for curing vanilla and they can be characterized into four stages. The first operation is to halt the plant's physiological functions by stopping vegetative development so that various enzymes can come into contact with cell substrates. The most practical methods are sun- or oven-wilting or hot water scalding. The second operation that is the most crucial step is sweating. Bean moisture is initially allowed to escape rapidly to get to a level that will reduce the risk of microbial spoilage during the subsequent processing. The curing enzymes are most active during this stage and vanillin and related compounds are released during this step that takes 7-10 days. At the end of the sweating process, the cured beans are still 60-70 per cent moisture: too high for safe storage. The third operation is drying to protect the beans from microbial spoilage and also to inhibit further enzyme activity and biochemical changes. At the end of drying the beans should be at 25-30 per cent moisture. The fourth step is storage in closed boxes for several months while chemical reactions take place. There are two main methods of curing, the Mexican process (sun method) or the Madagascar process (Bourbon method). The Mexican method begins with shed storage for some days until the beans start to shrink. The beans are then placed in sweating boxes, for at least 24 hours, where the temperature is maintained to encourage the slight fermentation and flavour-developing enzymatic reactions. This process of wilting and sweating may be repeated a number of times during the early part of curing and extend over a 2-3 week period. The moisture is reduced to the desired level of 25 per cent by frequent exposure to sunlight and sweating. The final curing occurs when the beans are matured in boxes for 2-3 months. The Mexican process of curing takes 5-6 months. The Madagascar process is faster and begins when the pods are dipped in hot water at 80°C for 7-15 min. The beans are then spread out on blankets in the sun and when well heated are rolled in blankets to stand and ferment overnight. This procedure is repeated over 10 days. At the end of this period the beans are placed on trays and allowed to further cure and dehydrate. The Madagascar process takes 35-40 days.

Allspice: Pimenta berry clusters are taken to a drying shed and left in heaps or sacks for up to 5 days to ferment. Berries are then spread on outdoor drying floors and turned daily to ensure even drying. The length of drying is dependent on the weather conditions (5-10 days) and the berries are covered at night and during rain. Moisture content close to 12 per cent is the target and the yield from 100 kg of green berries is about 55-65 kg. Microbial contamination often occurs during the fermenting and drying phase, especially during wet and humid weather. This contamination is almost impossible to counteract and a high microbial count results in berries unsuitable for human consumption. Correctly dried spice should have a pleasant characteristic odour and low microbial counts. These objectives are most easily met by artificial drying, under controlled conditions. Artificial driers (wood fired) are more common in areas where berries ripen in the wet season. Solar driers are more efficient but require a higher standard of postharvest berry management. Permanently placed large forced air driers, owned and operated by grower cooperatives or local government, are becoming more common. A maximum drying temperature up to 75° C can be used without loss of essential oil. The problems of microbial contamination can be kept to low levels with good drying and storage management. The dry berries are cleaned and bagged and kept in a clean dry store.

Drying to retain colour

Cardamom capsule colour and the retention of greenness are important for highest quality. Capsules cured immediately after picking retained greener colour and the loss of greenness was more significant if the capsules were stored for more than 12 hours. Bagging of the capsules in jute bags and cool storage, aids the retention of greenness. Cardamom capsules at picking contain 70-80 per cent moisture and to enable storage the moisture has to be taken down to 8-10 per cent by curing. Sun-drying is generally undesirable for cardamom because of bleaching and capsule splitting. The most widely-adopted drying system is a slow dry over 18-30 hours using a number of methods of artificial drying (solar, electric, kiln, bin) with various methods of hot air flow. A temperature between 40-50° C helps retain the green colour and an increase in temperature significantly increases the percentage of yellow capsules, split capsules and heat injury.

Drying leaves and stems

Drying is the transformation of the harvested leaf and stem, containing 80-90 per cent water, to a stable state containing 5-10 per cent water. It is important to minimize the holding time between harvest and drying and these activities should be co-ordinated to prevent delays in drying and the chance of spoilage. The temperature during drying is critical and if the drying temperatures are too hot the volatile components in the essential oils will be depleted and lost. In general terms drying temperatures should be below 40° C with forced air movement.

Drying flowers and buds

Clove buds are spread on matting and dried in the sun. Occasional gentle raking will aid even drying and it is important to overcome heating and mould formation. Damaged and spoiled buds are removed during this phase. Drying is completed in 4-6 days with buds losing about two-thirds of their weight. Saffron stigma should be dried the same day as picked from the field, following separation from the floral parts. The stigma is about 85 per cent moisture at field picking and to enable long-term storage of the filaments, a dry matter level close to 10.2 per cent is required after the drying. Drying temperatures above 30° C but below 70° C will provide good quality saffron and irrespective of the drying method, it is important not to over-dry to a brittle stigma. The duration of drying is dependent on the drying temperature used and the higher the temperature used, the shorter is the drying time.

Drying roots and rhizomes

Turmeric rhizomes have to be cured after harvest for both colour and aroma. The traditional method of curing rhizomes is to steam or boil fresh rhizomes in lime or a 0.1 per cent sodium bicarbonate solution. The 'curing' is to remove the raw odour, reduce the drying time, gelatinise the starch and produce a more uniformly-coloured final product. Traditionally the cooked rhizomes were spread in the sun to dry and this process takes 10-15 days. The final moisture level should be close to 6 per cent moisture content. Today the majority of internationally-traded turmeric rhizomes are artificially dried with hot-air drums, tray and continuous tunnel driers and in India a maximum temperature of 60° C is advised. An important factor in drying time is the preparation of the rhizome. Rhizomes to be dried can be sliced or whole, with slicing generally producing a more uniform and brightly coloured powder. The yield of the dried product varies from 10-30 per cent depending on the variety and the crop-growing environment. Ginger curing of the rhizome prior to drying directly affects the fibre and volatile oil content. Removal of the skin reduces the fibre content and also increases the oil loss. Peeling also affects the pungency as these compounds (gingerols) are in the skin. When sun drying is not an option wood-fired or solar driers can be used. The final dry matter should be in the 7-12 per cent range with a weight loss during drying of 60-70 per cent. Artificial drying minimizes the loss of quality and can also eliminate microbial contamination. Drying temperature, airflow and the length of drying all affect the flavour compounds in ginger.

Subsequent to the field harvest, the garlic bulb is broken into individual cloves and the loose paper shell removed by screening and airflow. The cloves are then washed and sliced. The sliced garlic cloves are dried down to about 10 per cent moisture, with a drying temperature below 60° C. In the final drying procedures care is needed to ensure there is no heat damage to the slices and the garlic is dried to 6.5 per cent moisture. Onions are dehydrated in the form of flakes, rings, kibbles and powder. Small-scale onion drying has been achieved by solar drying, but cabinet drying at 55-60° C for 10-15 hours gives a better product than sun drying or drying in solar huts. Commercial dehydration is achieved by forced hot air with an initial temperature of 75° C reducing to 55-60° C as the moisture content falls. Final storage moisture content close to 4 per cent is sought. Horseradish are very sensitive to wilting and immediately after harvest the roots should be pre-cooled to <5° C using forced air cooling or placed at 0° C with 90-98 per cent RH. Roots can be stored for 12 months under those conditions but pungency is rapidly lost at higher temperatures and roots dry out at lower RH. Perforated polyethylene bags and lined crates can maintain high RH during storage.

Cleaning/grading

The seed spices coming farm contain all sort of impurities besides varying in their fruit size. The types, shapes, characteristic of all these impurities are varied in nature hence cannot be separated by one process or by one machine. Cleaning the spice prior to packaging and sale is to ensure that the spice is of the highest quality and will obtain the highest price. Cleaning should remove all the foreign matter that lowers the quality and endangers the sale. Sieves, grading tables, flotation tanks and screens can all be used to ensure that the quality standards are met and an even line of high quality spice is obtained. The cleaning/grading operation is done at mandi level and at the level of processors. The cost of cleaning at mandi level is born directly or indirectly by farmers. Hence there is a need for simple and efficient cleaners cum grader, which can be used at farmers level only. The uniform size of seed spices is necessary for processing. Therefore grading is another important unit operation. The mixture of different size also fetches low price.

Table 141. Suggested cleaner/grading system for seed spices

S. No	Name of equipments	Capacity	Expected cost (Rs.)	Source
1	Hand operated cleaner	150 kg hr^{-1}	5000.0	CIAE Bhopal
2	Paddle and power operated cleaner	300-500 kg hr^{-1}	15000.0	CIAE, Bhopal
3	Seed cleaner cum grader	500-1000 kg hr^{-1}	-	Agrosaw
4	Hand operated drum type cleaner	100 kg hr^{-1}	5000.0	CIAE, Bhopal
5	Seed grader	300 kg hr^{-1}	20000.0	CIAE, Bhopal
6	Cleaner- cum - grader for cumin	100 kg hr^{-1}	30000.0	College of Agril Engg, Junagadh

Packaging

Processed plant materials should be packaged as quickly as possible to prevent deterioration of the product and as a protection against exposure to pest attacks and other sources of contamination. Continuous in-process quality control measures should be implemented to eliminate substandard materials, contaminants and foreign matter prior to final stages of packaging. Processed plant materials should be packaged in clean, dry boxes, sacks, bags or other containers in accordance with standard operating procedures and national and/or regional regulations of the producer and the end-user countries. Materials used for packaging should be non-polluting, clean, dry and in undamaged condition and should conform to the quality requirements for the plant materials concerned. Fragile plant materials should be packaged in rigid containers. Whenever possible, the packaging used should be agreed upon between supplier and buyer. Reusable packaging material such as jute sacks and mesh bags should be well cleaned (disinfected) and thoroughly dried prior to reuse, so as to avoid contamination by previous contents. All packaging materials should be stored in a clean and dry place that is free from pests and inaccessible to livestock, domestic animals and other sources of contamination.

A label affixed to the packaging should clearly detail the product name of the spice, the plant name, the place of production, the harvest date and the names of the grower and the processor and quantitative information. The label should also contain information indicating quality approval and comply with other national and/or regional labeling requirements. The label should bear a number that clearly identifies the production batch.

Packaging for seeds and fruits: Vanilla, for most of the world trade, has grades of whole and split beans which are subdivided according to length into various sizes between 12 and 26 cm and then put into bundles, each containing 70-100 beans and weighing between 150-200 g. These are packed into wax- lined cardboard boxes that hold 20-40 bundles. Beans remaining from the grading and sorting, such as short or broken beans, are bulked together and loosely packed into boxes. Cardamom, when dried, needs to be kept in polypropylene double lined polythene bags to ensure the flavour components are retained and to prevent re-adsorption of moisture, mould growth and reduce insect pests. Cloves when bagged should be stored in cool, clean and dry storege. Clove oils should be stored in full, airtight containers and remain sealed until required. Bulk oil is usually transported in 200 litre metal drums. After drying, the nutmeg seeds are put into bags and transported to the processing factory for kernel separation and grading. After grading, nutmeg mace can be classified, bagged, labeled and can be fumigated prior to export. Nutmegs are usually packed in double-layered linen, jute, sisal or woven bags. Mace follows a similar process but there is a need for three months curing before bagging and fumigation and top quality mace is packaged in plastic bags. Chillies (Capsicums, Chilli peppers and Paprika) that have been dried should be stored in airtight containers away from sunlight as air and light accelerates the rate of bleaching. Chillies should be stored in polyethylene bags or packs and can be stored for a year in a cool, dry and dark place.

Packaging for leaves and stems: It is essential that all material is dry to below 10 per cent. The leaves and stems of dried herbs should be stored in cool, dark and dry areas of low humidity and polyethylene bags or packs used.

Packaging for flowers and buds: Saffron stigma, when dried and graded, should be stored in airtight containers, in a cool dry place out of the light.

Packaging for roots and rhizomes: Fresh ginger should be stored in polyethylene bags with 2 per cent ventilation to prevent dehydration and mould development. Ginger rhizome for bulk shipping can be packed in jute sacks, wooden boxes or lined corrugated cardboard boxes. Processed, dry ginger should be packaged in laminated bags that have low oxygen permeability and stored in a cool dry environment. Turmeric is mostly traded as whole rhizome and then processed into powder or oleoresin. The colour constituents of turmeric deteriorate with light and to a lesser extent under heat and oxidative conditions. Ground turmeric should be stored in UV protective packaging and a cool dry environment.

Packaging for bark, wood and resins: Cinnamon quills are graded based on colour and pressed into cylindrical bales in jute cloth or corrugated cardboard cartons. Cinnamon and cassia, especially if ground, require polypropylene packaging and polyethylene is not recommended as the flavour components diffuse through it.

Transport of Spices

The harvested raw plant material of the spice crop should be transported promptly in clean, dry conditions. They may be placed in clean baskets, dry sacks, trailers, hoppers or other well-aerated containers and carried to a central point for transport to the processing facility. All containers used at harvest should be kept clean and free from contamination by previously-harvested plant products and other foreign matter. If plastic containers are used, particular attention should be paid to any possible retention of moisture that could lead to the growth of mould. When containers are not in use, they should be kept in dry conditions, in an area that is protected from insects, rodents, birds and other pests and inaccessible to livestock and domestic animals. Conveyances used for transporting bulk plant materials from the place of production to storage for processing should be cleaned between loads. Bulk transport, such as ship or rail cars, should be cleaned and, where appropriate, well ventilated to remove moisture from plant materials and to prevent condensation.

Storage

There is a need for quality storage both on-farm and off-farm, with cool stores and warehousing facilities linked to postharvest crop management. Spices deteriorate rapidly in adverse conditions and should be stored in well-prepared and maintained storage facilities. It is essential the moisture level of the spice to be stored is at a safe level (usually below 10 per cent) prior to storage. The storehouses should be damp-proof, vermin- proof and bird-proof and where possible have controlled ventilation and devices to control humidity and temperature. A dehumidifier fitted to a storage room, by keeping the atmosphere always dry, can eliminate mould and insect attacks. The room should be fumigated before storage, the walls whitewashed regularly and the facility kept dry. Only registered chemicals, agents authorized by the regulatory authorities of the source country and the countries of intended end-use should be used. All fumigation dates, chemicals agents and method of application should be documented. Polyethylene-lined gunny bags or laminated HDPE (high density polyethylene) are ideal for pepper storage. Prolonged storage of cloves reduces the oil content and may also dull the colour, which are both detrimental for the sale of the spice. Oil composition and the ratio of constituents of cloves can change in storage, with the rate and extent of change dependent on the production method. Nutmeg and mace should not be ground until required because of the loss of volatile constituents when in uncontrolled storage. Incorrectly stored nutmeg oil may also undergo significant composition changes if exposed to high ambient temperatures. Unprotected powders and oil can also absorb unpleasant odours and powders should be stored in sealed containers. To ensure storage without mould, Chillies (Capsicums, Chilli peppers and Paprika) fruit should be close to 10 per cent moisture or below. Dried rhizomes of turmeric are usually chemically treated or fumigated before storage to prevent insect and fungal infestation. Rhizomes stored in a cool chamber have less disease, rot and pest damage and less sprouting, compared to underground pit storage or heaps out doors.

PROCESSING OF SPICES

Grinding and Separating

Grinding an herb or spice to a specified particle size using standardized sieve apertures is a normal processing activity. Grinding gives easier mixing in the final food product and aids the dispersion of flavour throughout the food. A food manufacturer will have specific particle size requirements and the processor will have to mill the herb or spice through sieves that will obtain the fineness required. White pepper is produced by removing the pericarp (fruit wall) from ripe pepper fruits and then grinding. White pepper is traditionally prepared by steeping ripe fruits in water for a few days, rubbing to remove the pericarp, washing and drying. The berries that are beginning to ripen are put into bags and soaked in water to soften the pericarp, which is then washed off and the seeds dried in the sun. The quality of the water used for soaking the berries is important, as pepper soaked in muddy or dirty water may be discoloured or have foreign odours. Soaking in tanks of clean water produces pepper of higher quality. For black pepper, berries are harvested when mature green, de-spiked and dried in the sun. For both black and white pepper, the quality of the product is dependent on the weather, as almost all the pepper is sun-dried. Efforts to improve quality by cleaning out foreign matter, stalks and light berries by simple sieves and blowers reduce the amounts transported and increase the value. Ground pepper is obtained by grinding pepper with equipment like a hammer-mill. The pepper is first fluidized for the removal of extraneous matter and then passed through screens.

Extraction of Essential Oils

The extraction of essential oils from plant material can be achieved by a number of different methods and these are shown in the generalized flow diagram. There are five main methods of extraction:

- Expression
- Hydro or water distillation
- Water and steam distillation
- Steam distillation
- Solvent extraction

Value addition in seed spices: An opportunity for high profit

Seed spices are known for their aromatic and medicinal values. Aroma in these seed spice crops is due to essential oil present in the seed. Besides raw seed the market value of seed spice crops is of grounded powder or masala, essential oil and oleoresins. The global market is increasingly shifting away from commodity to value added products. India exports raw seed spices as well as their value added items to around 70 countries. There is immense scope of value addition in processing industry of seed spice crops. The value addition in case of seed spices should be done at two levels viz. at farm level and at processors level. The value addition at farm level or catchment areas enhances the rural employment and income and provide quality raw material for processors so that they can produce high quality end products.

There are excellent facilities in India for making spices oil, oleoresins, curry powder, ground spices and organic spices for export market. More value added products from seed spices, needed to be developed and marketed globally. Existing value added products needed to be strengthen to gain additional market advantage. The value added products are going to become an important contribution to India's foreign exchange earnings. The Government of India is also providing many benefits and incentives like advance license, duty draw back and market development assistance to exporter.

Value addition at farm level

The value addition at farm level is must because of major deterioration and post harvest losses take places at this stage. The objective of value addition at farm level is to enhance the processable character of farm produce and to achieve the quality and safe raw material for processing. This will also provide better income to the farmer, less losses in transport to urban areas for processing as well as creation of gainful employment at rural level. The major unit operations of processing conducted at farmers level are 1) Open sun drying and 2) Threshing. After threshing the produce is sold in bulk without cleaning and grading and hence the overall quality is low and fetches low price.

Storage/packaging

Farmers generally sell seed spices immediately in the same season. Traders and processors do the storage for further processing for value addition. The whole dried seeds are usually packed into sacks and stored in a cool dry place. The seed spices can be stored in jute bag with polyethylene line-bags. The poly and cotton bags help to retain volatile compounds. The fennel can be stored for one year but coriander is recommended for storage of six months only. Since storage of seed spices is simple it can be advocated at farmers level to avoid distress sale. Different storage system is presented in Table 142.

Table 142. Suggested improves storage structures for seed spices

S. No.	Name of equipments	Capacity (q)	Expected cost (Rs.)	Source
1	Hapur bin	2.5-7.5	5000.0	IGMRI, Hapur
2	Domestic metal bin	2.5-5.0	5000.0	IGMRI, Hapur
3	Evopartively cooled room	2000 - 5000	50000.0	CIPHET Ludhiana

Value addition at processors level

The value added products from seed spices at the level of processors are

Ground spices: These are the whole spices milled to a certain degree of fineness. The efficient grinding of spices prevents changes with respect to flavour and pungency. The optimum size of the grind for each spice depends on its end use and accordingly the conventional methods normally employ a hammer mill (coarse grinding), plate mills (domestic use) or pin mills (fine grinding). The particle size may vary from 850 microns to 50 microns of soft to medium hard materials according to the standards of comparison of Mohr scale ranged from 1 to 6. During conventional grinding that is grinding without cooling the spices, the temperature increases to over 90° C can occur in fast rotating mills which leads to substantial losses in volatile oils which leads to loss in quality. In an effort to decrease the loss in volatile oils by using cooling agents, an indirect water-cooling system has been developed. The indirect cooling methods have not achieved the desired expectations since cooling is limited to machine parts only and oil losses remained high. Conventional grinding with slow rotating roller mills, loss of volatile oils is low but results in lower production rate. Also these conventional methods suffer from various disadvantages like: (i) low efficiency (ii) not suitable for heat sensitive, high fat content materials (iii) gumming of grinding walls and sieves resulting in frequent stoppage of mill for cleaning (iv) reduction of grinding rate (v) consumption of enormous energy especially during spice grinding. Recently, low temperature grinding of spices using liquid nitrogen has achieved increasing importance. Nitrogen is a colourless, odourless and tasteless gas, which is not especially reactive to other elements. Most of the difficulties and degradations experienced in conventional grinding can be overcome by grinding spices in liquid nitrogen. Some of the advantages of cryogrinding are: (i) reduced oxidation of spice oils and increased stability due to pasteurizing effect on spices (ii) extreme fine grinding due to brittleness and fine grinding of difficult materials (iii) more uniform dispersal of flavour, reduced visual speckling and settling rates in liquid preparations (iv) increased throughput and reduced grinding power consumption (v) reduced loss of volatile oils and increased flavour strength (vi) reduction in microbial load (vii) lower cost of powder when the flavour strength is taken into account (viii) absolute control of particle size, shape and distribution. The major disadvantage of cryogenic grinding is its cost.

Table 143. Suggested equipment for grinding of seed spices

S. No.	Name of equipments	Capacity (kg hr^{-1})	Detail	Source
1	Low cost multipurpose mill	10-20	Suitable for coarse grinding	CIAE Bhopal
2	Hammer mill for spice	10-75	Suitable for coarse grinding	Commercially available
3	Pin mill	30-500	Used for medium hard material	Commercially available
4	High speed pulverizers	30-2000	A built-in water circulation jacket retains grinding temperature	Commercially available
5	Cryo grinders	100- 2000	Due to low boiling point and inert characteristic, liquid nitrogen and carbon-dioxide are ideal for cryogenic reactive cooling and subsequent milling with homogenous heat transfer.	Commercially available

Intervention in grinding industry: Present grinding technology applicable in seed spice industry is using traditional grinders which crush the seed under ambient temperature causing a loss of 15-43 per cent of essential oil. To reduce the loss there can be two approaches:

» **Cool grinding**: The grinder should be designed in a way that the temperature should not go beyond 25° C for reducing the loss of essential oil from the product. High aromatic value will enhance quality of the produce. Innovative technologies should also be developed to have low cost house hold or low capacity mixture grinders with cool grinding options.

» **Cryo-grinding**: Technology for cryogenic grinding using liquid nitrogen as cooling agent is available. Cryogenic grinding improves texture and aroma of powder by minimizing essential oils loss by approximately 3-10 per cent. The technology is applicable in all commodities having their value for essential oil, among seed spice cumin and coriander processing have high scope of cryogenic grinding. Chilly and turmeric processing can also be brought in cryogenic grinding because of better natural colour and aroma compared to conventional process. A 50 kg per hour capacity plant can be installed at a cost of Rs. 20 lakh, additional cost of liquid nitrogen is incurred in the technology but the net return is high compared to traditional system of grinding.

Seed spices extraction: Spices extracted compound serve as the alternate to whole and ground spices. They are used in various food products like sauces, confectioneries, beverage, and pharmaceutical and perfume industries. The essential oils and oleoresin are principal extracted compounds for value added modules. The essential oils are the aromatic, volatile compounds present in seed spices. In coriander seed it ranges from 0.1 to 0.4 per cent, in cumin seeds 2.3 to 4.8 per cent, in fennel 1.5 to 1.8 per cent, celery 5.0 to 6.0 per cent, ajowain 2.0 to 2.5 per cent. The essential oils extracted mainly by the method of distillation. The oil yield and quality of essential oil depend on the nature of feed material and pretreatments. The methods used for distillation are: 1. Water distillation 2. Water and steam distillation 3. Steam distillation. Oleoresin represents the overall flavours profile of the spices. It consists of the essential oil and non-volatile fraction constituents present in seed spices. The oleoresin is therefore designated as the true essence of spice and mostly used flavouring agents in food products.

Supercritical extraction

Spice extraction using supercritical fluids is a recent trend. It involves the use of gas above its critical temperature and pressure as extraction medium. The most popular extraction medium is carbon dioxide.

Curry powder: Curry powder is a ready-made mixes of various spices in ground form. The number of spices varies from 5 to 20 depending upon the producer and its end use. The ingredient of curry changes with the need for different foods. Curry powder is used to flavour foods and coriander is as a major ingredient in curry powder where it can be as much as 24 per cent. The export trade in curry powder at present is dominated by India. The packaging of curry powder has gained considerable importance as it increases the shelf life and fetches the consumer preference. The development of new and improved plastic film laminates and high-speed form fill sealing machines has created new opportunity for packaging the spices as instant spice powder. Exporting consumer packed pack spices can earn more unit value for the same quantity. It is also important to note that with stiff competition that India is facing in the value added market is particularly in the area of packaging.

Other products: Coriander dal (dhania dal) is mainly used as an adjunct in supari or pan masala. Coriander dal is obtained from the seeds of coriander. The coriander seeds are dehusked, flaked and given a mild heat treatment. The treated seeds are highly flavoured and consumed as a digestive chew. The traditional method of preparing dhania dal from the whole seeds of coriander is quite laborious and time consuming, besides being less efficient with high amount of brokens. In this method, the coriander dal is soaked overnight in salt water and partially dehydrated and husk removed manually. CFTRI has developed a technology with 2 t day^{-1} capacity to obtain dal from coriander seeds with higher yield and less brokens. This process can easily be adapted in rural areas where the raw material is available in plenty and the product can be marketed with suitable strategies.

Coriander

Manual ⇩

Coriander splits

Soaking for over night

Partial dehydration

Manual dehusking

Coriander dhal

Traditional method

Coriander

Milling ⇩

Coriander splits

Polishing

Coriander dal

Roasting and flavouring

Final products

Packaging

CFTRI method

Post harvest losses of spices:

The major losses in postharvest and production are dependent on many factors. Poor harvesting methods with immature crop product, disease or pest-contaminated material, or rotten and damaged material, all encourage crop losses. There is a need to have facilities such as artificial driers and dry storage to minimize the problem of rainfall interrupting the crop drying. The consequences of poor drying and storage multiply into microbial invasion which can have disastrous results on the potential sale and can lead to rejection of the crop. Poor processing methods, creating damage, can lead to loss of quality and losses, while poor storage facilities can also lead to losses to pests and to quality. Failure to comply with legislative requirements at the port of entry can also lead to rejection of a consignment. The cost of these losses is progressively greater as the products move through the postharvest chain. This is why it is important to keep information flowing with the marketplace so that an informed decision (such as, not to harvest a crop) can be made before investing in processing, packaging and transport.

References

Abeles F B, Morgan P W and Saltveit Jr M E 1992. Ethylene in plant biology. Second edition. San Diego: Academic Press, pp. 414.

Agrawal S, Divakara Sastry E V and Sharma R K 2002. Seed spices: Production, quality, and export. Pointer Publication, Jaipur, pp. 1-315.

Ahmad M S, Nayyer A, Aftab A, Nayak B, and Siddiqui M W 2014. Quality prerequisites of fruits for storage and marketing. *Journal of Postharvest Technology*, 2(1): 107–123.

Andersen Craig 1998. Postharvest handling workshop. Huntsville, AR. February, pp. 22.

Anonymous 1970. Ethylene for colouring matured fruits, melons and vegetables. New York: Union carbide corp. Product information, pp.11.

Anonymous 1989. Fresh produce manual: Handling and storage practices for fresh produce. Australian United Fresh Fruit and Vegetable Association Ltd, Footsore, Victoria.

Anonymous 1992. Put it on ice. American Vegetable Grower. June, pp. 17-18.

Anonymous 2014. Annual Report, Agricultural and Processed Food Products Export Development Authority (APEDA), pp. 135.

Anonymous 2018. Indian Horticulture Database, pp.1-165.

Anonymous, 1999. Plant resources of South-East Asia, No.13. Spices, in plant resources of South-East Asia, No.13, pp 1-15.

Armitage A M 1991. The georgia report stage of flower development at harvest. *The Cut Flower Quarterly,* 3(1):13.

Armitage A M 1993. Specialty cut flowers. Timber Press, Portland OR–USA.

Axtell P B and Dillon M 1995. Quality assurance for small-scale rural food industries. FAO Agricultural services bulletin No. 117, Rome: Food and Agriculture Organization of the United Nations.

Aylesworth Jean 1992. Deliver quality. American Vegetable Grower. June, pp. 19-20.

Barman K, Ahmad M S and Siddiqui M W 2015. Factors affecting the quality of fruits and vegetables: Recent understandings. In M. W. Siddiqui (Ed.), Postharvest biology and technology of horticultural crops: Principles and practices for quality maintenance (pp. 1–50). Waretown, NJ: Apple Academic Press.

Barth M M, Zhou C, Mercier J and Payne F A 1995. Ozone storage effects on anthocyanin content and fungal growth in blackberries. *J. Food Sci.*, 60: 1286- 1288.

Beattie, B S, McGlasson, W B and Wade, N L 1989. Postharvest diseases of horticultural produce, Volume 7: Temperate Fruits, CSIRO, Melbourne.

Beuchat L R 1991. Surface disinfection of raw produce. *Dairy, Food Environ. Sanitation*, 12(1): 6-9.

Bhati M S, Dixit V D and Bhati D H 1988. *Indian Perfumer*, 32: 4: 306 -309.

Blanpied G D 1985. Symposium on Ethylene in postharvest biology and technology of horticultural crops. *HortScience*, 20: 39-60.

Blanpied G D, Bartson J A and Turk J R 1985. A commercial development programme for low ethylene controlled atmosphere storage of apples. In. A. Roberts and G. A. Tucker, eds., Ethylene and plant development. London: Butterworths, 393-404.

Bleeker A B 1999. Ethylene perception and signalling: an evolutionary perspective. *Trends in Plant Sci.*, 4: 269-274.

Bott T R 1991. Ozone as a disinfecting in a process plant. *Food Control*, 2(1): 44-49.

Burg S P and Burg E A 1966. Fruit storage at sub atmospheric pressure. *Science*, 153: 314-315.

Byczynski Lynn 1997. Storage crops extend the season. Growing for Market. September, pp. 1, 4-5.

Carawan R E 1999. Personal communication. InFood Inc., Raleigh, N.C.

Cena A 1998. Ozone: Keep it fresh for food processing. Water Conditioning Purification, Sept. pp 112-115.

Chadha K L, Singh S K, Ram Asrey, Patel VB and Sethi S 2005. Maturity standards in fruits and vegetables. Proc. First Indian Horticulture Congress- 2004. Vo. II pp. 794-817. Published by The Horticulture Society of India, New Delhi.

Champ S R, Highley E and Johnson G L 1984. Post harvest handling of tropical fruits. Proceedings of an international conference, Chiang Mai, Thailand, 19-23 July 1993, Australian Centre for International Agricultural Research, Canberra, ACT.

Chand R 2002. WTO, Trade liberalization and India Agriculture, Mittal Publications, New Delhi.

Chand R 2005. WTO and its implication of Horticultural Trade in India. Proc. First Indian Horticulture Congress- 2004., pp. 1111-24. Published by The Horticultures Society of India, New Delhi.

Charkraverty A 2003. Handbook of postharvest technology: cereals, fruits, vegetables, tea and spices, pp. 1-25.

Coates L, Cooke A, Persley D, Beattie B, Wade N and Ridgeway R 1995. Post harvest diseases of horticultural produce, Volume 2: Tropical Fruits. DPI, Queensland.

Dhatt AS 2005. Post-harvest management technologies for horticultural crops- An over view, Proc. First Indian Horticulture Congress- 2004., Vol. II pp. 765-81, Published by The Horticulture Society of India, New Delhi.

Dhatt AS, Mahajan BVC, Sandhu KS, Garg A and Sharma SR 2007. Handbook on Post harvest Handling of Fruits and Vegetables. 3rd edition, PHPTC, PAU, Ludhiana

Dyduch J 1996. Studies on optimum density of celery (*Apium graveolens* L.) var. Dulce pers grown for seed. Annals University Marie Curie Scalodowska Sectio Horticulture, 4: 29-36. Malhotra S K 2004. Nigella in handbook of herbs and spices volume 2 Editor K.V. Peter. Wood Head Publisher Ltd. Cambridge England, pp. 206-214.

Ekechukwu O V and Norton B 1999. Review of solar-energy drying systems II: an overview of solar drying technology. *Energy Conversion and Management*, 40(6): p. 615-655.

Ewell A W 1940. Ozone and its applications in food preservation. Refrigeration Application Data Book, 2nd ed., Sect. II. Cold Storage Practice, pp. 199- 203. Am. Soc. Refrig. Eng., Menasha, Wis.

Ezekiel R.2005. Recent development in on farm storage of fresh fruits and vegetables. Proc. First Indian Horticulture Congress- 2004., Vo. II pp. 832-43. Published by The Horticultures Society of India, New Delhi.

Finger JM and Nogues J J 2004. The unbalanced round outcome. The new area in future, WTO negotiations. *The World Economy*, 25: 321-40.

Gane R 1936. The respiration of bananas in presence of ethylene. *New Physiologist*, 36: 170-178.

Gast K L B and Flores R 1991. Precooling produce. Kansas State University Cooperative Extension, Manhattan, Kansas, pp. 130-146.

Geering F 1999. Ozone applications, The state- of the-art in Switzerland. *Ozone Sci. Eng.*, 21(2): 99-118.

Gooch Jamie 1998. Getting into the 'o' zone. Fruit Grower. January, pp. 10-11.

Gopal Krishna Rao KP 2005. Influence of pre-harvest treatments on post-harvest quality of horticultural produce, Proc. First Indian Horticulture Congress- 2004, Vol. II pp. 782-93. Published by The Horticultures Society of India, New Delhi.

Graham D M 1997. Use of ozone for food processing. *Food Technol.*, 51(6): 72-75

Green C 2002. Export development of essential oils and spices by Cambodia. The International Trade Centre, Geneva [SSA, 2001-468] and The Ministry of Commerce, Kingdom of Cambodia. pp. 100.

Green C, Steele J and Willemse J 1999. The development and prospects for minor industrial crops in selected Southern African countries. World Bank Technical Paper, 424: 139-166.

Grierson W 1987. Postharvest Handling Manual Commercialization of Alternative Handling Crops Project. The Belize Agribusiness Company/ Chenomics International/USAID.

Halevy A H and Mayak S 1979. Senescence and postharvest physiology of cut flowers, Part 1. Horticultural Reviews, 1:204-236.

Halevy A H and Mayak S 1982. Senescence and postharvest physiology of cut flowers, Part 2. *Horticultural Reviews*, 3:59-143.

Hampson B C and Fiori S R 1997. Applications of ozone in food processing operations. Proc. of 1997, IOA PAG Conf., Lake Tahoe, Nev., pp. 261-267.

Hardenburg R E, Watada A E and Wang C Y 1986. The commercial storage of fruits, vegetables and florist and nursery stocks. USDA, Agriculture Handbook 66, pp 130-136.

Hartulistiyoso E, Rusli M S and Lucke W 1998. Postharvest technology and processing of spices in Indonesia. *Landtechnik*, 53(4): 252-253.

Heindl A 2000. Solar drying of medicinal and aromatic plants. *Zeitschrift fur Arznei and Gewurzpflanzen*, 5(2): 80-88.

Heindl A 2003. Problems of and suggested solutions for the processing and drying of medicinal and spice plants. *Zeitschrift fur Arznei- & Gewurzpflanzen*, 8(1): pp. 33-36.

Herner Robert C 1989. Sources of loss cited during post harvest handling. Great Lakes Vegetable Growers News. May, pp. 16.

Hoff J C 1997. Strengths and weaknesses of using Ct to evaluate disinfection practice. Proc. AWWA Seminar, Assurance of Adequate Disinfection, or Ct or Ct. AWWA, Denver, pp. 49-65.

Howell John C 1993. Postharvest handling. Vegetable Notes: Growing and Marketing Information for Massachusetts commercial growers, pp. 1-5

Huybrechts C, Deckers T and Valcke R 2003. Predicting fruit quality and maturity of apples by fluorescence imaging: Effect of ethylene and AVG. *Acta Horticulture*, 599, 243–247.

Isenberg E M R, Kasmire R E and Parson J E 1982. Vacuum cooling vegetables. Cornell Univ. Coop. Ext. Bull. 186. pp 10.

Jeffrey J J 1977. Engineering principles related to the design of systems for air cooling of fruits and vegetables in shipping containers. Proc. 29th Intl. Conf. on Handling Perishable Agricultural Commodities. East Lansing: Mich. State Univ. 151-164.

Kader A 1982. Postharvest technology of horticultural crops. Publication 3311, University of California, Davis, California.

Kader A A 1986. Biochemical and physiological basis for effects of controlled and modified atmosphere on fruits and vegetables. *Food Technology*, 40(99–100): 102–104.

Kader A A 2002. Postharvest biology and technology: An overview. In A. A. Kader (Ed.), Postharvest technology of horticultural crops. University of California, Division of Agriculture and Natural Resources, Special Publ. 3311, pp. 39–47.

Kader Adel A 2002. Post harvest Technology of Horticultural Crops. 3rd Edition, University of California, Agricultural and Natural Resources.

Kader Adel A and Merita Cantwell 2006. Produce quality rating scales and colour charts. Postharvest Horticulture Series No 23. Post harvest Technology Research & Information, Centre University of California, Davis.

Kays S J and Beaudry R M 1987. Techniques for inducing ethylene effects. *Acta Hort.*, 201: 77-116.

Kende H 1993. Ethylene biosynthesis. *Annu. Rev. Plant Physiol. and Plant Mol. Biol.*, 44: 283-308.

Kim J G, Yousef A E, and Chism G W 1999. Use of ozone to inactivate microorganisms on lettuce. *J. Food Safety*, 19: 17-33.

Kitinoja L 2002. Making the link: Extension of postharvest technology. In A. A. Kader (Ed.), Postharvest technology of horticultural crops. Publication 3311 (3rd ed., pp. 481–509). Oakland, CA: University of California.

Kitinoja L and Gorny J 1998. Post-harvest technology for fruits and vegetables Produce marketers: Economic opportunities. Quality and Food Safety by, Department of Pomology, University of California, Davis. A joint publication of UC Post harvest Outreach Program and Punjab Horticultural Post harvest technology Centre, USAID/ ACE.

Kondo F, Utoh K and Rostamibashman M 1989. Sterilizing effect of ozone water and ozone ice on various microorganisms. Bull. Faculty of Agric., *Miyazaki Univ.,* 36(1): 93-98.

Krishnamoorthy V, Meenakshi N and Madalageri M B 2002. Herbal spices - A review. *Journal of Medicinal and Aromatic Plant Sciences*, 24(1): 123-131.

Ladamia M S 2000. Response of Nagpur Mandarin to degreasing by ethylene trickle method. *J. Food Sci. Tech.*, 37: 103-10.

Lafuente M T, Cantwell M, Yang S F and Rubatzky V 1989. Isocoumarin content of carrots as influenced by ethylene concentration, storage temperature and stress conditions. *Acta Hort.*, 258: 523-534.

Larson R A 1980. Introduction to Floriculture. Academic Press, Inc., New York, pp. 607.

Lawrence B M 1993. A planning scheme to evaluate new aromatic plants for the flavour and fragrance industries, in New crops, J. Janick and J. E. Simon, Editors. Wiley: New York. pp. 620-627.

Lelievre J M, Latche A, Jones B and Bouzayen M 1997. Ethylene and fruit ripening. *Physiol. Plant.*, 101: 727-739.

Liu F 1970. Storage of bananas in polyethylene bags with an ethylene absorbent. *HortScience*, 5: 25-27.

Lopez E G 1983. Conservation de la Production Agricola. Barcelona Editorial Aedos, pp.188.

Lutz J M and Hardenburg R E 1968. The commercial storage of fruits, vegetables, and florist and nursery stocks. U. S. Department of Agriculture. Handbook No. 66. pp. 94.

Maini S B 2001. Post-harvest management of fruits and vegetables. In Hand Book of Horticulture (Ed. K.L. Chadha) ICAR, New Delhi pp. 930-40.

Malhotra S K 2004. Ajowan in handbook of herbs and spices volume 2 editor K.V. Peter. Wood Head Publisher Ltd. Cambridge England, pp. 107-116.

Marcelle G B 1995. Production, handling and processing of nutmeg and mace and their culinary uses: FAO Regional office for Latin America and the Caribbean, pp. 217.

Mayak S and Halevy A H 1980. Flower Senescence. IN: (ed. K.V. Thimann) Senescence in Plants, CRC Press, Boca Raton Fl., pp. 131-156.

McGregor B M 1987. Manual del transport de products tropical's. Manual de Agriculture 668. USDA, p. 148.

McGregor B M 1989. Tropical Products Transport Handbook. USDA, Office of Transportation, Agricultural Handbook Number 668.

Mckeon T A, Fernandez-Maculet J C and Yang S E 1995. Biosynthesis and metabolism of ethylene. In P. J. Davies, ed., Plant hormones. 2nd ed. Dordrecht: Kluwer, pp. 118-139.

Meena R K and Yadav JS 2001. Horticulture Marketing and Post-harvest Management. Pointer Publisher Jaipur, Rajasthan.

Melnick Rick 1998. Safety sets the table. American Vegetable Grower. February, pp. 9-11, 13, 15.

Mitchell F G 1972. Commercial cooling of fruits and vegetables. California Agricultural Experiment Station Extension Service, Manual, pp. 43.

Mitchell F G and Kader A A 1992. Postharvest technology of horticultural crops. University of California, Division of Agriculture and Natural Resources, Publication 3311. pp.296.

Nayyer M A, Siddiqui M W and Barman K 2015. Quality of fruits in the changing climate. In M. L. Choudhary, V. B. Patel, M. W. Siddiqui, & R. B. Verma (Eds.), Climate dynamics in horticultural science: Impact, adaptation, and mitigation (Vol. 2, pp. 269–278). Waretown, NJ: Apple Academic Press.

Newenhouse Astrid 1998. Line harvest containers with mesh bags for washing. Growing for Market. December, pp. 8.

Norton J S, Charig A J and Demoranville I E 1968. The effect of ozone on storage of cranberries. *Proc. Am. Soc. Hort. Sci.*, 93: 792-796.

Nowak J and Rudnicki R M 1990. Postharvest handling and storage of cut flowers, florist greens, and potted plants. Tiber Press, Portland OR–USA, pp. 210.

Ong K C, Cash J N, Zabik M J and Siddig M 1995. Chlorine and ozone as postharvest wash in the removal of pesticides from apples. In Book of Abstracts, Ann. Mtg., Inst. of Food Technologists, pp. 28.

Pal R K, Roy S K and Srivastava S 1997. Storage performance of 'Kinnow' mandarin in evaporative cool chamber and ambient condition. *Journal of Food Science and Technology*, 34: 200–203.

Pandey U B 2004. Onion and Garlic storage; status, constraints and challenges. In: Souvenir, Proc. First Indian Horticulture Congress- 2004., pp. 95-104. Published by The Horticultures Society of India, New Delhi.

Pandey U B 2005. Present Status, constraints and opportunities in export of fresh vegetables from India. Proc. First Indian Horticulture Congress- 2004., Vo. II pp. 965-77. Published by The Horticultures Society of India, New Delhi.

Pantastico Er B 1975. Post-harvest physiology, handling and utilization of Tropical and Subtropical fruits and vegetables. The AVI publishing Co. Inc: West port Connecticut.

Panwar K, Badiyala D and Jangpo B 1989. *Indian Cocoa, Areca nut and Spices Journal*, 12:146-147.

Parsons R A and Kasmire R F 1974. Forced-air unit to rapidly cool small lots of packaged produce. University of California Cooperative Extension, OSA pp.272.

Paul Sanjib K 2020. Edible films and coatings for fruits and vegetables. In: Hashmi, Saleem and Choudhury, Imtiaz Ahmed (eds.). Encyclopaedia of Renewable and Sustainable Materials, vol. 5, pp. 363–376. Oxford: Elsevier. http://dx.doi.org/10.1016/B978-0-12-803581-8.11509-7.

Perkins M 1997. Ozone in food processing applications, past experience, future potential and regulatory issues. Presented at Connec TECH' 97, Atlanta, Ga.

Piper J 1998. Tomato washing using ozonated water. Presented at 1998 EPRI Ozone Workshop, Memphis, May.

Pratt H K and Goeschl J D 1969. Physiological roles of ethylene in plants. *Annu. Rev. Plant. Physiol.* 20: 541-585.

Pruthi J S 1987. Cryo milling the latest technology of cold grinding of spices condiments, Proceedings of symposium, AFST (I), New Delhi, pp.38.

Pruthi J S 1992. Postharvest technology of spices: pre-treatments, curing, cleaning, grading and packing. *Journal of Spices and Aromatic Crops*, 1(1): 1-29.

Pryor A 1998. Ozone toxicology and guidelines for safe use in food processing ozonation systems. Presented at 1998 EPRI Ozone Workshop, Memphis, May.

Putievsky E, Ravid U and Dudai N 1986. The influence of season and harvest frequency on essential oil and herbal yields from a pure clone of sage (*Salvia officinalis*) grown under cultivated conditions. *Journal of Natural Products*, 49(2): 326-329.

Ray R C and Ravi V 2005. Post harvest spoilage of sweet potato in tropics and control measures. *Critical Reviews in Food Science and Nutrition*, 45: 623–644.

Reid M S 1995. Ethylene in plant growth, development and senescence. In P. J. Davies, ed., Plant hormones. 2nd ed. Dordrecht: Kluwer, pp. 486-508.

Rice R G 1999. Ozone in the United States of America, State of the art. *Ozone Sci. Eng.* 21(2): 99-118.

Rice R G, Farquhar W, and Bollyky L J 1982. Review of the application of ozone for increasing storage time for perishable foods. *Ozone Sci. Eng.*, 4(1): 147-163.

Rij R E, Thompson J F and Farnham D S 1979. Handling, precooling and temperature management of cut flower crops for truck transportation. Oakland: Univ. Calif. Coop. Ext. Leaflet 21058, pp. 1-12.

Roberts A W 1998. Particle technology reflections and horizons: an engineering perspective, Trans. Icheme, 76 A, pp. 775-796.

Roy S K and Khurdiya D S 1986. Studies on evaporative cooled zero-energy input cool chamber for the storage of horticultural produce. *Indian Food Packer*, 40: 26–31.

Rumpf H 1990. Particle technology (English edition), Powder Technology series, London, Chapman and Hall, pp. 128.

Sacalis J N 1989. Fresh (cut) flowers for designs. Post production guideline. D. C. Kiplinger Chair in Floriculture, Ohio State University, pp.128.

Sacalis JN 1993. Cut Flower Prolonging Freshness—Postproduction Care and Handling, 2nd ed., edited by Joseph L. Seals. Ball Publishing, Batavia, IL–USA.

Saha NC 2005. Latest Development in packing of Horticulture produce. Proc. First Indian Horticulture Congress- 2004, Vo. II pp. 854-59. Published by The Horticultures Society of India, New Delhi.

Salami P, Ahmadi H, Keyhani A and Sarsaifee M 2010. Strawberry post-harvest energy losses in Iran. *Researcher*, 4: 67–73.

Saltveit M E 1999. Effect of ethylene on fresh fruits and vegetables. *Postharvest. Biol. Technol.*, 15: 279-292.

Sapers G M 1998. New technologies for safer produce, Chemical-based treatments and decontamination by washing. In Proc. of Fresh Fruits and Vegetables: Food Safety Challenges, sponsored by Natl. Ctr. for Food Safety and Technology, Chicago, May pp.12-14.

Saraswathy S, Preethi T L, Balasubramanyan S, Suresh J, Revathy N, and Natarajan S 2013. Post harvest management of horticultural crops. Agrobios (India), Johpur, pp 1-543.

Sargent S A, Ritenour M A and Brecht J K 2000. Handling, cooling and sanitation techniques for maintaining postharvest quality. HS719. Horticultural Sciences Department, Florida Cooperative Extension Service, Institute of Food and Agricultural Sciences, University of Florida.

Sargent S A, Talbot M T and Brecht J K 1989. Evaluating precooling methods for vegetable packing house operations. *Proc. Fla. State Hart. Soc.* 101:175-182.

Sasseville David N 1988. Harvesting and handling produce: Plan now for high quality. Missouri Farm. May-June, pp. 19-21.

Scott K J and Wills R B H 1973. Atmospheric pollutants destroyed in an ultraviolet scrubber. *Lab. Pract.*, 22: 103-106.

Sherman M and Gull D D 1981. A flow-through system for introducing ethylene in tomato ripening rooms. Univ. Florida Veg. Crops. Fact Sheet VC-30. pp. 4.

Shikhamany SD 2004. Strategies to improve productivity, quality and post-harvest management in fruit crops. In: Souvenir, First Indian Horticulture Congress-2004., pp. 5-14. Published by The Horticultures Society of India, New Delhi.

Siddiqui M W and Dhua R S 2010. Eating artificially ripened fruits is harmful. *Current Science*, 99(12): 1664–1668.

Siddiqui M W, Patel V B and Ahmad M S 2015. Effect of climate change on postharvest quality of fruits. In M. L. Choudhary, V. B. Patel, M. W. Siddiqui, and S. S. Mahdi (Eds.), Climate dynamics in horticultural science: Principles and applications (Vol. 1, pp. 313–326). Waretown, NJ: Apple Academic Press.

Singh I S and Medhi G 1985. Biochemical changes during storage of beauty seedless grape. Proc. National Workshop on Post Harvest Management of Grapes, pp. 242-48.

Singh I S and Pathak R K 1988. Packaging of Jamun, Aonla and Bael fruits. Souvenir, packaging of fruits and vegetables in India. Agri-Horticultural Society Publication Hyderabad pp. 108-11.

Singh I S and Weaver R J 1982. Effect of pre-harvest application of Ethephon (during stage III) and SO2 generation packing device on ethylene production and quality of Tokay Grapes during cold storage. *Haryana J. Hort. Sci.*, 11: 149-53.

Singh IS 1984. Post-harvest handling of fruits (in Hindi). Directorate of Publication Haryana Agricultural University, Hisar.

Singh IS 2000. Manual on post-harvest handling and processing of fruits. N.D. University of Agri. and Tech. Publication, Faizabad.

Singh IS 2009. Post harvest handling and processing of fruits and vegetables. Westville Publishing House, New Delhi, pp. 1-194.

Singh K 2004. Strategies for increasing productivity, quality and post-harvest management of vegetables. In Souvenir, First Indian Horticulture Congress-2004., pp. 15-22. Published by The Horticultures Society of India, New Delhi.

Singh K K and Goswami T K 1999. Design of a cryogenic grinding system for spices. *Journal of Food Engineering*, 39(44): 359-368.

Sisler E C, Serek M, Dupille E and Goren R 1999. Inhibition of ethylene responses by 1-methylcyclopropene and 3-methylcyclopropene. *Plant Growth Regulators*, 27: 105-112.

Snowden A L 1990, 1991. Post-harvest diseases and disorders of fruits and vegetables, Volume 7 General introduction of fruits and vegetables, Volume 2. Wolfe Scientific, London.

Staby G L and Thompson J F 1978. An alternative method to reduce ethylene levels in coolers. *Flor. Rev.*, 163: 31, 71.

Staby G L, Robertson J L, Kiplinger D C and Conover C A 1976. Proc. of national floricultural conference on commodity handling. Ohio Flor. Assoc., Columbus, Ohio, pp.71.

Suslow Trevor 1997. Microbial food safety: an emerging challenge for small-scale growers. Small Farm News. June-July, pp. 7-10.

Tainter D R and G A T 2001. Spices and Seasonings: a food technology handbook. John Wiley and Sons Inc., New York. pp. 248.

Tauxe R, Kruse H, Hedeerg C, Potter M, Madden J and Wachsmuth K 1997. Microbial hazards and emerging issues associated with produce- A preliminary report to the National Advisory Committee on Microbiological Criteria for Food. *J. Food Protect.* 60: 1400-1408.

Thompson A K 1996. Post harvest Technology of Fruits and Vegetables. 1st Edition, Blackwell Science, Inc. USA.

Thompson A K 1998. Controlled atmosphere storage of fruit and vegetables. Wallingford, England: CAB International.

Thompson J F 2002. Cooling horticultural commodities. pp. 97-112. In: Kader, A. A. (ed). Postharvest Technology of Horticultural Crops. Univ. of California, Div. of Agriculture and Natural Resources, Publication 3311.

Thompson J F and Kasmire R F 1981. An evaporative cooler for vegetable crops. *Calif. Agric.* 35(3-4): 20-21.

Thompson J F, Bishop C F H, and Brecht P E Patrick 2004. Air transport of perishable products. University of California, ANR publication, 21618, pp. 22.

Thompson J F, Mitchell F G, Rumsey T R, Kasmire R F and Crisosto C H 1998. Commercial cooling of fruits, vegetables and flowers. Oakland: Univ. Calif. Div. of Ag. and Nat. Res. Publ. 21567, pp. 61.

Thompson J F, Mitchell F G, Rumsey T R, Kasmire R F and Crisosto C H 2003. Commercial cooling of fruits, vegetables and flowers, Revised Edition, University of California, ANR publication , 21567, pp.59.

Thompson J, Kader A and Sylva K 1999. Compatibility chart for fruits and vegetables in short-term transport or storage. University of California, Publication 21560. http:// postharvest. ucdavis.edu/Pubs/postthermo.html.

Vedamuthu G V Peter, Khader M D, Abdul M D and Rajan F S 1989. Yield components in coriander (Coriandrum sativum L.). *South Indian Hort.*, 37(5):287-90.

Verlet N 1993. Overview of the essential oils economy. *Acta Horticulture*, (333): 65-72.

Walls M M and Corgan J N 1994. Post-harvest losses from delayed harvest and storage of onion. *Hort. Sci.*, 20: 802-4.

Wardlaw C W and Leonard E R 1936. The storage of West Indian Mangoes, Low Temperature Research Station Memoir 3, pp.47.

Waskar D P and Roy S K 2002. Zero energy cool chamber storage of fruits. A review. *Indian Food Packer*, pp. 144-47.

Watada A E 1986. Effects of ethylene on the quality of fruits and vegetables. *Food Technol.*, 40(5): 82-85.

Watkins C B 2003. Principles and practices of postharvest handling and stress. In D. Ferree & I. Warrington (Eds.), Apples-botany, production, and uses (pp. 585–615). Boston: CABI.

Watkins J B and Ledger S 1990. Forced-air cooling. 2nd ed. Brisbane, Australia: Queensland Dept. of Primary Industries, pp. 64.

Weiss E A 1997. Essential oil crops. CAB International, pp.160.

Weiss, E. A. 2002. Spice crops, in Spice crops. De Silva, K. T. 1995. A manual on the essential oil industry. Vienna: UNIDO, pp.232.

Wilhelm Luther R 1992. Effect of cooling treatment on the quality of sweet corn. Tennessee Farm and Home Science, pp. 30-35.

Williams D W, Montecalvo J, Mueller E, Earls D, Swanson K and Petersen L 1995. Ozonation as an alternative disinfectant for carrot wash water. In Book of Abstracts, Ann. Mtg., Inst. of Food Technologists, pp. 8.

Wills R B H, Ku V V V, Shonet D. and Kim G H 1999. Importance of low ethylene levels to delay senescence of non-climacteric fruit and vegetables. *Austral. J. Exp. Agr.*, 39: 221-224.

Wills R B H, McGlasson W E, Graham D, Lee T H and Hall E G 1989. Postharvest: an introduction to the postharvest and handling of fruit and vegetables, 3rd edition. University of New South Wales Press, Sydney.

Wills R, McGlasson B, Graham D and Joyce D 1998. Post harvest: An Introduction to the physiology and handling of fruit, vegetables and ornamentals. 4th Edition, CAB International, UK.

Wilson L G, Boyette M D and Estes E A 1995. Post harvest handling and cooling of fresh fruits, vegetables and flowers for small farms. Leaflets 800-804. North Carolina cooperative extension service, pp. 17.

Yang S F 1987. Regulation of biosynthesis and action of ethylene. *Acta Hort.*, 201: 53-59.

For Product Safety Concerns and Information please contact our EU representative GPSR@taylorandfrancis.com Taylor & Francis Verlag GmbH, Kaufingerstraße 24, 80331 München, Germany

Batch number: 10726364

Printed by Printforce, the Netherlands